Introduction to Security

Introduction to Security

Ninth Edition

Robert J. Fischer

Edward P. Halibozek

David C. Walters

AMSTERDAM • BOSTON • HEIDELBERG • LONDON
NEW YORK • OXFORD • PARIS • SAN DIEGO
SAN FRANCISCO • SINGAPORE • SYDNEY • TOKYO
Butterworth-Heinemann is an imprint of Elsevier

Butterworth-Heinemann is an imprint of Elsevier
225 Wyman Street, Waltham, MA 02451, USA
The Boulevard, Langford Lane, Kidlington, Oxford, OX5 1GB, UK

Notices
Knowledge and best practice in this field are constantly changing. As new research and experience broaden our understanding, changes in research methods, professional practices, or medical treatment may become necessary.

Practitioners and researchers must always rely on their own experience and knowledge in evaluating and using any information, methods, compounds, or experiments described herein. In using such information or methods they should be mindful of their own safety and the safety of others, including parties for whom they have a professional responsibility.

To the fullest extent of the law, neither the Publisher nor the authors, contributors, or editors, assume any liability for any injury and/or damage to persons or property as a matter of products liability,negligence or otherwise, or from any use or operation of any methods, products, instructions, or ideas contained in the material herein.

Library of Congress Cataloging-in-Publication Data
Fischer, Robert J.
  Introduction to security / Robert Fischer, Edward Halibozek, David Walters. -- 9th ed.
    p. cm.
  Includes index.
  ISBN 978-0-12-385057-7
  1. Private security services. 2. Private security services--Management. 3. Industries--Security measures.
4. Retail trade--Security measures. I. Halibozek, Edward P. II. Walters, David, 1963- III. Title.

HV8290.G74 2013
363.28'9--dc23

                                                                        2012020446

British Library Cataloguing-in-Publication Data
A catalogue record for this book is available from the British Library.

For information on all Butterworth–Heinemann publications
visit our Web site at http://store.elsevier.com

Printed and bound by CPI Group (UK) Ltd, Croydon, CR0 4YY
Transferred to digital print 2012

Contents

Preface

As we write this ninth edition, we are reminded of how fast our profession has changed and how global responsibilities of a security leader have now become commonplace. A new author, David C. Walters, has joined our team, as we say a grateful goodbye to Gion Green, the original author. A complete revolution in the security industry has occurred since Gion passed over 30 years ago. The aftermath of the September 11, 2001 attacks on the World Trade Center and the Pentagon shocked the nation. Government involvement in certain areas of the security business is now a daily reality.

Coupled with the security industry's response to the threat of world terrorism is the absolute need to keep up with an always rapidly changing technology used by the industry. Unfortunately, this same technology is also used against persons and organizations by criminals. Protection of information, which was traditionally handled by placing it in vaults and marking it proprietary or Top Secret, has become ever more complex as the information is migrated to computer files and networks. Computer systems that now contain all types of information, from personal identification to inventory records, are making life both easier and more complex. Information exchange is made easy by the click of a computer key or a touch on a smart phone. This exchange of information relies on the Internet, and there are plenty of examples that clearly demonstrate the computer network is subject to attack by outside hackers and other criminal enterprises.

This ninth edition continues to maintain the basic concepts developed by Gion Green, covering the total picture and giving the reader a glimpse of various, diverse components that make up the security function. However, much of the security industry is undergoing continuous change. Former President George W. Bush's call for an international coalition to wage war against world terrorism put the world's citizens on an alert status. The tension in the Middle East and parts of Europe, focusing on religious conflicts, economic interests, and differences in political beliefs, continues to provide both opportunity and fear for those in the security industry. The world of security that changed dramatically on September 11, 2001 continues to evolve. This new edition utilizes the same basic concepts that have made this text a basic primer in the security field, while also focusing on current and future problems within the basic framework of security theory.

This updated edition has one less chapter, but no less material. Two chapters are still devoted entirely to the security issues created by the continuing presence of world terrorism. In addition, the material on information security, identity theft, transportation, contingency planning, retail security, and piracy has been completely updated.

Even as we put the final touches on this 2012 edition, the reality of the times may make some of the materials already outdated. However, we maintain the belief that the basic principles of loss prevention and security remain intact. The tools that allow us to apply the principles have become more complex and constantly changing. We still lock our valuables. Some use traditional locks and keys, others have sophisticated electronic tools that operate electronic locks, while still other information is locked into databases using information technology security including encryption and dynamic passwords.

Acknowledgments

We wish to thank those individuals who took time to complete surveys distributed by the publisher providing us with suggestions for updating the book for this ninth edition. These individuals include:

Robert B. Iannone, CPP
President
Iannone Security Management
Fountain Valley, California

Lennart E. Long
Adjunct Professor
Criminal Justice and Criminology Department
University of Massachusetts, Lowell
Lowell, Massachusetts

Dr. James S.E. Opolot
Professor
Administration of Justice
Texas Southern University
Houston, Texas

Chelsey Rennard
West Palm Beach, Florida

Rafael Rojas, Jr., DPA
Assistant Professor
Justice Studies Department
Southern New Hampshire University
Manchester, New Hampshire

Allen R. Sondej, Esq.
Adjunct Professor
New Jersey City University
Jersey City, New Jersey

Michael H. Witt
Director, Program Security
Ball Aerospace & Technologies Corp.
Boulder, Colorado

We also appreciate the occassional note we receive from those who use our book. Thank you to Klas Nilsson, Security Manager, Vanadisvagen 24, Stockholm, Sweden, for taking the time to comment on the materials on "Deciding on a Contract Security Firm."

In particular, special thanks to the content experts who reviewed and at times made substantial updates to materials contained in various chapters. These experts include: Dr. Sharon

Larson, Department of Health Science, Western Illinois University, who reviewed the information on drugs in the workplace; James Falk, Security Director, Ace Hardware, who reworked the Retail chapter; Steven C. Babb, retired Deputy General Counsel for Northrop Grumman Corporation, who reviewed and updated our chapter on security and the law; Deon Chatterton, Sr. Manager Systems & Technology at Cisco as well as Whitney Stein, Esq, of the Insurance Law Group, who helped with the rewriting of the insurance materials for this ninth edition.

We also thank Pam Chester, Acquisitions Editor, Butterworth-Heinemann, for her encouragement to update this classic text. Pam's understanding of the changing security climate inspired Ed and Bob to tackle a ninth edition and bring David in as a third author. Acknowledgments also go to Gregory Chalson and Amber Hodge at Elsevier, who kept us on task during the production of the manuscript.

Bob thanks those at Assets Protection Associates, Incorporated, who provided support services during his work on this revision, as well as comments and edits of various drafts. David would like to give special thanks to Deon Chatterton, Sr. Manager Systems & Technology at Cisco as well as Whitney Stein, Esq, of the Insurance Law Group.

Finally, we thank our dear wives Kathy and Phillis, who have always been understanding of the time needed to complete this project.

Chapter Acknowledgments

Terrorism

While this chapter has been thoroughly updated by the authors, the outline and original composition is primarily the work of Dr. Vladimir Sergevnin, Editor, *Illinois Law Enforcement Executive Form Journal*, and professor of Law Enforcement and Justice Administration at Western Illinois University.

Retail

This completely revised chapter is primarily the work of James Falk, Director, Loss Prevention/Property Administration, Ace Hardware Corporation, and his colleagues, Theresa Tapella, Manager of Retail LP and BC Planning at Ace Hardware and Bill Cafferty, Retail Loss Prevention Consultant. James and his team rewrote 90 percent of the chapter to reflect the many changes that have occurred in retail security over the past decade. These include the predominant use of "chain management" principles and the growing use of many types of technology, including the growing trend for web-based sales.

Violence/Drug Use

This chapter is a combination of two chapters. The violence materials are the work of David, who wrote the original chapter titled "Violence in the Workplace." The drug materials have been completed reworked by Dr. Sharon Larson, Health Sciences Department, Western Illinois University. Dr. Larson is an experienced practitioner in the health field, having served as a nurse in the U.S. Army, and a Health Education Specialist at the Beau Health Center in Macomb, Illinois before joining the faculty at Western.

Security and the Law

This chapter was thoroughly reviewed and updated by Steven C. Babb, retired Deputy General Counsel for Northrop Grumman Corporation. Steve's expertise in Human Resources and Labor Law, along with his extensive work with the Northrop Grumman Security Council on all security related legal issues and problems, made him the perfect contributor to this chapter.

Introduction

The chapters in Section I provide an overview of the security and loss prevention industry. Chapter 1 is a brief history of the development of the field in Europe and America, ending with a quick, crisp summary of the status of security in the 21st century. Chapter 2 identifies the roles of security, whether contract, proprietary or hybrid. Chapter 3 covers career options. Chapter 4 discusses the development of security as a profession. Issues discussed include training, certification and regulation. Chapter 5 provides the reader with an overview of the development and the important role of the Department of Homeland Security as it relates to the private sector.

As noted in the preface, the events of September 11, 2001 changed the face of security operations. Security is a common theme considered by almost every person in the developed and developing world. Yet, as security professionals know, the basic concepts and theories of security and loss prevention are not changed by a single event. The most significant changes are in the innovative tools that professionals use to achieve their goals. What is important in looking at the past is that we learn the lessons that are presented in the development of past security operations. We can then apply those lessons to the present situation, modify those that have potential to assist us in our efforts, and discard outdated and outmoded ideas and technology.

The information presented in Part I, along with recommendations just presented, will serve as a basis for understanding and applying the specific materials presented in Parts II and III.

Origins and Development of 21st Century Security

OBJECTIVES

The study of this chapter will enable you to:

1. Outline the historical development of security in America.
2. Discuss changing crime trends over the past decades.
3. Understand the role of professional associations/organizations in the development of a professional security industry.
4. Consider the changing role of security in our 21st century world.

Introduction

Security implies a stable, relatively predictable environment in which an individual or group may pursue its ends without disruption or harm and without fear of disturbance or injury. The concept of security in an organizational sense has evolved gradually throughout the history of Western civilization, shaped by a wide variety of institutional and cultural patterns.

In examining the origins and development of security, it should be noted that security holds a mirror up, not to nature, but to society and its institutions. Thus in medieval England there were programs to clear brush and other concealment on either side of the king's roads as a precaution against robbers, and to protect citizens from night thieves there were night watchmen. In the United States in modern times, these rudimentary security measures find their counterparts in the cleared areas adjoining perimeter fences and buildings, in security patrols, and in intrusion alarms. Throughout history it is possible to trace the emerging concept of security as a response to, and a reflection of, a changing society, mirroring both its social structure and its economic conditions, its perception of law and crime, and its morality. Thus, security remains a field of both tradition and dramatic change. The introduction of high-tech systems and computers has changed the nature of the job of the 21st century security professional. Security today must be directed toward modern problems including computer crime and world terrorism, yet we cannot forget the basic foundations on which the field has developed.

Security in England

The development of systems of protection (security) and law enforcement in England began to come with greater rapidity and sophistication beginning in the 14th through the 18th centuries. Seeds for this development were planted during the social revolution that heralded the end of the remaining elements of the feudal structure in the latter half of the 13th century.

Security was one thing in a largely rural society controlled by kings and feudal barons; it was another thing entirely in a world swept by enormous changes. The voyages of exploration, which opened new markets and trade routes, created a new and increasingly important merchant class whose activities came to dominate the port cities and trading centers. Concurrently, acts of enclosure and consolidation drove displaced small tenants off the land, and they migrated to the cities in great numbers.

By 1700 the social patterns of the Middle Ages were breaking down. Increased urbanization of the population had created conditions of considerable hardship. Poverty and crime increased rapidly. No public law enforcement agencies existed that could restrain the mounting wave of crime and violence, and no agencies existed that could alleviate the causes of the problem.

Different kinds of police agencies were privately formed. Individual merchants hired men to guard their property. Merchant associations also created the merchant police to guard shops and warehouses. Night watchmen were employed to make their rounds. Agents were engaged to recover stolen property, and the people of various parishes into which the major cities were divided hired parochial police.

Attention then turned to the reaffirmation of laws to protect the common good. Although the Court of Star Chamber, which gave the English monarchy all control over decisions of law, had been abolished in 1641, its practices were not officially proscribed until 1689 when Parliament agreed to crown William and Mary if they would reaffirm the ancient rights and privileges of the people. They agreed, and Parliament ratified the Bill of Rights, which for all time limited the power of the king as well as affirming and protecting the inalienable rights of the individual.

The 18th Century

By the 18th century, it is possible to discern both the shape of efforts toward communal security and the kinds of problems that would continue to plague an increasingly urban society into modern times.

In 1737, for instance, a new aspect of individual rights came to be acknowledged: For the first time, tax revenues were used for the payment of a night watch. This was a significant development in security practice because it was a precedent-setting step that established for the first time the use of tax revenues for common security purposes.

Eight years later, Parliament authorized a special committee to study security problems. The study resulted in a program employing various existing private security forces to extend the scope of their protection. The resulting heterogeneous group, however, was too much at odds. It proved ineffective in providing any satisfactory level of protection.

In 1748 Henry Fielding, magistrate and author (most notably of the unforgettable *Tom Jones*), proposed a permanent, professional, and adequately paid security force. His invaluable contributions included a foot patrol to make the streets safe, a mounted patrol for the highways, the famous "Bow Street Amateur Volunteer Force" of special investigators, and police courts. Fielding is credited with conceiving the idea of *preventing* crime instead of seeking to control it.

The Impact of Industrial Expansion

The Industrial Revolution began to gather momentum in the latter half of the 18th century. Like the migrations off the land 200 years earlier, people again flocked to the cities—not pushed this time, as they had been earlier, by enclosure and dispossession, but rather lured by promises of work and wages.

The already crowded cities were choked with this new influx of wealth seekers. What they found were long hours, crippling work, and miserly wages. Family life, heretofore the root of all stability, was virtually destroyed in this environment. Thievery, crimes of violence, and juvenile delinquency were the order of the day. All the ills of such a structure, as we see in analogous situations today, overtook the emerging industrial centers.

Little was done to alleviate the growing problems. Indeed the prevailing philosophy of the time argued against doing anything. In this new age in which statements of laissez-faire were generally accepted, industrial centers became the spawning grounds for crimes of all kinds. At one time, counterfeiting was so common that it was estimated that more counterfeit than government-issue money was in circulation. More than 50 false mints were found in London alone.

The backlash to such a high crime rate was inevitable and predictable. Penalties were increased to deter potential criminals. At one time, more than 150 capital offenses existed, ranging from picking pockets to serious crimes of violence. Yet no visible decline in crime resulted. It was a "society that lacked any effective means of enforcing the criminal laws in general. A Draconian code of penalties that proscribed the death penalty for a host of crimes failed to balance the absence of efficient enforcement machinery."[1]

Private citizens resorted to carrying arms for protection, and they continued to band together to hire special police to protect their homes and businesses.

For a more detailed consideration of the development of security and policing in England see an introductory text on law enforcement.

Security in the United States

Security practices in the early days of colonial America followed the patterns that colonists had been familiar with in England. The need for mutual protection in a new and alien land drew them together in groups much like those of earlier centuries.

As the settlers moved west in Massachusetts, along the Mohawk Valley in New York, and into central Pennsylvania and Virginia, the need for protection against hostile Indians and other colonists—French and Spanish—was their principal security interest. Settlements generally consisted of a central fort or stockade surrounded by the farms of the inhabitants. If hostilities threatened, an alarm was sounded and the members of the community left their homes for the protection of the fort, where all able-bodied persons were involved in its defense. In such circumstances, a semi-military flavor often characterized security measures, which included guard posts and occasional patrols.

Protection of people and property in established towns again followed English traditions. Sheriffs were elected as chief security officers in colonial Virginia and Georgia; constables were appointed in New England. Watchmen were hired to patrol the streets at night. As *Private Security: Report of the Task Force on Private Security* notes, "These watchmen remained familiar figures and constituted the primary security measures until the establishment of full-time police forces in the mid-1800's."[2]

Such watchmen, it should be pointed out, were without training, had no legal authority, were either volunteer or else paid a pittance, and were generally held in low regard—circumstances that bear a remarkable similarity to observations in the RAND report on private security in 1971.[3]

Development of Private Security

The development of police and security forces seems to follow no predictable pattern other than that such development was traditionally in response to public pressure for action.

Outside of the establishment of night watch patrols in the 17th century, little effort to establish formal security agencies was made until the beginnings of a police department were established in New York City in 1783. Detroit followed in 1801, and Cincinnati in 1803. Chicago established a police department in 1837; San Francisco in 1846; Los Angeles in 1850; Philadelphia in 1855; and Dallas in 1856.

New York, influenced by the recent success of the police reforms of Sir Robert Peel, adopted his general principles in 1833. By and large, however, police methods in departments across the country were rudimentary. Most American police departments of the early 19th century, as a whole, were inefficient, ill trained, and corrupt.

In addition, the slow development of public law enforcement agencies, both state and federal, combined with the steady escalation of crime in an increasingly urban and industrialized society, created security needs that were met by what might be called the first professional private security responses in the second half of the 19th century.

In the 1850s Allan Pinkerton (see Figure 1-1), a "copper" (police officers who were identified by the copper badges they wore) from Scotland, and the Chicago Police Department's first detective, established what was to become one of the oldest and largest private security operations in the United States, Pinkerton. Pinkerton's North West Police Agency, formed in 1855, provided security and conducted investigations of crimes for various railroads. Two years later the Pinkerton Protection Patrol began to offer a private watchman service for railroad yards and industrial concerns. President Lincoln recognized Pinkerton's organizational skills and hired the agency to perform intelligence duties during the Civil War. Pinkerton is also credited with hiring the first woman to become a detective in this country, well before the women's suffrage movement had realized its aims.[4]

FIGURE 1-1 Allan Pinkerton, President Lincoln, and Major General John McClellan. *(Photographed October 1862, Antietam, Md. Courtesy of National Archives.)*

In 1850 Henry Wells and William Fargo were partners in the American Express Company, chartered to operate a freight service east of the Mississippi River; by 1852 they had expanded their operating charter westward as Wells Fargo and Company. Freight transportation was a dangerous business, and these early companies usually had their own detectives and security personnel, known as "shotgun riders."

Washington Perry Brink in Chicago founded Brinks, Inc., in 1859 as a freight and package delivery service. More than 30 years later, in 1891, he transported his first payroll—the beginning of armored car and courier service. By 1900 Brinks had a fleet of 85 wagons in the field.[5] Brinks, Wells Fargo, and Adams Express were the first major firms to offer security for the transportation of valuables and money.

William J. Burns, a former Secret Service investigator and head of the Bureau of Investigation (forerunner of the Federal Bureau of Investigation [FBI]), started the William J. Burns Detective Agency in 1909. It became the sole investigating agency for the American Bankers' Association and grew to become the second largest (after Pinkerton) contract guard and investigative service in the United States. For all intents and purposes, Pinkerton and Burns were the only national investigative bodies concerned with nonspecialized crimes in the country until the advent of the FBI.

Another 19th-century pioneer in this field was Edwin Holmes, who offered the first burglar alarm service in the country in 1858. Holmes purchased an alarm system designed by Augustus Pope. Following Holmes, American District Telegraph (ADT) was founded in 1874. Both companies installed alarms and provided response to alarm situations as well as maintaining their own equipment. Baker Industries initiated a fire control and detection equipment business in 1909.

From the 1870s, only private agencies had provided contract security services to industrial facilities across the country. In many cases, particularly around the end of the 19th century and during the Great Depression of the 1930s, the services were, to say the least, controversial. Both the Battle of Homestead in 1892, during which workers striking that plant were shot and beaten by security forces, and the strikes in the automobile industry in the middle 1930s are examples of excesses from overzealous security operatives in relatively recent history.

With few exceptions, proprietary, or in-house, security forces hardly existed before the defense-related "plant protection" boom of the early 1940s. The impetus for modern private security effectively began in that decade with the creation of the federal Industrial Security Program (today named the Defense Industrial Security Program [DISP]), a subordinate command within the Department of Defense. The National Industrial Security Program (NISP) is the nominal authority in the United States for managing the needs of private industry to access classified information. The National Industrial Security Programs' Operating Manual (NISPOM/DoD 5220.22-M) today consists of 11 chapters and 3 appendices totaling 141 pages. The most recent 2006 revisions include The Intelligence Reform and Terrorism Prevention Act of 2004 and other changes taking effect since September 11, 2001.[6]

The Beginning of Modern Security

By 1955 security took a major leap forward with the formation of the American Society for Industrial Security (ASIS). Today the organization is the American Society for Industrial Security International, reflecting the global emphasis on security operations. For most practitioners, 1955 signifies the beginning of the modern age of security. Before 1955 there were no professional organizations of note, no certifications, no college programs, and no cohesive body to advance the interests of the field.

Today's changed climate for increased security services came as businesses undertook expanded operations that in turn needed more protection. Retail establishments, hotels, restaurants, theaters, warehouses, trucking companies, industrial companies, hospitals, and other institutional and service functions were all growing and facing a serious need to protect their property and personnel. Security officers were the first line of defense, but it was not long before that important function was being overchallenged by the increasing complexity of fraud, arson, burglary, and other areas in which more sophisticated criminal practices began to prevail. Security consulting agencies and private investigation firms were founded in increasing numbers to handle these special types of cases. From among these, another large contractor was to emerge and join the field alongside Pinkerton, Burns, Globe, and Brinks. In 1954 George R. Wackenhut formed the Wackenhut Corporation in company with three other former FBI agents. Today some of these giants, such as Guardsmark, continue, while others have changed branding. Today Wackenhut is G4S Secure Solutions. Pinkerton is now part of Securitas, the largest security firm today and operating globally. Burns has become a subsidiary of Borg Warner (Well Fargo). In addition, many regional firms like Per Mar Security, located in Davenport, Iowa, continue to develop and provide quality security services.

Early Security Information Sharing

The private sector entered security in another form during the 1960s and 1970s. Common businesses and industries created central repositories of security information deemed important to all of their common interests nationwide and made it available in various ways to their separate groups. Their purpose was to decrease loss by networking information that would prevent criminals from victimizing members of the group once anything was known that could be used to alert them.

Variously called "alliances," "bureaus," or possibly security or loss prevention "institutes," these groups became deeply entrenched as providers of valuable information and services. Their methods of dissemination vary with what is appropriate to the business for which they were founded but include circulating "hot" lists, newsletters with "wanted" pictures and descriptions of characteristic modes of operation, telephone chain calling to alert merchants within an area, and so on. Nationally available repositories of other types of industry-specific data are usually maintained also and can be accessed by members. These groups serve the private sector in its effort to survive against crime, and make their collected intelligence available to law enforcement.

Some presently existing groups are the National Insurance Crime Bureau (NICB) (www.nicb.org), the International Association of Arson Investigators (I.A.A.I.) (http://firearson.com), the Property Loss Research Bureau (PLRB) (www.plrd.org), and the Jewelers Security Alliance (JSA) (www.jewelerssecurity.org), to name a few. Still other groups serve similar functions by collecting records of insurance claims and spotting fraud, issuing periodic records of defaulted or dubious credit cards, and so on. The measures taken by these and other business associations to limit their losses and protect their members have spread to other areas in which there is today an increasing concern about excessive risk. Some of these areas include computer and other high-tech industries, and antiterrorism and executive protection alliances. The need for information for employment background checks has also led to the creation of information bureaus. The Internet has added its own twist in providing fast service for those looking for information ranging from criminal histories to credit checks. A number of Web sites sell information for fees ranging from a low of $15 to more than $100 for each search request.

The Costs

Expenditures in private security exceeded $100 billion annually in 2000.[7] The expenditures continued to grow especially following the September 11, 2001 attack on the World Trade Center with spending on home and security at $35 billion in 2001 and $69.1 billion in 2011.[8] Even as the anti-Vietnam War protest created a demand for additional security services during the 1970s, the threat of terrorism against U.S. business throughout the world, the kidnapping of executives assigned outside the United States by various extremist groups, and drugs and violence in the workplace create a demand for the 21st century. With this dynamic growth have come profits, problems, and increasing professionalism. Each is a significant part of the picture of security today.

Crime Trends and Security

During the 25 years roughly spanning the mid-1950s to the late 1970s, the United States became the victim of what the Task Force on Private Security of the National Advisory Committee on Criminal Justice Standards and Goals has called "a crime epidemic." The FBI's annual Uniform Crime Report Program (UCRP) documented the continuing steady increase in crimes of all types until 1981. Then, for the first time, the UCR Program reflected a modest overall decrease that has continued through the beginning of the 21st century.[9]

In 2005, however, 14,094,186 arrests occurred nationwide for all offenses (except for traffic violations), of which 603,503 were for violent crimes and 1,609,327 for property crimes. Although the number of arrests in 2005 increased only a slight 0.2 percent from the 2004 figure, arrests for murder rose 7.3 percent. An examination of the 2- and 10-year trends shows that the estimated number of property crimes in 2005 decreased 1.5 percent from the 2004 estimate and declined 13.9 percent when compared with the estimate for 1996. Preliminary figures for 2006 show a continuing trend with violent crime up 1.3 percent over 2005 and property crime down 2.9 percent over the previous year.[10]

Why this two-year increase occurred will be a topic of discussion for many years. However, the downward trend resumed in 2007 and continues through preliminary reports of 2011. Preliminary data for 2011 shows a 6.4% decrease in violent crime during the first 6 months of 2011, while property crimes decreased 3.7% during the same period.[11]

Gallup polls taken each year indicate that the fear of crime is an even greater problem than the crime rate itself would indicate. The consistency of survey results indicating that crime touched 25 percent of all American households during the year preceding each survey led Gallup to conclude that "the actual crime situation in this country is more serious than official governmental figures indicate." The most recent National Crime Victimization Survey (NCVS) data (for 2008) continue to show a 25-year decline in crime with the lowest reported levels in over 30 years. Over the past decade violent crime has declined 41% while property crime rates fell by 32%.[12] Although the NCVS indicates a decline in the number of offenses, the cost of business crime continues to be a major concern. The estimated figures on the extent of crime against business, ranging from $67 billion to $320 billion, have not been adequately studied since the mid-1980s and dramatize the absence of consistent hard data indicating the exact size of the problem today. Variations of billions of dollars in estimates are the result of educated guessing, interpolation, and adjustment for inflation. Still, some progress has been made in the last decade.

According to a 2002 Brookings Institute report, the Enron and WorldCom scandals alone cost the U.S. economy approximately $37 billion to $42 billion off the gross domestic product during the first year.[13] More recently, one individual, Bernard Madoff, has been charged with over 40 billion dollars in theft from investors.[14] A 2002 joint conference of the National White Collar Crime Center and the Coalition for the Prevention of Economic Crime identified the following as the most serious of economic crime problems:

- Money laundering
- Identity fraud
- E-commerce crime
- Insurance crime
- Victim services
- Terrorism.

The conclusion of the conference was that the amount of "dirty money" worldwide tops $3 trillion.[15]

The figure is double the $1.5 trillion dollar estimate for money laundering suggested by the International Monetary Fund. Sadly the United States Department of Treasury estimated that 99.9% of foreign criminal and terrorist money the U.S. attempts to stop gets deposited into secure accounts.[16] Obviously, figures vary, principally because satisfactory measures of many crimes against business and industry have not yet been found, but also because much internal crime in particular is never reported to the police, either because internal disciplinary action has already been taken, or to avoid bad publicity and management embarrassment that could result from exposing to the public the business's lack of security controls. Nevertheless, such questions as may exist concern only the degree, not the fact, of the dramatic escalation of crimes against business in our society. Security concerns remain constant for employee theft, property crime, and issues related to life safety. The newest problems revolve around fraud, computer crime, workplace violence, and terrorism.

As this brief history of security has indicated, there is always an intimate link between cultural and social change and crime, just as there is between crime and the security measures adopted to combat the threat. A bewildering variety of causes, both social and economic, are cited for criminal behavior in this era. Among them are an erosion of family and religious restraints, the trend toward permissiveness, the increasing anonymity of business at every level of commerce, the decline in feelings of worker loyalty toward the company, and a general decline in morality accompanied by the pervasive attitude that there is no such thing as right and wrong, but rather only what feels good.

In addition, the rapidly changing technology of business and personal lives is often far out ahead of security measures used to protect personal and business intellectual property. The dominance of the computer and related technology in business has improved worldwide business efficiency, but not without a price. The Internet, while providing the path for information transfer, has also provided unheard of opportunities to steal or manipulate intellectual property. Who had heard of a computer virus in the 1970s?

These changes in attitudes, personal values, and technology have created a new problem for security managers. In 1990 McDonnell Douglas Corporation fired 150 employees who allegedly used interest-free company loans intended for the purchase of computers to buy stereo equipment and other luxury items. A data-processing employee reportedly processed the $4,000 loans by printing phony invoices for computers.[17] These problems are dwarfed by the problems created with accounting practices at Arthur Andersen, WorldCom, and Enron, and, as noted above, the Bernie Madoff scandal.

In the wake of the Enron scandal, the government passed the Sarbanes-Oxley Act of 2002 to combat corruption in public companies. This regulation has aided the Chief Security Officer (CSO) by hopefully minimizing scandal and requiring greater financial disclosures, better scrutiny by corporate boards and their audit committees, and tighter overall accounting controls. Oversight by SEC regulators coupled with stronger internal control mechanisms clearly define white collar crimes as being prevalent and a security challenge that cannot be overlooked. Since the stock market crash of 2008, additional regulation of the financial business by the federal government has been growing. New legislation, as well as major amendments to existing regulatory rules, has been ostensibly enacted to make sure that further financial instability in the markets will not happen in the future. It is far beyond the scope of this book to attempt to analyze or even to catalog all of the factors involved in the trend toward increasing crime even were we to restrict such a study to crimes against business and industry. What is important here is to make clear note of the fact of such increases—and of their impact on society's attempts to protect itself.

Most significant is the realization that "the sheer magnitude of crime in our society prevents the criminal justice system by itself from adequately controlling or preventing crime."[18] In spite of their steady growth, both in costs and in numbers of personnel, public law enforcement agencies have increasingly been compelled to be reactive and to concentrate more of their activities on the maintenance of public order and the apprehension of criminals. Even community-oriented policing rests on the need for a cooperative approach to law enforcement. The approximately 650,000 local law enforcement personnel in this country cannot possibly provide protection for all those who need it.[19]

Growth of Private Security

Society has in recent times relied almost exclusively on the police and other arms of the criminal justice system to prevent and control crime. But today the sheer volume of crime and its cost, along with budget cutbacks in the public sector, have overstrained public law enforcement agencies. Private security must play a greater role in the prevention and control of crime than ever before. The Institute for Law and Justice, authors of the 1990 *Hallcrest II* report, indicated that by 2000 there would be more than 1.9 million persons employed in private security, with total expenditures for its products and services estimated at $100 billion.[20] This compares with police protection expenditures for federal, state, and local governments of only $45 billion. This gap has narrowed as the government agencies responded to the events of September 11, 2001. The Bureau of Labor Statistics reported that there were 1.2 million protective services employees in 2009 compared to public police officers who numbered 641,590.[21]

Growing Pains and Government Involvement

Inevitably, the explosive growth of the security industry in the second half of the 20th century was not without its problems, leading to rising concern for the quality of selection, training, and performance of security personnel. The hijackings that led to the destruction of the World Trade Center in 2001 were blamed on poorly trained contract security screeners at U.S. airports and governmental intelligence agencies. Whether the blame is fair may be debatable, because screeners were not looking for box cutters or other implements used by the hijackers. Within months the U.S. government had established federal control over this segment of security. Even within the industry itself, there is increasing pressure for improved standards, higher pay, and greater professionalism. The American Society for Industrial Security (ASIS) International has developed industry standards that are regularly being discussed by representatives in the security industry and federal government.

Considering the importance of private security personnel in the anticrime effort and their quasi-law enforcement functions, it is ironic that they receive so little training in comparison to their public-sector contemporaries. According to a 2005 study by Associated Criminal Justice and Security Consultants, the median number of hours of basic police training is 720 hours prior to licensing or certification by state police training boards. The same study found that many security officers, on average, receive less than 8 hours of pre-job training.[21] And to complicate these figures, often this training is completed through an orientation video. Still, there are contract and proprietary security operations that provide very good training programs. Some contract security companies, such as Wackenhut for example, have client contracts established that provide from 40 to 120 hours of pre-post assignment training requirements, dependent upon the designated officer's position. This does not include on-the-job (OJT) training. In addition, a specified 16-hour annualized training requirement to refresh officers and avoid complacency can be established. Security managers must not overlook the values of maximizing training opportunities and requirements.

This situation was debated during the 1990s. The Gore Bill, introduced in 1991 by then-Senator Al Gore (D-TN), recommended minimum training for all security personnel without setting a minimum standard. The Sundquist Bill, introduced in 1993 by Representative Don Sundquist (R-TN), spelled out specific training requirements, adding to the 1991 Senate bill. The Sundquist Bill recommended 16 hours of training for unarmed officers and 40 hours for armed personnel. Also in 1993, Representative Matthew Martinez (D-CA) reintroduced a bill mandating 12 hours of training for unarmed security personnel and 27 hours for armed officers. What is obvious is that the federal government has started to take an active interest in setting minimum standards for the security profession. (For a discussion of these bills, see "Why Is Security Officer Training Legislation Needed?" by John Chuvala III, CPP, and Robert J. Fischer, *Security Management*, April 1994.) Still, it is important to note that *no* federal legislation regulating private security had been passed until 2002 following the World Trade Center disaster. With the support of ASIS, the Private Security Officer Employment Standards Act of 2002 was passed, allowing all security employers access to federal employment background checks through National Crime Information Center.

Professionalism

Today private security has moved toward a new professionalism. In defining the desired professionalism, most authorities often cite the need for a code of ethics and for credentials including education and training, experience, and membership in a professional society.

This continuing thrust toward professionalism is observable in the proliferation of active private security professional organizations and associations. It is promoted by such organizations as ASIS (which has a membership of more than 30,000 security managers), the Academy of Security Educators and Trainers (ASET), the International Association for Healthcare Security and Safety (IAHSS), the National Association of School Security Safety and Law Enforcement Officers (NASSLEO), and the Security Industry Association (SIA). It finds its voice in the library of professional security literature—magazines, Internet sites, and books. And it looks to its future in the continued development of college-level courses and degree programs in security.

ASIS has adhered to a professional code of ethics, one mark of a true profession, since its inception. The group established the Certified Protection Professional (CPP) program, which requires security managers desiring certification to be nominated by a CPP member and to complete a rigorous test. This program and others are discussed in more detail in Chapter 4.

Despite the many efforts to professionalize the field of private security, there are still many who feel that major obstacles need to be overcome. The most persistent one has to do with the training and education of the contract security officer. (A distinction between contract and proprietary officers needs to be made. Proprietary officers—those hired directly by a company—are generally better trained and better paid than are their contract counterparts.) Many officers—no matter whether they are contract or proprietary—are underpaid, undertrained, undersupervised, and unregulated. Minimal standards do exist in some places, but there is still a reluctance to train, educate, and adequately compensate the security force. Business considerations in making a product for profit can make it difficult for companies to see the need for paying for costly security programs. Thus they often opt for the lowest-priced solution whether it affords real protection or not. Fortunately, this kind of thinking is undergoing a change as industry realizes that the adage "you get what you pay for" very definitely applies to the quality of security. This realization should in turn add pressure to industry to upgrade the position of the security officer. Current standards, codes of ethics, and educational courses need to be supported by industry participation.

One development in the evolution of training for line security staff is the Certified Protection Officer (CPO) program, established in 1986 by the International Foundation for Protection Officers, a nonprofit organization. The CPO program is being offered at a number of colleges in the United States and Canada. Additional information on this program is presented in Chapter 4.

New Thinking

A systematic approach to security is appropriate today, as more and more businesses are giving the responsibility for protecting all aspects of company assets to the security and loss prevention department. Security and loss prevention has evolved well beyond the officer at the gate. Though that post is still vital, today's business assets comprise an almost infinite variety of protection needs. Moreover, security increasingly includes protection against contingencies that might prevent normal company operation from continuing and from making a profit. And as the concept of risk management is further integrated into a comprehensive loss-prevention program, the security function focuses less and less on enforcement and more on anticipating and preventing loss through proactive programming. Such challenges indisputably require high-level security management and an increasingly well-credentialed group of security professionals.

The systems approach, as outlined by C. West Churchman in 1968, is the process of focusing on central objectives rather than on attempting to solve individual problems within an organization. By concentrating on the central objectives, the management team can address specific problems that will lead toward the accomplishment of the central objective. As noted earlier, these central objectives for the 21st century include protection from terrorism and control of economic crimes as well as continuing to combat traditional security problems. Today, we talk about integrated security systems and total assets protection and contingency plans.

These problems must be approached from a team perspective. Public law enforcement at local, state, and federal levels, along with security interaction and operations, must work together, sharing intelligence to control these problems and reestablish a sense of security in the world's citizens. Security, therefore, is the safety of reassurance.

The establishment of joint councils within ASIS and the International Association of Chiefs of Police (IACP) has increased communication between the public and private sectors. In addition the National Sheriffs' Association (NSA) has its Private Security Industry Committee. These groups have developed numerous cooperative programs, of which only a few will be mentioned. In the aftermath of September 11, 2001, cooperation among various law enforcement agencies—local, state, and federal as well as private security organizations—has been enhanced. With the establishment of the Department of Homeland Security, the federal government has attempted to increase the interoperability of all areas of the criminal justice system in an effort to eradicate terrorism in this country.

Summary

Although modern security relies heavily on technology, the basic theory of protection has changed little over the past centuries. Only the tools to implement the theory have changed. Where moats were used, we have high-tech sensors and fences. Where warded locks once protected buildings and rooms, we see state-of-the-art, computer-controlled, electronic locking mechanisms. Where watchmen walked the beat, we now find sophisticated camera systems.

Still, not every security measure has kept pace with the development of technology. Old techniques and technology are still commonplace in many operations. The one thing that has changed as we enter the 21st century is the need to consider terrorism as a significant threat to our country and its businesses. Homeland Security and Terrorism are discussed in Chapters 5 and 16, respectively. Our government and public businesses are now targets of individuals who choose to use terrorist tools to make their positions known. The potential has always been present, as noted by many security experts. However, the use of terrorist tools was not seen as likely given the ability of most individuals to find other means to express their positions.

We have entered a new era where the security professional must give full consideration to potential terrorist threats, just as they would to theft of intellectual property, burglary, robbery, shoplifting, fire, and other loss risks commonly associated with security/loss-prevention strategies. Add to this list the threats against the computer, now one of business's most prominent tools, and the mix becomes very interesting!

In sum, the security industry of the 21st century is moving toward a model of single solution security integrating all areas and operation of the firm. Although there are many parts to the plan, in the final plan there is convergence. Technology, information, human components, design, etc. must all be considered in the final operation. The following chapters discuss specific areas of interest, but in the end a good security program must consider all elements in order to succeed in today's world.

■ ■ CRITICAL THINKING ■

Why should a security professional have any interest in the historical development of the discipline?

Review Questions

1. What events in medieval England brought about the creation and use of private night watches and patrols?
2. How did World War II affect the growth of modern private security?
3. How do you believe the events of September 11, 2001 impacted the changes occurring in the private security/law enforcement relationship?
4. Discuss the extent of security's growth in this country. What are some of the reasons for the professionalization of the field of private security?
5. What do you believe are the greatest challenges facing private security in the 21st century?

References

[1] Van Meter C. Executive Director. In: Private security: report of the task force on private security. Washington, DC: National Advisory Committee on Criminal Justice Standards and Goals; 1976. [p. 30]

[2] Kakalik JS, Wildhorn S. In: Private police in the United States: findings and recommendations. (RAND Report R-869-DOJ)Santa Monica, CA: The RAND Corporation; 1971.

[3] Levine SA. In: Allan pinkerton: America's first private eye. New York: Dodd, Mead; 1963. [p. 33]

[4] Kakalik, Wildhorn. Private police. p. 94–5.

[5] Cunningham W, Strauchs JJ, Van Meter CW. In: *The hallcrest report II: private security trends 1970–2000.* Boston: Butterworth-Heinemann; 1990. [p. 295]

[6] www.dss.mil.odda/mspm06.htlm, downloaded 9/17/2010.

[7] Institute of Law and Justice, downloaded 1/10/2000.

[8] Dancs A. Homeland Security: Spending Since 9/11. costofwar.org, downloaded 3/10/2012.

[9] Federal Bureau of Investigation, U.S. Department of Justice. Uniform crime reporting program. Washington, DC: Government Printing Office.

[10] Ibid.

[11] www.fbi.gov/about-us/cjis/ucr/crime-in-the-u.s./2011/preliminary, downloaded 3/10/2012.

[12] Criminal Victimization 2008, BJS Bulletin, September 2009.

[13] Brookings Institute. Brookings Study Details Economic Cost of Recent Corporate Crises. www.brookings.edu/comm/news/20020725graham.htm.

[14] Three New Federal Lawsuits Against Madoff Family Business. www.dailyfinance.com/story/investing/court-appointed-trustee-madoff-family-lawsuit, downloaded 9/4/2010.

[15] Funny Money. The Economist Global Agenda, May 3, 2002. www.economist.com/agenda/displayStory.cfm?story_id+1116239.

[16] Dealing with Dirty Money. Hellenic Communication Service. www.helleniccomserve.com/dirtymoney.html, downloaded 9/4/2010.

[17] 150 Scheming Employees Fired. Peoria J Star May 27, 1990.

[18] Van Meter. p. 18.

[19] BLS Occupational Employment Statistics, 33-3051 Police and Sheriff's Patrol Officers, May 2009.

[20] Ibid.

[21] Associated Criminal Justice & Security Consultants, PLACE Project, Kaplan University, Evaluation of Basic Training Programs in Law Enforcement, Security and Corrections for Academic Credit, June 2005.

2

Defining Security's Role

OBJECTIVES

The study of the chapter will enable you to:

1. Define the concept of private security.
2. List various services offered by private security operations.
3. Understand the differences among proprietary, contract and hybrid security operations.
4. Discuss the issues that contribute to continued relations issues between private security operations and public law enforcement.

Introduction

During the 19th and 20th centuries, public police operated only on a local basis. They had neither the resources nor the authority to extend their investigations or pursuit of criminals beyond the sharply circumscribed boundaries in which they performed their duties. When the need arose to reach beyond these boundaries or to cut through several of these jurisdictions, law enforcement was undertaken by such private security forces as the Pinkerton Agency, railway police, or the Burns Detective Agency.

As the police sciences developed, public agencies began to assume a more significant role in the investigation of crime and, through increased cooperation among government agencies, the pursuit of suspected criminals. Concurrent with this evolution of public law enforcement, private agencies shifted their emphases away from investigation and toward crime prevention. This led to an increasing use of security services to protect property and to maintain order. Today, in terms of numbers, surety forces are by far the predominant element in private security.

But what other protective measures are available? Who provides them? Who is responsible for planning and executing these procedures? Where do the roles of private and public police overlap, and where do they diverge? What are the particular hazards for which private security is now held responsible, and how is it determined that threats are sufficient to justify the adoption of protective procedures? To answer these questions, it is necessary to define private security and its role more exactly.

What Is Private Security?

Although the term *private security* has been used in previous pages without question, there is no universal agreement on a definition or even on the suitability of the term itself. Cogent arguments have been made, for example, for substituting the term *loss prevention* for security.

The RAND report defines private security to include all protective and loss-prevention activities not performed by law enforcement agencies. Specifically,

> *the terms private police and private security forces and security personnel are used generically in this report to include all types of private organizations and individuals providing all of security-related services, including investigation, staffing key posts, patrol, executive protection, alarm monitoring and response, and armored transportation.*[1]

The Task Force on Private Security takes exception to this definition on several grounds. The task force argues that "quasi-public police" should be excluded from consideration on the grounds that they are paid out of public funds even though they may be performing what are essentially private security functions. The task force also makes the distinction that private security personnel must be employees of a "for-profit" organization or firm as opposed to a nonprofit or governmental agency. The complete task force definition states:

> *Private security includes those self-employed individuals and privately funded business entities and organizations providing security-related services to specific clientele for a fee, for the individual or entity that retains or employs them, or for themselves, in order to protect their persons, private property, or interests from varied hazards.*[2]

The task force argues that the profit motive and the source of profits are basic elements of private security. While this definition might be suitable for the specific purposes of the report, it hardly seems acceptable as a general definition. Many hospitals and schools, to name only two types of institutions, employ private security forces without for-profit orientation. Yet it would be difficult to contend, for example, that the members of the International Association for Healthcare Security and Safety (IAHSS) are not private security personnel.

The Hallcrest reports never formally defined the terms *security* or *loss prevention* but relied on the earlier definitions of these terms. These reports consider, however, the security or loss-prevention field in its broadest application and thus avoid getting bogged down in discussions of profit motive or specific tasks. The reports focus on the functional aspects of security, recognizing that the functions of security and loss prevention are performed by both the public law enforcement sector and private agencies.

Thus, neither the profit nature of the organization being protected nor even the source of funds by which personnel are paid holds up as a useful distinction. A night watchman at a public school is engaged to protect a nonprofit institution and is paid out of public funds. His function, however, is clearly different from that of a public law enforcement officer. He is—and is universally accepted as—a private security officer. How then should private security be defined for the purpose of this text?

The opening lines of Chapter 1 suggest that "*security* implies a stable, relatively predictable environment in which an individual or group may pursue its ends without disruption or harm and without fear of disturbance or injury." Such security can be effected by military forces, by public law enforcement agencies, by the individual or organization concerned, or by organized private enterprises. Where the protective services are provided by personnel who are paid out of public funds and also charged with the *general* responsibility for the public welfare, their function is that of public police. Where the services are provided for the protection of *specific* individuals or organizations, they normally fall into the area of private security.

Protection of Life and Property

The hazards against which private security seeks to provide protection are commonly divided into man-made and natural. Natural hazards may include fire, tornado, flood, earthquake, hurricane, blizzards, and other acts of nature that could result in disruptions or damage to the enterprise or organization such as to cause building collapse, equipment failure, accidents, safety hazards and other events that interrupt normal business processes causing work delays, stoppages and loss of revenue. It should be noted that fire is also quite often man-made, intentionally or unintentionally.

Man-made hazards may include crimes against the person (for example, robbery or rape) or crimes against property (theft and pilferage, fraud and embezzlement). In addition, man also creates problems through terrorism (domestic and international), espionage and sabotage, civil disturbances, bomb threats, fire (as noted above), workplace violence and accidents.

The degree of exposure to specific hazards will vary for different facilities. The threat of fire or explosion is greatest in a chemical plant; the potential of loss from shoplifting or internal theft is greatest in a retail store. Each organization or facility must ideally be protected against a full range of hazards, but in practice, a particular protection system will emphasize some hazards (those most likely to occur) more than others.

In some organizations, the area of accident prevention and safety has taken on such importance, primarily because of state and federal occupational safety and health legislation, that this responsibility has become a full-time objective in itself, in the charge of a director of safety. Security can then devote its energies to other areas of loss. Similarly, some large industrial facilities have full fire brigades. In most situations, however, both fire, accident prevention and disaster preparedness and business recovery are part of the responsibility of the security department. Disaster preparedness and business recovery are sub-processes of the macro processes of contingency planning and business resiliance. Each is discussed further in Chapter 11.

Security Functions

Security practices and procedures cover a broad spectrum of activities designed to eliminate or mitigate the full range of potential hazards (loss, damage, or injury). These protection measures may include but are by no means limited to the following:

1. Building and perimeter protection, by means of barriers, fences, walls, and gates; protected openings; lighting; and surveillance (security officers).
2. Intrusion and access control, by means of door and window security, locks and keys, security containers (files, safes, and vaults), visitor and employee identification programs, package controls, parking security and traffic controls, inspections, and security posts and patrols.

3. Alarm and surveillance systems to detect unauthorized intrusion attempts and irregularities.
4. Fire prevention and control, including evacuation and fire response programs, extinguishing systems, and alarm systems.
5. Emergency response, crisis management and disaster recovery planning.
6. Protection of intellectual property/data to include protection of information systems.
7. Prevention of theft and pilferage by means of vetting employees through the use of personnel screening, pre-employment background investigations, procedural controls, screening for promotions and positions of fiduciary responsibilities, and polygraph and PSE (psychological stress evaluator) investigations as needed.
8. Accident prevention and safety.
9. Enforcement of occupational crime- or loss-related rules, regulations, and policies.
10. Prevention of workplace violence.

In addition to these basic loss-prevention functions, security services in some situations might provide armored car and armed courier service, personal (also referred to as bodyguard) protection, management consulting, security consulting, and other specific types of protection.

These services may be *proprietary,* or in-house, in which case the security force is hired and controlled directly by the protected organization, or they may be *contract* security services, in which case the company contracts with a specialized firm to provide designated security services for a fee. Contract security employees are actually employees of the contract security firm. Most security functions may be provided by either proprietary or contract forces or services, and in practice, it is common to find a combination of such services used. This combination of proprietary and contract security is referred to as a *hybrid system.*

Security Services

Security services employed more than 1.1 million officers in 2000.[3] *Security* reported that after the September 11, 2001 bombing of the World Trade Center, 13 percent of respondents to the magazine's annual security survey indicated adding in-house staff or hiring more security officers through an outside contract service.[4] These predictions may have come true. According to the Bureau of Labor Statistics, there were 1.2 million protective service officers in 2001, reflecting at least a 10 percent increase over the 2000 figures. However, by 2005, there were still 1.2 million protective service officers.[5]

Security services in 1990 totaled approximately 107,000 companies doing an estimated $51 billion in business, primarily providing guard, investigative, central station alarm, and armored car and courier services. It has been said that six large, publicly owned firms dominate the industry, accounting for approximately half of all revenues generated by contract security services.[6] In 2000 *Security Services Report* indicated that 14 firms totaled more than $1.5 billion in corporate sales.[7]

In 1999 Securitas Group entered the contract security business in the United States with its purchase of Pinkerton's, Inc. By 2003 the Group had purchased other major guard firms, including Burns International, to become Securitas Security Services USA, Inc. Securitas provides security services to 80 percent of the Fortune 1000 companies producing annual revenues of over $2.5 billion. In 2012 Securitas Security Services USA, Inc. is by far the largest contract security firm operating in the United States.[8] In 2011 Securitas reported, in its Annual Report, that it remains the market leader in the USA and one of the leading security service providers in Canada and Mexico (See http://www.securitas.com/en/About-Securitas/Continued-global-growth/Market-position/or the Securitas Annual Report - 2011).

Many firms, particularly the smaller ones, specialize in specific types of services offered to a client. The larger the firm, the more likely it is to provide a full range of security services. The major categories of these services are security forces, patrols, consulting services, investigative services, alarm response, and armored car delivery and courier services. Moreover, many of the larger firms offer executive protection and event security as part of their service capabilities.

According to various sources, security officer services, whether proprietary or contract, are still in demand today despite the growth in the use of technology. People and companies turn to officers because psychologically they feel that technology or hardware may not be enough. *Security*'s reported prediction that the 1990s would be marked by diminishing in-house staff, redistribution of security decision making inside and outside, and an increased reliance on equipment[9] is only partially true. As noted above, in the aftermath of the September 11, 2001 terrorist attacks, security staffs increased by at least 10 percent. Although some proprietary firms are relying more on technology to reduce security-cost overhead, three basic trends are apparent. First, as the number of legal problems associated with inappropriate actions of officers increases, public outrage may eventually force states to regulate training and standards. However, as noted in Chapter 1, based on the June 2005 report by Associated Criminal Justice and Security Consultants, standards for security have not changed much since 2001. There have been several federal attempts to pass legislation mandating minimum standards for security personnel. In 2002 federal legislation to allow private security access to FBI records for screening of private security personnel was passed with the help of ASIS. Second, as the field grows, it will continue to attract better-qualified individuals. Of the 2,205 individual responding to the 2005 Annual Salary survey conducted by ASIS, 63 percent held at least an undergraduate degree. Almost a quarter had a master's degree or higher degree.[10] And, third, there will be a trend to disarm security personnel. This trend was true until the destruction of the World Trade Center. The presence of armed security personnel in some venues appears to be increasing.

Contract versus Proprietary Services

Before beginning a discussion of contract and proprietary services, it is important to make a clear distinction between the two types of operations. Proprietary security operations are those that are "in-house," or controlled entirely by the company establishing security for its operations. The company—for example, Jones Distributing—hires a chief security officer (CSO) and all the necessary support personnel (making them employees of the company) and equipment to operate a security department. Contract security services, on the other hand, are those operations provided by a professional security company that contracts its services to a company. In this case, Jones Distributing would contract with Fischer Security Services for specific security services. In most cases there would not be a CSO employed directly by Jones. Rather, the contract manager would work for Fischer Security Services. As we shall note later, the latest trend is to have a combination of proprietary and contract operations.

Early researchers perceived more rapid growth in contract security services than in proprietary security. Although many firms are considering contract services, some existing proprietary security operations are converting to hybrids with proprietary management and contractual line services. *Security* reported that the trend would be toward increased use of contract employees, products, and services, causing the employee numbers in the contract area to double by the year 2000.[11] This prediction was at least partially accurate as we entered the new millennium. Several of the largest firms adopted contract security services to replace their proprietary systems. However, the change has not been a clear departure from company control to contract. Three Fortune 500 companies have made the move to hybrid systems utilizing proprietary oversight, contract officers, and increased reliance on electronic advancements to replace outdated equipment and guards.

Since the various contract functions described in the preceding pages can be undertaken as proprietary (or in-house) activities, how is the choice to be made between the two types of services? The subject of contract services versus proprietary security has been debated in most of the major security periodicals for 20 years or more. Some of the conclusions are reflected in the following discussion of the relative merits of the two approaches to security services. The question of which is the most sensible approach, however, is best answered by the manager of the firm or organization contemplating security services. His or her decision will rest on the particular characteristics of the company. These characteristics will include the location to be protected, the type of assets requiring protection, the size of the force required, its mission, the length of time the officers will be needed, and the quality of personnel required.

Advantages of Contract Services

Cost

Few experts disagree that contract officers are less expensive than is a proprietary unit. In-house officers typically earn more because of the general wage rate of the facility employing them. In many cases, that wage level has been established by collective bargaining.

Contract officers generally receive fewer fringe benefits, and their services can be provided more economically by large contract firms by virtue of savings in costs of hiring, training, and insurance because of volume. Short-term security service on a proprietary basis can create such large start-up costs that the effort is impractical.

Liability insurance, payroll taxes, uniforms and equipment, and the time involved in training, sick leave, and vacations are all extra cost factors that must be considered in deciding on whether to establish a proprietary force.

Administration

Establishing an in-house security service requires the development and administration of a recruitment program, personnel screening procedures, and training programs. It will also involve the direct supervision of all security personnel. Hiring contract officers solves the administrative problems of scheduling and substituting manpower when someone is sick or terminates employment.

There is little question that the administrative workload is substantially decreased when a contract service is employed. At the same time, the contracting customer is obliged to continually check the supplier's performance of contracted services, including personnel screening procedures, on an ongoing basis. The customer must also insist on a satisfactory level of quality at all times. To this extent, the management of the client firm is not totally relieved of administrative responsibilities.

Staffing

During periods when the need for officers changes in any way, it may be necessary to lay off existing officers or take on additional staff. Such changes may come about fairly suddenly or unexpectedly.

In-house forces rarely have this flexibility in staffing. If they have extra people available for emergency use, such staff are an unnecessary expense when they are idle. Similarly if there were a temporary decrease in the need for officers, it would hardly be efficient to dismiss extra people only to rehire additional officers a short time later when the situation changed again.

Unions

Employers in favor of nonunion security officers support their position by arguing that such officers are not likely to go out on strike, are less apt to sympathize with or support striking employees, and can be paid less because they receive few if any fringe benefits.

Since most unionized officers are proprietary personnel, anyone subscribing to the arguments listed here would clearly favor hiring contract officers. Only a fraction, if any, of the officers employed by contract security agencies—for example Securitas, Wackenhut, and Guardsmark—are unionized. Although efforts to unionize contract security services remain, the relationship between contractors and unions is mixed. According to Stuart Deans, a veteran security officer and staff representative for the United Steelworkers union, the entire security industry may be about to take a huge step backward. It appears that security firms are pushing for an unregulated climate.[12] Contract firms make their profit on providing security services; because wages and other labor costs account for most of a company's expenses, having workers with high salaries makes bidding new contracts difficult.

Impartiality

It is often suggested that contract officers can more readily and more effectively enforce regulations than can in-house personnel. The rationale is that contract officers are paid by a different employer and because of their relatively low seniority have few opportunities to form close associations with other employees of the client. If properly managed, this produces a more consistently impartial performance of duty.

Expertise

When clients hire a security service, they also hire the management of that service to guide them in their overall security program. This can prove valuable even to a firm that is already sophisticated in security administration. A different view from a competitive supplier trying to create goodwill with a client can always be illuminating.

Advantages of Proprietary Officers

Quality of Personnel

Proponents of proprietary security systems argue that the higher pay and fringe benefits offered by employers as well as the higher status of in-house officers attracts higher-quality personnel. Such employees have been more carefully screened and show a lower rate of turnover.

Control

Many managers feel that they have a much greater degree of control over personnel when they are directly on the firm's payroll. The presence of contract supervisors between officers and client management can interfere with the rapid, accurate flow of information either up or down.

An in-house force can be trained to suit the specific needs of the facility, and the progress and effectiveness of training can be better observed in this context. The individual performance of each member of the force can also be evaluated more readily.

Loyalty

In-house officers are reported to develop a keener sense of loyalty to the firm they are protecting than do contract officers. The latter, who may be shifted from one client to another and who have a high turnover rate, simply do not have the opportunity to generate any sense of loyalty to the specified—often temporary—client-employer.

Prestige

Many managers simply prefer to have their own people on the job. They feel that the firm gains prestige by building its own security force to its own specifications rather than by renting one from the outside.

Obviously, in weighing the various factors on either side of this debate, prudent managers will carefully study the quality and performance of the security firms available to service their facility. They will make sure of the standards of personnel, training, and supervision in the security firm. They will also make a careful analysis of the comparative costs for proprietary and contract services, and will estimate both services' relative effectiveness in a particular application.

In situations where the demand for officers fluctuates considerably, a contract service is probably indicated. If a fairly large, stable security force is required, an in-house organization might be favored. However, as noted earlier, the trend of the future is toward hybrid systems, with proprietary supervision and contract officers.

Deciding on a Contract Security Firm

Seven of ten security directors from America's largest firms report that one of their top security concerns is "finding and retaining a really quality-driven contract security agency."[13] A variety of issues must be considered when a company or organization decides to hire a contract security firm. According to Minot Dodson, a former vice president of operations and training, California Plant Protection, Inc., the following areas should be analyzed:

The scope of the work
Personnel selection procedures
Training programs
Supervision
Wages
Licensing and insurance
Benefits
Operating procedures
Contractor data
Terms and conditions[14]

The analysis of the scope of the work should include, at a minimum, locations, hours of coverage, patrol checkpoints, and duties. The security firm should be aware of, and prepared to enforce, all applicable corporate security policies—particularly those dealing with access control, personnel identification, documentation procedures for removal of materials from the facility, handling customers and employees, and emergency procedures. Proprietary security objectives and priorities should be stated clearly. The employing (client) firm should also include references to expansion plans and determine how the security firm will handle the expansion. The client firm should also spell out expectations for security goals (for example, a 20 percent reduction of shrinkage) and determine how the security firm plans to meet these goals. Client performance criteria should be spelled out to include such things as what and when to report.

The organization choosing a contract service should also be able to set standards for the employees who will be protecting their facility. Standards for general appearance, rules of conduct, licenses needed, physical condition, educational levels, reporting skills, background, and language ability are certainly worth listing. According to Dodson, "You should check the personnel file on each prospective officer. Look for pre-employment background and police record checks, and verify application information."[15] The client firm should check training records and test scores as well as psychological test results (that is, pen and pencil tests when they are available). It is even advisable in some operations to interview both the security company and the prospective officer. The security company should agree to remove any officer for reasonable cause (that is, violating regulations).

In 1975 the Task Force on Private Security set the minimum recommended level of training for security officers at 120 hours.[16] Lobbying efforts on the part of cost-conscious CSOs, however, have reduced the generally recommended time to only 40 hours. Despite this recommendation, many security companies still maintain only an 8-hour indoctrination period. Client firms should review the security service's training procedures to make sure they meet specific company requirements. Some areas to consider are patrol techniques, first aid, liability and powers, fire fighting, public relations, and report writing. Client firms also have the right to request additional special requirements. Whether the decision is to contract or not to contract, the firm will generally get what it pays for.

Supervision of the contract is another concern. Employing firms should understand the entire organizational structure in the security company. Supervisors from the contract service should maintain regular contact with officers (including alternate sites) and make random checks on all shifts and workdays. The response time of supervisors can be critical, and radio or telephone contact should be possible at all times. Direct contact with a supervisor, however, should also be available within no more than one hour.

Wages are tremendously important. The quality of personnel is often directly related to the wage level. The client company, not the security service, should establish the minimum wage to be paid to security employees. What is a good minimum wage? The Bureau of Labor Statistics reported that 50 percent of all guards made between $14,930 ($7.80/hour) and $21,950 ($11.43/hour) in 2002.[17] The 2005 ASIS Security Survey reported that the range for 2005 was $10 to a high of $22.50 per hour.[18] In 2012, PayScale, a market leader in global online compensation data, reported national average salary rates for "Security Guard Jobs" between $17,089 and $33,692.[19] While better than the 2002 levels, there are still some officers making less than a survival wage. One implication of this is that underpaid officers might take advantage of their employment and steal from the contracting firm. At a minimum, officers should be paid at least what semi-skilled labor in the area is earning.

Although fringe benefits offered by security firms might not seem an area of much concern to the employing companies, they should be. In a field where the turnover rate of some security services is 200 percent annually, fringe benefits become very important in retaining quality personnel. Benefits might include cash bonus plans, sick leave, health insurance, and overtime pay. Other perks might be life insurance, pension funds, and paid education and training. Perhaps the best fringe benefit for many contract officers is paid uniforms and equipment. While the cost of these perks is usually reflected in the cost to the buyer, it should not be taken out of the officer's already-meager wages.

The preceding discussion is just one way of viewing a security service; other factors can also be considered. Howard M. Schwartz, former vice president of the mid-Atlantic division of Securitas International Security Services, Inc., suggests evaluating:

1. The security agency's understanding of the psychological factors that influence security's effect on business and industrial environments and the firm's ability to incorporate these tactical measures into its services.
2. The agency's understanding of the essential difference between security and law enforcement.
3. The agency's ability to apply creative solutions to security problems.
4. The agency's ability to involve all of the client firm's employees in a positive effort supporting the overall security program.
5. The agency's willingness and ability to be flexible and modify tactical approaches to meet changing needs.[20]

For one suggested format, consider the security firm evaluation analysis presented in Table 2-1.

Table 2-1   Security Firm Evaluation Analysis

Consideration	Score
1. Bid package	
a. All requested information provided in proposal	—
b. Quality of proposal presentation	—
c. Timelines of submission of proposal	—
Subtotal	—
2. Personnel	
a. Past employment checks (pre-employment)	—
b. Reference checks (pre-employment)	—
c. Psychological testing (pre-employment)	—
d. Polygraph testing (pre-employment)	—
e. Prehire evaluation of personnel	—
f. Prehire evaluation of personnel files	—
g. Basic qualifications	—
h. Security aptitude testing (pre-employment)	—
i. Management quality (top level)	—
j. Management quality (midlevel)	—
k. Supervision (first line)	—
l. EEO Program	—
m. Average length of employee service	—
Subtotal	—
3. Training	
a. Prehire classroom training (8 hours minimum)	—
b. Prehire classroom training testing	—
c. Training manual	—
d. Training manual testing	—
e. Training film usage	—
f. Training film testing	—
g. Training facilities	—
h. On-the-job training program	—
i. On-the-job training tests	—
j. Continuing training program	—
k. Continuing training tests	—
l. Advanced training program	—
m. Advanced training tests	—
n. State-certified training school	—
o. College-certified training/trainers	—
Subtotal	—
4. Supervision	
a. Selection	—
b. Training and testing	—
c. Site supervision	—
d. Field supervision	—
e. Visits and assistance	—
f. Employee evaluation reports	—
g. Response capabilities	—
Subtotal	—

(Continued)

Table 2-1   (Continued)

Consideration	Score
5. Employee Wages and Benefits	
a. Wage distribution by grade and scale	—
b. Longevity rewarded	—
c. Merit pay proposed	—
d. Health insurance	—
e. Life insurance	—
f. Holidays	—
g. Vacation	—
h. Sick pay	—
i. Bereavement pay	—
Subtotal	—
6. Insurance	
a. General liability	—
b. Care, custody, and control	—
c. Errors and omissions	—
d. Employee dishonesty	—
e. Excess liability (umbrella form)	—
f. Workmen's compensation	—
g. Automobile liability	—
h. Policy exclusions	—
i. Policy availability	—
j. Cancellation notification	—
k. Self-insured on any portion of insurance	—
Subtotal	—
7. Operational Considerations	
a. 24-hour, 365-day operations department	—
b. Management response	—
c. Additional services ability	—
d. Uniforms provided	—
e. Uniform cleaning and maintenance provided	—
f. Emergency response capabilities	—
g. Post orders	—
h. Client control of operations	—
i. Financial stability	—
j. Standard of performance for guards	—
k. Service agreement	—
l. Periodic polygraph testing	—
Subtotal	—
8. Cost of Service	
a. Cost factor detail	—
b. Standard time fee	—
c. Overtime fees	—
d. Holiday fees	—
e. Effective rate	—

(Continued)

Table 2-1   (Continued)

Consideration	Score
f. Equipment fee	—
g. Billing periods	—
Subtotal	—
9. Other Considerations	
a.	—
b.	—
c.	—
Subtotal	—
Grand Total	—

Source: "How to Select a Guard Company," *Security World* (November 1983): 39. Copyright 1983. A Cahners Publication.

Instructions: Rate each proposal topic and other observation as follows: high = +1; average = 0; low = −1. Add all +1s and subtract all −1s in each section to obtain subtotals. Then add and subtract subtotals to obtain the overall rating. Ratings can be ranked as: 50 to 77, excellent; 25 to 49, above average; 0 to 24, average; −1 to −34, below average; and −35 to −77, poor. Retain this rating sheet to verify that you have done everything possible to select a competent security company.

Hybrid Systems

If the relationship between the client company and contractor, whether a straight security provision contract or a hybrid program incorporating both proprietary and contract security services, is to be successful, all parties to the contract must be willing to communicate openly with each other. This means that both the contractor and contractee must be willing to share in successes and mishaps. The in-house CSO has the power to optimize a contract.

A good hybrid security operation consists of four components:

- An engaged corporate liaison.
- Consistent contract management support.
- Periodic reviews.
- Accurate quality measurements.

Engaged Liaison

The right person for this job is someone who already knows and understands the basics of loss prevention/security. The company should not assume that the contract security firm would run itself. The liaison should monitor, but not micromanage, the security contract. The liaison should review security logs and follow-ups daily. The liaison should also be available to the contract manager to discuss any incidents or issues that need immediate attention.

Support

The contractor is also obliged to provide a responsive and interested manager. The manager must be able to juggle personnel, provide adequate training, satisfy customers, and still return a profit for the security company. The three keys to a successful contract are accessibility, meetings, and proper resource management.

Reviews

A periodic review of services being provided to determine if the contract is being fulfilled is essential. Officers assigned to the contract should continue to meet the expectations set forth in the contract. A company should periodically audit the contractor's records for compliance.

Measurement

Attainable and realistic activities should be expected. The contractor should make sure that officers know these expectations and comply. The liaison should expect compliance and when it is not forthcoming, action plans should be developed to remedy the situation.[21]

The growth in the hybrid security operation, while forecasted by the *Hallcrest II* report, is expanding at a rate greater than had been anticipated. According to Bill Cunningham, president of Hallcrest Systems, Inc., "the reduction in proprietary officers (was) dropping at 2% [per] annum compounded."[22] There are currently no statistics on whether this trend continues.

Private Security and Public Law Enforcement

It should be noted that public and private police might, in certain circumstances, perform the same functions for the same individuals or organizations. A law enforcement officer might in some circumstances be assigned to protect a threatened individual; a private bodyguard frequently is hired to perform the same protective function. Public police commonly perform patrol functions, which include checking the external premises of stores or manufacturing facilities. But patrol is also one of the major activities of private security. The activity itself then is not always a differentiating factor. Private security functions are essentially client-oriented; public law enforcement functions are society- or community-oriented.

Another key distinction is the possession and exercise of police powers—that is, the power of arrest. The vast majority of private security personnel have no police powers; they act as private citizens. In some jurisdictions, "special officer" status is granted, in most cases by statute or ordinance, which includes limited power of arrest in specified areas or premises. The limitations on the exercise of special police powers and the fact that their activities are client-oriented and client-controlled (as opposed to being directed primarily by public law enforcement agencies) make it reasonable to include such personnel as part of the private security industry. (This discussion omits the situation of the law enforcement officer who is moonlighting as a part-time private security officer because police powers in that situation derive from the public rather than the private role.)

As early as 1975 the Task Force on Private Security stated that "public law enforcement and private security agencies should work closely together, because their respective roles are complementary in the effort to control crime. Indeed, the magnitude of the nation's crime problem should preclude any form of competition between the two."[23] As noted in the previous section, however, even though the roles of the two groups are similar (in fact, overlapping in many areas), they are not identical. The roles should be complementary, but in reality the two groups interrelate and interact. Most contact between public and private agencies is spontaneous and cooperative, but far too often the contact is negative, to the detriment of both groups.

According to experts in the field, the relationship between the two groups continues to be strained (although personal contacts may be warm) because of several key issues:

1. Lack of mutual respect.
2. Lack of communication.
3. Lack of law enforcement knowledge of private security.
4. Perceived competition.
5. Lack of standards for private security personnel.
6. Perceived corruption of police.
7. Jurisdictional conflict, especially when private problems (that is, corporate theft, arson) are involved.
8. Confusion of identity and the issues flowing from it, such as arming and training of private police.
9. Mutual image and communications problems.
10. Provision of services in borderline or overlapping areas of responsibility and interest (that is, provision of security during strikes, traffic control, shared use of municipal and private fire-fighting personnel).
11. Moonlighting policies for public police and issues stemming from these policies.
12. Difference in legal powers, which can lead to concerns about abuse of power, and so on (that is, police officers working off duty may now be private citizens subject to rules of citizen's arrest).
13. False alarm rates (police resent responding to false alarms), which in some communities are more than 90 percent.

Historically, public police have often accused the private sector of mishandling cases, breaking the law to make cases, being poorly trained, and generally being composed of those who could not meet the standards for police officers. The private security sector often views the public sector as being self-centered and arrogant. Moreover, public law enforcement officers often moonlight, thereby taking work away from the private security sector. Even today the private sector is still considered by public police as only somewhat effective in reducing direct-dollar crime loss, and its contributions to reducing the volume of crime, apprehending criminal suspects, and maintaining order are judged ineffective. Public law enforcement has given private security low ratings in ten areas, including quality of personnel, training, and knowledge of legal authority. The feelings about the lack of training may be justified; only 50 percent of the states presently have mandated provisions for training security officers, although regulation through licensing has been the norm in most states.[24]

On the other hand, the employment of police officers as private security personnel during their off-duty hours has also caused much criticism. Some say that moonlighting police are only "hired guns," and that such police officers take jobs away from security firms. Other problems include the question of who is liable for the officer's actions. Is the employer of the off-duty officer liable, or does the liability stay with the police department who trained the officer?

Another source of conflict is the high rate of false alarms. While improved technology has reduced the number of false alarms, there are still problems associated with the human element, critter infringement and the occasional electronic failure. When an alarm sounds, an alarm company employee may respond, or the police department may be called. In some jurisdictions, police report that 10 percent to 12 percent of all calls for police service are from false alarms. Many experts believe that 95 percent to 99 percent of all alarms are false. Some police departments have reacted to this high rate by fining alarm companies or businesses and even delaying their response pending internal confirmation of an actual intrusion.

Yet much of the conflict between private and public agencies is the result of misconception. There is a general misunderstanding of the roles played by the respective agencies. Perhaps this is understandable because even within their own areas police and private security officers often fail to understand the common goals of other agencies.

Complementary Roles

Despite the misconception that working together for a common goal is difficult to achieve, police departments and private security agencies are beginning to work together, at times unknowingly. The idea that only the public police protect public property is another misconception. The federal government has more than 10,000 contract security officers patrolling federal offices and buildings. In many cities, police departments have turned to private agencies to protect courts, city buildings, airports, and museums. In other areas, it is generally accepted that the protection of private property is the responsibility of the owner. When a crime occurs, however, it is the local police who are usually called.

A third misconception is that the private security sector is primarily concerned with crime prevention and deterrence rather than with investigation and apprehension. In reality, store detectives in many major cities make more arrests each year than do local police officers. In addition, certain types of crime are no longer investigated by local police departments but have instead become the job of private security personnel; these include credit card fraud, single bogus checks, and some thefts. The growing problems of drugs and violence in the workplace have also added to security's role in law enforcement. Cooperation between security and public law enforcement is vital in dealing with both these problems and the threat of terrorism.

Obviously some degree of complementary activity already exists. What can be done to improve the perceptions the two areas have of each other and foster cooperative efforts? A variety of methods could improve cooperation between the two areas. The formation of joint private- and public-sector task forces to study responses to terrorism and major crime issues and recommend strategies is crucial. Data files from both sectors should be more freely exchanged. Private security personnel are often not allowed access to information on criminal cases even as a follow-up on data originally entered by them. Third, joint seminars on terrorism and business crime have been developed to help the two areas better understand their respective roles.

It appears that the private sector will become increasingly involved in the crime prevention area; public law enforcement will then be free to concentrate more heavily on violent crimes and crime response. Most CSOs are willing to accept more responsibility for minor criminal acts within their jurisdictions. The new activities most likely include intrusion alarm response, investigation of misdemeanors, and many preliminary investigations of other criminal offenses. The public sector is also willing to give up some areas of responsibility because it is "potentially more cost effectively performed by private security."[25] Some areas—for example building security—will need to be shared if we are to be successful in our war on terrorism.

However, other areas of conflict remain that may take some time to resolve. Often CSOs do not report many criminal offenses. This is a source of concern for many police managers. CSOs may fail to report crimes for any of the following reasons: lax charging policies of prosecutors; administrative delays in prosecution; court proceedings that might reveal more about their organizations than management wants known, which could influence public, customer and investor perceptions of the company; and a perception that courts are unsympathetic to business losses.

Just how effective private security can be depends to a large degree on whether public law enforcement and private-sector professionals are able to form a close partnership. *Hallcrest II* recommended:

1. Upgrading private security. Statewide regulatory statutes are needed for background checks, training, codes of ethics, and licensing.
2. Increasing police knowledge of private security.
3. Expanding interaction. Joint task forces are needed, and both groups should share investigative information and specialized equipment.
4. Experimenting with the transfer of police functions.[26]

Progress continues to be made in all these areas. Nationally recognized law enforcement/private security partnerships include the NYPD Shield (Area Police/Private Security Liaison) and the Nassau County PD Security/Police Information Network (SPIN).[27]

Relationships after September 11, 2001

The events of September 11, 2001 have brought private security operations and public enforcement closer. In the United States, cooperation between public law enforcement and private security was evident during the anthrax scare that followed the attack on the World Trade Center. A study of the cooperation between the two areas is presented in Chapter 5.

The events of September 11th have helped build private security/public policing partnerships to prevent and respond to terrorism and public disorder. This was demonstrated by the 2004 National Policy Summit project supported by the U.S. Department of Justice, Office of Community Oriented Policing Services (COPS). This joint partnering effort involved the International Association of Chiefs of Police and the Security Industry Association (SIA), ASIS International (ASIS), the National Association of Security Companies (NASC) and the International Security Management Association (ISMA).

Through their working groups, summit participants made five recommendations. The first four are national-level, long-term efforts. The fifth recommendation relates to local and regional efforts that could begin immediately.

1. Leaders of the major law enforcement and private security organizations should make a formal commitment to cooperation.
2. The Department of Homeland Security and/or Department of Justice should fund research and training on relevant legislation, private security, and law enforcement–private security cooperation.
3. The DHS and/or DOJ should create an advisory council composed of nationally prominent law enforcement and private security professionals to oversee day-to-day implementation issues of law enforcement–private security partnerships.
4. The DHS and/or DOJ, along with relevant membership organizations, should convene key practitioners to move this agenda forward.
5. Local partnerships should set priorities and address key problems as identified by the summit:
   - Improve joint response to critical incidents
   - Coordinate infrastructure protection
   - Improve communications and data interoperability
   - Bolster information and intelligence sharing
   - Prevent and investigate high-tech crime
   - Devise responses to workplace violence

Execution of these recommendations should benefit all concerned:

- Law enforcement agencies will be better able to carry out their traditional crime-fighting duties and their additional homeland security duties by using the many private security resources in the community. Public–private cooperation is an important aspect—indeed, a potent technique—of community policing.
- Private security organizations will be better able to carry out their mission of protecting their companies' or clients' people, property and information, while at the same time serving the homeland security objectives of their communities.
- The nation as a whole will benefit from the heightened effectiveness of law enforcement agencies and private security organizations.[28]

Summary

Security/loss-prevention functions, while diverse, have a common goal. As noted earlier the definition of security has been debated, but the bottom line is clear. Security services protect both private and public places. Law enforcement protects both public and private property. The difference is found in their primary goals. Law enforcement agencies are charged with the protection of government interests, representing the people. Private security is charged with protecting a specific interest, whether public or private.

The old distinction between public law enforcement and private security will continue to exist. However, the story is different when considering the relationship between contract and proprietary security. As business continues to evolve, it appears that the hybrid systems will become a dominant organizational scheme for many businesses establishing security operations.

Finally, all people concerned with security, whether federal government agencies, state law enforcement organizations, local law enforcement, or private security, will need to learn to work together to focus resources needed to successfully combat threats created by the potential of terrorist attacks and cyber crimes.

■ ■ CRITICAL THINKING ■

If you were to develop a security operation within a major company—for example, The Boeing Company—would you favor having a contract, proprietary or hybrid security organization?

Review Questions

1. What does the term *private security* mean?
2. What are the differences between proprietary and contractual security services?
3. What are the basic services typically performed by contractual security personnel?
4. What are the advantages and disadvantages of using contractual security services?
5. What are the advantages and disadvantages of using proprietary security services?
6. What factors should be considered when deciding on a security firm?
7. Describe the relationship between public law enforcement and private security. What are the major problems?

References

[1] Kakalik JS, Wildhorn S. In: Private police in the United States: findings and recommendations. (RAND Report R-869-DOJ) Santa Monica, CA: The RAND Corporation; 1971. [p. 3]

[2] Van Meter C Executive Director. In: Private security: report of the task force on private security. Washington, DC: National Advisory Committee on Criminal Justice Standards and Goals; 1976. [p. 4]

[3] Protective service occupations. U.S. Department of Labor, Bureau of Labor Statistics; February 2002, Bulletin 2540-11.

[4] Zalud B. 2002 Industry Forecast Study Security Yin-Yang: Terror Push, Recession. Security January 2002.

[5] U.S. Department of Labor, Bureau of Labor Statistics; 2001, 2005, 2006a.

[6] Siatt W, Matteson S. Special report: trends in security. Security World January 1982:25.

[7] Security services—private companies report. Freedonia. October 2000, www.ecnext.com.

[8] Securitas in the United States. http://www.securitas.com/us/en/About-Securitas1/Securitas/

[9] Zalud B. What's happening to security. Security September 1990:42.

[10] 2005 ASIS Survey of U.S. Security Salaries. ASIS International; 2005.

[11] Zalud B. p. 44.

[12] Lyons T. Labour looks out for security guards: the private-security industry rolls back union gains of the '90s. www.eye.net/eye/issue/issue 03.02.00/news/guards.html.

[13] Dalton. Looking for the quality-oriented contractor. Secur Technol Des September 1994:6.

[14] Serb TJ. How to select a guard company. Secur World November 1983:33.

[15] Ibid.

[16] Cunningham W, Strauchs JJ, Van Meter CW. In: The hallcrest report II: private security trends 1970–2000. Boston: Butterworth-Heinemann; 1990. [p. 290]

[17] Protective service occupations.

[18] 2005 ASIS salary survey.

[19] See http://www.payscale.com/research/US/Job=Security_Guard/Hourly_Rate.

[20] Serb. p. 33.

[21] Harne EG. Partnering with security providers. Secur Manage March 1996:36–9.

[22] Hybrid staffing grows as contract replaces proprietary. Security April 1996: 86.

[23] Van Meter. p. 19.

[24] Associated Criminal Justice & Security Consultants, PLACE Project, Kaplan University, Evaluation of Basic Training Programs in Law Enforcement, Security and Corrections for Academic Credit, June 2005.

[25] Cunningham W, Strauchs JJ, Van Meter CW. In: The hallcrest report II: private security trends 1970–2000. Boston: Butterworth-Heinemann; 1990. [p. 290]

[26] Cunningham W, Taylor TH. In: Private security and police in America: the hallcrest report. Portland, OR: Chancellor Press; 1985. [p. 275]

[27] Security-police information network. Study of law enforcement-private security partnerships. COPS Office funded project, 2007.

[28] National policy summit: private security/public policing. Olhausen Research, Inc.; 2004: 3–4.

3

Career Opportunities in Loss Prevention

OBJECTIVES

The study of this chapter will enable you to:

1. Discuss key factors that affect the security job market.
2. Know the various job opportunities open to individuals pursuing careers in security.

Introduction

Security is a critical management function in most businesses. Where security strategies and operations were almost unheard of 35 years ago, there are now vice presidents of loss prevention and even chief security officers (CSOs) at most multi-national corporations who report directly to the chief executive officer (CEO) or the chief operating officer (COO). In most instances, the security position has become an integral part of management decisions similar to, for example, the vice presidents of operations, manufacturing, sales and distribution. In the 21st century, security is without a doubt playing a major role in protecting companies in a wide array of areas including physical and personnel assets, cyber, sales, health and safety, as well as a company's brand reputation.

Security career opportunities in areas of business, industry and government vary but all have a common theme—to protect the enterprise while mitigating risk. Security has also become more international in scope as many companies have operating entities in most if not all continents around the world. This has become important for companies to compete in a global marketplace. Security managers of the future will need to "think global" in order to respond local.

Key Factors Increasing Security Opportunities

Among the factors that create inviting career opportunities in security, none is more significant than the explosive growth of the protection function, as was briefly described in Chapter 1. The events of September 11, 2001 also added to an already healthy growth of the security industry. *Security*, in its 2002 annual report on the status of security, indicated that three factors were at work impacting the security field. First, in the aftermath of September 11th, organizations were forced to reevaluate protection efforts and this spurred spending on electronic security and hiring of additional security personnel. Second, the cost of insurance continued to spiral upward. This trend forced companies to shift more emphasis to security or accept uninsured losses. Third, the recession in the economy put pressure on security executives to cut costs.[1] In 2007 the economy was improved and the cost of insurance was once again under control, and companies were investing in the protection of company assets and hiring "professional" security staff. However, given the meltdown in the economy in 2008 and its steady rebound, security managers will need to remain flexible in addressing coverage models and adjusting budgets accordingly.

Positive considerations for careers in security, including individual advancement or growth within the field, include the following:

- The continuing professionalism of security as reflected in higher standards of educational and certification criteria, and experience resulting in higher salaries, especially at management levels. In essence, security has become more than a job; it has become a career for most entering the field today.
- The rapid growth of the loss prevention function has created a shortage of qualified personnel with management potential, meaning less competition and greater opportunities for advancement for those who are qualified. Those with proven international experience will have "a leg up" in the industry.
- The shift in emphasis to programs of prevention and service rather than of control or law enforcement has broadened the security function within the typical organization. This broadening of responsibility often necessitates a "generalist" approach, which may require security practitioners to be knowledgeable in the traditional areas of security operations and investigations, and in specialized areas. This would include emergency response, crisis management, OSHA and other regulatory agency compliance, various audit functions (with overlapping duties touching on information technology (IT) or information security (InfoSec)) as well as the employee relations aspect of human resources (HR).
- The presence of both 2- and 4-year degree programs as well as master's-level study in criminal justice and/or security has created a rising generation of trained security personnel at the corporate management level. Many companies, especially larger corporations, are actively pursuing candidates with a degree. In fact, for most companies, a 4-year degree has become the minimum requirement to be considered for a career opportunity.
- The demographics of an aging police and security force at the management level create opportunities for advancement.

The Security/Loss-Prevention Occupation

Whether you recognize the protection function by titles such as loss prevention, security administration, or industrial security, the basic function of modern security remains the same. Security helps protect people and prevents losses to the enterprise it is commissioned to protect. For every product manufactured, someone is waiting in the wings to make an illegal profit by stealing, counterfeiting or manipulation of processes and records. For every security device installed, some "would be" criminal is determined to find a method to defeat it. And as the events of September 11th have made crystal clear, for every business and political entity, there is now a potential for attacks based on political and philosophical reasons or a martyr's call to action.

Much like law enforcement itself, the security profession is basically a recession-proof occupation, particularly at the security officer level. The need for educated and trained security officers and administrators is increasing with the need to counteract terrorism, computer crime, embezzlement, employee theft, drugs and violence in the workplace, fraud, and shoplifting. The U.S. Department of Labor indicated that employment in security is expected to grow faster than the average for all occupations through 2010 as concerns about crime, vandalism, and terrorism continue to increase the need for security.[2] The Freedonia Group, Inc. projects private security services will increase at a rate of 4.3 percent annually through 2012 while security equipment demand will rise by 7.4 percent annually through 2014.[3] These projections are based on a perceived risk of crime and low expectations of public resources to respond in a timely manner.

Security professionals are hired by many different organizations and at all levels—line; lower, middle, and upper management; corporate; and so on. Among organizations that have security operations are banks, colleges, government agencies, hospitals, public utilities, restaurants, hotels, retail stores, insurance companies, museums and other cultural institutions, mining firms, oil companies, supermarkets, telecommunications and Internet companies, transportation companies, and office buildings. Within each of these broad areas, security personnel perform many different functions, including personnel protection, computer security, coupon security, disaster management, crime prevention, proprietary information security (intellectual property), white-collar crime and employee investigations, counterterrorism, guard force management, investigations, physical security, crisis management, plant security, privacy and information management, fire prevention, environmental health and safety, and drug abuse prevention and control.

The American Society for Industrial Security International (ASIS) Committee on Academic Programs has suggested that students seeking careers in security should pursue course work in security, computer science, electronics, business management, law, police science, personnel, and information management.[4] This suggestion is supported by security educators and practitioners in *Suggested Preparation for Careers in Security/Loss Prevention*.[5] Building on the ASIS committee statements, the editors developed a book of readings that have a common thread: the need for security/loss-prevention personnel to have a broad understanding of various disciplines and specific skills for certain specialties. The editors suggest that specific skills are needed for all students of security—communication, management, and law. Other subject areas, such as fire and computer security, will depend on the students' interests.

Security Manager

Salaries of security directors (CSOs) average $108,000 per year, according to a 2010 ASIS salary survey.[6] While the average salary is $108,000, many security directors make more than

$250,000 per year, which includes incentive compensation. This is the fifth year where compensation has increased from its $88,000 2006 base. The average 2001 security executive had 10 or more years of experience and at least a bachelor's degree.[7] Over 70% of the managers have at least an undergraduate degree with over 28% holding a master's.[8] The 2010 data indicates the value companies are placing on education, as the author reports that many of these executives are young with experience in areas other than security. One possible answer to this shift is the principle of convergence in the evolving role of security professionals. The 2010 survey also notes the importance of professional certification for the 21st century security executive. Approximately 50% of the respondents hold a professional certification, of which 32% reported having achieved the ASIS Certified Protection Professional status.[9]

Personal Security

With the danger of kidnapping and threats from other areas—including disgruntled employees—the demand for executive protection specialists, or bodyguards, is increasing.

According to the Professional Bodyguard Association (PBA), today's bodyguard can be anyone who has the desire to complete the appropriate training. The professional bodyguard is not just the "tough guy" portrayed in Hollywood movies. The bodyguard of the 21st century must be well-trained and professional. Training includes familiarity with new technology, including threats of electronic bugging. However, the PBA also notes that traditional self-defense skills and firearms proficiency are still critical and necessary skill sets.[10]

What protective specialists need is common sense, the ability to pay attention to detail, and patience. Bodyguards should also know about laws and customs in different places (countries, states, cities) where they might be living or traveling with their principals. In a field once dominated by men, women are becoming more prominent as protection specialists. Most employers of executive protection specialists want a person who can fit into the executive's work and personal schedule.

Bodyguard schools are now becoming prominent, and many of the students are former law enforcement or military personnel. Skills taught include use of weapons and hand-to-hand combat. In addition, schools might also prepare the specialists with skills in protocol, dress, and specialized knowledge of alarm systems and closed-circuit TV (CCTV). Superior communication skills and an understanding of the principal's business are important skill sets to have to foster the best chemistry with the principal you are charted to protect. While the salaries and benefits are generally excellent, burnout is high. Long hours and time spent away from friends and family eventually take their toll.

Brand Protection

A new area of security is evolving called Brand Protection. Brand Protection in some cases falls under the security department and in other instances is a stand-alone speciality function within an organization. Brand Protection works to mitigate damage to a company's brand reputation, which includes reputational risk reviews and the mitigation of counterfeiting and grey marketing of a company's products. Most consumers choose a product based on its reputation for quality, and come back time and time again to that quality, reliability and safety in a broad area of consumer products such as fast food, electronics, games, and pharmaceuticals, to name just a few. Protecting a company's brand and its products directly protects the consumer while capturing and protecting revenue.

Private Investigators

The private investigator (PI) is a gatherer of facts or an information finder for evaluation purposes. In essence, the PI is a researcher who spends the greatest portion of time collecting background information for pre-employment checks on personnel, background checks of applicants for insurance or credit, and investigation of insurance claims. Much of this work is non-crime related although it may involve either part- or full-time undercover investigations of employee theft, workman's compensation fraud or detection of shoplifting.

Investigations in divorce-related matters are declining as divorce laws become more liberal. Investigations of "significant others," however, are increasing as people become more interested in what their love interests do in their spare time. Tracing missing persons or investigating criminal matters on behalf of the accused is a very small part of a typical investigator's work. Web searches are also becoming more commonplace, with individuals acting as their own private investigators. Many websites offer an information service for a fee that provides information such as criminal convictions on their love interests. Although there are situations, such as the long-term relationship between the American Bankers' Association and the Burns Agency, in which private investigators are called upon to supplement the work of public police, the great majority of private investigative work is complementary to the public law enforcement effort.

A growing number of investigators are engaged in litigation investigations—that is, gathering data for defense and plaintiff law firms preparing for trial in civil court. In general, PIs are involved in locating missing persons, obtaining confidential information, and solving crimes. Many PIs work for businesses and lawyers whereas others work independently. Independent offices may be only one-person operations or may employ several operatives or contract work to part-time investigators. These independent investigators charge from $35 to more than $150 per hour. Long hours and 7-day work weeks are the norm when a PI is on a case.

Good PIs develop skills that include the ability to conduct thorough surveillance and background checks. Some cases involve undercover investigations, which require a complete understanding of a sometimes dangerous assignment, i.e., including developing a cover and dealing with people who must not know who you are. The private investigation business requires the investigator to learn the law, as well as interviewing and investigative techniques. The best PIs also possess good verbal and written skills as well as analytical skills. Attendance at or membership in the National Association of Legal Investigators is an excellent aid to improving basic skills. Generally speaking, the outlook for jobs in this area is good. According to the Bureau of Labor Statistics, the market is stable and good in all states, with an estimated 40,000 private investigators in the United States.

A person interested in this field should consult individual state laws, which vary greatly, to determine what criteria must be met for licensing. Criteria vary from virtually none to extensive experience, completion of a written and/or oral examination, and interview by a State Board. Various certifications are available for those wishing to specialize. For example, the Certified Fraud Examiner (CFE), the Professional Certified Investigator (PCI), or the Physical Security Professional (PSP) certifications may add credibility and value to services offered by investigators. More will be said about certification in Chapter 4.

Consultants

Many firms that seek professional assistance in determining their vulnerabilities and risks look not to the police but to those in the private security sector, who are professionally trained to assess security needs and make appropriate recommendations. As with any profession, some people emerge as experts, who then sell their expertise for a fee. These people are commonly referred to as *consultants* because they are paid for their professional opinions and recommendations. In short, security consultants provide advice for a contractually agreed upon fee.

Security consultants are the specialists of the field. They generally operate as sole proprietors of a business that sells specific security expertise. Some security consultants charge more than $100 per hour and bill more than $200,000 annually. They do not work for specific equipment companies or firms. Advice commonly purchased from the consultant consists of information from three general areas:

1. Number, quantity, and use of security personnel
2. Direction and content of security policies and procedures
3. Alternatives in security hardware

Consultants also offer training seminars on specific problem areas, such as executive protection, computer security, and disaster planning. Technical consultants generally recommend, and in some cases install, security systems. Security managers often rely on these technically experienced individuals to recommend CCTV/monitoring systems, biometric or other access controls, alarms systems or software for a variety of purposes including systems integration. Security consultants are generally people who have paid their dues working in investigation or security management. Many have published articles and books. Today the completion of a Ph.D. is helpful since the title "Doctor" carries added weight if the consultant is called to testify in court. As civil litigation increases, the demand for security experts who can testify as credible witnesses also increases. It takes years of experience to achieve the levels of knowledge to operate in this arena.

Opportunities in Sales

With the tremendous growth in the use of technology, the opportunities to sell all types of security products are evident. For the person with the personality for sales, the opportunities to represent major security product manufacturers and distributors are excellent. Visit any major trade show, such as ASIS or the America's Security Expo (ASC), to get an idea of the number of well-versed sales representatives in the field.

Opportunities in Industry

Typically the greatest opportunities in industrial security exist in larger companies that employ proprietary security forces. Some of these proprietary security forces are unionized, which are collectively bargained under agreements with the United Plant Guards Workers of America (UPGWA). Many firms actively recruit the career-oriented person with a certificate or degree in a recognized security or criminal justice program.

However, due to cost-cutting measures, the trend toward *hybrid* security operations has reduced the demand for proprietary security officers in many companies. Large employers, such as General Motors, John Deere, and Caterpillar, have made a transition to a system of contract security officers with proprietary oversight. In addition these industrial giants have been actively installing state-of-the-art technology that has changed and in some cases supplanted the need for security officers.

Despite this reduction in demand, the security director at one major manufacturing facility with a highly progressive security program reports being interested in hiring only those applicants with at least a bachelor of science degree in security from a recognized security or criminal justice program.

Opportunities in Retail

The retail field provides a diversity of job opportunities in security, from the entry-level position of the uniformed security officer (or blazer-jacketed "host") to the internal theft investigator. Positions are available both with retail stores and chains and with security service companies, which provide such services as undercover and shopping investigations. There are many openings for those without experience but with the education, ambition, and aptitude that might make them successful in retail security. Many companies today provide their own training for shoppers and other investigators even though the employees have no investigative experience. Alertness, resourcefulness, courage, and self-assurance are often more important than specific experience.

There are many different types of operations in the retail industry. Security has had its impact in virtually every operation—from the discount store through the department store to the supermarket. The recognition of the importance of inventory shrinkage to the company's profit picture and the necessity for loss prevention is, or soon will be, almost universal. Those companies that do not accept this necessity, in the words of one ranking retail executive, simply will not remain in business.

The entry-level position in one company includes many students recruited from criminal justice and security programs as well as sales personnel crossing over to security. Many of these employees work part-time while they are going to school.

It will be interesting to follow this career field as traditional retail stores continue to see their markets diminished by Internet companies selling the same products. While traditional retail security jobs may decline, this change in sales opportunities opens new jobs for persons interested in security.

Opportunities in Health Care

Hospital security officers make up the vast majority of persons employed in hospital security. According to Russell Colling, a nationally known health-care security authority and author of *Hospital Security*, 5th edition, the officer who prepares for advancement (through a combination of education beyond high school and field experience) can look to numerous supervisory, investigative, training, fire prevention, and safety positions in the field.[11]

Hospital security officers generally earn more than do their counterparts in other industries because of the variety of duties requiring a higher-than-average amount of training. The officer must also be able to interact effectively with the medical community as well as with patients and visitors under conditions of frequent stress. Salaries, however, do vary by location.

Security director positions generally require at least 4 years of college preparation and considerable field experience. Like so many other areas of security, as Colling observes, hospital security is just coming into its own.

Airport and Airline Security/Sky Marshals

As a result of the events of September 11, 2001, airport and airline security have undergone tremendous scrutiny and change. The verdict was clear. Old ways of handling airport security measures were not adequate. Something needed to be done. With the passage of the Airport Security Federalization Act of 2001, airport security in the United States has changed forever. What had been the domain of contract and proprietary security organizations is now under the control of the federal government as an arm of the Department of Homeland Security. To become a screener the following criteria must be met:

- High school diploma or a combination of education and experience determined to have equipped the individual to perform the duties of the screening position.
- Basic aptitudes and physical abilities including color perception, visual and aural acuity, physical coordination, and motor skills.
- Command of the English language.
- Be a citizen of the United States.

Preference will be given to individuals who are former members of the armed forces or former employees of an air carrier whose employment was terminated as a result of a reduction in workforce.

An older federal program, the Federal Air Marshals, was given new life following September 11. The number of air marshals deployed on selected passenger flights was increased significantly.[12] Standards are high, as is the rigor of the training program. The mystique of this position often clouds the fact that the job requires long periods of travel and much boredom. The number of marshals is classified, but estimates are that the current number falls between 2,500 and 4,000, up from the 23 agents employed in the program prior to September 11. The average salary is approximately $62,000 annually. While the position is prestigious, marshals often complain of fatigue and boredom. There are few chances for promotion. Many question the value of this costly program. Over the past five years sky marshals have made only 51 arrests and none were related to terrorism.[13]

In addition to airport security operations, now controlled by the federal government, there are other opportunities with major airlines. However, most major airlines still prefer to hire former federal agents in managerial positions. This situation is not unique to airlines, of course. Former Federal Bureau of Investigation (FBI) agents can be found in a great many corporate security jobs in business and industry around the globe. Both the experience and qualifications required by many federal law enforcement jobs have generally been highly regarded in the private sector. The ambitious, career-minded security aspirant could do far worse than consider a period of service with a federal law enforcement agency as a springboard to a promising position in industry, including airline security.

Qualifications are high. Almost all employees in this field have college degrees. Many have law degrees and 5 or more years of federal experience.

Hotel Security

The hotel and motel industry was at one time characterized by serious neglect of many security responsibilities, an attitude that has now changed due to a number of very large awards by the courts against hotels or motels charged with negligent security, particularly in the area of protecting guests. This past neglect, coupled with court-mandated responsibility, has created opportunities for security professionals. In addition, as travelers express safety concerns regarding potential terrorist attacks, hotel security managers will need to find ways to assure interested travelers that the facility has ample protection against such attacks.

Opportunities in the hotel industry exist in great numbers, both for on-the-site positions and at the corporate home office level. Except at the corporate level or the management level in large hotels, however, the salary range is relatively low in relation to the security industry as a whole. On the other side of the coin, the entry level for the person with any combination of hotel experience and security education or experience can be quite good, with clear opportunities for advancement.

In a related area, the future of security in high-rise apartment buildings and housing complexes offers great potential for the security professional. These opportunities exist because of the growing emphasis on the concept of total environmental protection and the threat of civil suits by residents or guests who are victimized on building property.

Campus Security

The rapid progress of campus security during the past 25 years has created excellent opportunities for career positions in the field. Openings in many progressive and professional campus departments provide a challenging work environment, but offer good salaries and fringe benefits, as well as chances for advancement. Such departments are looking for young, career-minded individuals, with particular interest in those enrolled in, or graduates of, a criminal justice degree program. They may also require more specialized training or certifications, such as various modules from the National Incident Management System (NIMS) offered through the Federal Emergency Management Agency (FEMA). Interestingly—unlike many areas of modern security—campus security has generally evolved from a low-visibility operation to a highly visible, police-oriented image in response to rising crime problems on campuses.

This highly visible, police-oriented upgrade in image is due in part to the mandate of the Campus Security Act of 1990 and through subsequent amendments by the Higher Education Opportunity Act of 2008.[14] Recent shooting events on college campuses have also led to increased interest in providing better security for all those attending our college campuses.

A good-sized department will include line officers, field supervisors, shift commanders, a coordinator of line operations, and a director. Many departments also have specialized positions such as investigator and training officer. Salaries vary from department to department, and from one area of the country to another.

Banking Security

Banks must comply with minimum federal regulations on security, as promulgated in the Bank Protection Act of 1968. The act mandates that there must be a security manager. Banks comply with this requirement in a variety of ways. Small operations often delegate security responsibilities to one of the senior bank officers. Larger banks, however, hire security managers who often are former FBI agents with an understanding of federal regulations regarding currency and fraud. There is a heavy reliance on electronic technology and physical security rather than on large staffs.

The bank guard may become a casualty of technological advances, as many banks are finding it more cost-effective to eliminate guards in favor of physical security improvements.

Security Services

In general, security personnel at the lower operational levels earn less in contract security organizations than they do in proprietary guard forces. But this is not necessarily true for investigators and other personnel at higher levels.

Young people should seek opportunities in security service organizations, since the growth of security services has been meteoric and there is no leveling off in sight. The demand for good executives is insatiable.

Because the good loss-prevention or security executive is much less a police officer than a systems expert, auditor, and teacher, security experts recommend broad-based education and experience in such areas as accounting, industrial engineering, management, personnel, law, statistics, labor relations, and report writing.

On another level, there is a demand for technically qualified individuals capable of providing specialized security services ranging from alarm sales, installation, and service, to alarm systems consulting. Continuing changes in the application of security hardware and systems will bring an increasing demand for the services of those who can advise users on their selection and implementation.

Guard Services

Guard supply represents the major service provided in the industry today. The majority of guards work for contract security agencies, but many firms hire their own security staff (proprietary security). Only part of the guard's job is crime-related. Whenever it is possible or necessary, guards are required to prevent major crimes and to report those that have been committed. In short, the job of a guard is to observe and report. But their major role may be to direct traffic, to screen persons desiring access to a facility, and generally to enforce company rules. In many modern applications, the role is more helpful, rather than regulatory. They may direct or escort persons to their destination within a facility, act as receptionists or sources of information, or be primarily concerned with safety.

Since many guards are concerned for only a small percentage of their time with crime-related activity, there is some effort in various quarters to adjust the guards' appearance to fit their roles by outfitting them in blazers and slacks rather than in a uniform with its police or coercive connotation.

A guard is, however, a guard. Even if, in a particular assignment, he or she is never confronted with criminal activity, the guard is still charged with certain responsibilities in that area and is responsible for protecting the interests of his or her employer on the employer's property.

According to the Bureau of Labor Statistics guard salaries range between $10.00 per hour to $16.00 per hour. Private guards differ from public police both in their legal status and in that they perform in areas where the public police cannot legally or practically operate. The public police have no authority to enforce private regulations, nor have they the obligation to investigate the unsubstantiated possibility of crime (such as employee theft) on private property. The job of the private guard is to provide specific services under the direction and control of a private employer who wants to exercise controls or supervision over the company's property or goods, or to provide additional services that the public police as a practical matter simply cannot provide.

Patrol Services

Private patrol services offer a periodic inspection of various premises by one or more patrols operating either on foot or in patrol cars. The tour of such patrols may cover several locations of a single client or include several establishments owned by different clients within a limited neighborhood. Inspections of patrolled premises may be visual perusal from the outside, or they may require entering the premises for a more thorough inspection. Typically the arrangement made with a client specifies that a certain number of inspections will be made within a given period of time or with specified frequency.

The patrol differs from the guard by operating through a tour covering various locations, whereas the guard stands a fixed post or walks a limited area. The patrol service is more economical because the guard maintains a post for the full period during which a danger exists. Due to predictability factors, the patrol has the possible disadvantage of being circumvented by an intruder who knows that there will be some period of time between inspections of a given premise.

Armored Delivery Services

Armored delivery services provide for the safe transfer of money, valuables, or any goods the employer may wish to move from one location to another. By far the widest use of this service is to transport cash and negotiables from a receiving point to a bank or other depository. Payrolls, cash receipts, or cash supplies for daily business are the principal traffic of the armored delivery service.

Personnel employed by such services are not concerned with the general security of the premises they serve; their responsibility is confined to the safe transport of sensitive items as directed by the customer. Courier services perform a similar function in the safe transport of valuables. They are distinguished from armored car services principally by using means other than special armored vehicles.

Locksmithing

This is a classic profession requiring a lengthy apprenticeship. Locksmiths in the United States sell, install, and repair locking devices, safes, and vaults. Some also sell and install various alarm and electronic access-control systems. Positions are available in shops, where apprentice locksmiths spend their time learning the trade under a master craftsperson. Much of the work is done on an emergency basis, thus the hours are often long and irregular. The best jobs involve keying new facilities or rekeying older structures such as office buildings and motels.

Alarm Response Services/Technology Experts

Central station alarm systems consist essentially of alarm sensors located in the protected premises and a communication line from the sensors to a privately owned central station alarm board that is monitored and responded to by private security personnel (see Figure 3-1). Many city codes require that alarm systems be tied into a central station operation.

FIGURE 3-1 A security guard walks through an emergency housing site set up for hurricane victims. *(FEMA photo by Leif Skoogfors.)*

In some cases, central station systems do not dispatch personnel to respond to an alarm but rather relay the alarm received to public police headquarters. But in most such systems, the alarm is relayed and someone is also sent to the scene.

Alarms connected to a central station are usually designed to detect intrusion, but they can also be used to monitor industrial processes or conditions.

Certainly, central station coverage of a facility is cheaper than full-time security employees performing essentially the same function. A drawback is that the false-alarm rates for many intrusion systems are still very high, resulting in resistance to the use of direct connection systems to police headquarters. Central station operators have some flexibility in checking the validity of an alarm before notifying the police, so they may to some degree reduce the incidence of false alarms demanding police response.

In cases where central station personnel actively investigate the intrusion, and even take steps to apprehend a suspect before the arrival of the police, they supplement the public police effort.

With a dominant use of electronics in security, the demand for professionals who understand the applications of alarm technology, CCTV, and other high-tech applications within security continues to grow (Figure 3-2). Alarm installation is an excellent skill to learn. While most positions in this business are through distributors and contract security services, there is a trend toward proprietary positions. Certification for this area of study is available through the National Alarm Association of America (NAAA).[15]

With the increased use of satellite technology, the alarm industry is rapidly changing to one with great reliance on integrated systems controlled by computers at remote locations. Major alarm firms are marketing their expertise in integration and support systems, creating an entirely new area of employment that incorporates knowledge of alarms systems with computer programming.

Computer/Intellectual Property Security

This is an important example of the new frontiers opening in the loss-prevention field in response to social and technological changes. Computer security executives and investigators call for a blend of education and experience in computer science and security. Today virtually all organizations have computers that need protection due to a growing number of cybercrimes. In fact, the Department of Justice (DoJ) now has a Computer Crimes and Intellectual Property (CCIP) section that is dedicated to working with industry to combat the growing number of cybercrimes, including counterfeiting of electronic gear.[16]

Although the first response was to protect the computer and related hardware from attack, it soon became clear that of greater importance was protecting the information stored in the computers. With the advent of information transfer through intranet and Internet systems, the problems associated with the protection of company and private information being transferred among various locations has become an even bigger security concern. In today's e-commerce world many firms generate information (customer lists, research projects, and so on) that has become a major target of identity thieves, new age bank frauds, credit thefts, and other scams. In addition, as security protection has come to rely on computer backbones to support

FIGURE 3-2 A security officer on foot patrol with handheld device. *(Permission of Hirsch Velocity.)*

21st century technology, criminals are looking to exploit weaknesses in both hardware and software designs to compromise security operations.

A person with an interest in and aptitude for computer software, computer technology, and related industries could do well in considering using these talents in the protection of computer information and information transfer.

Government Service

While most persons who study the loss-prevention/security field are contemplating work in the private sector, there are opportunities with public law enforcement and the military.

Individuals with technical training/education in crime prevention are valuable additions to government organizations that are combating both traditional criminal activities (burglary, robbery, computer crimes) and acts of terrorism. Given the growing cooperative nature among agencies in regards to the war on terrorism, opportunities for specialists appear to be good.

Other Services

In addition to the areas described above, private security firms also provide such services as crowd control, canine patrol, polygraph examination, psychological stress evaluation, drug testing, honesty testing, employee assistance services, and other related loss-prevention assistance to business and industry. Various firms have entered the area of drug testing to meet the demand for a drug-free work environment. Consultants are also providing programs on how to deal with workplace violence. Still other firms are specializing in electronic sweeps because espionage activities have increased. And as would be expected there are now many experts who offer advice on protection from terrorist attacks.

Summary

Although salary levels and security applications vary in different parts of the world as well as within the different areas of business and industry—or even within the same type of business or industry—it is nevertheless possible to perceive the coming of age of security throughout the 21st century.

Still, more universally accepted standards of training and applicant screening and higher wage scales are needed. The opportunity for vertical movement within the security structure must be both present and perceived. But even in these areas there are encouraging signs. Security is seen as more professional than in years prior not only by traditional law enforcement agencies but also department peers within the organization it seeks to protect. One thing is for certain—security is not going away anytime soon and will continue to grow given a turbulent economic environment, rising crime levels and geo-political, social and civil unrest around the globe.

The use of outside investigators and security consultants will increase as security functions become more specialized. The current trend in hiring security executives is to find someone with a broad background, as convergence is one of the key factors in the changing role of the security executives.

■ ■ CRITICAL THINKING ■

If you were giving advice to someone contemplating a career path in security/loss prevention, would you guide that person to a program in criminal justice or some other field such as computer science? Does the educational path have any impact on the type of positions that are available?

Review Questions

1. What factors are increasing career opportunities in security?

2. Pick one area of security that interests you and discuss the career opportunities that presently exist.

3. How important is a college education in obtaining a position as a security manager?

References

[1] Zalud B. 2002 Industry forecast study security Yin-Yang: terror push, recession drag. Security January 2002.

[2] Protective service occupations. Occupational outlook handbook. 2002–2003 edition, U.S. Department of Labor, Bureau of Labor Statistics; February 2002, Bulletin 2540-11, p. 14–17.

[3] Freedonia Group. Industrial Business Research Company, website review @ <www.freedoniagroup.com>, Security, Electronics and Communication. December 2010.

[4] Career opportunities in security and loss prevention. Washington, DC: ASIS Foundation. p. 1.

[5.] Chuvala III J, Fischer RJ, editors. Suggested preparation for careers in security/loss prevention. Dubuque, IA: Kendall/Hunt; 1991. p. iii.

[6] Mike Moran, You Earned It. <www.securitymanagement.com/print/5965>, and Mike Moran, Median Compensation Up 6 percent for Security Professionals in 2010.<www.securitymanagement.com/print/7512>, downloaded 1/17/2011.

[7] Compensation in the security/loss prevention field. 13th ed. Abbot Langer & Associates, Inc.; December 2001.

[8] Moran.

[9] Ibid.

[10] Professional Bodyguard Association, downloaded from <http://myersmith.tripod.com/home.html>, 8/15/2007.

[11] Colling RL. In: Hospital security, 4th ed. Boston: Butterworth-Heinemann; 1992.

[12] Airport Federalization Act of 2001. November 6, 2001, Sections 104, 105.

[13] Meckler L, Carey S. U.S. air marshal service navigates turbulent times. Wall St J (02/09/07) p. A1.

[14] U.S. Department of Education, website review @ <www.ed.gov>, Campus security: lead and manage my school. December 2010.

[15] Keller SR. Technology: unlocking the future for security practitioners. In: Chuvala III J, Fischer RJ, editors. Suggested preparation for careers in security/loss prevention. Dubuque, IA: Kendall/Hunt; 1991. p. 102.

[16] U.S. Department of Justice, website review @ <www.justice.gov>, Computer crimes and intellectual property section. December 2010.

4

Security Education, Training, Certification, and Regulation

OBJECTIVES

The study of this chapter will enable you to:

1. Understand the past and present situation regarding the training of security personnel.
2. Identify the key professional organizations and their efforts to provide professional guidance and certification programs.
3. Trace the efforts of the federal government to create legislation mandating security screening and training standards.
4. Discuss the role of higher education in providing a foundation for professional security.
5. Identify the professional and academic journals available to security practitioners.

Introduction

Few security officers in the United States receive adequate pre-job or on-the-job training to perform the tasks so often assigned to them. While a few companies provide good pay, benefits, and training, they are too often the exceptions. In addition, until recently, the industry leaders did little to establish criteria for the business through professional certification. While some industries did develop certification, such as the Certified Protection Professional (CPP) designation developed by the American Society for Industrial Security International (ASIS), there was little done for the line security officer, although the International Foundation for Protection Officers, based in Alberta, Canada, did develop a Certified Protection Officer (CPO) program.

Concerns over the quality and regulation of training were first recognized at the federal level. In 1991, then-Senator Al Gore introduced the first of several pieces of legislation aimed at setting minimum standards for the security profession. Ten years later, the events of September 11, 2001 brought the issue of training and credentials for security personnel to the forefront. In the aftermath of September 11, airport security procedures, security personnel, and training were called into question. As a result of the government and public focus on airport security, the federal government took over security at the nation's airports in an attempt to assure the public that high-quality, highly trained persons are staffing our airport security operations.

It now appears that the security industry is undergoing major changes in the areas of training, regulation, and certification. These efforts are being led by the federal government and ASIS.

Adequacy of Private Security Training

The status of private security training has traditionally been low. There were no uniform standards for courses—content, length, method of presentation, instructor qualifications, or student testing.[1] In 1976, the Report of the Task Force on Private Security found a lack of quality programs and for the first time made specific recommendations.[2] Unfortunately, many of these recommendations have yet to be implemented, although progress has been made.

Although federal government studies have called for attention to training issues and some standardization of training, training continues to be regulated by individual states, each with its own standards.

A 2005 review of state regulation and training found that 43 states now license or regulate the security industry. While this is impressive, this study found that mandated training occurs in only 13 states and the variety in hours ranges from four to just over 40.[3] It thus appears that training for private security personnel is still less than adequate. This may be one reason why the public law enforcement sector, mandating a median basic training program of 720 hours, has for many years held a poor opinion of the private security profession. The occupation must be professionalized. A listing of licensing and training requirements for security officers in the United States can be found through the National Association of Security Investigative Regulators at www.iasir.org/Security.htm.

Proposed Federal Regulation

In 1991, the first effort by the federal government to pass legislation to regulate the private sector was introduced by former Vice President Al Gore, then a senator from Tennessee. The bill proposed minimum standardized training for essentially all security personnel, although it would only be mandatory for those involved in government security operations either directly or as contractors. The Gore Bill proposed training in the following areas:

- Fire protection and fire prevention
- First aid
- Legal information relevant to providing security services
- Investigation and detention procedures
- Building safety
- Methods of handling crisis situations
- Methods of crowd control
- Use of equipment needed in providing security services
- Technical writing for reports

The bill mandated examination and commensurate certification procedures to ensure the quality of the basic training, but specifics were not spelled out.

The second initiative was made in 1992 under the direction of Representative Matthew Martinez (D-CA). The initial Martinez package was much more specific than the Gore Bill. The Martinez proposal provided for a minimum of 8 hours of basic classroom instruction and successful completion of a written examination, plus a minimum of 4 hours of on-the-job training. Individual states would set standards for individuals or entities conducting the classroom instruction.

The bill also stated that the classroom portion of the training must include, but may be expanded beyond (at the discretion of the instructor or state licensing agency), the following:

- Legal powers and limitations of a security officer, including law of arrest, search, seizure, and use of force
- Safety and fire detection and reporting
- When and how to notify public authorities
- Employers' policy, including reporting incidents and preparing an incident report
- Fundamentals of patrolling
- Deportment and ethics
- General information, including specific assignments and equipment use.

In 1993 House Bill 2656 was introduced by Representative Don Sundquist (R-TN). This bill was similar to the Gore Bill, mandating that states have screening, training, and other requirements and procedures for issuing licenses to security personnel. The Sundquist Bill also stipulated that security employees would need to pass a drug screening test that met the guidelines of the National Institute on Drug Abuse, as well as physical and psychological fitness tests. The bill also required a check of records with the National Crime Information Center.

A major difference in the Sundquist Bill was that it required a minimum of 16 hours of initial training, of which 8 must be preassignment with the balance occurring as on-the-job training. Armed personnel would need to complete a mandatory 24-hour program above and beyond the 16 hours already stipulated. In perhaps the most dramatic departure, Sundquist's bill also mandated annual training requirements, including a 4-hour refresher course. Armed personnel would have to complete additional hours of refresher courses on firearms and requalify in the use of their duty weapon.[4]

Although the above bills show movement in the right direction, they would not have led to uniform federal standards. The federal government, sensitive to states' rights, is suggesting a minimum standard that states will be free to enhance. The most current effort was S.2238, "The Private Security Officer Employment Standards Act of 2002." The key provision of this bill requires all security employees (contract and proprietary) to undergo a criminal history check.[5] This legislation is now law, allowing security employers access to criminal history records. However, this key legislation focuses on criminal history checks rather than training and development standards.

The Role of Higher Education

In November 1978 a seminar titled "Meeting the Changing Needs of Private Security Education and Training" was held at the University of Cincinnati as a follow-up to the report of the Task Force on Private Security and the first National Conference on Private Security. The majority of participants were academics. The interest of the academic world in security education at first increased and then leveled off. However, the interest is certainly not new. The demand for improved training and education in the field of security has existed since 1957.[6]

With the 21st century renewed interest in security, brought on by increasing problems with security in our new technologically changing world, as well as the threat of terrorism, has brought about a renewed emphasis on security education. Security programs will likely get a new "lease on life" as they find their place in the 21st century struggles against world terrorism and techno-crimes.

As more and more private security managers receive their degrees, the overall quality of private security employees will increase because college-educated people will do their best to see that the private security occupation becomes worthy of the term *professional*. And perhaps of more importance is the recognition that many of the new world security jobs require more education.

The problem for many colleges and universities continues to be defining security education. Most of the programs developed in the '70s and '80s were shaped by law enforcement education where these new programs were housed. From a review of the literature, it is apparent that educators are not of like minds on the placement of security curricula in colleges and universities. One view shows preference for equal status with law enforcement programs because the fields are very interrelated. A second view is that security should be a completely independent major with alliances to departments of business. Yet a third view indicates that the placement of the program is not as important as an interdisciplinary approach to the curriculum. The degree designation is of little importance.[7]

What is evident is that there are well-established security education programs at respected colleges and universities. Almost 20 years ago, Fischer and Chuvala presented information on security education at the 1993 ASIS annual meeting in Washington, D.C. They reported that in 1993, 60 programs had been identified as offering baccalaureate or higher levels of security education. In surveying this identified group, Fischer and Chuvala determined that there were only 21 institutions actually offering security degrees. As Chuvala noted in his presentation, most of the other programs offered criminal justice degrees with security courses.[8] While the field has increased to over 75 programs at the baccalaureate or higher level, many of these programs offer related degrees that have an emphasis on loss prevention/security.[9] Although only one-third of the institutions of varying size and administrative organization were identified as offering degrees in security, certain generalizations can be made about security education at the baccalaureate level. In general, programs are small and are staffed by faculties who have more experience in public law enforcement than in security. Despite the small size of programs, most institutions express support for the programs.

The final determinant of program success or failure is the program's ability to deliver a product that is attractive to the security industry. If the graduates of a program are not of adequate quality, the program will fail. And while criticisms are many, there are programs that have been able to identify problems and develop successful degree plans. The future of security education is excellent when one considers the growth evident in the field.

The involvement of the ASIS Foundation in sponsoring a master's program indicates the growing interest of professional security managers in providing graduate education in loss prevention/security for its membership. Today ASIS tries to fill the role of a security institute.

Over the past few years, the ASIS Foundation has made tremendous advances in providing innovation within the security profession. Perhaps the most visible success of the ASIS Foundation has been the establishment of *Security Journal,* which is published jointly with Butterworth-Heinemann. The Foundation also offers scholarship funds to students interested in pursuing an education in security.[10]

In addition to the work through the ASIS Foundation, ASIS has also been actively involved with Webster University in preparing a graduate certification program in security management. The program is designed for those individuals seeking additional education beyond the bachelor's degree, but not wanting a master's degree.[11] ASIS in cooperation with Wharton University of Pennsylvania also offers a program for security executives on basic business principles. In 2011 ASIS also partnered with Northeastern University to develop a four-day program, "Business Concepts for the Effective Security Manager." Also to be considered is the availability of online courses. The World Wide Web has allowed for the distribution of many different types of information on the Internet.

Following the events of September 11, 2001 and with the support of the federal government, a number of programs offering certification and various degrees in homeland security or emergency management have been developed. Whether these programs are simply "knee jerk" reactions to the terrorist threat or viable and sustainable programs is yet to be seen.

While ASIS and other organizations have in recent decades started to focus on training and education, there is still an inadequate amount of attention from government and industry leaders in the establishment of a uniform approach to standards for education and training, particularly for uniformed officers.

Training

Development and training of security personnel must be a continuing concern of management. Indeed the lack of adequate training in the past has been the major criticism leveled against private security, both within the industry and outside it. Today wages for contract guards are still generally low and training has not improved substantially.[12]

While the majority of all guards (both proprietary and contract) receive some preassignment training, in the contract area 40 percent of the guards had completed only on-the-job training. In general, it is apparent that proprietary security personnel report more training than do contractual personnel. While the Private Security Task Force recommended that contract security personnel complete a minimum of 8 hours of formal preassignment training, as well as a basic training course of at least 32 hours within 3 months of assignment,[13] this recommendation has not been implemented. Even the federal initiatives through the Gore, Martinez, and Sundquist bills have not resulted in any substantive changes to date.

It is clear that adequate training can and must be an important aspect of security planning in the proprietary organization. The need is as great in contract security services, of course, where the problem is compounded by the competitive pressures of the marketplace. The onus for low training standards must be borne by employers whose overriding consideration in selecting security services is the lowest bid. Proficiency in security is largely a product of the combination of experience and a thorough training program designed to improve the officers' skills and knowledge and to keep them current with the field. The recommendations of the Task Force on Private Security included the following:

1. A minimum of 8 hours of formal preassignment training.
2. Basic training of a minimum of 32 hours within 3 months of assignment, of which a maximum of 16 hours can be supervised on-the-job training.[14]

The merits of training will be reflected in the security officer's attitude and performance, improved morale, and increased incentive. Training also provides greater opportunities for promotion and a better understanding on the part of the officers of their relationship to management and the objectives of the job.

It should not be presumed that former law enforcement officers require no training. They do. In order for them to be successful in security, they must develop new skills and—not incidentally—forget some of their previous training.

A training program should cover a wide variety of subjects and procedures, some of them varying according to the nature of the organization being served. Among them might be:

- Company orientation and indoctrination
- Company and security department policies, systems, and procedures
- Operation of each department
- Background in applicable law (citizen's arrest, search and seizure, individual rights, rules of evidence, and so forth)
- Report writing
- General and special orders
- Discipline
- Self-defense
- First aid
- Pass and identification systems
- Package and vehicle search
- Communications procedures
- Techniques of observation
- Operation of equipment
- Professional standards, including attitudes toward employees

At least one contract firm has also recognized changes in the field brought about by the technology revolution. Barton Protective Services has developed several technology training programs. The program relies on computer-based e-learning. The firm is also installing "Tech-Knowledge-y" labs in branch offices, affording an opportunity for all officers to participate in the e-learning.[15]

As mentioned earlier, the Private Security Officer Employment Standards Act of 2002 was to be the beginning step for the federal government imposing uniform standards on the security industry. However, the version of the Act that actually passed did not address training standards. Rather the Act focused on employment standards by allowing security employers access to criminal histories of prospective employees.

Certification and Regulation

An issue closely tied to training issues is regulation and certification. The authors applaud ASIS for the development of its CPP program and its new initiatives for specialized certification. Given the current efforts by the federal government and ASIS, it is likely that a balanced approach between industry-imposed standards and preemptive state legislation will be the model for the United States in the 21st century. Industry-imposed standards can be successful, as noted by the success of the British Security Industry Association (BSIA). BSIA industry-imposed standards reportedly cover 90 percent of Britain's security industry. The BSIA has adopted standards pertaining to personnel screening, wage levels, supervision, training, liability insurance, and physical facilities.[16]

In 2003 ASIS International ventured into the area of industry regulation, creating an ASIS Commission on Guidelines. The goal of this commission is to "advance the practice of security through the development of risk mitigation guidelines." The Commission has written General Security Risk Assessment Guidelines (see www.asisonline.org).[17]

Regulation

Considering the importance of private security personnel in the anticrime effort and their quasi–law-enforcement functions, it is ironic that they receive so little training in comparison to their public-sector counterparts: 0 to just over 40 hours compared to a median of 720 for police officers. While it is ironic, the reason is obvious. Legislation mandates training for public law enforcement personnel, whereas this is not the case for security personnel. A look at licensing standards for private security companies reveals that little has changed with regard to regulation of this already huge and still growing giant. Considering the lack of progress in establishing uniform training standards, it is difficult to support a contention that the "best regulator" is the marketplace. Still, the 2003 initiatives by ASIS offer a hope that the United States will have some success with industry-led standards as in Great Britain. Federal and industry leadership is essential because it is doubtful that the states will provide any guidance.

In the states that do have legislation, the key words that might be used to describe the composite package of legislation are "lack of uniformity." Terminology is not uniform but, more importantly, there is no consensus on the degree to which the state should regulate training, licensing, and education/experience.

It is also interesting to note that, of those states that do attempt to regulate security, only a few include proprietary security forces in their regulatory statutes. This has established a double standard for in-house and contract employees performing essentially the same functions.

There is still a need for the following:

1. *Standards, codes of ethics, and model licensing.* The efforts of the Task Force on Private Security and the PSAC have stood the test of time, and both groups were well represented by law enforcement, business, and all facets of the security field. Statewide licensing should be required for guard and patrol, private investigation, and alarm firms. The profound effects on upgrading private security relationships with law enforcement will occur as

a result of the cooperative action of the security industry, law enforcement, and state governments in implementing the measures encompassed by the Task Force on Private Security and PSAC efforts.

2. *Statewide preemptive legislation.* Although law enforcement agencies seek closer local control over private security, a proliferation of local licensing ordinances deters adoption of minimum standards and imposes an unnecessary financial burden on contract security firms with redundant licensing paperwork and fees. Some latitude might be granted local law enforcement agencies to impose tighter control on some aspects of private security operation, but the controls should not be unduly restrictive and should withstand tests and measures of cost-effectiveness.

3. *Interstate licensing agency reciprocity.* Interstate operation of contract security can be unnecessarily hampered by having the same personnel comply with different personnel licensing requirements in adjacent states—and sometimes in cities and counties. The same standards of state-level licensing and regulation in all states and reciprocity (e.g., recognition of other states' regulatory provisions) would facilitate more efficient delivery of security services and decrease state regulatory costs.[18]

HR 2092, the Martinez/Barr compromise bill, would have made it easier to determine whether a security candidate had a criminal record.[19] The new Private Security Officer Employment Standards Act of 2002 accomplished what Martinez/Barr set out to accomplish.

Certification

The growth of programs leading to certification is an indication of the professionalization of the security field. Today it is possible to receive several certification designations, each of which has its special appeal. An indication of the level of heightened interest in the broad-based security professional can be seen in the results of the diligent efforts of ASIS International. This society has long been interested and involved in creating standards of competence and professionalism to identify those security practitioners who have shown a willingness to devote their attention to achieving higher goals of education and training in their chosen career. As was noted earlier, the security profession has, in the past, been characterized by the transitory nature of much of its personnel. Training standards have frequently been low, and even many executives in the field were generalists without either specific work-related experience or specific training in security. Many factors have been brought to bear on this problem, and changes have been and are being made.

One ASIS program is designed to upgrade those career security persons who are willing and able to qualify for certification as CPPs. The certification board in this program was organized in 1977 and since that time has provided sufficient evidence of professional performance capability through certification to stress the importance of the CPP. "Positions available" announcements in the *Wall Street Journal,* the *Chronicle of Higher Education,* and other publications have included requirements that state "Certification as a Protection Professional by the ASIS [is] desirable," or the candidate "must have certification as a CPP." This trend will continue as employers and the public become more aware of the CPP program.

Certification in this program is far from pro forma. Both educational and work experiences are required before a candidate can be considered. If candidates meet the basic standards, they must then take an examination on both mandatory and optional subjects. It is through this program and those given by colleges and universities across the country that the goal of professionalism in the practice of security will be achieved.

In 2003 ASIS International began offering two technical certifications—Professional Certified Investigator (PCI) and Physical Security Professional (PSP). The PCI is for those whose primary responsibility is to conduct investigations, while the PSP is designed for those individuals whose primary responsibility is to conduct threat surveys and design integrated security systems or for those who install, operate, and maintain security systems.[20]

Similar efforts have been made to improve the professional image of the security officer through the Certified Protection Officer (CPO) program. The program was founded in 1986 by the International Foundation for Protection Officers (IFPO). The first CPOs were granted in 1986, and that certification is now available through several colleges in the United States and Canada. The program is "designed to provide theoretical educational information to complement the field experience of Security Officers."[21] Topics of study include:

- Introduction to security
- Officer and the job
- Physical security
- Legal aspects
- Human relations
- Security as a career
- First aid and CPR
- Preventive security

Candidates must complete an application, obtain nominations from two security or police professionals, and complete the training program before certification is granted.

With the growth of integrated systems and the need to protect intellectual property, at least three certifications are available for security/information systems specialists. These are the Certified Information Security Auditor (CISA) offered by the Information Systems Audit and Control Association, the Certified Information Systems Security Professional (CISSP) from the International Information Systems Security Certification Consortium, and the Globally Certified Intrusion Analyst (GCIA) from the Sans Institute.

Other certifications are available in specific fields. The International Association for Healthcare Security and Safety (IAHSS) offers the Certified Health Care Protection Administrator (CHCPA) designation; the United Security Professionals Association (USPA), Inc., offers the Certified Financial Security Officer (CFSO) designation; and the Academy of Security Educators and Trainers (ASET) offers the Certified Security Trainer (CST) program. With the increase in computer-based fraud the Association of Certified Fraud Examiners was established in 1996 with their own certification program (CFE). And, as mentioned earlier, the events of September 11, 2001 foster the development of certification in Homeland Security. Of course, other groups have also developed various programs to identify competence in specific areas.

Magazines and Periodicals

Any discipline that claims to be a profession must have its own mechanism for distribution of information. In most disciplines or professions, information is distributed through professional publications. Security and loss-prevention publications have changed dramatically in the last 30 years as a reflection of the growing professionalization of the field. At one time *Industrial Security* (now *Security Management*) and *Security World* (now *Security*) magazines were about the only publications in the security field. Today the number has grown substantially.

Studies conducted over the years indicate that *Security Management* and *Security* dominated the field but that other publications are also being more widely read.[22] Other publications often mentioned included: *Security Products, Hospital Security, Security Journal,* and *Journal of Security Administration* (as of 2009 the *Journal of Security Administration and Homeland Security*). With the growing need to understand the electronic age, magazines like *Info Security* offer security professionals up-to-date information on the latest computer security technology and problems. *Access Control and Security Systems Integration,* along with *Security Technology and Design (ST&D),* offers the latest in security technology advances. Appendix A lists the major publications in security.

With the advent of the technology age, a focus on publications cannot ignore the growing use of the World Wide Web. ASIS has developed its own information-sharing Web presence as ASIS Online and ASISNET. *Security* magazine has also developed its own Web site. As the use of the technology associated with the computer continues to improve, it is more than likely that other publications will join the trend toward Web dissemination. Many corporations and government units already make extensive use of the computer and its power to distribute information via the Internet. Appendix B provides a list of useful Internet sites.

Summary

Security education has undeniably undergone tremendous growth in the last 25 years. Academic programs in security, with a few exceptions are, relative to traditional subject areas, young. Most were established within the last 40 to 45 years. In general, most have been reasonably successful, as the demand for college-educated security managers continues to grow. Leaders in the field, both academics and practitioners, indicate that security should seek recognition as its own distinct area of study. While some believe that the programs can find this autonomy within the criminal justice field, others believe that the field would be better off in colleges of business. The concept of convergence of fields for security managers may push this argument in favor of a broad education rather than a specialized, focused degree.

Security education is here to stay whether it is housed in criminal justice programs, colleges of business, or independent programs. In 1994 the World Institute for Security Enhancement (WISE) was established. WISE offers training, consulting, research and development, and "think tank" services.[23] (See www.worldinstitute.org/wise/nav_left.html.) Much research on various security topics is currently being produced at colleges and universities. Much of this research is academic in orientation, however, and perhaps of little value to practitioners. In this respect, an institute much like the ASIS Foundation, which would support practical research, is certainly desirable.

While the status of education in security is good news, the same cannot be said for training. It is truly unfortunate that until 2003 the federal government had not taken an active part in establishing minimum requirements for security personnel, who often perform the same duties as do police officers. Even the states that now regulate the security industry in reality pay little attention to it. Considering the fact that the private police outnumber the public sector by more than a two-to-one margin and that the field is growing at a rate of approximately 12 percent each year, it is time for the federal government to take an active role in requiring states to develop adequate legislation for security training or provide an impetus for a British model in the United States. This industry model may become reality as ASIS continues to develop industry standards.

The National Association of Private Security Industries (NAPSI), Inc., reports that 61 percent of all guard companies surveyed conduct continuing education courses for their guards and are continually looking for new training materials.[24] Let us take a lesson from the issue of police training. It was not until the establishment of the Law Enforcement Assistance Administration and federal legislation that the states began to require adequate training for police officers. Today they receive a median of 720 hours of basic training. In addition, most states also have an ongoing training program once the basic course has been completed. Given the improved quality of police education and training after federal involvement, it is likely that similar results would occur in the private sector should the federal government decide to become involved in the regulation of training and education. While the Gore, Sundquist, Martinez, and Barr initiatives indicated a growing interest in federal regulation, more needs to be done to get similar bills out of committee and onto the floors of the House of Representatives and Senate.

On the bright side the recent ASIS International efforts at establishing industry-based guidelines for security and the introduction of Senate Bill 2238, "The Private Security Officer Employment Standards Act of 2002," which was signed into law, are evidence that perhaps we are beginning to understand the need for professional security services.

■ ■ CRITICAL THINKING ■

What are the future implications for regulating the training and certification of security personnel through federal mandates?

Review Questions

1. How does the presence of college degree programs in security enhance the field?
2. What is your view of regulation in the security field? Should the government play a role, or should the industry regulate itself?

3. How does the level of training and regulation for security personnel compare with that of the police?

4. Do you observe any differences in the level of professionalization between security management and line security officers? If so, what are they?

References

[1] Hallcrest report: model security guard training curricula. Private security advisory council to LEAA, U.S. Department of Justice. McLean, VA: Hallcrest Press; 1978. p. 1.

[2] Van Meter Executive Director C. In: Private security: report of the task force on private security. Washington, DC: National Advisory Committee on Criminal Justice Standards and Goals; 1976. p. 88–9.

[3] Associated Criminal Justice and Security Consultants, LLC. Evaluation of basic training programs in law enforcement, security and corrections..., June 2005.

[4] Chuvala III J, Fischer RJ. The role of regulation. Secur Manage March 1993:61–3. Chuvala III J, Fischer RJ. Why is security officer training legislation needed? Secur Manage April 1994:101–2.

[5] ASIS International. ASIS efforts result in change to security guard screening bill, downloaded 11/14/2002, <www.asisonline.org/newsroom/newsreleases/100802guard.html>.

[6] Wathen TW. Careers in security—one professional's view. Secur Manage July 1977:45–8.

[7] Cunningham W, Taylor TH. In: Private security and police in america: the hallcrest report. Portland, OR: Chancellor Press; 1985. p. 264.

[8] Fischer RJ, Chuvala III J. "Security education: an update," a paper presented at the Annual Meeting of ASIS, Washington, DC; September 1993.

[9] Academic institutions offering degrees and/or courses in security. ASIS International IRC; July 2007.

[10] Working for security. ASIS Dyn January/February 1991: 8–9.

[11] Webster offers graduate certificate in security management. Secur Manage July 1996: 119.

[12] Cunningham et al. p. 141–56.

[13] Cunningham, Taylor. p. 264.

[14] Van Meter. p. 99–106.

[15] Cunningham, Taylor. p. 263–64.

[16] Ibid. p. 152–53.

[17] ASIS International. ASIS international develops framework for creating security guidelines and standards, downloaded 1/12/2003, <www.asisonline.org/newsroom/newsreleases/010803guidelines.html>.

[18] Cunningham, Taylor. p. 265.

[19] Secur Manage January 1996: 114.

[20] ASIS International. ASIS international launches two additional professional certification programs, downloaded 1/12/2003, <www.asisonline.org/newsroom/newsreleases/120902certs.html>.

[21] The protection officer: training manual. Cochrane, Canada: The Protection Officer Publications; 1986.

[22] Bottom NR. Periodical literature in security and loss control. J Secur Adm June 1985:9. Palmiotto MJ, Travis III LF. Faculty readership of security periodicals: use of the literature in a new discipline. J Secur Adm June 1985:30. Fischer RJ. The development of baccalaureate degree programs in private security 1957–1980, unpublished dissertation, Ann Arbor, MI: University Microfilms.

[23] Bottom N. Editorial: a security institute is born. J Secur Adm December 1994:ii–iv.

[24] Guard companies specify officer training needs. Security January 1991: 9.

5

Homeland Security: Security Since September 11, 2001 and Beyond

OBJECTIVES

The study of this chapter will enable you to:

1. Understand the response of the private sector to the events following September 11, 2001 and key developments since, especially in the area of cyber security.

2. Discuss the federal government's response to the events of September 11, 2001.

3. Consider the cooperation of government and private security in the effort to protect the United States and businesses from potential terrorist attacks.

4. Identify the efforts of selected state and local governments in preparing to protect their constituents from potential terrorist threats.

5. Consider the positive achievements and efforts to protect individual liberties, businesses and citizens from terrorism as well as potential pitfalls and the corresponding erosion of individual rights.

Introduction

There are few tragedies in history that create a lasting impression. However, no one of cognizant age will likely forget where he or she was the morning of September 11, 2011. This is a day that affected life in the United States and throughout the world. Security as we had come to know it would begin an evolution that will continue for many years. The most telling change in the United States was the creation of the Department of Homeland Security (DHS).

President George W. Bush made a commitment to the American people and others in the world community to fight a war on terrorism. More will be said about terrorism later in this text but for the purpose of setting the stage, terrorism is generally defined as a systematic use of violence and intimidation to achieve a political goal.

On January 24, 2003, former Pennsylvania Governor Tom Ridge was sworn in as the first Secretary of the DHS. This was a watershed moment in the federal government's response to terrorism against the United States. The Presidential cabinet-level department merged 22 federal agencies with over 180,000 employees. To put this landmark move into perspective, the establishment of the DHS is the largest reorganization of the United States federal government since the creation of the Department of Defense in 1947 under President Harry S. Truman. President Bush remarked at the swearing-in ceremony that this "begins a vital mission in the defense of our country." While it appears that the Department is here to stay, it is apparent that there are successes and failures since its creation. Disappointment with DHS's Federal Emergency Management Agency (FEMA) during Hurricane Katrina is just one symptom of a larger problem of handling a large organization with multiple layers of bureaucracy and sometimes competing agendas.

> The security field continues to undergo the most significant changes it has seen since World War II!

Its current secretary, Janet Napolitano, is only the third person to hold the cabinet level position and with her recent announcement of the "If you see something, say something" public awareness campaign, the DHS continues to evolve in its messaging to the public. It has also evolved to secure the nation from both terror threats and natural disasters. The significance of this move cannot be overstated. The security industry, which for the most part operated independently of federal mandates, has undergone dramatic changes. For example, one of the most radical changes was the federalization of the national airport security and passenger screening system, which was traditionally handled by the private sector—an overreaction to the events of September 11, 2001 perhaps. Federal involvement is also visible in other transportation fields such as maritime activities and trucking.

The American people are also experiencing greater federal involvement in the establishment and maintenance of the nation's security. The DHS also created the Homeland Security Advisory System (HSAS), designed to provide the American public with an ongoing indication of the level of potential terrorist threat against the nation. In February 2003, the federal government raised the terrorism alert to orange, the second-highest level of concern. Many unduly distressed citizens purchased plastic and duct tape to prepare their home for a possible biological/chemical attack by a terrorist. The color-coded threat level of HSAS seems to have outlived its usefulness and was replaced with the National Terrorism Advisory System (NTAS).

The new NTAS according to Secretary Napolitano, "…counts on the American public as a key partner in securing our country." In essence, the color-coded system has been retired for NTAS bulletin alerts.[1] Public law enforcement continues to wonder what role it will play along with private enterprise in maintaining public safety in the face of terrorism. While DHS has dispersed billions of dollars in grants to states and local law enforcement agencies to assist them in preparation for terrorist incidents, the DHS recently changed strategies to target only the major cities for significant funding over the next several years.

Global Security

While the focus of this chapter is on the United States and its Homeland Security programs, it would be foolish to believe that the destruction of the World Trade Center has not impacted other countries. President Bush's declaration of war on terrorism resulted in a worldwide coalition in this battle. It was estimated that governments around the world spend an estimated $550 billion on homeland security each year. The figures reached $572 billion by 2005 and billions more through various Congressional appropriations through 2011. In February 2007, the Pentagon Comptroller reported that the war on terrorism had cost Americans $545 billion to date. Compared to the $56 billion figure from 2003, the cost for 2008 was estimated at $141.7 billion.[2]

Private Enterprise Response

The response by the private sector has been varied. While it is clear from some of the following examples that private sector response is often prompted by federal mandates, private industry has also initiated many security enhancements on their own to reassure its employees, partners and customers. The American Society for Industrial Security (ASIS) worked closely with the United States federal government in an attempt to pass industry standards such as the Private Security Guard Act of 2002. Ultimately, this effort resulted in portions of the 2002 Act being included in the Intelligence Reform Act signed by President Bush in December 2005. As a result, the Department of Justice (DOJ) now allows employers in all 50 states to receive FBI criminal background checks on persons applying for or holding jobs as private security officers.[3] In addition, ASIS has initiated an effort to establish industry-wide standards for security operations and new certification programs for security professionals, such as the Professional Certified Investigator (PCI) and Physical Security Professional (PSP). ASIS is also a thought leader in sponsoring seminars on terrorism issues around the world. To learn more, reference: www.asisonline.org.

ASIS International proposed uniform standards for the security industry.

Corporate executives have mixed reactions to the impact that terrorism has had on security operations, with some getting directly involved in security planning for their enterprises. In a poll by Booz Allen Hamilton conducted in late 2002, a year after the bombing of the World Trade Center, 80 percent of the CEOs of 72 firms with more than 1 billion dollars in annual revenues believed that security was more important then than prior to September 11, 2001, but that 3 percent did not expect any increase in security spending.[4] Since September 11, 2001, due in part to a recalibration of the need for security spending, the average increase in security spending has been only 4 percent. Some professional security experts strongly believe and ominously warn that spending is small in comparison to the real threat that terrorism poses to their corporation.[5]

Transit Security

According to the Government Accountability Office (GAO), transit agencies have taken many steps to improve security. These include traditional security vulnerability assessments, revising emergency plans and emergency response training for employees. However the GAO indicates that there are still many challenges to be addressed. Chief among these is funding. In an effort to reduce this barrier, the Federal Transit Administration is working on obtaining increased funding for transportation security. They have been successful in gaining attention for airport, maritime cargo, and some land transportation systems. Still, the real problem is endemic in the system itself. The transportation network must be open and accessible if it is to remain a viable means of public transport.

Airport Security

Airport security and screening itself was dramatically impacted by the hijackings on September 11, 2001. In 2006, the Transportation Security Administration (TSA) reported that it confiscated over 13 million prohibited items at airport passenger security checkpoints; 1.6 million were knives and in one week during April 2011 18 firearms were confiscated.[6]

Detailed confiscation statistics can be found on the TSA website at www.tsa.gov/research.

Bus Security

The bus transportation system, represented by the American Bus Association, has established an Anti-Terrorism Action Plan designed to improve safety and security for bus operations. The four goals of this plan include:

- Promoting security vigilance among operators through training and partnerships with law enforcement agencies
- Assisting companies in the development of plans
- Preserving the bus industry as a strategic transportation reserve
- Protecting the transportation infrastructure[7]

Port/Shipping Security

In the arena of port security and the shipping industry, security professionals are encouraging the United States government to extend U.S. boundaries to foreign ports. That would put much of the emphasis on security at loading points in these foreign ports. One plan, the Container Security Initiative (CSI), first proposed in early 2002 by the U.S. Customs Office, has major international components. The CSI calls for international security criteria to identify high-risk cargo containers. These containers would be pre-screened at their point of shipment. Of course, such security measures might be difficult to implement considering the need for international understanding and additional security personnel. By mid-2004, 20 of the world's largest seaports had become partners with the United States in the CSI.[8]

Still, a study by BDP International found that 30 percent of shippers are factoring in additional time to comply with the Advanced Manifest System. The DHS rule requires the filing of complete import manifest documentation at least 24 hours before U.S. bound ships are loaded at foreign ports.[9]

The maritime shipping industry will have more federal involvement as the Coast Guard increases patrols at U.S. ports and waterways (Figure 5-1). Sea marshals from the Coast Guard

FIGURE 5-1 Members of a U.S. Coast Guard security unit patrol the harbor near a vehicle cargo ship. *(DoD photo by PA1 Chuck Kalnbach.)*

are assigned to "high interest" vessels arriving and departing from U.S. ports. The Coast Guard is also providing increased protection around the nation's critical petro-chemical facilities.[10]

One possible model, which already works, is the Federal Aviation Administration's (FAA) foreign airport security assessment program. Following this model U.S. Customs would identify ports that fail to meet standards set by the United Nations' International Maritime Organization. What has hindered this approach is a general lack of direction.[11]

Still the International Maritime Organization, composed of over 100 governments, agreed to a security plan that would impose significant regulations on ports and seagoing vessels. The International Ship and Port Facility Security Code took effect in July 2004. The code would eventually require ship operators to develop security plans, appoint ship and company security officers and maintain a minimum level of *on board* security. Port officials would be required to develop similar plans and hire a port facility security officer. The code allows the state controlling ports to deny access to ships not meeting security standards.[12]

While the code is certainly a paper victory, the reality of cooperation often has a blanketing effect on agreements. In recent testimony from the GAO before the Subcommittee on National Security, Veterans Affairs, and International Relations, House Committee on Government Reform, the Director of Physical Infrastructure Issues testified that U.S. efforts to widen security to exporting countries is often slowed by lack of follow-through by foreign governments.

In a case, Estonia delayed installing detection equipment for seven months while finalizing an agreement and monitors sat for two years while Lithuanians argued over the correct power supply.[13]

Rail Security

In the area of rail transit, the sheer number of daily travelers and cargo transported makes extensive security measures a challenge, to say the least. However, Amtrak and regional metropolitan systems have increased security, in some cases requiring rail staff to check tickets prior to boarding the train. In large rail terminals, taxi stands have been moved to the street and away from underground rail locations. States and local governments have been asked to provide either police or National Guard protection for selected rail bridges. The rail companies themselves have been asked to increase security at major facilities and key rail hubs. At the request of the Department of Transportation (DOT), railroad operators are required to monitor shipments of hazardous material and increase security measures on trains carrying such materials. Furthermore, rail operators must report any incidents through the hazardous materials incident reduction program. There are approximately 140,000 miles of rail line throughout the United States. On average the system carries approximately 2 million loads of hazardous materials and chemicals each year. Additional details regarding DOT's Federal Railroad Administration Program can be found at www.dot.gov.

While the United States rail system has yet to be a victim of any large-scale attack, systems in other countries like England, Spain and India have been targeted over the past 10 years. While the TSA is responsible for transportation security and ranks an attack against chemicals in transit and stored as among the most serious risks facing the United States, budget allocations for rail and other surface transportation pale in comparison to the budget for aviation.[14]

Over-the-Road

Over-the-road security is without a doubt the most difficult area to regulate, with over 4 million miles of interstate, national and other roads in use by the trucking industry in the United States.[15] Still the government and trucking industry are working to monitor hazardous material carriers. Companies are suggesting increased security to include:

- Employee identification checks
- Communication plans including increased use of Global Positioning Systems (GPS)
- Operator awareness training and incident reporting

The Freight Transportation Security Consortium, including businesses in asset tracking, vehicle monitoring, and the freight industry in general, are suggesting expanded use of GPS. In particular the groups would like to see all hazardous materials carriers monitored so that vehicles moving away from their predetermined route will be spotted early and law enforcement alerted to the possible problem. This would be akin to air traffic control and the filing of pre-designated flight plans. The cost associated with the proposal has drawn criticism.

The Transportation Security Administration (TSA) also randomly searches vehicles outside airport terminals during heightened security alerts established by the DHS.[16] Supreme Court and other legal precedents are cited in establishing their authority for the random searches, but it is likely that such warrantless, random searches will be challenged. There is also a zero tolerance enforcement policy on vehicles being left unattended or loitering too long at curbside check-in areas. Cell phone waiting lots for arriving passenger pickups are far removed from the terminal and have been established to mitigate the risk of a terrorist attack, while at the same time to reduce traffic congestion in the terminal area. This helps the TSA monitor suspicious behavior.

Electrical Grid Issues

The Pacific Northwest Economic Region (PNWER), a group of electric power companies, conducted a simulated terrorist attack on the region's power grid. The result showed clearly that the region was not prepared to handle such an attack. To deal with the lack of security, both private and government organizations developed guidelines for protecting electric facilities and distribution grids. On the private side, the Edison Electric Institute (EEI) developed guidelines, which have been passed on to the North American Electric Reliability Council (NERC), the U.S. Department of Energy's coordinator for the United States electrical infrastructure. The guidelines cover, among other things:

- Vulnerability/risk analysis
- Threat response
- Emergency planning
- Business continuity
- Communications
- Physical security
- Cyber issues
- Intrusion detection
- Backgrounding/screening

It should be clear that these are traditional security concerns. However, and a bit alarming, until the events of September 11, 2001, most power companies had not taken them as serious issues. While some companies do have model security programs, the majority of regional power firms have lagged well behind. This is to be expected, as the threat did not seem probable.

According to Michael Gips in an article for *Security Management*, "most experts… said that their utilities were either well on their way to meeting all the recommendations or were well beyond them."[17]

While the NERC guidelines are being considered, the Federal Energy Regulatory Commission (FERC) has also developed security standards for electric utilities. In developing standards FERC sought the assistance of the NERC. The result is an amalgamation of NERC recommendations and FERC standards. The purpose of the standards is designed to prevent anyone from disrupting the electrical market and to ensure the reliability of the electric grid.

FIGURE 5-2 A nuclear power facility.

Nuclear Facilities

In the meantime the Nuclear Regulatory Commission (NRC) is requiring additional security measures as a supplement to its significant control measures to protect radioactive materials that are already in place. There are currently 104 nuclear reactors operating at 65 sites in 31 states, which accounts for 19.6% of the United States' electricity generation as this revised edition goes to print (Figure 5-2).[18] Oversight of emergency preparedness is shared by the NRC and the Federal Emergency Management Agency (FEMA). Although the NRC has issued orders to increase standards, it was of concern that when asked, the NRC did not know how many foreign nationals were employed at nuclear reactors and that background checks were generally not adequate. According to a report by Representative Edward J. Markey (D-MA) the only reactor that was designed to withstand ramming by a large airliner is at Three Mile Island in Pennsylvania. Markey's report also expresses concerns over the handling of spent fuels. The NRC has taken Markey's report seriously and has been involved in a complete security review since September 11, 2001. The NRC has issued over 30 directives since including orders to increase guard presence, construction of barriers and increased surveillance. Contractors at nuclear facilities are being scrutinized for possible security risks. While the NRC works on security upgrades, many of the plans require cooperation with other federal agencies such as the FAA, FBI and DOD. Workload and budget limitations have also slowed progress in meeting their own mandated deadlines.[19] Although not the result of a terrorist act but rather a 9.0 magnitude earthquake, the recent tsunami off the coast of Japan, which forced the shutdown and crippling of four nuclear reactors, serves as a good reminder that the nuclear threat can be real. Countless lives were lost and many people are still missing.

For additional information, see the NRC website at: http://www.nrc.gov/about-nrc/emerg-preparedness.html.

Oil and Gas Facilities

There are over 2.2 million miles of gas, oil and hazardous material pipelines in the United States. Security of the lines of course is only part of the issue, as refineries and oil/gas fields are also potential targets. Prior to September 11, 2001 there were minimum security standards for pipelines that unfortunately exist today. The Department of Transportation's Office of Pipeline Safety (OPS), however, began looking at risk associated with pipeline safety and security. The OPS has required that pipeline operators proactively identify and address risks in areas where a rupture would have the greatest impact on populations or the ecological system. An example of the sheer devastation that can be caused by a ruptured gas line is what occurred in San Bruno, California during the early evening hours on September 10, 2010. Seven people were confirmed dead and countless more injured after the explosion and fireball. While not the result of a terrorist act but rather an aged and cracked section of a pipeline of 54 years, it does point out in human terms that the threat can be real.[20]

One of the most promising tools for dealing with pipeline monitoring and other oil and gas security challenges may rest in the application of unmanned aerial vehicles (UAVs). Since 2003, UAVs have been monitoring offshore oil fields checking for thieves and oil leaks. Infrared technology works well in detecting oil and gas leaks.[21]

While protection of the United States distribution and manufacturing facilities is important, terrorists recognize the potential for disrupting the American economy by attacking other suppliers. The Arabian Peninsula's e-magazine *Sawt al-Jihad* (*Voice of Holy War*) posted a notice that Venezuela and Mexico are big oil suppliers for the United States market. In addition, Canada, which exports over half its daily production of 2.5 million barrels of oil to the United States via pipeline, was targeted by al Qaeda in web postings in 2007.[22]

Water Utilities

The Department of Justice (DOJ) has notified water utilities that the enforcement of the Safe Drinking Water Act (SDWA) security provisions is a top federal government priority. The American Water Works Association (AWWA) Executive Director reports that the industry has been active trying to meet federal security deadlines.[23] According to Jack Hoffbuhr, "Water utilities throughout the nation have spent hundreds of millions of dollars in infrastructure costs including water monitoring, physical security systems and emergency training and planning to protect American's water supplies from terrorism." The Public Security and Bioterrorism Act mandated that water systems serving over 100,000 persons must meet vulnerability assessments by March 31, 2003. The Environmental Protection Agency (EPA) reported to Congress on these assessments. It concluded that $276.8 billion is needed in infrastructure improvement over the next 20 years to comply with current regulations. Of this amount, $1 billion is needed for security-related needs.

While most of the covered water utilities were in compliance by the March 31 deadline, it must be understood that most water systems serve populations well under 100,000. The AWWA estimates that it would cost over $450 million to bring these smaller systems into compliance with the Public Security and Bioterrorism Act. Communities between 50,000 and 100,000 had until December 31, 2003 to meet standards, while small communities (3,300 to 50,000) had until June 2004. The Act authorized $160 million in funding for fiscal year 2002 and funds as needed through 2005. Supplemental funding provided $90 million for assessing vulnerabilities and security planning.

In 2005, President Bush requested $5 million for state water security grants and $44 million for a new water security initiative called *Water Sentinel*. Water Sentinel is designed to establish early warning systems in several pilot cities through water monitoring and surveillance for chemical and biological contaminants. The House Appropriations Committee urged the EPA to clear up program management goals and provide better justification for the request in their fiscal year 2007 budget submission. The EPA requested $41.7 million for Water Sentinel for fiscal year 2007. In 2006 Congress allocated $837.5 million to help communities finance projects needed to comply with drinking water standards.[24] Implementing programs such as the Water Sentinel security initiative is challenged with current Congressional budget cuts among other competing and higher priority initiatives designed to protect the United States from another terrorist attack.

Retail

With the federal government sending messages that the next major attack by terrorists might be against retail establishments, mall security operations have come under much greater scrutiny. While many stores retain their proprietary and contract security forces to catch shoplifters and internal thieves, those located in malls also rely on mall security personnel to provide protection to shoppers in the mall commons and parking areas. The mall owners often contract these services. In response to concerns over targeting of mall properties, there was a brief interest in bringing out horse patrols. While the malls have not taken this direction, the entertainment business has increased security through the use of horse patrols. Horse patrols provide good mobility and a high observation point in crowd conditions. The lack of use in the retail environment is blamed on the current poor economy and weak consumer spending.[25]

In the aftermath of September 11, 2001 a number of malls began a 14-hour training program for security personnel on how to spot suicide bombers. The training also includes behavioral and situational awareness. They are receiving training that was at one time reserved for the Israeli police and the U.S. military, among other elite organizations. Training to spot possible terrorist activity is important and helps to reinforce to consumers that they can shop in a safe and secure environment. The International Council of Shopping Centers has held anti-terrorism classes since 2004. According to a spokesman for the Council, everyone has a key role to play, from mall managers to engineers and maintenance people, to keep the consumer safe.

Are malls good targets for terrorist attacks? According to anti-terrorism instructors, it is nearly twice as likely that a commercial establishment will be targeted over a government building or military installation.[26] This is due in part to the open environment, relative ease of access and the potential for mass casualties.

Yet, even with good intentions, there are still questions regarding the quality of security personnel. On February 14, 2007, six people were killed in Trolley Square, a major mall in Salt Lake City. A single armed gunman committed the murders and was eventually killed by the local SWAT team. Where were the security personnel? According to a DOJ report, "An Assessment of the Preparedness of Large Retail Malls to Prevent and Respond to Terrorist Attacks," 60.2 percent of the 120 mall directors surveyed reported that training for security staff had not improved since September 11, 2001 and 94 percent indicated that there had been no change in hiring requirements.[27]

City Centers

The concern that a terrorist organization might attack public gatherings focused interest on security measures at civic/cultural city centers. However, it was clear that many centers had been considering security an important issue long before September 11, 2001. A prime example of what security can accomplish is the Town Center Improvement District (TCID) just outside of Houston, Texas. The Center covers 1.5 square miles, employs approximately 15,000 people and draws over 15 million visitors each year. Security measures include the use of mounted patrols and contract services with local law enforcement departments. The visibility of security has made employees and shoppers feel safe, an issue of greater importance following the events of September 11, 2001.

A 2006 Rand study identified 39 proactive security measures that can substantially reduce the risk of terrorist attacks at mall shopping centers. Among these are the following:

- Public information campaigns encouraging people to report suspicious packages
- Placing vehicle barriers at pedestrian entrances to block suicide car bombers
- Searching kiosks for bombs and weapons
- Clearly labelling exits so shoppers can quickly find their way out in an emergency
- Searching all bags and requiring everyone entering shopping centers to remove their coats to check for explosives and weapons[28]

Public Events and Cultural Centers
SPORTS ARENAS

Sharpshooters, fighter jets, bomb dogs and 1,500 police officers secure Yankee Stadium! Immediately following the September 11, 2001 targeting of the World Trade Center, the World Series was being hosted at New York's Yankee Stadium. How do you protect thousands of fans? The Yankee management team decided to have 1,500 police officers on alert in and around the facility. They also had sharpshooters and fighter jets. Bomb dogs were employed to sniff the stadium. All flights over the stadium were banned. While this was probably a wise move considering predictions of additional terrorist attacks, such measures are extreme and no one expects that level of security when attending public functions. It was also important to show the world community that the United States was prepared to protect the event and send a clear message to any *would-be* terrorist of a symbol of force in order to mitigate an attack.

According to a *Security Management* Survey of 150 (from a total of 752) United States and Canadian sports facilities, all but one of the 47 respondents have increased security since September 11, 2001. In most cases these increased security measures have tightened restrictions on bags, coolers and other items. Patrons are carefully inspected at the gates, while over 90 percent of the organizations lock down the facilities between events. Also included in security upgrades is greater credentialing of staff. Almost 80 percent of the organizations also monitor air intakes, but only 33 percent monitor water systems. And, 66 percent have increased electronic surveillance.[29]

The increase in security for arenas is also the result of the rise in the American sports hooligan, which Europe has long dealt with in their soccer stadiums. In 2011 a Giants baseball fan was attacked and critically injured in the parking lot of the Dodger baseball stadium. A similar attack occurred in 2003 resulting in a death of a baseball fan.[30] Facility managers and security professionals alike are being proactive in attempting to assure the public of a safe, secure and enjoyable experience at their chosen arena event.

CONFERENCE SITES AND HOTELS

But what is reasonable? What expectations should the public have when attending a play, while at a convention, visiting museums or staying at hotels? These are real concerns of companies planning conventions. Some firms in the aftermath of September 11, 2001 avoided New York City altogether in favor of other locations like Tampa, Florida and Phoenix, Arizona. Security, which was often invisible, is now expected to be seen to quell perceptions of fear by visitors.

At a minimum, security professionals for hotels and resorts have been requested to review disaster and crisis management plans and in some cases conduct training drills for employees. Employees have been trained to watch for abandoned luggage, and there are ongoing discussions about screening hotel guests' baggage. Of course, subterranean parking facilities are considered potential targets, but there has been little change in most self-park hotel operations.

One conference planner discovered that nine applications to attend the conference were using credit cards that were not their own.

Conference planners are also concerned with the potential of foreign students and professors, who could be working for their government in an intelligence-gathering capacity or terrorist organization, to try to infiltrate seminars in larger numbers as the Department of Homeland Security's (DHS) SEVIS computer begins to deny student visas to many of these individuals. One conference planner that initiated increased background checks of attendees in 2002 discovered nine applications out of 350 attendees who were listed as university students, using credit card numbers listed to other persons. The firm became suspicious since their programs had never drawn attendees from Nigeria or Ghana. An FBI inquiry confirmed the firm's concerns.[31]

MUSEUMS

Museums and other cultural institutions are in a unique position of drawing inspiration, and sometimes mass protest, from current events. For example, at the Museum of Contemporary Art in San Diego, California, a recent exhibit depicted images from inside Abu Ghraib prison of bound and hooded prisoners attached to electrodes. These were controversial images that inflamed the Arab world. According to Justin Giampaoli, Chief of Security for the Museum of Contemporary Art in San Diego, California: "There is an inherent challenge in providing a safe ... environment in an open campus ... with over 200,000 annual visitors with 50 percent [being] international visitors."[32] Post September 11, 2001 museums and other cultural institutions have acknowledged the need for significant improvements in their security programs from stricter access control, package handling and identification management. It should also be noted that the DHS offers an annual Urban Area Security Initiative grant specifically for museums that are earmarked for the hardening of targets through installation of security systems, CCTV and access control measures, among other security enhancements.

Construction Industry

The construction industry has also become involved in security concerns. In 2002 the Construction Specifications Institute (CSI) announced plans to revise its MasterFormat system to include more specifications on security. The Format includes recommendations for contractors on such things as construction requirements, products and activities. According to Dennis Hall, Chair of the CSI's MasterFormat task team, "We're not opposed to creating a strict security division if the industry (security) comes back and says we need a security division."[33] According to architect and author Barbara Nadal, architects and building planners are even considering threats such as chemical and biological weapons in their designs and retrofits. Structural engineers are considering "progressive collapse design," where only the impacted area of a building collapses while the remainder of the structure remains intact.[34]

Agriculture

Although most Americans were not aware of the $50 million loss to the beef industry in 2001, concerns over perceived *agroterrorism*, as it has become known, has resulted in tighter security measures as recommended by federal agriculture inspectors. Experts agree that infecting cattle, pigs, sheep or other animals with disease would likely erode public confidence in U.S. food supplies. Source diseases are common and with the predominate use of feedlots and confinement farming, disease can infect thousands of animals at a time. The regular trading of animals at auctions and fairs would also move the disease rapidly from a source site to other areas of the country.

With the creation of the Department of Homeland Security (DHS), the United States Department of Agriculture's (USDA) department of Animal and Plant Health Inspection Service-Veterinary Services (APHIS-VS) has received support for expanded services to monitor borders.[35]

In another agricultural area, the concerns of the American Psychopathological Society (APS) focus on two broad areas: prevention and preparedness. Prevention efforts focus on traditional issues of security, border protection and secrecy. Preparedness is concerned about early detection of threats, rapid diagnosis and response/recovery. The private sector is concerned that the federal government seems to be focusing its efforts on only security, as shown by legislative initiative during 2002. For example, the USDA has already created a list of plant pathogens considered high risk for terrorist use against United States crops.

There are those who believe that the key element in the battle against potential agroterrorism has been overlooked. The farmer must learn about the potential of attack and what to look for. According to a recent study of farmers by the Extension Disaster Education Network, over 66 percent of the farming respondents said that they either lacked access to materials about *agroterrorism* or were unaware of whether they even had access to such materials; 75 percent said they had not made investments to make their farms biosecure. It cannot be ruled out that some farmers are financially unable to make their farms biosecure as well.

To encourage reporting of suspicious people, the American Farm Bureau Federation reports that it has been encouraging its members to track who is on their property and develop ties with local law enforcement agencies.

Hospitals

While hospitals have been sensitive to security issues for decades, the aftermath of September 11, 2001 provided a renewed emphasis on security issues. Many hospitals have been working to increase access control as a primary means of mitigating problems such as theft, terrorism, vandalism, narcotics problems, and handling the mentally ill and disgruntled.

In terms of increased access control, modern state-of-the-art controls are being used. While applying traditional access control theory, CCTV, identification systems, access control cards, metal detectors and other devices are being used and evaluated in an environment that requires public access to many areas of the hospital complex.

Parking facilities are being carefully monitored using CCTV. Security officers are also employed to patrol the facilities. Many facilities have also installed emergency call boxes.

High-risk areas such as the emergency room, obstetrics/pediatrics, psychiatric, cash handling, pharmacy, research, operating rooms, and locker rooms are controlled with increased access controls that include alarmed doors, detectors, card readers and CCTV.[36]

Still, according to Jeff Aldridge, a nationally known expert on hospital security, "Even today, many hospitals still have the same open door policy they have practiced for decades." The concept of "public access" is taken literally so that any person may enter the hospital to seek treatment or visit someone. They have the right to come and go as they please. Aldridge suggests that the open door policy is no longer safe. Everyone who enters the hospital must be identified through controlled access. Old and misunderstood fire codes and old open building designs contribute to problems of restricting access. However, access can and must be controlled. Aldridge suggests that all hospitals must conduct a security assessment or audit.[37] You can read more about managing security risks in Chapter 7.

Federal Response in the United States
Department of Homeland Security (DHS)

As noted in the introduction, DHS folds 22 law enforcement, security and intelligence agencies into one conglomerate organization. The overarching mission of DHS is to preserve freedoms while protecting the United States from potential land, sea, air or in-country terrorist attacks.

The most significant federal move following 9–11 was the creation of the new Department of Homeland Security (DHS).

DHS's primary concerns are:

- Prevention of terrorist attacks within the United States
- Reduction of America's vulnerability to terrorism
- Minimize damage and recover from attacks that do occur, which now includes natural disasters

While Congress sought to have a new organization with its own intelligence-gathering operation, the Homeland Office suggested that intelligence operations are best left with the existing agencies within the new department. The Department has set up a Terrorist Threat Integration Center (also known as *Fusion Centers*) that are staffed by analysts from the CIA, FBI, NSA, Pentagon and employees of DHS. Congressional leaders still worry that such an arrangement does not improve on the intelligence situation prior to September 11, 2001.[38] In 2007, President Bush created the Director of National Intelligence position to oversee intelligence coordination as Congressional fears regarding organizational failures and interagency competition proved to be all too true.

As mentioned in the introduction, DHS was the leader in creating the Homeland Security Advisory System (HSAS), which has been replaced with the National Terrorism Advisory System (NTAS). President Bush signed the system into law in March 2002 with now Secretary of DHS Janet Napolitano changing it to the NTAS in 2011. Both HSAS and now NTAS systems are designed to improve coordination and cooperation among all levels of government and the public. Both are intended to offer a common vocabulary to understand the threat of terrorism. Of perhaps more significance and less controversial is the Ready Campaign (www.ready.gov), designed to educate the public on potential threats and steps in proper planning. While the site is still active and provides good information on a variety of topics, the new warning system developed by DHS has been designed to be more effective than HSAS, which was seen as ineffective and overly alarmist. It is still too early to tell if the new NTAS will be effective. A key aspect of the new NTAS and overall emphasis of DHS is to share responsibility. More specifically, Napolitano commented: "...the Department of Homeland Security is focused on strengthening our country's defenses by getting all stakeholders—including the public—the information and resources they need in order to play their part in helping to secure the country."[39] Complementing the efforts of the NTAS advisory system is the Federal Emergency Management Agency's (FEMA) National Domestic Preparedness Consortium (NDPC). The primary mission of NDPC is to enhance the preparedness of federal, state, and local responders to reduce the United States' vulnerability to incidents involving weapons of mass destruction and terrorism among other preparedness training.[40]

The interest in protecting the United States from further terrorist attacks reflects several issues discussed at great lengths in the security business. First, what are the respective roles of the state and private enterprise? In the above section, the private response was discussed. In the following section, the response by the United States federal government and other government units is discussed. The federal government seems to be exploring efforts to control segments of American life that have not been subject to government scrutiny. Whether such efforts will be successful, or perhaps more importantly accepted, is yet to be seen.

Cyber Security and Computer Protection

Complementing Interpol's long-standing efforts to fight global terrorism, the USA Patriot Act was perhaps the first major tool developed to investigate and fight world terrorism at the local level. The Patriot Act increased the reach of the federal government in investigating computer crimes. The Computer Fraud and Abuse Act was the base, but the Patriot Act now allows the United States to reach out to foreign computer users who pass information through a server in the United States. Patriot Act II, first proposed by then Attorney General John Ashcroft, received such a negative reaction when leaked to the public that it was disassembled and parts inserted into other legislation. For example, the Intelligence Authorization Act was signed into law by President Bush on December 13, 2003. The new act increases the use of surveillance measures in the war against terrorism. The original Patriot Act had given the FBI authority to obtain client records from banks by requesting the records using a "National Security Letter." There was no need to appear before a judge or show probable cause. The letter also directed the institutions to be silent regarding the release of information on whose records were being sought. The new legislation expands the FBI power to include stockbrokers, car dealerships, casinos, credit card companies, insurance agencies, jewellers, airlines, the U.S. Post Office and any other business "whose case transactions have a high degree of usefulness in criminal, tax, or regulatory matters." The FBI will continue monitoring individuals suspected of terrorists' links. In particular, the FBI, along with the DOJ and DHS, are working cooperatively to identify organizations and individuals who facilitate terrorism through fundraising, logistical support and recruitment.

Opposition to the Patriot Act is plentiful and on-going. For many pundits, the Patriot Act violates judicial oversight and the Fourth Amendment protections that prohibit unreasonable search and seizure.[41] Early in his executive administration, President Barack Obama extended the Patriot Act without reform or modifications. As the Patriot Act has remained unchanged for two presidential administrations, arguably one can conclude the value the Patriot Act brings in terms of intelligence in the fight against global terrorism.

It is without question that securing critical information networks, computers and the cyber space environment in general is priority. To ignore it is to risk serious economic and national security implications. That is why DHS is securing the nation's computer networks through a variety of initiatives. The most significant suggestion is the development of a private, compartmentalized federal network for government agencies and private sector experts to share information during major events. The system would be part of the newly created Cyber Warning Information Network (CWIN), which includes federal offices responsible for the security of the federal computer systems as well as private sector interests.

The supporters of CWIN note that this is not a government regulation effort, with government mandates, but rather an industry-led activity to find solutions. According to Harris Miller, President of Information Technology Association of America, this is important since between 85–90 percent of the infrastructure used by government and industry is owned by private industry. However, there are those who feel that the approach, while worthy, lacks teeth. There is no system to enforce guidelines, as they are only "guidelines."[42]

Comprehensive National Cyber Security Initiative

Significant progress has been made since September 11, 2001 toward securing the nation's computer and digital infrastructure. However, there is still a lot of work to be done as computer systems and networks continue to evolve in sophistication. A proactive cyber security approach helps to ensure that the United States stays one step ahead of the criminal enterprise as well as the terrorist network bent on disrupting the nation's computer systems. For example, the Comprehensive National Cyber Security Initiative (CNCI) is one such proactive program that is definitely a step in the right direction. Shortly after taking office, President Obama ordered a full-scale review of the federal government's efforts to defend the United States information and communications infrastructure. The end goal of the review was to develop a comprehensive approach to secure the nation's digital infrastructure. Not surprisingly, the finding is one that the United States government and country are not adequately prepared.

To ensure the nation is prepared, CNCI consists of a number of key goals designed to secure the United States' cyberspace, as follows:

- **To develop a front line defense against today's immediate threats.** For example, sharing information regarding network vulnerabilities, threats and events within the federal government as well as state and local governments.
- **To defend against the full spectrum of threats.** For example, enhancing the United States government's counterintelligence capabilities and increasing the security of the supply chain for key information technologies. In March 2010, the GAO issued a report regarding the integrity of the DOD supply chain with recommendations that are designed to mitigate the risks of counterfeit parts.[43]
- **Lastly, to strengthen the future cyber security environment.** For example, expanding cyber education and research to deter hostile or malicious activity in cyberspace.

All of this will be done in coordination with the Executive Branch's Cybersecurity Coordinator who now has regular access to the President of the United States. Other key players include cyber security professionals in the government, and state and local governments and the private sector. In fact, the private sector and many information security professionals have served with distinction on many of the cyber security advisory boards. All of this is designed to ensure an organized and unified response to cyber incidents regardless of their origin and purpose whether terrorist, economic, social unrest or politically motivated.[44]

Outside the cyber security agenda but equally as important, many other federal government agencies such as the Department of Health and Human Services (HHS) has put local health departments, hospitals and medical care providers on alert to report any unusual disease patterns. The department also has enhanced its inspection of imported foods.

In addition, the HHS, in cooperation with the Centers for Disease Control and Prevention (CDC), issued regulations for procedures in laboratories handling certain toxins or pathogens. The regulations limit the handling of these agents to certain people. Each facility must develop and implement a plan to ensure security of designated agents. The plan must include inventory control, security training for non-security personnel, methods for reporting suspicious persons, loss of agents, and alteration of inventory documents. Access controls must be established for all employees and visitors. Visitors and employees who are with housekeeping or maintenance in areas of laboratory or security operations must be escorted and monitored at all times. Designated agents must be secure at all times as well.

The health care industry, reacting to the mandates of the DHS, is also impacted by President Bush's decision to have military, health services workers and other first responders vaccinated against smallpox. In December of 2002, President Bush ordered smallpox vaccinations for 500,000 soldiers. Many health officials ponder whether the risk of death associated with the vaccinations is worth the preventative value. Once again, only time will tell whether this program was over protective.[45]

The United States Department of Agriculture (USDA) has alerted food and agriculture community workers to monitor feedlots, stockyards, and import and storage areas. The USDA has also stepped up its efforts to provide resources to protect the $140 billion agriculture industry. In 2006 the USDA published its Pre-Harvest Security Guidelines and Checklist. The document provides farmers with guidelines for risk assessment, checklists of things to do to protect various types of agricultural activities and a list of websites for various agricultural resources.[46]

The Federal Emergency Management Agency (FEMA) developed the publication, *Are You Ready? A Guide to Citizen Preparedness*. This comprehensive guide to disasters provides information on a step-by-step outline of how to prepare for a disaster. The guidelines include information on disaster supply kits, emergency planning, how to locate and evacuate to a shelter, etc. A full copy is available online at: www.fema.gov/areyouready.

Following the problems associated with the lack of appropriate response from FEMA to the New Orleans–Katrina disaster of 2006, the DHS has established the National Advisory Council. The Council will advise FEMA on all aspects of emergency management in an effort to ensure close coordination with all involved.[47]

The Immigration and Naturalization Service (INS) has proposed more detailed information about persons who enter or leave the United States by boat or plane. Even American citizens may be required to fill out forms detailing their travel. With this request is a proposed system that will allow for quicker cross-checking of databases and matching records of arrivals and departures.

The proposed rules would not impact buses, trains or private transportation, but target commercial airlines (passengers and crews), cruise ships, cargo flights and vessels. Canadian officials who believe the system would bog down in the border areas between Canada and the United States have voiced concerns over the program.[48]

In May 2002, the Enhanced Border Security and Visa Entry Reform Act was signed into law. The act requires that foreign nationals wishing to enter the United States must have a tamper-proof, machine readable visa with biometric identifiers. The Act implementation deadline was October 26, 2004.[49] Beginning in January of 2008, all persons entering the United States must have a legal passport. This includes United States citizens who routinely cross the borders to Mexico and Canada.

The US-VISIT Program

According to the Department of Homeland Security this is a top priority because it:

- Enhances the security of citizens and visitors
- Facilitates legitimate travel and trade
- Ensures the integrity of the immigration system
- Protects the privacy of visitors[50]

The Student and Exchange Visitor Information System (SEVIS)

In the area of tracking potential terrorists the DHS initiated a $36 million computer tracking system to monitor student and exchange visitors at universities. The system, known as the Student and Exchange Visitor Program (SEVP), was designed to track approximately 500,000 foreign students who come to the United States each year to attend school. Two of the terrorists involved in the September 11, 2001 bombing were approved for student visas six months after the attack.

The Student and Exchange Visitor Information System (SEVIS) links approximately 70,000 schools admitting foreign students to the Customs and Border Protection (CBP), the agency that replaced Immigration and Naturalization Services in 2002. The rollout of the SEVIS system in March 2003 resulted in a multitude of problems. Forms were not available on the SEVIS website; printed information designated for one school might turn up on a printer thousands of miles away at another school. Foreign students and professors have been held up for days, and active students are being listed in the system as "dropped out."

While the intent of the system is good, the zero tolerance built into the system is oppressive. As one university official said, "You can't fight terrorism by terrorizing students."[51]

Despite early problems, the program has continued to grow. As of April 2007 there were 951,654 active non-immigrant student exchange visitors (counting dependents) in the United States. The SEVIS database includes over 2.5 million entries. The largest population of students are in the following states: California, New York, Texas, Massachusetts and Illinois.[52]

Federal Identification Cards

Federal identification cards have been discussed at some length since the destruction of the World Trade Center. While a hot button topic for many Americans, the federal government continues to discuss options that would allow for higher levels of trust in identification systems currently controlled by most states. These include driver's licenses, birth certificates, and death certificates. Such a coordinated system would require agreement and cooperation among all 50 states including funding. An effort to develop a federal identification card failed to receive support in the version presented in the Patriot Act II. The debate continues today, as some opponents are concerned that tracking technology would be embedded in the cards and the federal government could actively monitor your whereabouts. Setting the debate aside, the federal government issues all citizens Social Security numbers and passports who want them, so processes are in place to move forward with a federal identification card should the states and opponents concede to the proposal.

Homeland Security Presidential Directive/Hspd-12 (HSPD-12)

On August 27, 2004, President Bush released the Homeland Security Presidential Directive/ Hspd-12, commonly referred to as HSPD-12. The Presidential directive establishes a policy for a common identification standard for federal employees and contractors. In short, it is a mandatory government-wide standard for a secure and reliable form of identification issued by the federal government to employees, contractors and contract employees. Secure and reliable means identification that:

- is issued based on sound criteria for verifying an individual's identity.
- is strongly resistant to identify fraud, tampering, counterfeiting, and terrorist exploitation.
- can be rapidly authenticated electronically.
- is issued only by providers whose reliability has been established by an official accreditation process.

The system includes criteria for a graded security system.[53]

Total Information Awareness Program

The Defense Advanced Research Projects Agency (DARPA) working for the Department of Defense (DOD) is working, among many other projects, to develop what it calls the Information Awareness Program (IAP) in supporting the DOD war on terrorism. The program would utilize new surveillance and information analysis systems aimed at protecting American citizens. The system when completed would provide a computerized record of an individual's private life, supplying a paper trail of the person's entire life, including vital statistics, medical, financial, e-mail, Internet, phone and travel records. More information on this program can be found at www.darpa.mil/DARPATech2002/presentation.html.[54]

There are many questions about this initiative, including issues involving privacy safeguards, limitations on access, and protection of records already protected by existing legislation.

DARPA activities continue as the engine for radical innovation in the DOD. DARPA's mission is to prevent technological surprise for us *and* to create technological surprise for our adversaries. Stealth is one example of how DARPA created technological surprise.

The following is a list of ongoing DARPA research that promises major benefits to DOD (the Strategic Plan can be downloaded from DARPA's website, www.darpa.mil), and which may become icons of significant technical achievement by themselves.

- **Networks**: self-forming, robust, self-defending networks at the strategic and tactical level are the key to network-centric warfare.

FIGURE 5-3 The MQ-9 Reaper drone, a small unmanned aircraft system. *(U.S. Air Force photo by Senior Airman Larry E. Reid Jr.)*

- **Chip-Scale Atomic Clock**: miniaturizing an atomic clock to fit on a chip to provide very accurate time as required, for example, in assured network communications.
- **Global War on Terrorism**: technologies to identify and defeat terrorist activities such as the manufacture and deployment of improvised explosive devices and other asymmetric activities.
- **Air Vehicles**: manned and unmanned air vehicles that quickly arrive at their mission station and can loiter there for very long periods (Figure 5-3).
- **Space**: the U.S. military's ability to use space is one of its major strategic advantages, and DARPA is working to ensure the United States maintains that defense advantage.
- **High Productivity Computing Systems**: supercomputers are fundamental to a variety of military operations, from weather forecasting to cryptography to the design of new weapons; DARPA is working to maintain our global lead in this technology.
- **Real-Time Accurate Language Translation**: real-time machine language translation of structured and unstructured text and speech with near-expert human translation accuracy.
- **Biological Warfare Defense**: technologies to dramatically accelerate the development and production of vaccines and other medical therapeutics from 12 years to only 12 weeks.
- **Prosthetics**: developing prosthetics that can be controlled and perceived by the brain, just as with a natural limb.

- **Quantum Information Science**: exploiting quantum phenomena in the fields of computing, cryptography, and communications, with the promise of opening new frontiers in each area.
- **Newton's Laws for Biology**: DARPA's Fundamental Laws of Biology program is working to bring deeper mathematical understanding and accompanying predictive ability to the field of biology, with the goal of discovering fundamental laws of biology that extend across all size scales.
- **Low-Cost Titanium**: a completely revolutionary technology for extracting titanium from the ore and fabricating it promises to dramatically reduce the cost for military-grade titanium alloy, making it practical for many more applications.
- **Alternative Energy**: technologies to help reduce the military's reliance on petroleum.
- **High Energy Liquid Laser Area Defense System**: novel, compact, high-power lasers making practical small-size and low-weight speed-of-light weapons for tactical mobile air- and ground-vehicles.

Federal Building Security Initiatives

Even with federal government intervention, there have been problems. In the aftermath of the 1995 Oklahoma City bombing of the Muir Federal Building, the Department of Justice (DOJ) moved security at federal buildings to the top of its priority list. Following September 11, 2001 the GAO was asked to consider how well the concerns voiced after the Muir bombing had been addressed. Auditors found that only 50 percent of the agencies had completed security assessments. Security appears to be easy to discuss, but more difficult to implement. The GAO continues to monitor progress on security recommendations.

Protection of Chemical Manufacturing Sites

For many years now the Department of Homeland Security (DHS) has been spearheading efforts to set federal standards for the chemical manufacturing sector. The agency's rules will be designed to provide safeguards to prevent sabotage and require enhanced perimeter security at chemical plants. Some states—for example, New Jersey—while not opposed to legislation, do not want it to set a standard that the states cannot exceed. In the case of New Jersey, they already have standards in place that exceed those expected to be implemented by the federal government.[55]

Anti-Bioterrorism

DHS *Biowatch* program was established in 2003 in approximately 30 cities to monitor the air for possible biological attack. Equipment was installed quickly, but there was no detailed plan on how to respond in the event of positive alarms. The impact was felt in Houston in October 2003 when the alarms showed positive for tularemia, a naturally occurring pathogen. Too often well-intentioned plans are not tied to the reality of response by actual cities and their responders.[56]

State/Local Response

All areas of the country are seeing local response to the potential threat of terrorism. In Alaska, the Anchorage fire department has rekeyed its facilities. In Arizona, the Gilbert Police Department has assigned extra officers to schools, power plants and water supply operations. Rockford, Illinois has added a bomb-sniffing dog to its ranks. Orange County, Florida sheriff's deputies were among the first county law enforcement officials to receive smallpox vaccinations. The City of Los Angeles Police Department is working with DHS on public information and awareness *"if you see something, say something"* campaign. Overall, such added security measures are costly.

The Costs

As an example of cost associated with heightened security associated with the protection of the United States, the governor of Illinois reported that the orange alert terror status costs the state about $20,000 a day. Such costs can be devastating to state and local governments without some type of federal support.[57]

> In 2003 communities were spending $70 million a week on security.

At the 2002 U.S. Conference of Mayors, 192 cities leaders with populations of over 30,000 were polled. The pollster reports it is estimated that municipalities would spend approximately $2.1 billion on equipment in 2002. A 2003 study of cities of over 30,000 (1,185 in the United States) found that this group of communities was spending $70 million a week on security. New York topped the list, spending an extra $5 million each week. San Francisco, Los Angeles, and Atlanta were spending just over $2 million per week followed by Fresno at $1.5 million. Other cities report little added expense due to homeland security including Rockford, Illinois, Winston-Salem, North Carolina and Fort Collins, Colorado. San Antonio reported an increase of only $15,000 per week.[58]

This is a drop in the bucket compared to the $525 million spent between September 11, 2001 and January 1, 2002.[59] The significance of the need for local support in the federal government's battle against terrorism is exemplified by where the government is spending money. In March 2003, the DHS released approximately $600 million dollars to the states for local first responders out of an annual federal allocation of roughly $38 billion. The funds are designated for equipment, planning and training exercises. The problem, as with many federal initiatives, is that the money did not begin flowing until July. In addition the department provided another $750 million for firefighters.[60]

A quick check of the math will show that, while the funding is welcome, it is only a small portion of what local and state governments have already invested in the federal government's war on terrorism.

In 2007, while funds had been spent on first responder equipment and training, many state officials, particularly those in law enforcement, complained that there had been too little effort placed on improving communication of vital intelligence information and not enough emphasis on local intelligence operations. The total fiscal year budget for homeland security was estimated at $54.3 billion.[61]

School Safety

DHS has encouraged local school districts to implement or review crisis/emergency planning. In cooperation with the Department of Education, the two agencies have designed a "one-stop" shop to help local school districts plan for any emergency, including natural disasters, violence and of course terrorism.

In addition, the Department of Education announced that it would make $30 million available to help schools improve emergency response and crisis management plans. Additional information can be found at www.ed.gov/emergencyplan.[62]

Emergency Response Plans

Many cities have or are working on or re-evaluating emergency response plans. The plans fill a need in case of a terrorist attack, and are useful in cases of natural disasters such as the recent and devastating tornados that killed hundreds in Alabama. Pre-existing plans in Washington, D.C. and Dallas/Fort Worth are cited as examples of how having plans can reduce confusion and get the city back to its business, although, in after-event analysis, both plans were found lacking in the areas of coordination of specific communication.

What is evident in most of these plans is the multi-jurisdictional nature of the response and the need to coordinate/communicate actions taken by various team members. In addition, these plans have been developed utilizing a variety of sources including the Federal Emergency Management Administration (FEMA) and the Critical Incident Protocol developed by the School of Criminal Justice at Michigan State.[63]

Examples of Specific State Response

Mississippi has enacted a law making it illegal to possess or release harmful biological substances. The law also prohibits false claims of exposure. The punishment established by the law is a fine up to $100,000 and 20 years of prison.

Illinois mandated that all municipalities with populations over 1 million adopt an ordinance mandating emergency procedures for high-rise buildings by January 1, 2004. Plans included procedures for evacuating people with disabilities, the roles of building personnel, contact information and instructions for conducting drills at least once per year.[64]

In Colorado the Denver Department of Public Health has set out to monitor early symptoms of a bio terror attack. The program, called "Syndromic Surveillance," looks to identify changes in health patterns early, so that a bio terror attack may be identified before too many people fall ill. The program was featured on the Homeland Security website.

In December 2002, western governors agreed to work together to develop common standards for emergency communications systems and procedures for use in responding to terrorist attacks or other emergencies. Such systems require cooperation between state and local governments as well as private companies.[65]

The Port Authority of New York and New Jersey spent approximately $2.7 billion on security-related costs since the 9/11 attacks. Expenses were primarily for security personnel at cargo facilities located at Newark Liberty, Kennedy, LaGuardia and Teterboro airports. It was estimated that the Authority spent approximately $2.3 million weekly on protection of these air cargo facilities. In 2007 the Authority spent $394 million for police and emergency salaries, up from $180 million in 2000. The Authority also spent $18 million on black box tracking technology used in all shipping containers, which can then be monitored by satellite.[66]

The Chicago Transit Authority working with the Chicago Police Department has made it possible for Chicago police cars to watch live views from inside buses. The CTA is outfitting buses with radio equipment that can be picked up by nearby police cars. The city is studying a proposal by AT&T and EarthLink to create a Wi-Fi system to monitor the entire city.[67]

Private/Public Joint Initiatives

The need and critical importance of cooperation between public law enforcement and private security has never been greater. Yet, the difficulties of establishing effective cooperation between the two groups, particularly in the area of emergency planning and response, remain a major obstacle to effective responses. According to Penelope Turnbull, Director of Crisis Management and Business Continuity for Marriott International and Vice Chair of the ASIS International Disaster Management Council, "I do believe that there is a growing realization that greater cooperation between and within the two sectors is required..."[68] According to Turnbull, initiatives will need to overcome traditional issues of trust, jargon and objectives. It also goes to professional respect as traditional law enforcement at the federal, state and local levels still have a bias against the security professional as someone who lacks training and credentials. Their view? One could argue they view the security person and the profession in general as someone who flunked out of another profession.

However, a good example of how public and private cooperation can work was seen in the aftermath of the crash of American Airlines Flight 77 into the Pentagon. Through the cooperation of law enforcement and commercial communications companies, the problems usually associated with increased telephone and radio traffic during an incident of this magnitude were virtually eliminated. As further evidence and according to the Public Safety Wireless Network (PSWN)—a joint initiative between the Departments of Treasury and Justice—the response showed true interoperability despite the fact that many departments operate on different frequencies. Cooperation with the Arlington County dispatch center and commercial companies Nextel, Cingular and Verizon Wireless made the effort truly work. There were, however, lessons to be learned; most of these lessons had to do with proactive information sharing prior to an event that would make all known resources available to the users.[69]

The National Institute of Justice has also taken a "let's work together" approach. The Institute published flow charts, diagrams, matrices and other help aids for businesses in the chemical industry interested in analyzing their security systems.

The Austin, Texas public transportation system has increased its security by purchasing wireless digital cameras that are located both inside and outside city buses. The funding for this program is from the federal government and could reach over one-half million dollars. While the general public has often seen "big brother" in such projects, the Austin community seems to have accepted the cameras. The Austin Police Department has also played a significant role in specifying the camera system and establishing chain of custody in the event the cameras record criminal activities.

According to the DHS, companies, universities and government agencies are reporting cyber attack crimes to the U.S. Computer Emergency Readiness Team (US-CERT) in increasing numbers. In 2005, 5,000 incidents were reported. In 2006, 23,000 incidents were reported, an almost five-fold increase. In the first quarter of 2007 more than 19,000 incidents were reported.[70] For updated reporting metrics, go to www.us-cert.gov.

In an effort to assist local and state law enforcement agencies faced with the possible attack using nuclear or duty bombs, the United States National Nuclear Security Administration has been working on "Render Safe" devices that could be used to disarm such devices. The "Render Safe" program is in response to concerns that in the event of a terrorist threat involving nuclear/dirty bomb technology, the Nevada-based Nuclear Emergency Search Team (NEST) would not be able to get to the scene of the attack in time.[71]

While the DHS promotes investment in the "Render Safe" system, others criticize it as an extraordinarily costly program that will provide few security gains. Costs must certainly be a concern as New York officials are haggling over the burden of maintaining and operating a network of detection machines that the Domestic Nuclear Detection Office, DHS, is planning on installing in New York City.[72]

Concerns

No one should minimize the loss of 3,000 lives to a single terrorist attack or ignore significant attacks by terrorists throughout the world, but it is equally important to consider the impact of overreaction on the health of the world both mentally and economically. In the big picture, most businesses and institutions do not represent viable targets for terrorist attacks. The emphasis on protection from terrorism may be diverting attention and funding away from other more significant and real threats. This is something that risk analysis through vulnerability/probability studies may be able to answer. For additional information regarding vulnerability, probability and risk mitigation, refer to Chapter 7.

At the same level of thinking, it is worth noting that the United States government has established the largest new cabinet level department since World War II. Within the operations of this super-department are government plans to play larger roles in protecting the American public. The question that needs to be asked is, "Just how much protection do we need and at what cost?"

Arguably, some believe that al Qaeda may have achieved a portion of its goal in the establishment of the DHS. The impact of its attack may have caused government and industry to overspend on security. This argument is countered by a look at the percent of the gross domestic product (GDP) going toward homeland defense. The Federal Reserve Bank reports that, while expenditures on security have tripled, the total allocated in 2003 ($38 billion) in this area represents only .35% of the GDP. Even when added to the Department of Defense budget of $379 billion, the total still is less than 3.8% of the GDP. The Federal Reserve also indicates that the amount being spent by state and local governments is less than 1/1000 of state and local budgets. Estimates in a recent study indicate that the Fed figures were still accurate through 2005.[73]

Still state and local governments are complaining that expenses are in addition to already allocated funds. Fiscal health remains an important issue in many states where deficit spending continues to be a major problem.

On another level tension has been created between the need to monitor and evaluate people entering this country as well as those already here. Security needs to appear to be great. To accommodate those needs government agencies are asking for unprecedented access to personal information. On the other side is what many Americans consider a given right to privacy. The right to privacy requires that steps are taken to protect personal information. Recent legislation has increased the security of personal information, while legislation following September 11, 2001 has also increased the government's ability to monitor and control personal behavior in public places. While the public opinion immediately following September 11, 2001 was in favor of increased security, there seems to be a shift back to privacy as time passes. This shift is best exemplified in the change in public opinion concerning the use of federal identification cards with biometric identifiers. Polls conducted by Harris, Pew Research Center and the Washington Post found support for the national ID system ranged from a low of 44 percent to 70 percent. In a poll by Gartner Group (March 2002) only 26 percent of the respondents viewed the system as positive.[74] Departing House Majority Leader Dick Armey warned in 2002 that the nation should be careful in sacrificing freedom for safety in the fight against terrorism. American Civil Liberties Union cyber chief Barry Steinhardt has said that, "The surveillance monster is growing, but the legal chains to these monsters are weakened even when we should be strengthening them."[75] He refers to projects that would call high-tech security cameras to instantly match people with those in databases.

Although the establishment of the Department of Homeland Security (DHS) was touted as a progressive move, bringing together the intelligence and response capabilities of various and often competitive federal agencies, the truth is that the transition to a super-agency has not been smooth. Competitive egos are deep rooted and difficult to root out. The failure of FEMA in post Katrina reflects poorly on its supervising unit. In a recent GAO bulletin, the Department is criticized for failure to provide adequate support to its mission components. In particular, DHS has problems in acquiring goods and services and providing proper oversight of this function.[76]

In addition, some legislation passed soon after the September 11, 2001 attacks, while well intentioned, has failed to achieve any meaningful objectives. For example, the 2002 Federal Information Security Management Act (FISMA) mandated security planning for agencies, requiring a risk analysis of IT systems, and certification and accreditation of the systems. While the act is appropriately written, there is no way under the current legislation to measure if the plan actually improves security. According to Bruce Brody, vice president of information assurance at CACI International Inc., "Federal systems and networks are like Swiss cheese. FISMA over five years has not helped us to be appreciably more secure."[77]

Perhaps the most telling assessment of the new Department is the February 7, 2007 report by the GAO. While DHS is just over six years old, the importance of its stated mission requires that any problems associated with the consolidation of over 22 agencies into a cohesive unit must be overcome. As of the February 2007 report the GAO notes that DHS continues to face programmatic and partnering challenges. To help ensure its missions are achieved, DHS must overcome continued challenges related to cargo, transportation, and border security; systematic visitor tracking; efforts to combat the employment of illegal aliens; and outdated Coast Guard asset capabilities.[78]

Notwithstanding the death of Al Qaeda Chief Osama Bin Laden accomplished by the United States Seal Team 6 in Pakistan, the DHS is here to stay and will work collectively with other federal government agencies to keep the United States safe from the next generation of terrorists and those certainly to follow Bin Laden.

■ ■ CRITICAL THINKING ■

What are the potential liabilities associated with limiting individual rights? Are the potential protections afforded by greater government scrutiny worth the reduction in individual freedom?

CASE STUDY

The Bill of Right of the United States Constitution provides for the protection of life, liberty and the pursuit of happiness for all Americans. All Americans have a reasonable right and a certain expectation of privacy in their work and personal lives. You are the Security Manager for a large facility which employees 25,000 employees primarily in a cubicle and open work environment. The facility is surrounded by a very larger parking lot with little or no trees. You are asked by the CEO of Walsh Corporation to install closed circuit cameras (CCTV) in the facility and parking lot as a deterrence to would be criminals.

1. What challenges are presented by installing CCTV cameras?
2. What might you do to advertise the fact CCTV cameras are in use?
3. Where would CCTV cameras not be allowed under any circumstances?

Review Questions

1. What new initiatives are underway through the American Society for Industrial Security International?
2. What impact have federal regulations, rules and guidelines had on private business in the United States?

3. Summarize what you know about the Department of Homeland Security.
4. What measures if any have been taken to secure world ports and the shipping industry?
5. Discuss the problems associated with electrical power security.

References

[1] The Blog@Homeland Security, posted by Secretary Janet Napolitano, 4/20/2011.

[2] Garamone J. Top DoD Budget Official Outlines War on Terror Costs. American Forces Information Services, <www.defenselink.mil/news/NewsArticle.aspx?ID=2949>.

[3] ASIS press release, Workplace security legislation supported by ASIS international implemented by department of justice, February 7, 2006.

[4] Anderson T. A year of reassessment. Secur Manage January 2003 [61+].

[5] US Corporate Security Spending has increased only modestly since September 11th, Continuity Central online.

[6] Transportation Security Administration website <www.tsa.gov>, RESEARCH 2006 and April 2011.

[7] The American Motorcoach Industry's Anti-Terrorism Bus Association, 2002.

[8] Cargo security initiative established at world's 20 largest seaports. Current Issues, USINFO, July 27, 2004.

[9] Security rules adding time to shipments. Secur Beat. March 18, 2003, download 3/18/2003, <www.securitysolutions.com>.

[10] Ports vow to run a tight ship. Secur Manage, May 2002:16–7.

[11] Ibid.

[12] Maritime security. Secur Manage, March 2003:142.

[13] Container security. Secur Manage, February 2003:21.

[14] White Jr CH. Guest viewpoint: developing railroad security. Tempe, AZ: Institute for Supply Management.

[15] Ibid.

[16] TSA Cites authority to search vehicles; Gets passing grade from GAO. Secur Beat, February 25, 2003. downloaded 2/25/2003, <www.securitysolutions.com>.

[17] Gips MA. They secure the body electric. Secur Manage November 2002:77–81.

[18] Nuclear Energy Institute, US Nuclear Power Plant Statistics, <www.nei.org>. 2010.

[19] NRC struggling to address reactor shortcomings. Secur Manage June 2002:17.

[20] Dremann S, Schilling S. San Bruno's exploded gas line runs through Mountain View. MountView Voice, 9/20/2011.

[21] http://www.aeronautics-sys.com/Index.asp?CategoryIC=116&ArticleID=276&Page=16/20/2007.

[22] Jones J. Canada oil sector takes al Qaeda threat seriously. Reuters, <www.reuters.com/articleId=USN11433721020070214>.

[23] Water utilities scramble to meet federal deadlines. Secur Beat March 18, 2003, downloaded 3/18/2003, <www.securitysolutions.com>.

[24] Tieman M. Safe drinking water act: implementation and issues, Congressional Research Service for Congress, May 3, 2006.

[25] Apuzzo M. Mall security guards get anti-terrorism training. Dly Texan 12/03/2004.

[26] Mall shooting spree raises security questions. Retail Traffic February 14, 2007.

[27] Ibid.

[28] Premo R. City center solutions. Secur Manage April 2002:85–92.

[29] Survey assesses sports facility security. Secur Manage February 2003:14.

[30] Nachaman C. San Francisco Giants Fan Beaten and Critically Injured. Business Insider 4/1/2011.

[31] When attendees may be terrorists. Secur Manage March 2003:16–8.

[32] Giampaoli J. Chief of Security, Museum of Contemporary Art, San Diego, California, Phone Interview, 5/2/2011.

[32a] Gips MA. Open spaces in a tight spot. Secur Manage March 2003:22.

[33] Security can cement construction role. Secur Manage March 2003:22.

[34] Building security starts with planning and design. Secur Beat, February 25, 2003, downloaded 2/25/2003, <www.securitysolutions.com>.

[35] Gips MA. The first link in the food chain. Secur Manage February 2003:41–7.

[36] Leahy RF, Michelman BS. Healing access control woes. Secur Manage March 2003:88–96.

[37] Aldridge J. Hospital security: the past, the present, and the future, SecurityInfoWatch.com, August 31, 2005.

[38] Using the Same Old Eyes. Newsweek March 10, 2003:7.

[39] The Blog@Homeland Security, posted by Secretary Janet Napolitano, 4/20/2011.

[40] FEMA, National Domestic Preparedness Consortium bulletin, <www.dhs.gov/consortium>. 5/20/2010.

[41] Martin D. With a whisper, not a bang. San Antonio Current, 12/24/2003.

[42] Cyberspace protection plans get mixed reviews. Secur Manage November 2002:42.

[43] GAO Report. DOD Should Leverage Ongoing Initiatives in Developing Its Program to Mitigate Risk of Counterfeit Parts, March 2010.

[44] Department of homeland security. The comprehensive national cybersecurity intiative. Executive Office of the President of the United States memorandum, 2009.

[45] Widespread smallpox vaccination too dangerous for U.S. population. Secur Prod March 2003:54.

[46] Pre-Harvest Security Guidelines and Checklist 2006, USDA.

[47] Homeland Security Establishing a National Advisory Council, FEMA release.

[48] Ibid.

[49] Government Wants Detailed Information about Guests. Secur Prod March 2003:52.

[50] Ibid.

[51] Becker R. Glitches riddle database to track foreign students. Chicago Tribune Monday March 17, 2003:1.

[52] Student and Exchange Visitor Information System: General Summary Quarterly Review, April 24, 2007.

[53] Homeland Security Presidential Directive/Hspd-12, August 27, 2004 press release, <www.whitehouse.gov/news/releases/2004/08/print20040827-8.html>.

[54] DARPA, <http://www.dhs.gov/xtrvlsec/programs/content_multi_image_0006shtm>.

[55] Marsico R. An appeal on chemical plant safety. The Star-Ledger February 9, 2007.

[56] Lipton E. New York to test ways to guard against nuclear terror. New York Times February 9, 2007.

[57] Governor outlines more security preparations. Macomb J, Thursday, March 20, 2003:2A.

[58] Cities Spending Extra $70 Million a Week on Security, Survey Says, CNN.com, March 27, 2003, downloaded 4/4/2003, <http://cnn.allppolitics.printthis.clickability.com/pt/cpt?action=cpt&expire=-1&urlID=583092>.

[59] Municipal security. Secur Manage April 2002:19.

[60] Sarkar D. Hometown security gets funding boost. Fed Comput Week, March 17, 2003, downloaded 3/18/2003, <www.fcw.com/fcw/articles/2003/0317>.

[61] Hobijn B, Sager E. What has homeland security cost? An assessment: 2001–2005. Curr Issues Econ Finance February 2007;13(2).

[62] Paige, Ridge Unveil New Web Resources to Help Schools Plan for Emergencies. U.S. Department of Education, downloaded 3/18/2003, <www.ed.gov/PressReleases/03-2003/03072003.html>.

[63] What it took. Secur Manage March 2003:68.

[64] U.S. State legislation – illinois. Secur Manage February 2003:93.

[65] Western governors agree to work together for common emergency communications. Secur Prod February 2003:34.

[66] Port authority spending its own funds on security. Access Control Secur Syst January 23, 2007, <www.securitysolutions.com/ne\ws/port-authority-security/index.html>.

[67] Van J. In Chicago, Police to See Live Surveillance from Buses. SecurityInfoWatch.com, June 21, 2007.

[68] Public-private emergency planning. Secur Manage November 2002:39.

[69] Wireless works for first responders. Secur Manage April 2002:33.

[70] Lemos R. Companies Increasingly Reporting Attacks, <www.securityfocus.com/brief/430>.

[71] Davidson K. Devices could disable terror bombs. San Francisco Gate February 7, 2007:A-3.

[72] Lipton E. New York to test ways to guard against nuclear terror. The New York Times February 9, 2007.

[73] Homeland security costs. Secur Manage February 2003:20.

[74] ACLU Cyberchief Worried About Privacy, CNN.com March 30, 2003, downloaded 4/4/2003, <http://cnn.technology.printthis.clickability.com/pt/cpt?actions=cpt&expire=04%2F29%2F20>.

[75] Ibid.

[76] Observation on the Department of Homeland Security's Acquisition Organization and on the Coast Guard's Deepwater Program, GOA Highlights, GAO-07-453T, February 8, 2007.

[77] Jackson W. Experts: It's time to fix FISMA, GCN, <www.gcn.com/cig-bin/udt/id=gcn_daily&story.id=43103>.

[78] Management and Programmatic Challenges Facing the Department of Homeland Security, GOA highlights, February 2, 2007, GAO-07 452T.

Basics of Defense

In this section, the basic tools used in security and loss prevention are discussed. These range from a theoretical understanding of law, honesty, risk analysis, and surveys, to applied technology in locks, alarms, barriers and procedures. Other tools include an understanding of contingency planning, insurance, fire protection, and safety. All of these elements make up the foundation on which any good security program is based.

These are foundational tools, but it is important for security experts to realize that the applications of these tools can and do change with time. Technology changes, especially computing, have made the 18th-century lock a modern card-operated security device. Theories have continued to evolve, with contingency planning now a must for any organization. And even the basic law continues to change as new cases and events are reviewed by the courts.

6

Security and the Law

OBJECTIVES

The study of this chapter will enable you to:

1. Understand the basic principles in the application of law.
2. Identify the differences in law as it applies to private security and public law enforcement.
3. Discuss various trends in liability issues.
4. Know new legal issues that apply to private security matters and how to maintain currency.

Introduction

Although in the United States public police and protection services derive their authority to act from a variety of statutes, ordinances, and orders enacted at various levels of government, private police function essentially as private citizens. Their authority to so function is no more than the exercise of the right of all citizens to protect their own property. Every citizen has common-law and statutory powers that include arrest, search, and seizure. The security officer has these same rights, both as a citizen and as an extension of an employee's right to protect his or her employer's property. Similarly, this common-law recognition of the right of defense of self and property is the legal underpinning for the right of every citizen to employ the services of others to protect property against any kind of incursion by others.

The broad statement of such rights, however, in no way suggests the full legal complexities that surround the question. In common law, case law, and state statutes, as well as in the basic authority of the U.S. Constitution, privileges and restrictions further defining these rights abound. The body of law covering the complex question of individual rights of defense of person and property contains many apparent contradictions and much ambiguity. In their efforts to create a perfect balance between the rights of individuals and the needs of society, the courts and the legislatures have had to walk a narrow path. As the perception of society's needs changed or as the need for the protection of the individual became more prominent, a swing in the attitudes of the courts and the legislatures became apparent. This led to some confusion, especially among those with little or no knowledge of the law.

It is of enormous value, therefore, for everyone engaged in security to pursue the study of both civil and criminal law. Such a study is aimed neither at acquiring a law degree nor certainly at developing the skills to practice law. It is directed toward developing a background in those principles and rules that will be useful in the performance of the complex job of security.

Without some knowledge of the law, security officers frequently cannot serve their clients' interests. They may subject themselves or their employers to ruinous lawsuits through well-meaning but misguided conduct. In cases that must eventually go to court, handling of evidence, reports, and interrogations may be critical to the case; without an understanding of legal processes and how they operate, the cases could be lost.

In short, the pursuit of security itself involves contact with others. In each such contact, there is a delicate consideration of conflicting rights. Without an appreciation of the elements involved, the security officer cannot perform properly.

Because for the purposes of this book we are primarily interested in civil and criminal law—both have major implications for the security officer and for the industry—it is useful to distinguish between them. Criminal law deals with offenses against society (corporations are of course part of society and can be either criminals or victims). Every state has its criminal code, which classifies and defines criminal offenses. Criminal law is the result of a jurisdiction either using common law, which was adopted from English traditions, or passing specific legislation called statutory law. (In some jurisdictions both are used.) When criminal offenses are brought into court, the state takes an active part, considering itself to be the offended party.

Civil law, on the other hand, has more to do with the personal relations and conflicts between individuals, corporations, and government agencies. Broken agreements, sales that leave a customer dissatisfied, outstanding debts, disputes with a government agency, accidental injuries, and marital breakup all fall under the purview of civil law. In these cases, private citizens, companies, or government agencies are the offended parties, and the party found at fault is required to directly compensate the other party.

This chapter is intended as a guide to some of the intricacies of criminal and civil law, with primary emphasis on civil actions. It is aimed at those subjects with which a security officer would most likely be confronted. It will deal with substantive law (statutes and codes) or that portion of the law that concerns the rights, duties, and penalties of individuals in their relationships with each other. Procedural law is the other of the two divisions of the law. It deals with the rules of court procedure and the mechanisms of the legal machine.

Security, Public Police, and the U.S. Constitution

The framers of the U.S. Constitution, with their grievances against England uppermost in mind when they were creating a new government, were primarily concerned with the manner in which the powerless citizen was or could be abused by the enormous power of government. The document they created was concerned, therefore, not with the rights of citizens against each other but rather with those rights with respect to federal or state action.

Breaking and entering by one citizen on another may be criminal and subject to tort action (a civil wrong not involving a breach of contract), but it is not a violation of any constitutional right. Similar action by public police is a clear violation of Fourth Amendment rights and, as such, is expressly forbidden by the Constitution.

The public police have substantially greater powers than do security personnel to arrest, detain, search, and interrogate. Whereas security people are, as a rule, limited to the premises of their employer, public police operate in a much wider jurisdiction. At the same time, the public police are limited by various restrictions imposed by the Constitution. With some exceptions, private police are not as a rule touched by these same restrictions.

Public police are limited by the Constitution, which prohibits officials from denying others their constitutional rights. The Fourth and Fourteenth Amendments are most frequently invoked as the cornerstones of citizen protection against arbitrary police action. The exclusion of evidence from criminal proceedings is one penalty paid by public police for violation of the search provision of the Fourth Amendment. For the most part private police are not affected by these restrictions.

Sources of Law

All law, whether civil or criminal, has its source in constitutional law, common law (also referred to as case law), or statutory law. The following discussion can be applied to either civil or criminal law as it has developed over the past century. Although today's criminal law is primarily statutory, civil law, particularly tort law, is essentially judge-made and created in response to changing social conditions.

Common Law

At one time, the principal source of law in the United States was English common law. Although common law may also refer to judge-made (as opposed to legislature-made) law, to law that originated in England and grew from ever-changing custom, or to written Christian law, the term is most commonly used to refer to the English common law that has been changed to reflect specific U.S. customs.

Some states have preserved the status of common-law offenses for their criminal codes, while others have abolished common law and written most of the common-law principles into statutes. Some states are still using both common and statutory law.

Case Law

When a case goes to court, the outcome is usually governed by prior court opinions of similar nature. Those preceding cases have usually been resolved in such a way as to put to rest any doubts as to the meaning of the governing statutes or common-law principles as well as to clarify the attitude of the courts regarding the legal issue involved. The court opinions used have established precedent that will guide other courts in subsequent cases based on the same essential facts. Because the facts in any two cases are rarely precisely the same, opposing attorneys cite preceding cases whose facts more readily conform to their own theory or argument in the case at hand. They, too, build their case on precedents, or case law already established. It is up to the court to choose one of the two sides or to establish its own theory. This is a very significant source of our law essentially becoming common law.

Because society is in a constant state of change, it is essential that the law adapt to these changes. At the same time, there must be stability in the law if it is to guide behavior. People must know that the law as it appears today will be the same tomorrow, that they will not be punished tomorrow for behavior that was permitted today. They need to know that each decision represents a settled statement of the law and that they can conduct their affairs accordingly. So the published decisions of the appellate courts become guides to the meaning of the law and in effect become the law itself. Their judgments flesh out legislative enactments to give them clear outlines. Such interpretations based on precedents are never regarded lightly and in legal terms are *stare decisis*, or "let the decision stand."

This does not mean that each decided case locks the courts forever into automatic compliance. Conditions that created the climate of the earlier decision may have changed, rendering the precedent invalid. And there are cases decided in such a narrow way that they cannot be applied beyond that case. Further, there is nothing that prevents review of a decision at the time of a later case. If the reviewing court agrees that the earlier case was in error, it may not be bound by the earlier precedent.

So it can be seen that case law is an important source of the law; it provides a climate of legal stability without closing the law to responsiveness to changing needs.

Statutory Law

Federal and state legislatures are empowered to enact laws that describe crimes. The authority to do so emanates from the U.S. Constitution and from the individual state constitutions. These constitutions do not specifically establish a body of criminal law. In general, they are more concerned with setting forth the limitations of governmental power over the rights of individuals. But they do provide both for the authority of legislative action in establishing criminal law and for a court system to handle these as well as civil matters.

Much criminal law is, in fact, the creation of the legislatures. The legislatures are exclusively responsible for making statutes. The courts may find some laws unconstitutional or vague and thus set them aside, but they may not create statutes. Only the legislatures are empowered to do that.

The Power of Security Personnel

Security personnel are generally limited to the exercise of powers possessed by every citizen. There is no legal area where the position of a security officer as such confers any greater rights, powers, or privileges than those possessed by every other citizen. A few states go contrary to this norm and confer additional arrest powers for security personnel after the completion of a designated number of hours of training.[1] As a practical matter, if officers are uniformed they will very likely find that in most cases people will comply with their requests. Many people are aware neither of their own rights nor of the limitations of the powers of a security officer. Thus, security officers can obtain compliance to directives that may be, if not illegal, beyond their power to command. In cases where security officers have unwisely taken liberties with their authority, the officers and their employers may be subject to the penalties of civil action. The litigation involved in suing security officers and their employers for a tort is slow and expensive, which may make such recourse impossible for the poor and for those unfamiliar with their rights. But the judgments that have been awarded have had a generally sobering effect on security professionals and have probably served to reduce the number of such incidents. Criminal law also regulates security activities. Major crimes such as battery, manslaughter, kidnapping, and breaking and entering—any one of which might be encountered in the course of security activities—are substantially deterred by criminal sanctions.

Further limitations may be imposed on the authority of a security force by licensing laws, administrative regulations, and specific statutes directed at security activities. Operating contracts between employers and security firms also specify limits on the activities of the contracted personnel.

Classes of Crimes

A crime has been defined as a voluntary and intentional violation by a legally competent person of a legal duty that commands or prohibits an act for the protection of society.[2]

Since such a definition encompasses violations from the most trivial to the most disruptive and repugnant, efforts have long been made to classify crimes in some way. In common law, crimes are classified according to seriousness, from treason (the most serious) to misdemeanors (the least serious). Most states do not list treason separately and deal with felonies as the most serious crimes, and misdemeanors as the next in seriousness, with different approaches to the least serious crimes (those known as infractions in some jurisdictions, less than misdemeanors in others, and petty offenses in still others). It will become apparent why security specialists should understand the nature of a given crime and its classification because such considerations will be important in determining:

- Power to arrest
- The need to use force in making the arrest
- Whether and what to search
- Various other considerations that must be determined under possibly difficult circumstances and without delay

Serious crimes like murder, rape, arson, armed robbery, and aggravated assault are felonies. Misdemeanors include charges such as disorderly conduct and criminal damage to property.

Felonies and Misdemeanors

From the time of Henry II of England, there has been a general understanding that felonies comprised the more serious crimes. This is true in modern U.S. law as far as it goes, but clearly the definition of felony must be pinned down more precisely if it is to be used as a classification of crime and if courts are to respond differently to felons than they would to another type of lawbreaker. The definition of a felony is by no means standard throughout the United States. In some jurisdictions, there is no distinction between felonies and misdemeanors.

The federal definition of a felony is an offense punishable by death or by imprisonment for a term exceeding one year. The test, then, for a felony is the length of time that punishment is imposed on the convicted person.

A number of states follow the federal definition. In those states, a felony is a crime punishable by more than a year's imprisonment. The act remains a felony whatever the ultimate sentence may actually be. Other states provide that "[a] felony is a crime punishable with death or by imprisonment in the state prison." This definition hinges on the place of confinement rather than, as in the federal description, the length of confinement.

Some states bestow broad discretionary powers on a judge by providing that certain acts may be considered either a felony or a misdemeanor depending on the sentence. The penalty clauses in the statutes thus involved specifically state that if the judge should sentence the defendant to a state prison, the act for which he was convicted shall be a felony (under the state definition of a felony) but if the sentence be less than such confinement the crime shall be a misdemeanor.

The distinction can be very important. In states where arrest by private citizens (for example, security personnel) is covered by statute, an arrest may be made only where the offense is committed in the presence of the arrester. In the case of arrest for a felony, the felony must in fact have been committed (though not necessarily in the presence of the arrester), and there must be reasonable grounds to believe the person arrested committed it. In other words, security employees, unlike police officers, act at their own peril.

A police officer has the right to arrest without a warrant where he reasonably believes that a felony has been committed and that the person arrested is guilty, even if, in fact, no felony has occurred. A private citizen, on the other hand, is privileged to make an arrest only when he has reasonable grounds for believing in the guilt of the person arrested and a felony has in fact been committed.[3]

Some states, however, do allow for citizen arrest in public-order misdemeanors. Making a citizen's arrest, which must be recognized as the only kind of arrest that can be made by a security officer, is a privilege, not a right, and as such is carefully limited by law. Such limitation is enforced by the ever-present potential for either criminal prosecution or tort action against the unwise or uninformed action of a security professional.

Private Security Powers

Arrest

Arresting a person is a legal step that should not be taken lightly. A citizen's power to arrest another is granted by common law and in many jurisdictions by statutory law. In most cases, it is best to make an arrest only after an arrest warrant has been issued. Most citizens' arrests occur, however, when the immediacy of a situation requires arrest without a warrant. The exact extent of citizens' arrest power varies, depending on the type of crime, the jurisdiction (laws), whether the crime was committed in the presence of the arrester, or the status of the citizen (strictly a private citizen or a commissioned officer).

In most states, warrantless arrests by private citizens are allowed when a felony has been committed and reasonable grounds exist for believing that the person arrested committed it. Reasonable grounds means that the arrester acted as would any average citizen who, having observed the same facts, would draw the same conclusion. In some jurisdictions, a private citizen may arrest without reasonable grounds as long as a felony was committed.

Most states allow citizens' arrests for misdemeanors committed in the arrester's presence. A minority of states, however, adhere closely to the common-law practice of allowing misdemeanor arrests only for offenses that constitute a breach of the peace and that occur in the arrester's presence. A good source on state-specific legal issues related to security personnel is the National Association of Security and Investigative Regulators (www.iasir.org).

Although the power of citizen's arrest is very significant in the private sector because it allows security officers to protect their employer's property, there is little room for errors of judgment. The public police officer is protected from civil liability for false arrest if the officer has probable cause to believe a crime was committed, but the private officer (citizen) is liable if a crime was not committed, regardless of the reasonableness of the belief.

This distinction is illustrated by the case Cervantez v. J.C. Penney Company.[4] In this case, an off-duty police officer, moonlighting as a store detective for J.C. Penney Company, made a warrantless arrest of two individuals for misdemeanor theft. Later they were released because of lack of evidence. The plaintiffs sued the company and the officer for false arrest, imprisonment, malicious prosecution, assault and battery, intentional infliction of emotional distress, and negligence in the selection of its employee. The primary issue in the Cervantez case was whether the officer could rely on the probable-cause defense. The court's decision rested on whether the officer acted as a police officer in California or as a private citizen.

The store and officer argued that the probable-cause defense was sound because the detective was an off-duty police officer and thus could arrest on the basis of probable cause. The plaintiffs argued that the store detective should be governed by the rules of arrest applied to private citizens and that the officer was therefore liable for his actions because no crime had been proven. The plaintiffs contended that the officer was employed as a private security officer, and thus his arrest powers were only those of a private citizen. The California Supreme Court ruled that the laws governing the type of arrest to be applied depend on the arrester's employer at the time of the arrest. Since the officer was acting as a store detective when he made the arrest, his arrest powers were no greater than those of a private citizen. Thus probable cause could not be used as a defense against false arrest.

Some states have avoided the problem of the Cervantez case by extending the probable-cause defense to private citizens. The most common extension involves shoplifting arrests. Many states have a mercantile privilege rule that allows the probable-cause defense for detentions but not for arrests. The law permits a private citizen or his employees to detain in a reasonable manner and for a reasonable time a person who is believed to have stolen merchandise so that the merchant can recover the merchandise or summon a police officer to make an arrest. Some states have extended this merchant clause to cover public employees in libraries, museums, or archival institutions.

The exact extent of the protection afforded to merchants and their employees or agents depends on the individual state's statutes. Some states offer protection against liability for false arrest, false imprisonment, and defamation; others offer protection against false imprisonment but not against false arrest. It is interesting to note that very few states allow a merchant to search a detainee. The private citizen's authority to search is unclear and will be discussed later in this chapter.

Detention

Detention is a concept that has grown largely in response to the difficulties faced by merchants in protecting their property from shoplifters and the problems and dangers they face when they make an arrest. Generally, detention differs from arrest in that it permits a merchant to detain a suspected shoplifter briefly without turning the suspect over to the police. An arrest requires that the arrestee be turned over to the authorities as soon as practicable and in any event without unreasonable delay.

All the shoplifting statutes refer to "detain," not to "arrest," a terminology probably derived from the thought that a distinction could be made between the two. The distinction is based on the fact that an arrest is for the purpose of delivering the suspect to the authorities and of exercising strict physical control over that person until the authorities arrive. A detention, or temporary delay, would not be termed an arrest as commonly defined. The distinction is difficult to defend but the statutes are clear. In Illinois, for example:

> *Any merchant who has reasonable grounds to believe that a person has committed retail theft may detain such person, on or off the premises of a retail mercantile establishment, in a reasonable manner and for a reasonable length of time for all or any of the following purposes:*
> *a. To request identification;*
> *b. To verify such identification;*
> *c. To make reasonable inquiry as to whether such person has in his possession unpurchased merchandise and, to make reasonable investigation of the ownership of such merchandise;*
> *d. To inform a peace officer of the detention of the person and surrender that person to the custody of a peace officer.[5]*

California was one of the first states to establish merchant immunity in a 1936 supreme court decision; in Collyer v. S. H. Kress Co.,[6] the court upheld the right of a department store official to detain a suspected shoplifter for 20 minutes.

Most statutes include the merchant, employee, agent, private police, and peace officer as authorized to detain suspects, but they do not include citizens at large, such as another shopper. Most of the statutes also describe the purposes of detention and the manner in which they may be conducted. These purposes are to search, to interrogate, to investigate suspicious behavior, to recover goods, and to await a police officer. The manner in which the detention is to be conducted is generally described as "reasonable" and for "a reasonable period of time."

The privilege of detention is, however, subject to some problems. There must be probable cause to believe theft already has taken place, or is about to take place, before a merchant may detain anyone. Probable cause is an elusive concept and one that has undergone many different interpretations by the courts. It is frequently difficult to predict how the court will rule on a given set of circumstances that may at the time clearly indicate probable cause to detain. Second, reasonableness must exist both in time and the manner of the detention or the privilege will be lost.

Interrogation

No law prohibits a private person from engaging in conversation with a willing participant. For public law enforcement, should the conversation become an interrogation, the information may not be admissible in a court of law. The standard is whether the statements were made voluntarily.

A statement made under duress is not regarded as trustworthy and is therefore inadmissible in court. This principle applies equally to police officers and private citizens. A confession obtained from an employee by threatening loss of job or physical harm would be inadmissible and could also make the interrogator liable for civil and criminal prosecution.

The classic cases involving interrogation, generally applied to only public law enforcement officers, are Escobedo v. State of Illinois[7] and Miranda v. Arizona.[8] Today the Miranda case has become the leading case recognized by most American citizens in reference to "their rights." On March 13, 1963, Ernesto Miranda was arrested at his home and taken to a Phoenix police station. There he was questioned by two police officers who during Miranda's trial admitted that they had not advised him that he could have a lawyer present. After 2 hours of interrogation, the officers emerged with a confession. According to the statement, Miranda had made the confession "with full knowledge of my legal right, understanding any statement I make may be used against me." His confession was admitted into evidence over defense objections during his trial. He was convicted of kidnapping and rape. On appeal, the Arizona Supreme Court upheld the conviction, indicating that Miranda did not specifically request counsel. The U.S. Supreme Court reversed the decision based on the fact that Miranda had not been informed of his right to an attorney, nor was his right not to be compelled to incriminate himself effectively protected.

Although the principle behind the Miranda v. Arizona decision was the removal of compulsion from custodial questioning (questioning initiated by law enforcement officers after a person has been taken into custody or otherwise deprived of freedom), it generally only applies to public law enforcement officers. The police officer must show that statements made by the accused were given after the accused was informed of the facts that speaking was not necessary, that the statements might be used in court, that an attorney could be present, and that if the accused could not afford an attorney, one would be appointed for the accused prior to questioning. These Miranda warnings are not necessary unless the person is in custody or is deprived of freedom and is subject to interrogation. Based on this distinction, most courts agree that private persons are not generally required to use Miranda warnings because they are not public law enforcement officers.

In the case In re Deborah C.,[9] the California Supreme Court upheld the principle that private citizens are not required to use Miranda warnings and that statements made by the accused in citizen's arrests are admissible in a court of law. The court felt that the Miranda rationale did not apply to the retail store environment because store detectives lack the psychological edge that police officers have when the latter are questioning someone at a police station.

A few states require citizens to use a modified form of Miranda warnings before questioning, and some—a definite minority—prohibit questioning. Wisconsin law states, "[t]he detained person must be promptly informed of the purpose of the detention and be permitted to make phone calls, but shall not be interrogated or searched against his will before the arrival of a peace officer who may conduct a lawful interrogation of the accused person."[10]

In 1987 the case of State of West Virginia v. William H. Muegge[11] expanded the Miranda concept to private citizens in the state of West Virginia. Muegge was detained by a store security guard who observed Muegge place several items of merchandise in his pockets and proceed through the checkout aisle without paying for those items. The security guard approached Muegge, identified herself, and asked him to return to the store office to discuss the problem. The officer ordered Muegge to empty his pockets, which contained several unpaid-for items valued at a total of $10.65. The officer next read Muegge his "constitutional rights" and asked him to sign a waiver of rights. Muegge refused and asked for the assistance of a lawyer. The officer refused the request and indicated that she would call the state police. At some time either prior to the arrival or after the arrival of the state trooper, the defendant signed the waiver and completed a questionnaire that contained various incriminating statements. At the trial, the unpaid-for items were admitted without objection and the questionnaire was read aloud over the defendant's objection. Although the court felt that the specific Miranda warnings were not necessary, it ruled that whenever a person is in custodial control mandated by state statute (that is, merchant clauses) the safeguards protecting the constitutional right not to be compelled to be a witness against oneself in a criminal case apply.

In the case of interrogation involving employees, a special rule applies if the employees are represented by a labor union. Under the *Weingarten* rule adopted by the National Labor Relations Board (NLRB) and subsequently upheld by the US Supreme Court in 1975,[12] if the employee reasonably believes that the interrogation may result in disciplinary action, he is entitled to have a union representative present during the interrogation if he so requests. At one time the NLRB extended this rule to employees not represented by labor unions,[13] but it later determined that this was not appropriate.[14]

Some employers use polygraphs in connection with interrogations of employees. Use of polygraphs for this purpose has been greatly restricted by the federal Employee Polygraph Protection Act of 1988 (EPPA), which generally only permits an employer to request that an employee take a polygraph in connection with an ongoing investigation of economic loss or injury to the employer's business, where the employer has a reasonable suspicion that the employee was involved in the loss. In addition, the employer is required to provide the employee with a detailed signed written notice (as further described in EPPA) before making such a request.[15] Some state laws further restrict the use of polygraphs.[16]

Search and Seizure

A search may be defined as an examination of persons and/or their property for the purpose of discovering evidence of guilt in relation to some specific offense. The observation of items in plain view is not a search as long as the observer is legally entitled to be in the place where the observation is made. This includes public property and private property that is normally open to the public, for example, shopping malls, retail stores, hotel lobbies, and so on.

Common law says little about searches by private persons and is inconclusive. Searches by private persons, however, have been upheld by the courts where consent to search was given and where searches were made as part of a legal citizen's arrest. The best practice to follow is to contact police officials, who can then ask for a search warrant or search as part of an arrest. Because searches often need to be conducted on short notice without the aid of a police officer, however, it is important to understand several factors.

First, in a consent search, the searcher must be able to show that the consent was given voluntarily. Second, the search cannot extend beyond the area for which consent to search was given. It is advisable to secure a written agreement of the consent to search. Third, the person who possesses the item must give the consent. Possession, not ownership, is the criterion for determining whether a search was valid. Although many firms issue waivers to search lockers and other work areas, an officer must remember that the consent to search may be withdrawn at any time. If the consent is withdrawn, continuing a search might make the officer and the company liable for invasion of privacy. Some companies have solved this problem by retaining control over lockers in work areas. In this situation, workers are told that the lockers are not private and may be searched at any time. As will be discussed later in this chapter, a key element of proving invasion of privacy is to prove that the individual had a reasonable expectation of privacy. Thus, establishing and communicating policies stating that a company may search lockers, desks, etc. is often a key defense to such claims.

A search made as a part of an arrest is supported by case law. In general, the principle of searching the arrestee and the immediate surroundings, defined as the area within which one could lunge and reach a weapon or destroy evidence, has been repeatedly held as constitutional. The verdict on searches incident to arrest by security officers is still mixed. In People v. Zelinski,[17] the California court disapproved of searches made incident to an arrest but did approve of searches for weapons for protective reasons. New York courts tend to support searches, indicating that private officers, like their public counterparts, have a right to searches incident to an arrest. In general, it appears that unless the security officer fears that a weapon may be hidden on the arrestee, the officer should wait until the police arrive to conduct a search unless permission is given for such a search.

Even in the statutes governing retail shoplifting, the area of search is limited. Some states neither forbid nor condone searches; rather, they allow security personnel to investigate or make reasonable inquiries as to whether a person possesses unpurchased merchandise. In other states, searches are strictly forbidden, except looking for objects carried by the suspected shoplifter. Courts, however, generally favor protective searches where officers fear for their own safety.

Exclusionary Rule

In a historic decision, the U.S. Supreme Court ruled that any and all evidence uncovered by public law enforcement agents in violation of the Fourth Amendment will be excluded from consideration in any court proceedings. That means all evidence, no matter how trustworthy or indicative of guilt, will be inadmissible if it is illegally obtained. This landmark case (Mapp v.

Ohio)[18] was the most important case that contributed to the development of the "exclusionary rule," which states that illegally seized evidence (and its fruits) are inadmissible in any state or federal proceedings. Weeks v. United States[19] set the stage for the later, all-inclusive decision in Mapp by holding that evidence acquired by officials of the federal government in violation of the Fourth Amendment must be excluded in a federal prosecution. The Mapp case is clear in its application of the exclusionary rule to state and federal prosecutions. The question is, does the exclusionary rule apply to private parties? The determining case in this area is Burdeau v. McDowell.[20]

Unlike illegal searches conducted by public law enforcement officers, evidence secured by a private security officer conducting an illegal search is still admissible in either criminal or civil proceedings. In Burdeau v. McDowell, the U.S. Supreme Court said, "[i]t is manifest that there was no invasion of the security afforded by the Fourth Amendment against unreasonable searches and seizures, as whatever wrong was done was the act of individuals in taking property of another." If such evidence is admissible, why should private-sector employees concern themselves with the legality of searches? Even though the evidence is admissible, security officers who conduct illegal searches may be subject to liability for other actions, including battery and invasion of privacy.

There is considerable controversy over the Burdeau case because some people fear that constitutional guarantees are threatened by the acceptance of evidence illegally obtained by private security personnel. It is clear that any involvement by government officials constitutes "state action" or an action "under color of law" and is limited by the constitutional restrictions that apply to public police actions. In State v. Scrotsky,[21] the New Jersey court excluded evidence obtained when a police detective accompanied a theft victim to the defendant's apartment to identify and recover stolen goods. The court held that "[t]he search and seizure by one served the purpose of both and must be deemed to have been participated in by both." The exclusionary rule is applied in this case, as in many others, to discourage government officials from conducting improper searches and from using private individuals to conduct them.[22]

In cases where private parties act independently of government involvement, the courts have not been so clear. In a significant case, People v. Randazzo,[23] the California court admitted evidence obtained by a merchant in a shoplifting case. The court did not deal with any questions of Fourth Amendment violation since there was no state action involved. The court held that redress for the victim of an unreasonable search conducted by a private individual not under color of law is a tort action, and thus the exclusionary rule does not apply. In Thacker v. Commonwealth,[24] the Kentucky court held that a private party acts for the state when that party makes an arrest in accordance with the state's arrest statute and thus would be subject to the exclusionary rule. On the other hand, following the Burdeau precedent, a federal district court found no state action in a case where the plaintiff alleged she was wrongfully detained, slapped, beaten, harassed, and searched by the manager and an employee of the store.[25] The plaintiff sued, alleging among other things that the employee, a security officer, was acting "under color of law" because he was licensed under the Pennsylvania Private Detective Act. The court rejected this argument and found that the Pennsylvania law "invests the licensee with no authority of state law."

In summary, although public police are clearly limited by constitutional restrictions, generally private security personnel are not so limited. Provided that they act as private parties and are in no way involved with public officials, they are limited by criminal and civil sanctions but are not bound at this time by most constitutional restrictions.

Use of Force

On occasion, security personnel must use force to protect someone or to accomplish a legitimate purpose. In general, force may be used to protect oneself or others, to defend property, and to prevent the commission of a criminal act. The extent to which force may be used is restricted; no more force may be used than is reasonable under the circumstances. This means that deadly force or force likely to create great bodily harm will not be allowable unless the force being used by the assailant is also deadly force or force likely to create great bodily harm. If the force exceeds what is deemed reasonable, officers and their employers are liable for the use of excessive force, which can range from assault and battery to homicide. This is the same degree of power extended to the ordinary citizen.

Self-Defense

In general, people may use reasonable force to protect themselves. The amount of force may be equal to, but not greater than, the force being used against them. In most states, a person can protect himself or herself, except when that person was the initial aggressor. Most states allow self-defense to be used by a person against whom force is being used.

Defense of Others

Security officers may protect others just as they protect themselves. However, two different approaches to defense of others are evident. In the first approach, the officer must try to identify with the attacked person. In this position, the officer is entitled to use whatever force would be appropriate if he or she were the person being attacked. If the officer happens to protect the wrong person—that is, the aggressor—the officer is liable regardless of his or her good intentions. In the second approach, the defender may use force when it is reasonable to believe that such force is necessary. In this case, the defender is protected from liability as long as he or she acted in a reasonable manner.

Defense of Property

In defense of property, force may be applied, but it must be short of deadly force, which is generally allowable only in cases involving felonious attacks on property during which loss of life is likely. As noted by Schnabolk, "one may use deadly force to protect a home against an arsonist but the use of deadly force against a mere trespasser would not be permitted."[26] Security officers acting in the place of their employers are empowered to use the same force that their employers are entitled to use.

Force Used during Arrest or Detention

Like the police, the private citizen security officer has the right to use reasonable force in detaining or arresting someone. Many states still follow the common-law principles that allow deadly force in the case of fleeing felons, but many others have restricted the use of deadly force. This restriction allows the use of deadly force only in cases where the felony is both violent and the felon is immediately fleeing. Figure 6-1 on the Use of Force Model has been widely used for training of police officers and has application for security personnel faced with various levels of suspect compliance.

Prevention of Crimes

To determine the amount of force a security officer may use to prevent crimes, the courts have considered the circumstances, the seriousness of the crime prevented, and the possibility of preventing the crime by other means. Under common law, a person can use force to prevent a crime. The courts have ruled, however, that the use of force is limited to situations involving felonies or a breach of the peace and that nonviolent misdemeanors do not warrant the use of force. Deadly force is justifiable in preventing a crime only if it is necessary to protect a person from harm. The ruling case is Tennessee v. Garner.

Use of Firearms

Most states regulate the carrying of firearms by private citizens. Almost all states prohibit the carrying of concealed weapons, whereas only half of them prohibit carrying an exposed handgun. Although all states excuse police officers from these restrictions, some states also exempt private security officers. Even in states that prohibit carrying concealed or exposed handguns, there are provisions for procuring a license to carry weapons in this manner.

In recent years a number of states have passed laws permitting employees to bring firearms onto a company's premises if such firearms are locked in a vehicle, and, for most laws, hidden from plain view, either being placed in a locked container or in the trunk of the car.[27] The idea behind most of these laws is that employees should be able to protect themselves with firearms if they are driving through dangerous areas on the way to work.

Civil Law: The Controller for Private Security

Tort Law: Source of Power and Limits

A tort is a civil action based on the principle that one individual can expect certain behavior from another individual. When the actions of one of the parties do not meet reasonable expectations, a tort action may result. In security applications, a guard may take some action to interfere with the free movement of some person. There is a basis for a suit no matter whether the guard knows those actions are wrong or is unaware that the actions are wrong, or is unaware that the actions are wrong but acts in a negligent manner.

Thus tort law may be invoked for either an intentional or negligent act. In some cases, liability may be imposed even though an individual is not directly at fault. One branch of this doctrine is "strict liability," and does not generally affect the security officer. Strict liability applies to a provider of defective or hazardous products or services that unduly threaten a consumer's personal safety. Vicarious liability, however, is of concern to enterprises that contract

USE OF FORCE SCALE

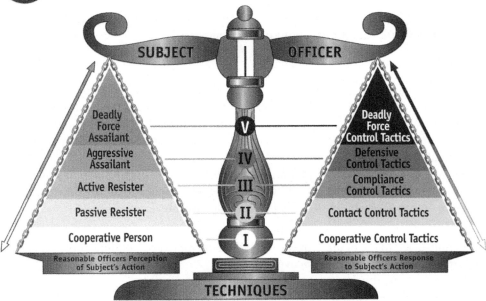

LEVEL V		
SUBJECT	**OFFICER**	**TECHNIQUES**
DEADLY FORCE ASSAILANT Is a person whose actions will probably cause death or great bodily harm.	DEADLY FORCE CONTROL TACTICS 1. Firearms 2. Other measures which could result in death or great bodily harm	1. Verbal Command (if time permits) 2. Knife Defense	3. Weapon Retention/Take Back (Service Weapon) 4. Firearms

LEVEL IV		
AGGRESSIVE ASSAILANT Is a person who performs physical actions, without weapons, that are aggressive and he/she demonstrates behavior that is likely to cause physical injury.	DEFENSIVE CONTROL TACTICS 1. Punches, kicks, & other strike/ blocking techniques/ to stop aggression 2. Intermediate tools 3. Other Non-Lethal tools (including firearms using ammunition defined in 720 ILCS 5/7-8(b) 1997	1. Verbal Command 2. Baton Retention 3. Weapon Retention/O.C. Spray 4. Blocks/Strikes (Baton and Empty Hands) 5. Escape Techniques	6. Baton Chops and Jabs 7. Intermediate Weapons (per department policy, other weapons authorized to carry stun-gun, taser, mace, etc.)

LEVEL III		
ACTIVE RESISTER Is a person who exhibits resistive movement to avoid physical control.	COMPLIANCE CONTROL TACTICS 1. Stunning techniques (with or without control instruments) 2. Take-downs 3. Control instruments techniques/leverage 4. Chemical agents 5. Canine deployment	1. Verbal Command 2. Pressure Points/Distractors 3. Same Techniques as Level II, but with Torque 4. Driver Removal Techniques (inside and outside hand) 5. Handshake Control	6. Baton Positions (on guard, strike ready, cross arm, and vertical) 7. O.C. Spray 8. Knifehand Transition 9. Clamp Technique

LEVEL II		
PASSIVE RESISTER Is a person who exhibits no resistive movement in response to verbal or other direction by the officer.	CONTACT CONTROL TACTICS 1. Joint manipulations/used to guide or direct the subject 2. Pressure sensitive area techniques	1. Verbal Command 2. High Gooseneck 3. Handshake Control	4. Rear Wristlock 5. Armlock 6. Control Applications

LEVEL I		
COOPERATIVE PERSON Is a person who is cooperative or can be developed into a cooperative individual.	COOPERATIVE CONTROL TACTICS 1. Officer presence 2. Verbal direction 3. Restraint devices	1. Verbal Command 2. Basic Alert Stance 3. Basic Escort Position 4. Standing Cuff - One on One	5. Cuffed Search 6. Uncuffed Search 7. Standing Cuff - Two on One

*Developed by Joe L. Smith, Police Training Specialist, Police Training Institute in 1999   Use of Force Scale, University of Illinois ©1999   10032

FIGURE 6-1 Use of force model. *(Used with permission from the Police Training Institute, University of Illinois, Champaign.)*

or employ security services. Vicarious liability is an indirect legal responsibility; for example, the liability of an employer for the actions of an employee.

Negligence

The Restatement of Torts[28] states that "[it] is negligence to use an instrumentality, whether a human being or a thing, which the actor knows, or should know, to be incompetent, inappropriate or defective and that its use involves an unreasonable risk of harm to others." This statement has particular importance to security employers and supervisors in hiring, supervision, and training of employees.

In all cases of negligence, the plaintiff (the person who brings an action, the party who complains or sues) must prove the case by a preponderance of the evidence (more than 50 percent or "more likely than not") in all of the following areas:

1. An act or failure to act (an omission) by the defendant
2. A legal duty owed to the plaintiff by the defendant, the person defending or denying, and/or the party against whom relief or recovery is sought
3. A breach of duty by the defendant
4. A foreseeable injury to the plaintiff
5. Actual harm or injury to the plaintiff

A relatively new concept in the area of negligence is comparative fault. This concept accepts the fact that the plaintiff may have contributed to his or her own injury, such as being in a restricted area or creating a disturbance or some hazard. In the past, the theory of contributory negligence prevented the plaintiff from collecting for injuries if he or she contributed somehow to his or her own injury. In comparative negligence, the relative negligence of the parties involved is compared, and the plaintiff who may have contributed to the injury may get some award for part of the injury for which he or she is not responsible. There are three types of comparative negligence statutes: (1) pure approach, (2) the 50/50 rule, and (3) the 51 percent rule. In the pure approach, the plaintiff may collect something for injuries even if he or she was primarily responsible for the injuries. In theory, the jury could award the plaintiff 1 percent of the damages if he or she is found to have contributed 99 percent to the injury. Under the 50/50 rule, the plaintiff can collect for damages if he or she was responsible for no more than 50 percent of the negligence. In the 51 percent situation, the plaintiff's acts must not have contributed more than 49 percent of the situation in order to collect damages. Regardless of the rule followed, the degree to which the plaintiff is responsible for the end result is considered before judgments are pronounced. One example of the 50/50 rule is the following Illinois statute:

> In all actions on account of bodily injury or death or physical damage to property, based on negligence … the plaintiff shall be barred from recovering damages if the trier of fact finds that the contributory fault on the part of the plaintiff is more than 50% of the proximate cause of the injury or damage for which recovery is sought. The plaintiff shall not be barred from recovering damages if the trier of fact finds that the contributory fault on the part of the plaintiff is not more than 50% of the proximate cause of the injury or damage

for which recovery is sought, but any damages allowed shall be diminished in the proportion to the amount of fault attributable to the plaintiff.[29]

Cases involving negligence in providing adequate security on the part of firms have been increasing. Recent cases have resulted in awards to plaintiffs in individual cases of more than $1 million. More will be said about this issue later in the chapter.

One relatively recent development in the area of negligence is claims of negligent hiring or negligent retention of employees. As summarized by a recent California court decision, *Phillips v TLC Plumbing, Inc.*, 172 Cal. App. 4th 1133, 1139 (2009): " 'An employer may be liable to a third person for the employer's negligence in hiring or retaining an employee who is incompetent or unfit.' 'Liability for negligent hiring… is based upon the reasoning that if an employer hires individuals with characteristics which might pose a danger to customers or other employees, the enterprise should bear the loss caused by the wrongdoing of its incompetent or unfit employees.' Negligence liability will be imposed on an employer if it 'knew or should have known that hiring the employee created a particular risk or hazard and that particular harm materializes.' "

Because of the risk of negligent hiring claims and for other business reasons, many employers now require applicants to successfully complete a background investigation prior to being hired. Such investigations are subject to a number of legal restrictions, which are discussed later in this chapter.

Intentional Torts

An intentional tort occurs when the person who committed the act was able to foresee that the action would result in certain damages. The actor intended the consequences of the actions or at least intended to commit the action that resulted in damages to the plaintiff. In general, the law punishes such acts by compensatory judgments that exceed those awarded in common negligence cases.

The most common intentional torts are discussed in the following sections.

Assault
Assault is intentionally causing fear of imminent harmful or offensive touching but without touching or physical contact. In most cases, courts have ruled that words alone are not sufficient to place a person in fear of harm.

Battery
Battery refers to intentionally harmful or otherwise offensive touching of another person. The touching does not have to be direct physical contact but may instead be through an instrument such as a cane or rock. In addition, the courts have found battery to exist if "something" closely connected to the body but not actually a part of the body is struck.[30] In Fisher v. Carrousel Motor Hotel, Inc., the plaintiff was attending a conference that included a luncheon. The luncheon was a buffet, and while Fisher was in line, one of the defendant's employees snatched the plate from his hand and shouted that no Negro could be served in the club. The Texas court

of appeals held that a battery can occur even though the subject is not struck. It ruled that, so long as there is contact with clothing or an object identified with the body, a battery can occur. From a security point of view, the contact must be nonconsensual and not privileged. Privileged contact is generally granted to merchants who need to recover merchandise; privilege is generally a defense against charges of battery if the merchant's actions were reasonable. If the touching were unreasonable, however, the plaintiff would have a case for battery. The same argument holds for searches: if a search is performed after consent has been given, no battery has occurred. If consent is not given, however, the search is illegal, and a battery has probably occurred.

False Imprisonment or False Arrest
False imprisonment or false arrest is intentionally confining or restricting the movement or freedom of another. The confinement may be the result of physical restraint or intimidation. False imprisonment implies that the confinement is for personal advantage rather than to bring the plaintiff to court. This is one of the torts most frequently filed against security personnel.

Defamation
Defamation refers to injuring the reputation of another by publicly making untrue statements. Slander is oral defamation, while libel is defamation through the written word. The classic case of a security officer yelling "Stop, thief!" in a crowded store has all the necessary elements for slander if the accused is not a thief. Although it is generally true that truth is an absolute defense in defamation issues, the courts may also look at the motivation of the defendant. True statements published with malicious intent can be prosecuted in some jurisdictions. In this age of high technology, the courts have now included statements made on television or other broadcasts in the libel category. It is apparent that the courts view these types of statements as being more permanent and as reaching broad audiences.

Malicious Prosecution
Malicious prosecution is groundlessly instituting criminal proceedings against another person. Malice is an essential element. To prove malice, the plaintiff must show that the primary motive in bringing about criminal proceedings was not to bring the defendant to justice. Classic cases include proceedings brought about to extort money or to force performance on contracts. Although there is no liability for reporting facts to the police or other components of the criminal justice system, if the prosecution resulted from biased statements of fact, incomplete reports, or the defendant's persuasion (political, sexual, religious, and so on), liability for malicious prosecution might be proved.

Invasion of Privacy
Intruding on another person's physical solitude, disclosing private information about another person, or publicly placing someone in a false light is called invasion of privacy. Four distinct actions fall into this category: (1) misappropriation of the plaintiff's name or picture for commercial advantage, (2) placing the plaintiff in a false light, (3) public disclosure of private facts, and (4) intrusion into the seclusion of another. For security purposes, the last type of claim is

the most common during observation of an individual. Concern over liability for invasion of privacy is increasing; this liability may be the result of background investigations, searches of work areas, video surveillance, or the use of truth detection devices.

The common law tort of invasion of privacy has two elements. "First, the defendant must intentionally intrude into a place, conversation, or matter as to which the plaintiff had a reasonable expectation of privacy. Second, the intrusion must occur in a manner highly offensive to a reasonable person."[31]

In determining whether the plaintiff has a reasonable expectation of privacy, one factor the courts look at is the work area itself—at one extreme, it would be hard to argue that there is a reasonable expectation in a crowded company cafeteria, but few would deny that employees in restrooms have such an expectation. Another factor is employer policies—if the employer has clearly communicated that it reserves the right to search desk drawers or undertake video surveillance of an area, then it makes it harder for a plaintiff to claim a "reasonable" expectation.

The California Supreme Court case in *Hernandez v Hillsides, Inc*[32] is instructive. In that case the defendant operated a private nonprofit residential facility for neglected and abused children. Plaintiffs were two female employees who shared the same office. The director of the facility learned that late at night, after plaintiffs had left the premises, an unknown person had repeatedly used a computer in plaintiffs' office to view pornographic material. In an effort to catch the culprit, the defendant used a hidden video surveillance camera that could be operated from a remote location to view the work area around the computer. The camera was not operated during normal work hours. Nevertheless, when plaintiffs found out about the camera, they sued for invasion of privacy. Among other things, they alleged that they occasionally used the office to change or adjust their clothing. After weighing all the factors, including the defendant's policies, the Court concluded that although the plaintiffs had established a reasonable expectation of privacy in the work area, they had not met their burden of proving an invasion that would be highly offensive to a reasonable person.

A related area to video surveillance is tape recording of audio conversations as is sometimes done in workplace investigations. Some states have restrictions on such tape recording, such that it may not be permissible to tape record a conversation (either in person or by telephone) without the consent of the other party if the conversation can reasonably be deemed to be confidential.[33]

Trespass and Conversion

Trespass is the unauthorized physical invasion of property or remaining on property after permission has been rescinded. Conversion means taking personal property in such a way that the plaintiff's use or right of possession of chattel is restricted. In simpler terms, conversion is depriving someone of the use of personal property.

Intentional Infliction of Mental Distress

Intentional infliction of mental distress refers to intentionally causing extreme mental or emotional distress to another person. The distress may be either mental or physical and may result from highly aggravating words or conduct.

Security and Liability

In the past few years, the number of suits filed against security officers and companies has increased dramatically. Predictions for the next 10 years indicate no further increase but that the number of suits will continue at the present levels. One possible reason for the leveling off of suits is that security management has a better understanding of the problems associated with liability situations today. The earlier increase may be partly attributed to the growth of the security industry and to the public's demand for accountability and professionalism in the security area. Most of the cases filed against private security officers and operations belong in the tort category, as was mentioned earlier in this chapter. The individual who commits a tort is called a tortfeasor, while the injured party is called the plaintiff. The plaintiff may be a person, a corporation, or an association. Torts are classified as intentional, negligent, or strict liability. An intentional tort is a wrong perpetrated by someone who intends to cause harm to another. In contrast, a negligent tort is a wrong perpetrated by someone who fails to exercise sufficient care in doing what is otherwise permissible. Strict liability imposes responsibility on a defendant for inherently dangerous products and acts regardless of intent.

In most cases of negligence, the jury considers awarding damages to compensate the plaintiff. The awards generally take into account the physical, mental, and emotional suffering of the plaintiff, and future medical payments may be allowed for.

Punitive damages are also possible but are more likely to be awarded in cases of intentional liability. Punitive damages are designed to punish the tortfeasor and to deter future inappropriate behavior. Punitive damages are also possible in negligence cases in which the actions of the tortfeasor were in total disregard for the safety of others.

Duty to Protect from Third-Party Crime

The area of civil liability is of great importance to the security industry because the courts have been more willing to hold the industry legally responsible for protection in this area than in others. This trend is particularly noticeable in the hotel and motel industry, where owners are liable for failure to adequately protect guests from foreseeable criminal activity. In some circumstances, a hotel or motel owner might be held accountable for failure to provide adequate protection from criminal actions. In Klein v. 1500 Massachusetts Avenue Apartment Corporation,[34] a tenant who was criminally assaulted sued the corporation. The decision centered on the issue that the landlord had prior notice of criminal activity (including burglary and assault) against his tenants and property. In addition, the landlord was aware of conditions that made it likely that criminal activities would continue. The court ruled that the landlord had failed in an obligation to provide adequate security and was thus liable. A similar case was made against Howard Johnson's by the singer and actress Connie Francis.[35] Francis alleged that the hotel had failed to provide adequate locks on the doors. The jury awarded Francis more than $1 million.

Recent decisions (Philip Aaron Banks, et al. v. Hyatt Corporation and Refco Poydras Hotel Joint Venture and Allen B. Morrison, et al. v. MGM Grand Hotel, et al.) have followed earlier landmark cases.[36] In the Banks case, a federal court held the hotel liable for foreseeable events that led to the murder of Banks by a third party. Banks was shot only 4 feet from the hotel door. The suit alleged that the hotel failed to provide adequate security and to warn Banks of the danger of criminal activity near the hotel entrance. The jury awarded the plaintiffs $975,000, even though evidence was introduced that showed that the hotel had made reasonable efforts to provide additional protection in the area. The court stated that "the owner or operator of a business owes a duty to invitees to exercise reasonable care to protect them from injury," noting that "the duty of a business to protect invitees can extend to adjacent property, particularly entrances to the business premises, if the business is aware of a dangerous condition on the adjacent property and fails to warn its invitees or to take some other reasonable preventive action."

In the Morrison case, a robber followed Morrison from the hotel desk into the elevator after Morrison had cashed in his chips and withdrew his jewelry and cash from the hotel's safe. The robber took Morrison's property at gunpoint and then knocked him unconscious. Morrison brought suit against the hotel for failing to provide adequate security, noting that a similar robbery had recently occurred. The federal appellate court supported Morrison's contention, saying, "a landowner must exercise ordinary care and prudence to render the premises reasonably safe for the visit of a person invited on his premises for business purposes." In McCarthy v. Pheasant Run, Inc.,[37] however, another federal court recognized that invitees who fail to take basic security precautions may not have cause for action against the hotel. The difference in the rulings in McCarthy and Morrison points out the need for security managers to be aware of decisions within their own states and federal jurisdictions.

The foreseeability issue has been applied to other areas of business in recent years. In Sharpe v. Peter Pan Bus Lines,[38] a Massachusetts court awarded $550,000 for a wrongful death attributed to negligent security in a bus terminal. The same basic concept of foreseeability was applied in Nelson v. Church's Fried Chicken.[39] In fact, the concept of foreseeability has been expanded beyond the narrow opinion that foreseeability is implied in failure to provide security for specific criminal behavior. This concept implies that, since certain attacks have occurred in or near the company, the company should reasonably be expected to foresee potential security problems and provide adequate security. In a recent Iowa Supreme Court decision, the court abolished the need for prior violent acts to establish foreseeability. In Galloway v. Bankers Trust Company and Trustee Midlands Mall,[40] the court ruled foreseeability could be established by "all facts and circumstances," not just prior violent acts. Therefore, prior thefts may be sufficient to establish foreseeability because these offenses could lead to violence. In another case, Polly Suzanne Paterson v. Kent C. Deeb, Transamerica Insurance Co., W. Fenton Langston, and Hartford Accident & Indemnity Co.,[41] a Florida court held that the plaintiff may recover for a sexual assault without proof of prior similar incidents on the premises.

Nondelegable Duty

Another legal trend is to prevent corporations from divesting themselves of liability by assigning protection services to an independent contractor. Under the principle of agency law, such an assignment transferred the liability for the service from the corporation to the independent contractor. The courts, however, have held that some obligations cannot be entirely transferred. This principle is called nondelegable duty. Based on this principle, contractual provisions that shift liability to the subcontractors have not been recognized by the courts. These contractual provisions are commonly called hold harmless clauses.

Take for example, Dupree v. Piggly Wiggly Shop Rite Foods Inc. The court decided that:

Public policy requires [that] one may not employ or contract with a special agency or detective firm to ferret out the irregularities of his customers or employees and then escape liability for the malicious prosecution or false arrest on the ground that the agency and/or its employees are independent contractors.[42]

Imputed Negligence

Imputed negligence simply means that "by reason of some relation existing between A and B, the negligence of A is to be charged against B, although B has played no part in it, has done nothing whatever to aid or encourage it, or indeed has done all that he possibly can to prevent it. This is commonly called 'imputed contributory negligence.'"[43]

Vicarious Liability

One form of imputed negligence is vicarious liability. The concept of vicarious liability arises from agency law in which one party has the power to control the actions of another party involved in the contract or relationship. The principal is thus responsible for the actions of a servant or agent. In legal terms, this responsibility is called *respondeat superior*. In short, employers are liable for the actions of their employees while they are employed on the firm's business. Employers are liable for the actions of their agents even if the employers do nothing to cause the actions directly. The master is held liable for any intentional tort committed by the servant when the servant's purpose, however misguided, is wholly or partially to further the master's business.

Employers may even be liable for some of the actions of their employees when the employees are neither at work nor engaged in company business. For example, consider the position of an employer who issues a firearm to an employee. The employee, at home and therefore off duty, plays with the firearm, which discharges and injures a neighbor. The neighbor may sue the employer for negligently entrusting a dangerous instrument to an employee or for the negligence in selecting a careless employee.

The principle of *respondeat superior* ("let the master respond") is well established in common law. It is not in itself the subject of any substantial dispute, and at those times when it becomes an issue in a dispute, the area of contention is factual rather than the doctrine itself. As was noted earlier, in the doctrine of *respondeat superior*, "[a] servant is a person employed by a master to perform service in his affairs, whose physical conduct in the performance of the service is controlled or is subject to the right of control by the master. The Minnesota court in Graalum v. Radisson Ramp has stated that the right of control and not necessarily the exercise of that right is the test of the relationship of master and servant. Basically, the issue revolves around the distinction between a person who is subject to orders as to how he does his work and one who agrees only to do the work in his own way."[44] There is no question that an employer (master) is liable for injuries caused by employees (servants) who are acting within the scope of their employment. This is not to say that the employees are relieved of all liability. They are in fact the principal in any action, but since the employee rarely has the financial resources to satisfy a third-party suit, an injured person will look beyond the employee to the employer for compensation for damages.

Clearly the relationship between master and servant under *respondeat superior* needs definition. Under the terms of the Graalum v. Radisson Ramp ruling, in-house security officers are servants, whereas contract security personnel may not be. In the latter case, as discussed previously, contract personnel are employees of the supplying agency, and in most cases, the hiring company will not be held liable for their acts. The relationship is a complex one, however.

In some cases the "joint employer" doctrine may apply. Under this concept, an employee can have more than one employer. For example, one employer may pay his wages and set his hours of work, but the other may give day-to-day instruction on work assignments, such as typically occurs where a temporary staffing agency provides employees to another employer. If a joint employer relationship is found, each of the employers may be fully liable. This issue most frequently arises in the context of employment claims.[45]

If security officers are acting within the scope of their employment and commit a wrongful act, the employer is liable for the actions. The matter then turns on the scope of the officer's employment and the employer–employee relationship. One court described the scope of employment as depending on:

1. The act as being of the kind the employee is employed to perform.
2. The act occurring substantially within the authorized time and space limits of the employment.
3. The employee being motivated, at least in part, by a purpose to serve the master.[46]

This is further refined in Hayes v. Sears, Roebuck Co., in which the court found that if the employee is in pursuit of some purpose of his own, the defendant is not bound by his conduct, but if, while acting within the general scope of his employment, he simply disregards his master's orders or exceeds his powers, the master will be responsible for his conduct.[47]

Liability then is a function of the control exercised or permitted in the relationship between the security officer and the hiring company. If the hiring company maintains a totally hands-off posture with respect to personnel supplied by the agency, it may well avoid liability for wrongful acts performed by such personnel. On the other hand, there is some precedent for considering the hiring company as sharing some liability simply by virtue of its underlying rights of control over its own premises, no matter how it wishes to exercise that control. Many hiring companies are, however, motivated to contractually reject any control of security personnel on their premises in order to avoid liability. This, as was pointed out in *The Private Police*, works to discourage hiring companies from regulating the activity of security employees and the company that exercises controls (e.g., carefully examines the credentials of the guard, carefully determines the procedures the guard will follow, and pays close attention to all his activities) may still be substantially increasing its risk of liability to any third persons who are, in fact, injured by an act of the guard.[48]

It is further suggested in this excellent study that there may be an expansion of certain non-delegable duty rules into consideration of the responsibilities for the actions of security personnel. As was discussed previously, the concept of the nondelegable duty provides that there are certain duties and responsibilities that are imposed on an individual and for which that individual remains responsible even though an independent contractor is hired to implement them. Such duties currently encompass keeping the workplace safe and the premises reasonably safe for business visitors. It is also possible that the courts may find negligence in cases where the hiring companies, in an effort to avoid liability, have neglected to exercise any control over the selection and training of personnel, and they may further find that such negligence on the part of the hiring company has led to injury to third-party victims.

Vicarious liability requires a direct employer–employee relationship; it does not apply to cases in which an independent contractor is working for a firm. This is because the employer has no way of controlling the way an independent contractor performs the work. There are many exceptions to this rule, however. For example, the employer may be liable for the negligent selection of the contractor, or the employer may have exercised some day-to-day control over the employee.

Criminal Liability

Criminal liability is most frequently used against private security personnel in cases of assault, battery, manslaughter, and murder. Other common charges include burglary, trespass, criminal defamation, false arrest, unlawful use of weapons, disorderly conduct, extortion, eavesdropping, theft, perjury, and kidnapping. Security officers charged with criminal liability have several options in defending their actions. First they might try to show that they were entitled to use force in self-defense or that they made a reasonable mistake, which would negate criminal intent. Other defenses include entrapment, intoxication, insanity, consent (the parties involved concurred with the actions), and compulsion (the officer was forced or compelled to commit the act). As has been already noted in previous discussions, a corporation or an association could be charged with criminal liability as well as an individual officer.

The reporting of crime is an area in which security officers are liable for criminal prosecution. In general, private citizens are no longer obliged to report crime or to prevent it. But some jurisdictions still recognize the concept of misprision of felony—that is, concealing knowledge of a felony. Such legislation makes it a crime to not report a felony. To be guilty of misprision of felony, the prosecution must prove beyond a reasonable doubt that (1) the principal committed and completed the alleged felony, (2) the defendant had full knowledge of that fact, (3) the defendant failed to notify the authorities, and (4) the defendant took affirmative steps to conceal the crime of the principal.

Security officers may also be liable for failure to perform jobs they have been contracted or employed to perform. If guards fail to act in a situation in which they have the ability and obligation to act, the courts suggest that they could be criminally liable for failure to perform their duties, assuming a criminal act is committed. At the minimum the guard is subject to tort liability.

Another issue in security work involves undercover operations. Many times security operatives are accused of soliciting an illegal act. Where security officers clearly intended for crimes to be committed, they may be charged with solicitation of an illegal act or conspiracy in an illegal act. This is in contrast to the public sector, where most police officers are protected by statute from crimes they commit in the performance of their duty. Thus only the private citizen may be charged with such an offense, and the only issue that can be contested is the defendant's intent.

Entrapment, which is solicitation by police officers, is another charge that may be leveled against security officers. While entrapment does not generally apply to private citizens (the case of State v. Farns[49] is frequently cited to prove that entrapment does not apply to private citizens), several states have passed legislation that extends entrapment statutes to cover private persons as well as police officers. Until the issue is resolved in the courts in the next few years, security officers involved in undercover operations should be careful to avoid actions that might lead to entrapment charges.

Recent Trends in Liability

As is evident from the listing of cases used to discuss security issues, case law continues to evolve. Although the basics attested to in the preceding discussions remain relatively constant it is the duty of security managers to follow current case-law thinking.

Recent cases in the area of vicarious liability include Durand v. Moore (Texas Court of Appeals, 1997), Kirkman v. Astoria General Hospital (New York Court of Appeals, 1994) and Iglesia Cristiana LaCasa Del Senor Inc. v. L. M. (Florida Court of Appeals, 2001). Cases out of California in 2000–01 focused on the issue of whether the acts were committed in furtherance of the employer's business or on the employee's subjective motivations (University Mechanical v. Pinkerton's (2003 Cal. App. Unpub. LEXIS 2418) and Maria D. v. Westec Residential Security Inc. (84 Cal. App. 4th 125 [2000])).

A recent study of premises liability cases over the past 10 years was conducted by Liability Consultant, Inc. and reported in the October 2002 issue of *Security Management*. Figure 6-2 contains a table showing the average settlements for the cases examined between 1992–2001 by type of crime. The highest number of premises liability cases was filed in New York, followed by Texas and Georgia. Most cases were filed against condominiums, retail stores, and bars, with the largest number dealing with parking lot facilities. Still, it appears that business owners have learned from past cases because the chances of prevailing in a liability suit of this type are just barely in favor of the defendant. Cases of interest include Doe v. Kmart Corp. (Circuit Court of Charleston County, South Carolina, 1997), Smith v. Sparks Regional Medication Center (Arkansas Court of Appeals, 1998). In Shadday v. Omni Hotels Management Corporation (U.S. Court of Appeals Seventh Cirucuit, No. 06-2022, 2007) the court held that the Omni Hotel was not liable for the rape of a hotel guest by another guest. The court stated that the hotel could "hardly be required to have security guards watching every inch of the lobby every second of the day and night."

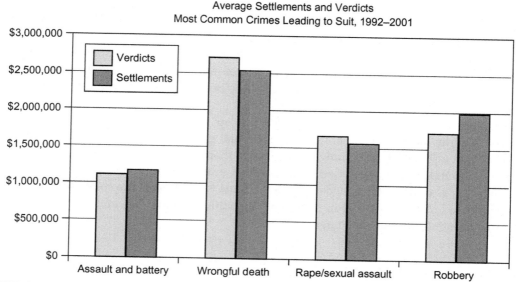

FIGURE 6-2 Settlement and verdict table. *(Source: Developed by Liability Consultants, Inc.)*

Recent Trends in Privacy

With the growing use of software programs that monitor use and content of employees using company computers, there are those who have cited invasion of privacy as a legal concern. Most courts are following early precedents supporting employers and their right to monitor use. Decisions clearly conclude that the employer owns the software and hardware and thus has the right to monitor its use. In particular, the courts have ruled that some company information such as medical records, legal files, and financial statements must be protected from unauthorized viewing and possible manipulation. To ensure the security of these records, monitoring of usage is essential. The courts have also determined that computer use logs are in fact legal records and may be admitted into court proceedings to substantiate company actions. The courts are clearly saying to employees that if you have something private to do, do it on your own time and with your own computer.[50]

New Laws

There are many new laws impacting on security and loss prevention passed each year (some have been mentioned in earlier chapters). Some of the more significant ones are noted here.

Background Investigations

Because of concerns over negligent hiring claims and for other reasons, many companies now require that applicants satisfactorily pass a background investigation prior to being hired.

Typically, such a background investigation would include checking on prior criminal convictions, verifying prior employment, and, in some cases, verifying educational degrees and job-related licenses/certificates. It might also include credit checks. There are a number of legal requirements in this area. In particular, under 1996 amendments to the federal Fair Credit Reporting Act (FCRA),[51] if an outside agency is used to obtain such background information the employer must comply with FCRA requirements, which primarily involve providing the applicant with information about his rights under FCRA and getting a written authorization to obtain background information about the applicant. If the employer then decides not to hire the applicant because of information in the background investigation report, it is required to inform the applicant at least five business days in advance of any such action and advise him of his right to receive a copy of the report. The applicant thus has an opportunity to contact the provider of the report as to any disputed information before a final no-hire decision is made. In addition to FCRA, several states have analogous laws with their own special requirements.[52]

One of the most common reasons for disqualifying an applicant based on a backgound investigation report is prior criminal convictions. There are a couple of legal issues to bear in mind in this area. First, a number of states restrict the types of convictions that can be used to disqualify an applicant from employment. A common restriction is to preclude the employer from considering certain crimes if the conviction is too old.[53] Some states also do not permit employers to ask about or consider arrests that did not result in a conviction.[54] Second, the federal Equal Employment Opportunity Commission (EEOC) has expressed a concern that a blanket policy precluding all persons with criminal convictions from being hired at a company has a disparate impact against African Americans and other protected groups, and a number of cases have recently been filed challenging such blanket exclusionary policies.[55] For these reasons, many companies consider each conviction on a case by case basis in relation to the job at issue.

Social Security Numbers

With increased concern over identity theft, a number of states have enacted legislation protecting employees' social security numbers. Typically such laws prohibit disclosing such numbers to the general public or, in some cases, showing such information on a paystub.[56] In addition, employers in some states can't require individuals to transmit social security numbers over the Internet unless the transmission is secure or the number is encrypted.[57]

Privacy of Medical Information

In 1996 the Health Insurance Portability and Accountability Act (HIPAA) was passed.[58] The act includes strong protections to ensure the security and privacy of medical records. It provides for both civil and criminal penalties for breach of its provisions.

Privacy of Financial Information

The U.S. Financial Services Modernization Act was passed in 1999.[59] This act establishes a duty for all financial services to protect the privacy of their customers.

These early efforts were the beginning of concern for ensuring personal privacy of records maintained by legitimate business, but often compromised by criminals. More is said about this issue under the topic of identity theft in Chapter 18.

The Courts

The process of adjudication varies between civil and criminal courts. The differences occur in that the civil courts have no accused; rather there are plaintiffs and defendants who believe that there is a cause for action rather than a violation of the law. The courts operate on many of the same rules, but the verdict in a civil case is by the preponderance of the evidence rather than by proof beyond a reasonable doubt.

The Procedure

In the event of an arrest, the accused must, by law, be taken without unnecessary delay before the nearest judge or magistrate. The court may proceed with the trial in the case of a misdemeanor charge unless the accused demands a jury trial or requests a continuance, and such is ordered by the court.

If the charge is a felony, the judge or magistrate conducts a preliminary hearing, an informal process designed to determine if reasonable grounds exist for believing that the accused committed the offense as charged. If such grounds do not appear, the accused will be discharged. If the judge finds that there are reasonable grounds for believing that the accused may have committed the offense as charged, he or she will "bind the accused over" for the action of the grand jury. The accused will be held in jail in the interim unless bond is paid, if the offense is bailable.

The grand jury is required by many states to consider the evidence in any felonious matter. Grand juries, usually consisting of 23 citizens, of whom 16 constitute a quorum, do not conduct a trial. They hear only the state's evidence. The accused may not be accompanied by an attorney into the hearing room and may not, in most cases, offer evidence in his or her own behalf. Misdemeanors are not handled by grand jury action but are usually prosecuted on an information, a document filed by the prosecuting attorney on receipt of a sworn complaint of the victim or a witness or other person who is personally informed about the circumstances of the alleged incident.

For a felony charge, the jury proceedings must result in a vote of at least 12 members for the accused to be indicted, and an indictment must be obtained in those jurisdictions that require it even if it is determined that there is reasonable grounds for prosecution at the preliminary hearing. This procedure was instituted as a constitutional guarantee in federal cases as a safeguard against arbitrary prosecutorial action. Those states that have the same requirement are also motivated to provide protection at the state and municipal level.

If an indictment (also known as a true bill) is voted, the next step is the appearance of the accused before a judge who is empowered to try felony cases. At this time, the accused is confronted with the charges and is asked to plead. If the plea is guilty, the accused may be sentenced without further court action; if the plea is not guilty, the trial is set for some future date.[60]

Development of Case Law

Much of the law, or more accurately its interpretation and hence its application, comes from the continuing judgments of the courts in cases all over the country. Most of the hundreds of cases heard daily are routine and, however significant they are to the participants, represent no particularly startling legal principle nor any significant upheaval in the day-to-day conduct either of the courts or of the average person. Patterns in jurisprudence emerge, however, and landmark cases that may have routine beginnings do appear.

It is essential for a lawyer to keep up with this flood of information because legal practice is being constantly reshaped by events in courtrooms around the country. Even the casual student with only a sporadic interest in the dynamics of the legal world should have some way of researching areas of immediate concern.

Legal Research

Many sources of information are available to the researcher: legal encyclopedias, dictionaries, legal periodicals, and code books setting forth the statutes. The encyclopedias focus on legal principles and theories along with cases in which such principles predominate. There are also digests that index cases. While the following presents information on these sources, the Internet and access to various legal research sites has made the area of legal research much easier for most casual learners.

Reporters

Perhaps the most useful sources are the bound volumes of reported cases called reporters that list the decisions of the appellate court. These decisions establish the precedents that are the cornerstone of the judicial system. Despite the large amount of available information, legal research is only as effective as the resources in the library or those obtainable through interlibrary loan. Legal research generally involves locating (1) the applicable statute, (2) the applicable case law, and (3) related articles in professional journals. Once the issue has been narrowed, the search for statutory law begins. Annotated criminal codes contain not only the statute but also brief notes and citations on court decisions that will be valuable in interpreting legal decisions. Federal statutes related to criminal law are found in The United States Code Annotated (USCA), Title 18, Crimes and Criminal Procedures.[61]

With the advent of the computer age much of the before-mentioned legal research is simplified through the use of various computer sources. Westlaw is a common computer-based subscription service available for a fee. Other information is available through the World Wide Web.

Case Reports

In legal writing, cases are frequently cited to show how a court applied a legal principle. Each such citation is followed by certain figures and abbreviations that are simply a convenient way to indicate the location of a description of the elements of the case in a reporter. Cases are arranged with volumes for cases in each state. In addition to state reporters, a private publisher

(West Publishing, St. Paul, Minnesota) has established what is termed the "national reporter system," in which blocks of states by geographical area are combined in various volumes. For example, appellate decisions from courts in Illinois, Massachusetts, Indiana, Ohio, and New York are contained in the Northeastern Reporter (N.E.) United States Reports. These are the official case reports of the United States Supreme Court and contain full transcripts of the majority opinions as well as other concurring or dissenting opinions in the Court's cases. The specifics of cases are summarized, and lists of the principals involved are given.

For a certain 1966 wiretapping case in Illinois, the citation is "People v. Kurth, 34 IL. 2d 387, 216 NE 2d 154 [1966]." Translated, this means that the decision of the appellate court in the case of People v. (versus or against) Kurth can be found in the Illinois Reporter, second series, volume 34, on page 387; this same decision can be found in volume 216 of the Northeastern Reporter, second series, on page 154.

The reported decision indicates the contending parties, a synopsis of the case up to the time it appeared before the reviewing court, the decision of the court, the relevant points of law considered and decided by the court (in the opinion of the legal experts employed by the publisher), the majority opinion, and the minority opinion if there is one included. Other information includes dates, names of justices and contending attorneys, and even citations of the case if it has passed through prior appeals before the current one. For this wealth of information, the reporters are invaluable aids to any research of points of law and of the cases in which they are found.

Digests and Summaries

Another useful set of sources in researching legal issues are the digests and summaries. One of the most helpful is *Shepard's Acts and Cases by Popular Names*,[62] a digest that gives the references and citations necessary to find the legislation or court decisions. Another source is *Corpus Juris Secundum*,[63] an encyclopedic compilation of criminal and civil law based on reported cases. The *Criminal Law Digest*[64] is a one-volume digest of leading court decisions. Each annual volume is cumulative from 1965 when the digest was first published. The *Digest* indexes the *Criminal Law Bulletin* (CLB),[65] which gives specific information on cases. Still another source is the *Criminal Law Reporter*,[66] which has an alphabetical index of cases by subject matter and a straight alphabetical listing.

Once a case has been found, it is easy to find other related cases using *Shepard's Citations*.[67] This publication allows researchers to gather all the case citations that relate to issues in the known case. The legal road is filled with bumps and potholes. There is no easy way to deal with it. Alert security managers will keep abreast of the climate in their jurisdictions and of the latest developments. There are rewards for the knowledgeable professional with an acquaintance with the law and its changes. For the unwary or uncaring, the road can be troublesome indeed.

Liability costs, while remaining relatively high, will continue the downward trend that began in 1986. As was noted earlier, the downward trend may be attributable to better-prepared security officers and managers, particularly to security management, which is learning the value of risk management and making good use of proper training.

Computer-Based Legal Sources
While the previous hard copy sources have provided many legal scholars with invaluable assistance, today's casual student will find that much of the information contained in these sources has been made accessible through various computer-based sites. The two best-known and most popular sites are LexisNexis (www.lexisnexis.com) and West Group (www.westpub.com). Another good site is Legal Resources (www.paralegals.org/LegalResources). In today's world of electronic communication, information on state laws may also be found through a computer search.

Summary

Although the area of law may appear complicated to many, it is one of the most important tools used by security personnel. A failure to understand offender rights, legal obligations of security personnel, and other legal matters can result in serious litigation involving employers as well as the employee.

The preceding materials present an overview of legal issues, but by no means should they be considered an adequate substitute for courses on criminal law, criminal procedure, contracts, and other related legal topics.

■ ■ CRITICAL THINKING ■

Why do the courts continue to make distinctions between public law enforcement powers and private sector enforcement issues?

Review Questions

1. Why is a practical knowledge of the law important to the security officer and the security manager?
2. What impact does tort law have on the private security industry?
3. What makes an arrest different from a detention?
4. What are the major legal differences between public police and private security officers?
5. Why is the legal term *respondeat superior* important to the contract security industry?
6. What is the difference between criminal and civil law?
7. The private security industry may enforce criminal law but it is restricted by civil law. How is this possible?

References

[1] Michigan Revised Statutes, Section 338.1051–338.1083.
[2] Pursley RD. In: Introduction to criminal justice. 5th ed. New York: Macmillan; 1991. p. 35.
[3] U.S. v. Hillsman, 522 F. 2d 454, 461 (7th Cir. 1975).

[4]   Cervantez v. J.C. Penney Company, 156 Cal. Rptr. 198 1978.

[5]   720 Illinois Compiled Statutes 5/16A-5.

[6]   Collyer v. S. H. Kress Co., 5 Cal. 2d 175, 54 p. 2d 20 1936.

[7]   Escobedo v. Illinois, 378 U.S. 478, 84 S.Ct 1758, 12L.Ed.977 1964.

[8]   Miranda v. Arizona, 384 U.S., 436, 86 S.Ct 1602, 16L.Ed.2d.691 1963.

[9]   In re Deborah C., 1977 Rptr. 852 1981.

[10] Wisconsin Statutes Annotated, Section 943.50.

[11] West Virginia v. William H. Muegge, 360 SE 2d. 216 (W.Va 1987).

[12] NLRB v. J. Weingarten, Inc., 420 U.S. 251 1975.

[13] Materials Research Corp., 262 NLRB 1010 1982.

[14] Sears, Roebuck & Co., 274 NLRB 230 (1985). See also E.I. Du Pont de Nemours & Co. (lll), 289 NLRB 627 1988.

[15] EPPA is set forth in 29 U.S.C. Sections 2001 et. seq. The exception for investigations is set forth in Section 2006(d).

[16] See, for example, California Labor Code Section 432.2; Connecticut General Statutes Section 31-51g.

[17] People v. Zelinski, 594 P 2d 1000 1979.

[18] Mapp v. Ohio, 367 U.S. 643 1961.

[19] Weeks v. United States, 232 U.S. 383 1914.

[20] Burdeau v. McDowell, 256 U.S. 465 1921.

[21] State v. Scrotsky, 39 NJ 410, 416 189 A.2d 23 1963.

[22] People v. Jones, 393 NE 2d. 443 1979.

[23] People v. Randazzo, 220 Cal. 2d 268, 34 Cal. Rptr. 65 1963.

[24] Thacker v. Commonwealth, 310 Ky. 701, 221 SW 2d 682 1949.

[25] Weyandt v. Mason Stores, Inc., 279 F. Supp. 283, 287 (W.D. Pa. 1968).

[26] Schnabolk C. In: Physical security: practices and technology. Boston: Butterworth-Heinemann; 1983. p. 74.

[27] See, for example, Arizona Revised Statutes Sections 12-781 and 13-3112; Indiana Code Section 34-28-7.2; Louisiana Revised Statutes Annotated Sections 32:292.1 and 40:1379.3.

[28] Restatement of Torts, Second Section 307.

[29] Illinois Revised Statutes, Ch. 110, sections 2–116.

[30] Fisher v. Carrousel Motor Hotel, Inc., 424 SW 2d 627 (TX 1976).

[31] Hernandez v Hillsides, Inc., 47 Cal. App. 4th 272, 286 2009.

[32] 47 Cal. App. 4th 272 2009.

[33] See, for example, California Penal Code Section 632; Florida Statutes Annotated Sections 934.03, 934.15.

[34] Klein v. 1500 Massachusetts Avenue Apartment Corporation, 439 F 2d 477 (D.C. Cir. 1970).

[35] Garzilli v. Howard Johnson's Motor Lodge, Inc., 419 F Supp. 1210, (D.CT. E.D.N.Y. 1976).

[36] Philip Aaron Banks, et al. v. Hyatt Corporation and Refco Poydras Hotel Joint Venture, 722 F 2d 214 1984; Allen B. Morrison, et al. v. MGM Grand Hotel, et al., 570 F. Supp. 1449 1983.

[37] McCarthy v. Pheasant Run, Inc., F 2d 1554 1987.

[38] Sharpe v. Peter Pan Bus Lines, No. 49694, Suffolk County, MA.

[39] Nelson v. Church's Fried Chicken, 31 ATLA L. Rep 84 1987.

[40] Galloway v. Bankers Trust Company and Trustee Midlands Mall, No. 63/86-1879 Iowa Supreme Court 1988.

[41] Polly Suzanne Paterson v. Kent C. Deeb, Transamerica Insurance Co., W. Fenton Langston, and Hartford Accident & Indemnity Co., 472 S. 2d 1210.

[42] Dupree v. Piggly Wiggly Shop Rite Foods Inc., 542 SW 2d 882 (Texas 1976).

[43] Prosser WL. Hornbook Series. In: Handbook of the law of torts. 4th ed. St. Paul, MN: West; 1970. p. 458.

[44] Graalum v. Radisson Ramp, 245 Minn. 54, 71 NW 2d 904, 908 1955.

[45] See, for example, Watson v Adecco Employment Services, Inc., 252 F. Supp 2d 1347, 1354-55 (M.D. Fla. 2003).

[46] Fornier v. Churchill Downs-Latonia, 292 Ky. 215, 166 SW 2d 38 1942.

[47] Hayes v. Sears, Roebuck Co., 209 P. 2d 468, 478 1949.

[48] Kakalik JS, Wildhurn S. In: The private police: security and danger. New York: Crane, Russak & Co.; 1977.

[49] State v. Farns, 542 P.2d 725 (Kan. 1975).

[50] Levine DE. Content monitoring and filtering. Secur Technol Des March 2003:74.

[51] 15 U.S.C. Section 1681 et. seq.

[52] See, for example, California Civil Code Section 1785.1 et. seq.; Colorado Revised Statutes Sections 12-14.3-101, et. seq.

[53] See, for example, California Labor Code Section 432.8 (employer cannot inquire about marijuana conviction more than 2 years old); Hawaii Revised Statutes Section 378-2.5 (employer can only consider convictions 10 years old or less).

[54] See, for example, California Labor Code Section 432.7; New York Exec. Law Section 296(15) and (16).

[55] Michelle Natividad Rodriguez and Maurice Emsellem, 65 Million Need Not Apply: The Case for Reforming Criminal Background Checks for Employment (National Employment Law Project, March 2011).

[56] California Civil Code Section 1798.85.

[57] Maine Statutes Annotated Section 325E.59.

[58] 42 U.S.C. Sections 300gg et. seq.

[59] 15 U.S.C. §§ 6801, 6809, 6821, and 6827.

[60] Neubauer D. In: America's courts and the criminal justice system. Pacific Grove, CA: Brooks/Cole; 1988.

[61] U.S. Code Annotated, Title 18 Crimes and Procedures (St. Paul: West), 50 vols., updated annually.

[62] Shepard's acts and cases by popular names. Colorado Springs, CO: Shepard's/McGraw-Hill; updated annually.

[63] Corpus juris secundum. New York: American Law Book Co.; 101 vols., updated annually.

[64] Douglas JA, Benton DS. Criminal law digest. Boston: Warren, Gorham & Lamont; updated annually.

[65] Criminal law bulletin. Boston: Warren, Gorham & Lamont.

[66] Criminal law reporter. Washington, DC: U.S. Bureau of National Affairs.

[67] Shepard's citations. Colorado Springs, CO: Shepard's/McGraw-Hill; updated annually.

7

Risk Analysis, Security Surveys and Insurance

OBJECTIVES

The study of this chapter will enable you to:

1. Understand the use of risk management tools to determine the probability of an event occurring and the potential cost to the company should that event occur.
2. Identify traditional alternatives for optimizing risk management strategies.
3. Identify the role of insurance as a risk management strategy.
4. Know the basic types of security-related insurance.

Introduction

Once security goals and responsibilities have been defined and an organization has been created to carry them out, as discussed in earlier chapters the ongoing task of security management is to identify potential areas of loss and to develop and install appropriate security countermeasures to mitigate those losses. This process of study is called *risk analysis*. Implicit in this approach is the concept of security as a comprehensive, integrated function of the organization. One part of this comprehensive, integrated job is the security survey, which is used to identify potential problem areas. More will be said about the survey later in this chapter.

A small business, particularly one with minimal loss potential or relative ease of defense, might adequately be served (as many are) by basic physical security controls such as a good lock on the door, an alarm system or by a contract guard patrol.

This comprehensive view of the loss-prevention function might be contrasted with more limited security responses such as:

- *One-dimensional security*, which relies on a single deterrent, such as guards or simple insurance coverage.
- *Piecemeal security*, in which ingredients are added to the loss-prevention function piece by piece as the need arises, without a comprehensive plan.
- *Reactive security*, which responds only to specific loss events.
- *Packaged security*, which installs standard security systems (equipment, personnel, or both) without relation to specific threats, either because "everybody's doing it" or on the

theory that packaged systems will take care of any problems that might arise. This is akin to prescribing a remedy without diagnosing the illness.

An integrated or systems approach to security is not always the desired solution. A small business, particularly one with minimal loss potential or relative ease of defense, might adequately be served (as many are) by basic physical security controls such as a good lock on the door and an alarm system or by a contract guard patrol. Include some basic insurance and you might have a reasonable security package. But as the areas of loss increase and become more complex and as the ability to protect a growing company against those losses with one-dimensional responses decreases, it becomes increasingly necessary to adopt a more comprehensive security program. If security is not to be one-dimensional, piecemeal, reactive, or prepackaged, it must be based on analysis of the total risk potential. In other words, in order to set up defenses against losses from crime, carelessness, accidents, or natural disasters, there must first be a means of identification and evaluation of the risks.

Risk Management

The first step in risk analysis involves recognizing the threats. In essence, the security manager's job is one of a professional worrier by trade. It costs a company great sums of money to erect buildings, protect assets, systems and information as well as personnel. But compared to putting money into research and development, which is investing in the company's future, putting money into preventing loss is only spending money to prevent something undesirable from happening. Although both investments involve risks, spending money on a product is dynamic and speculative and thus is a more interesting risk to take. Moreover, investments in new products provide the potential for increased revenue. Investment in security controls may prevent or mitigate losses but always at some cost. Risks to real and intellectual property are generally overlooked, underestimated or considered a "cost of doing business." Even when the risk is recognized, managers prefer to operate under the calculated risk theory. What is often overlooked in this process is the word "risk." However, over the past several decades, businesses have been forced to come to terms with the potential consequences of taking security risks. The two alternative solutions, which should be complementary, are (1) investment in loss-prevention techniques and (2) insurance.

Today the progressive manager recognizes that property risks are formidable and that they must be managed. Risk management may thus be defined as making the most efficient before-the-loss arrangement for an after-the-loss continuation of business. As a consequence, good insurance programs and security or loss-prevention programs are in demand. The concept of *risk management* presents a sensible approach to this complicated problem: It allows risks to be handled in a logical manner, using long-held management principles. Insurance in and of itself is no longer able to meet the security challenges faced by major corporations. To meet this challenge, insurance companies have found loss-prevention techniques and programs invaluable. A good risk-management program involves four basic steps:

1. Identification of risks through the analysis of threats and vulnerabilities.
2. Analysis and study of risks, which includes the probability and severity of an event.

3. Optimization of risk-management alternatives:
   a. Risk avoidance
   b. Risk reduction
   c. Risk spreading
   d. Risk transfer
   e. Self-assumption of risk
   f. Any combination of the above
4. On-going study of security programs.

The approach must be total; there can be no shortcuts.

Asset Assessment

However, before considering threats, it is important to have a clear understanding of what assets are being protected. Moreover, it is essential to have a thorough understanding of the nature and type of business (or organization) being protected. Protecting assets may be as simple as protecting a person (a "star" athlete or actor), or as complicated as protecting billions of dollars of materials, including people, physical assets (buildings, machines, raw materials) and information stored in complex computer systems. More will be said about assets when we discuss the issue of criticality later in this chapter.

Threat Assessment

The first step in risk analysis is identifying the threats and vulnerabilities. Many threats to business are important to security, but some are more obvious than others. The key is to consider the specific vulnerabilities in a given situation. Each individual firm has problems and threats that are unique. For example, a retailing company may be less concerned about fire hazards than is a manufacturing firm that operates a foundry. A retailer will be concerned with shoplifting, whereas a software developer is more concerned with code being stolen by a competitor or employee, essentially protecting its cyber environment. Employee theft, on the other hand, is a real problem. Today it appears that drug use and abuse and workplace violence, along with computer integrity issues, may be security problems found to some extent in all organizations. More will be presented on these and other topics in later chapters.

Specific threats are not always obvious. Although it seems to be common sense to check doors, locks, and gates to control access, accessibility through walls made of inferior materials or through a poorly constructed door or doorframe is a less obvious consideration. Awareness of all the possibilities is the mark of a good security manager. The best manager can think like a thief and thus is able to consider policies to reduce the vulnerability of company property. Therefore a manager must develop the ability to analyze vulnerabilities. A thorough analysis is comprehensive and accurate and leads to effective countermeasures. Once it has been completed, a vulnerability analysis—also called a *security survey* or *audit*—should be repeated on a regular basis.

> Threat × Vulnerability × Impact on Asset Value = Risk

The Security Survey

In the process of risk analysis that proceeds from threat assessment (identifying risk) to threat evaluation (determining the criticality and dollar cost of that risk) to the selection of security countermeasures designed to mitigate or prevent that risk, one of management's most valuable tools is the security survey (see Appendix C for sample surveys).

A security survey is essentially an exhaustive physical examination of the premises and a thorough inspection of all operational systems and procedures. Such an examination or survey has as its overall objective the analysis of a facility to determine the existing state of its security, to locate weaknesses (vulnerabilities) in its defenses, to determine the degree of protection (mitigation measures to reduce vulnerabilities) required, and ultimately to lead to recommendations for establishing a total security program (reduce risks) that also seeks to limit legal liability.

Motivation to set the survey in motion should come from executive management to ensure that adequate funds for the undertaking are available and to guarantee the cooperation of all personnel in the facility. Since a thorough survey will require an examination of procedures and routines in regular operation and an inspection of the physical plant and its environs, executive management's interest in the project is of the highest priority.

> Whoever undertakes the survey should have training in the field and should also have achieved a high level of ability.

The survey may be conducted by staff security personnel or by qualified security specialists employed for this purpose. Some experts suggest that outside security people could approach the job with more objectivity and would have less of a tendency to take certain areas or practices for granted, thus providing a more complete appraisal of existing conditions. Others suggest that outsiders do not have a clear picture of the internal workings of the organization. These opposite opinions may point to an operation that incorporates both external oversight utilizing internal security staff. Utilizing both internal and external resources for the survey does lend a great degree of credibility to the effort. When seeking funding to implement recommended improvements, support from an independent expert can be very useful.

Whoever undertakes the survey should have training in the field and should also have achieved a high level of ability. It is also important that at least some members of the survey team be totally familiar with the facility and its operation. Without such familiarity, it would be difficult to formulate the survey plan, and the survey itself must be planned in advance to make the best use of personnel and to study the operation in every phase.

Part of the plan may come from previous studies and recommendations. These should be studied for any useful information they might offer. Another part of the survey plan will include a checklist made up by the survey team in preparation for the actual inspection. This list will serve as a guide and reminder of areas that must be examined, and once it has been drawn up, it should be followed systematically. In the event that some area or procedure has been omitted in the preparation of the original checklist, it should be included in the inspection and its disposition noted in the evaluation and recommendation.

Since no two facilities are alike—not even those in the same business—no checklist exists that could universally apply for survey purposes. The following discussion is intended only to indicate those areas where a risk may exist. It should be considered as merely a guide to the kinds of questions or specific problems that might be handled.

The Facility

When analyzing security risks, the security manager should look at a number of aspects of the company, giving consideration as potential security problems to the following:

- *The perimeter.* Check fencing, gates, culverts, drains, access hatches, lighting (including standby lights and power), overhangs, and concealing areas. Can vehicles drive up to the fence?
- *The parking lot.* Are employees' automobiles adequately protected from theft or vandalism? How? Is the lot sufficiently isolated from the plant or office to prevent unsupervised back-and-forth traffic? Are bollards or other barriers used? Are there gates or turnstiles for the inspection of traffic, if that is necessary? Are these inspection points properly lighted? Can packages be thrown over or pushed through the fence into or out of the parking lot?
- *All adjacent building windows and rooftops.* Are spaces near these adjacencies accessible to them? Are they properly secured? How?
- *All doors and windows less than 18 feet above ground level.* How are these openings secured?
- *The roof.* What means are employed to prevent access to the roof?
- *The issuance of main entrance keys to all tenants in a building.* How often are entrance looks changed? What is the building procedure when keys are lost or not returned? How many tenants are in the building? What businesses are they in?
- *Any shared occupancy, as in office buildings.* Does the building have a properly supervised sign-in log for off-hours? Do elevators switch to manual, and can floors be locked against access outside of business hours? When are they so switched? By whom can they then be operated? Who collects the trash, and how and when is it removed from the building? Are lobbies and hallways adequately lighted? What guard protection does the building have? How can guards be reached? Are washrooms open to the public? Are equipment rooms locked? Is a master key system in use? How are keys controlled and secured? Is there a receptionist or guard in the lobby? Can the building be accessed by stair or elevator from basement parking facilities?
- *All areas containing valuables.* Do safes, vaults, or computer rooms containing valuables have adequate intrusion detection systems (alarms)? What alarms are in place to protect against burglary, fire, robbery, or surreptitious entry? Are computers protected from hackers and unauthorized employee use? How are hardware, software and media protected?
- *The off-hours when the facility is not in operation and all nighttime hours.* How many guards are on duty at various times of day? Are guards alert and efficient? How are guards

equipped? How many patrols are there, and how often do they make their rounds? What is their tour? What is the guard communication system? Are post orders (for fixed/mounted and mobile posts) documented and up-to-date?

- *The control and supervision of entry into the facility.* What method is used to identify employees? How are applicants screened before they are employed? How are visitors (including salespeople, vendors, and customers) controlled? How are privately owned vehicles controlled? Who delivers the morning mail and when? How are empty mail sacks handled? Do you authorize salespeople or solicitors for charity in the facility? How are they controlled? Are their credentials checked? Who does the cleaning? Do they have keys? Who is responsible for these keys? Are they bonded? Who does maintenance or service work? Are their toolboxes inspected when they leave? Are their credentials checked? By whom? Are alarm technicians and other technical and telephone company people allowed unlimited access? Is the call for their service verified? By whom? How is furniture or equipment moved in or out? What security is provided when this takes place at night or on weekends? How is the movement of company property controlled? Are messengers permitted to deliver directly to the addressee? How are they controlled? Which areas have the heaviest traffic? Are visitors claiming official status, such as building or fire inspectors, permitted free access? Are their credentials checked? By whom?

- *Keys and key control (traditional or electronic).* Are keys properly secured when they are not in use? Are locks replaced or recorded when a key is lost? Are locks and locking devices adequate for their purpose? Are all keys accounted for and logged? What system is used for the control of master and submaster keys? Is there adequate security to prevent unauthorized access and use of computer keying systems? Can the use of hardware (locks and keys) be replaced with electronic access control systems?

- *Fire.* Are there sufficient fireboxes throughout the facility? Are they properly located? Are the type and number of fire extinguishers adequate? Are they frequently inspected? How far is the nearest public fire department? Have they ever been invited to inspect the facility? Does the building have automatic sprinklers and automatic fire alarms? Are there adequate fire barriers in the building? Is there an employee fire brigade? Are fire doors adequate? Are "no smoking" signs enforced? Are flammable substances properly stored? Is there a program of fire-prevention education? Are fire evacuation drills conducted on a regular basis?

- *Computer access.* What is the potential for loss of equipment? What is the impact to the business if computer systems are rendered inoperable? What information is stored on computer systems that if lost or compromised could cause the organization damage, to include loss of profits and reputation, if compromised?

- *Video surveillance.* Is there a video surveillance system in place? Are cameras properly located and protected? Who monitors the system or is it monitored? What type of system has been designed for review and destruction of old video footage or digital records? What controls assure that images cannot be manipulated?

- *Computer systems and network.* Are computer equipment and network (including network switching and router rooms) properly protected from fire, water damage and physical attacks? Is there a backup system? Are computer files properly backed up? Where are backups stored? Does the computer system have a backup power system or protection for power surges? What types of access measures exist? How is the system protected from unauthorized access? Are firewalls in place? How is the system protected from hackers? Is there a system to protect the computer from viruses?

- *Landscaping.* This is an important aspect of the security survey that should never be overlooked. Are bushes, overgrowth and trees pared back enough to ensure there are no hiding places for a potential criminal?

For all its seeming length, this list contains only a sample of the kinds of questions that must be asked when conducting a comprehensive survey of any facility. The list is only a general overview of some of the aspects to be covered.

General Departmental Evaluations

Each department in the organization should be evaluated separately in terms of its potential for loss. These departmental evaluations will eventually be consolidated into the master survey for final recommendations and action. Basic questions might be as follows:

Is the departmental function such that it is vulnerable to embezzlement?
Does the department have cash funds or negotiable instruments on hand?
Does the department house confidential records?
What equipment, tools, supplies, or merchandise can be stolen from the department?
Does the department have heavy external and/or internal traffic?
Does the department have target items in it such as drugs, jewelry, or furs?
What is the special fire hazard in the department or from adjacent areas?

These are questions that may serve to guide the survey in focusing on particular areas of risk in each department examined. Where particular risks predominate, special attention must be paid to providing some counteraction to remove them. Key areas of internal concern are: pilferage and theft, sabotage, corporate espionage, money storage and handling, drug storage, mail/postal operations, high-value item storage, shipping and receiving, storage of chemicals and explosives, fuel pumps and storage, utilities including gas, electrical and water telecommunications distribution rooms.

> It is well to remember here, especially when it appears that questions concerning the probity of company officers are posed, that the job of the security survey is not to make judgments on whether a criminal act is likely to occur, but rather on whether it could.

Human Resources Organization

Particular security problems are associated with human resource organizations. Protection of personal identifiable information has become a high priority in recent years, with states and the federal government passing laws requiring stringent protection of such information. Ensuring personal information is properly protected drives the security manager to consider the following:

- Can the human resources area be isolated from the rest of the facility and/or building after hours?
- How are door and file keys secured? How is access control to human resource areas managed? If human resource records are stored on computer systems, are proper controls in place? Can computer files be accessed from remote locations?
- Are hard copy files kept locked during the day when they are not in use?
- What system is followed with regard to the payroll department when employees are hired or terminated?
- What are the relationships between personnel and payroll staff?
- What are the employment procedures? How are applicants screened?
- How closely do personnel work with security on personnel employment procedures?
- Are new employees given a security briefing? By whom?
- Does the company have an incident reporting system? Are employees aware of the program? Does the company have a follow-up security awareness training program?

Security of human resources records, electronic and paper is of extreme importance. Normally these files contain information on every employee, past and present, from the chief executive to the newest employee. This information is highly confidential and must be handled that way. There can be no exceptions to this firm policy.

Accounting

The accounting department has total supervision over the firm's money and will generally be the area most vulnerable to major loss due to crime. Certainly protective systems have been in operation in this area from the company's founding, but these systems must be reevaluated regularly in light of ongoing experience to find ways of improving both their efficiency and security:

- Cashier
- Accounts Receivable
- Accounts Payable
- Payroll
- Company Bank Accounts

Information Systems

Computer-related security problems have become more critical to companies as they increase their dependence on information systems and criminals continue to become more sophisticated with information technology and more persistent with their criminal intentions. Even the smallest companies depend upon information systems and information technology to run their business. Protecting information systems and mitigating risks is essential for any company wanting to survive, never mind flourish. The importance of this area is stressed in

Chapter 17, Computer Technology and Information Security Issues. Until then, the security manager should be thinking about the following fundamental issues:

- Are adequate auditing procedures in effect on all programs and systems?
- What are the protocols governing system access?
- How is computer use logged? How is the accuracy of this record verified?
- How is remote access tracked for LANs, WLANs or WANs?
- Are adequate firewall controls in place?
- What is done to determine access from outside sources through the Internet, if appropriate?
- Are there audits of downloads to laptop computers?
- What is the off-site storage procedure? How are such files updated?
- Who has keys to computer spaces and how often is the list of authorized key holders evaluated?
- What controls are exercised over access and how often is the list of those authorized to enter updated?
- What fire prevention and fire protection procedures are in effect? What training is given employees in fire prevention and protection?
- Is there off-site backup hardware? How is it secured?
- How are printouts of confidential information handled?

Purchasing

This is an area subject to many temptations. Graft is often freely offered in the form of cash, expensive gifts, lavish entertainment, and luxurious vacations—all in the name of seeking the goodwill of the purchasing agent. Generally speaking, this is not a security matter unless the agent succumbs to the extent of paying for goods never delivered or paying invoices twice. If all the attention from vendors causes a purchasing agent to buy unwisely, that is a management concern in which security plays no role.

There are, however, some areas in the purchasing function in which security might be involved:

- What are the procedures preventing double payment of invoices? Fraudulent invoices? Invoices for goods never received? How often is this area audited?
- Are competitive bids invited for all purchases? Must the lowest bid be awarded the contract?
- What forms are used for ordering? For authorizing payment? How are they routed?
- Since purchasing is frequently responsible for the sale of scrap, waste paper, and other recoverable items, who verifies the amount actually trucked away? Who negotiates the sale of waste or scrap material? Are several prospective buyers invited to bid? How is old equipment or furniture sold? What records are kept of such sales? Is the system audited? What controls are placed on the authority of the seller?

Having in place a company policy, documenting and defining the rules of operation and behavior for company purchasing agents will provide a foundation for minimizing such events.

Shipping and Receiving

Freight and merchandise handling areas are particularly troublesome as there is a great potential for theft in these areas. Close attention must be paid to current operations and efforts must continually be directed to their improvement.

- What inspections are made of employees entering or leaving such areas?
- How is traffic in and to such areas controlled? Are these areas separated from the rest of the facility by a fence or barrier?
- Where is merchandise stored after receipt or before shipment? What is the security of such areas? What is the nature of supervision in these areas?
- What is the system for accountability of shipments and receipts?
- Is the area protected with guards, surveillance equipment or other means?
- What losses are being experienced in these areas? What is the profile of such loss (type, average amount, time of day)?
- Is merchandise left unattended in these areas?
- Are truck drivers provided with restroom facilities separate from those of dock personnel? Are they isolated from them at all times to prevent collusion?
- How many people are authorized in security storage areas, and who are they?

Miscellaneous

Other general security concerns include:

- What records are kept of postage meter usage? What controls are established over meter usage?
- How is the use of supplies and materials controlled? (Figure 7-1)
- How are forms controlled?

Report of the Survey

After the survey has documented the full scope of its examination, a report should be prepared indicating those vulnerable areas with weak security controls in place and recommend measures that might reasonably bring the security of the facility up to acceptable standards, mitigating potential areas of risk. On the basis of the status in the survey and considering the recommendations made, a security plan can now be developed.

In some cases, compromises may have to be made. For example, operational considerations may make an ideal security program with full coverage of all contingencies too costly to be practical. In such cases, the plan must be restudied to find the best approach for achieving acceptable security standards within these limitations. Keep in mind, the objective is to mitigate and manage risks. It is almost always too costly to completely eliminate risks.

It must be understood that security directors will rarely get all of what they want. As in every department, they must work within the framework of the possible. Where they are denied extra personnel, they must find hardware that will help to replace people. Where a request for more coverage by closed-circuit television (CCTV) is turned down, they must develop inspection

*"They're sure fussy about checking out
tools. The next thing you know they'll
want them returned"*

FIGURE 7-1 Loss prevention.

procedures or barriers that may serve a similar purpose. If on the other hand (CCTV) is
approved, a delicate balance must be achieved between coverage and the rights of workplace
privacy. The question must be asked: Is there a reasonable expectation of privacy? If at any
point they feel that security costs have been cut to a point where the stated objective cannot be
achieved, they are obliged to communicate that opinion to management who will then determine whether to diminish the original objective or to authorize more money. It is important,
however, that security directors exhaust every alternative method of coverage before going to
management with an opinion that requires this kind of decision.

Operational Audits and Programmed Supervision

The mechanism by which the security survey is administered is as important as is the survey form itself. A security survey may focus on physical security measures or procedures and
is conducted on a periodic basis; an operational audit (OA) considers all aspects of the security operation on a continuing basis. The operational audit is a methodical examination, or
audit, of operations. The purpose of the examination is threefold: (1) to find deviations from
established security standards and practices, (2) to find loopholes in security controls, and
(3) to consider means of improving the efficiency or control of the operation without reducing
security.

Because the audit is an ongoing process achieved through program supervision, it is relatively inexpensive. An OA is based on the concept of programmed supervision without which the
audit would become nothing more than a simple security survey. *Programmed supervision* (PS)

is a means of making sure that a supervisor or other employees go through a prescribed series of inspections that will ascertain that functions or procedures for which they are responsible are being properly executed. Supervisors are thus conducting OAs by evaluating their areas of responsibility on an ongoing basis. A truly successful OA requires the supervisor to make the necessary inspection and to record specific re-checkable findings, not just to record a simple checkmark as "yes" or "no." For example, a supervisor in a loading dock area should be required to check all steps in the shipping/receiving area. Where are truck drivers authorized to be during the loading/unloading of the truck? What procedures are followed to determine that the load count is accurate? How are broken parcels handled? The supervisor must answer each of these questions carefully and fully. As noted, the supervisor cannot simply respond with a "yes" or "no." The aggregate of several area OAs results in a divisional OA, and divisional OAs considered in aggregate are an entire company's OA.

The OA should be distinguished from its simpler sister, the security survey. A security survey begins by developing a checklist of items that the security team feels are important. For example, are there adequate locks, alarms, and guard patrols? Do security breaches (that is, doors and windows) have adequate protection, and are they built of substantial materials? Although some security surveys involve a check of procedures, many do not. For example, it would be wise to ensure that check-in procedures at the warehouse are being followed.

The OA builds on the security survey. For many operations, a security survey may be conducted once a year, or even less frequently. The OA, however, is conducted regularly and frequently. Once the OA begins, it continues until someone in a position of authority decides that it is no longer necessary. The audit, through the process of PS, requires that supervisors regularly report whether procedures are being followed and if those procedures are adequate. Some procedures might need to be amended as the work changes to meet new demands. The OA also requires supervisors to report physical conditions regularly (such as whether the doors are locked regularly as specified).

While the security survey is better than nothing, the OA goes beyond an occasional survey. The security survey relies heavily on either the proprietary security force or on a contractor. The OA uses the management resources of the company.

Using the information gained from vulnerability analysis, security surveys, and OAs, the security manager can develop a comprehensive security plan.

Probability

Once vulnerabilities have been identified through the use of the security survey or OA, it is essential to determine the probability of loss. For example, suppose that one vulnerability involves the theft of trade secrets. Within the area of trade secrets, subcategories of vulnerabilities are identified, including the loss of information from research and development through employee turnover or negligence. Should security dollars be spent to reduce the potential for such a loss? It is not possible to say until the probability has been assessed. Will a loss certainly occur if nothing is changed, or is the occurrence improbable? When security managers are confronted with a series of problems, they must determine which problems need immediate attention. *Probability* is a mathematical statement concerning the possibility of an event occurring. Is it possible to reduce security risks to a mathematical equation that can be used to determine probability? Unfortunately, such mathematical precision must wait until various subjective security measures can be turned into numerical values. This has not yet occurred.

The best that can be done today is to make subjective decisions about probability. Such decisions should be based on data like the physical aspects of the vulnerability being studied—for example, spatial relationships, location, and composition of the structure. Procedural considerations must also be studied. What policies exist? The history associated with the industry is of great importance, particularly the vulnerability being studied. Has the product been a target before? What is the current state of the art of thieving? Later in this book various physical security devices will be discussed. Each has its advantages, but the reality is that criminals, depending on their own education and level of determination, may find methods of overcoming each security device. How aware are potential thieves of the technology to defeat existing security devices?

Criticality

Probability cannot stand alone when the security manager analyzes which security problem to address first. For example, a certainty that someone will steal money from the company cafeteria may not warrant attention as immediate as the possibility that someone might tamper with software used to maintain company inventories, purchasing transactions, and quality assurance of products and services. To help separate vulnerabilities into still finer categories, security managers use the principle of *criticality*. The term has been defined to mean the impact of a loss as measured in dollars. As noted earlier in this chapter, the security manager must know what assets are being protected including their relative value to the overall health of the organization. The concept has also been expanded to include how important the area, practice, or whatever is to the existence of the organization. The dollar loss is not simply the cost of the item lost, but also includes:

1. Replacement cost
2. Temporary replacement
3. Downtime
4. Discounted cash
5. Insurance rate changes
6. Loss of marketplace advantage (includes time to market before the competition)
7. Impact to company reputation (damage to a company, whether self-inflicted or caused by external entities, can damage the reputation "brand name" of a company, causing current and future customers to lose faith and thus negatively impacting future revenue)

Consider the Enron Corporation and its collapse in late 2001 due to internal accounting fraud. Internal accounting practices not only led to the bankruptcy of Enron but caused a scandal bringing into question the accounting practices of many corporations. Congress shortly thereafter created the Sarbanes-Oxley Act, driving significant reform in corporate accounting practices.

Criticality is an extremely important concept for security managers to understand. In general, company executives who usually think in terms of cost–benefit analysis will not be interested in spending money for security if the cost is greater than the potential loss of money. It is essential for the security director to be able to explain that criticality is far more than just the direct cost of the items lost. Replacement costs include the new purchase price, the costs of delivery, installation costs, any additional materials needed during the installation, and other indirect costs.

A second major cost may be temporary replacement. Consider an attack on an information system. If the main computer is damaged by sabotage or fire, the company will most likely need to process its information by some other means to include use of a time-sharing arrangement with a computer firm or another company or the use of a hot-site backup. A hot-site backup processing capability is highly useful for contingencies, but it can be very costly. The cost of these temporary measures should be taken into consideration.

A third possible cost is downtime, the cost associated with not being able to continue business while the computer is inoperable. One possible cost in this category, depending on various company policies and union contracts, may be the wages for employees who are idled.

A fourth cost factor, discounted cash, is money lost when invested funds must be withdrawn from time certificates or other investments to pay for any of the above costs. For example, consider the loss of income on a $100,000 certificate of deposit, held at 12.9 percent interest, if the certificate is cashed early to pay delivery and installation costs.

A fifth cost involves the possible increase in insurance premiums associated with loss problems. Insurance rates will increase as losses go up.

Yet a sixth cost is the potential loss of marketplace advantage created by the loss of product markets due to sabotage, work slowdowns, and so forth. If the product is not available when consumers want to buy it, they will turn to alternatives. In some cases, they will stay with the competitor's product.

All six factors need to be added into the criticality cost. Many security managers are surprised to find that the criticality cost can be double the cost of an item. Likewise, company managers often fail to consider these indirect costs.

The Probability/Criticality/Vulnerability Matrix

Criticality, much like probability, is a subjective measure, but it can be placed on a continuum. Consider the continuum for criticality and probability in Table 7-1. By using the rankings generated for probability and criticality and by devising a matrix system for the various vulnerabilities, it is possible to quantify security risks somewhat and to determine which vulnerabilities merit immediate attention. Although some areas of importance may be obvious, some security executives may be surprised to find that other areas are more critical than they first surmised.

For example, consider the cash theft vulnerability matrix shown in Table 7-2. Cash theft has been chosen since it is simple to calculate criticality costs. By considering the gross sales of the firm and its current assets, the impact of the loss of cash from each of the areas listed can be determined. In addition, by considering the history of loss and the number and quality of security devices present, it is possible to estimate the probability of a cash theft.

Table 7-1   The Probability/Criticality Matrix

Probability	Criticality
1. Virtually certain	A. Fatal
2. Highly probable	B. Very serious
3. Moderately probable	C. Moderately serious
4. Probable	D. Serious
5. Improbable	E. Relatively unimportant
6. Probability unknown	F. Criticality unknown

Adapted from Richard J. Healy and Timothy J. Walsh, *Industrial Security Management* (New York: American Management Association, 1971), p. 17.

Using the system presented in Table 7-1, alphabetical and numerical values can be assigned to each vulnerability area. For example, the manager's office might be categorized as A4, which indicates that the loss of $200,000 in a company with total current assets of $300,000 could be "fatal" and that the probability of the loss occurring is "probable" based on the amount of money tempting the thief and the level of security present. Each area may be classified in the same fashion. Then it is usually possible to rank the importance of attacking each area using criticality as the most important variable, for example: A1, A5, B3, B4, C1, D4, E2 (see Table 7-3). The only exception to this order of ranking occurs in the cases of F (criticality unknown) and 6 (probability unknown). If the security director cannot assign a probability or criticality to a certain item the criticality should be assumed fatal and the probability virtually certain. To do otherwise is suicidal!

If a decision has to be made, criticality should take precedence over probability. The security director, however, should implement measures to reduce the threat to the improbable level whenever the measures are cost-effective.

Alternatives for Optimizing Risk Management

Once the security probability and criticality analysis has been completed and the security problems have been identified and ranked in importance, the security manager in cooperation with company executives must decide how to proceed. As was noted earlier, there are several risk-management alternatives: risk avoidance, risk reduction, risk spreading, risk transfer, and self-assumption of risk.

Risk avoidance is removing the problem by eliminating the risk. This can be accomplished by transferring responsibility to another area. For example, the manufacturing of a small transistor by company M may be a security problem. To avoid the risk, company M decides to subcontract the manufacturing process to another firm that is better suited to handling this type of product security. Thus, the risk for company M is avoided.

Risk reduction is decreasing the potential ill effects of safety and security problems when it is impossible to avoid them. For example, as a result of a security survey and vulnerability analysis, the security manager has determined that company N has a high risk of money loss in the central budget office because there are no positive admittance controls and no alarms.

Table 7-2 Cash Theft Vulnerability Matrix

Building Location	Amount of On-Hand Dollars		Accountability Records		Area Has Physical Bounds		Area Locked		Positive Control on Admittance		Alarm Protection		Surveillance Devices		Cash in Storage Container		Bait Money Kept		History of Cash Loss	
	NBH	OT	NBH	OT	NBH	OT	NBH	OT	NBH	OT	NBH	OT	NBH	OT	NBH	OT	NBH	OT	NBH	OT
Manager's office	200,000	40,000	Y	Y	Y	Y	N	Y	N	Y	Y	Y	N	Y	Y	Y	Y	Y	N	N
Manager's secretary	300	300	Y	Y	N	N	N	Y	Y	N	N	N	N	N	Y	Y	N	N	Y	Y
Cafeteria	2,000	0	Y	N	Y	Y	N	N	N	N	N	N	N	N	N	—	N	—	Y	N
Loading dock	1,500	500	Y	Y	Y	Y	N	N	N	N	N	N	N	Y	Y	Y	N	N	Y	N
Visitor reception	100	100	N	N	N	N	N	N	N	N	N	N	N	N	N	Y	N	Y	N	Y

Data for a company with gross sales per year of $310,000 and total current assets of $300,000.

NBH = normal business hours

OT = other times

Y = yes

N = no

Table 7-3 Probability/Criticality Assessment and Ranking

	Criticality		Probability	
	NBH	**OT**	**NBH**	**OT**
Manager's officer	A	B	2	4
Manager's secretary	D	D	2	2
Cafeteria	C	—	1	—
Loading dock	C	D	1	2
Visitor reception	D	D	2	1

Suggested Rank Order

	Manager's officer NBH	A2
	Manager's officer OT	B4
	Cafeteria NBH	C1
	Loading dock NBH	C1
	Visitor reception OT	D1
	Visitor reception NBH	D2
	Loading dock OT	D2
	Manager's secretary NBH	D2
	Manager's secretary OT	D2

Creating a policy for positive admittance and the installation of proximity alarm devices can reduce the risk. The risk is never totally eliminated since the old adage "where there's a will there's a way" applies to employees and outsiders who want to steal money.

Risk spreading is decentralizing a procedure or operation so that a security or safety problem at one location will not cause a complete loss. Suppose that company M is producing a microchip at a high risk of loss. It can spread its production risk by subcontracting some of the components to other companies or by producing the components at other sites owned by company M.

Risk transfer generally means removing the risk to the company by paying for the protection of an insurance policy. Insurance options will be discussed later in this chapter. *Self-assumption* of risk involves planning for an eventual loss without benefit of insurance. In all procedures to minimize risks, insurance should be considered as a valuable addition to safety and security procedures.

> Insurance should be considered as a valuable addition to safety and security procedures.

The Cost-Effectiveness of Security

It is unlikely that any evaluation will ever absolutely determine the cost-effectiveness of any security operation. A low rate of crime committed against the organization along with minimal loss of assets—whether compared to past experience, to like concerns, or to neighboring businesses—is an indication that the security department is performing effectively. But how much is being protected that would otherwise be damaged, stolen, or destroyed? This can be any figure, from the total exposure of the entire organization to some more refined estimate based on the incidence of criminal attack locally or nationally, the average losses suffered by the industry in general, or the reduction in losses by the organization over a given period.

An estimate based on such figures might well serve as a practical guide to the usefulness of the security function. On the other hand, if a security operation costing $400,000 annually were estimated, by some formula using a mix of the data mentioned previously, to have saved a potential loss in theft and vandalism of $300,000, would it be deemed advisable to reduce the department's operating budget by $100,000 or more? Obviously not! This would be roughly analogous to reducing or canceling insurance because damage or loss and subsequent insurance recovery for a specific period or incident were less than the cost of the premium. Security can be considered as insurance against unacceptable risks.

Studies on the role of security and related investments as part of corporate costs of risk conducted by the Risk and Insurance Management Society (RIMS) show that the share of the total cost for risk control in these areas has risen. The rise in percentage indicates a growing awareness by management of the role security can and must play in the total package of risk control as well as strategies by the insurance industry to recover from losses incurred in the 1990s and events following September 11, 2001 including the collapse of Enron, problems at WorldCom and catastrophic storms like Katrina. The changes following 2001 have resulted in a now relatively healthy property/casualty insurance industry. This in turn has resulted in declines in rates in all segments of the industry continuing into 2007.[1] By 2010, actual rates grew by nearly 5 percent.[2] However, this growth may be due to the depletion of capital needed to underwrite new business, caused by the economic downturn in 2008 more so than any action or inaction caused by Risk Managers.

Cost-effectiveness studies must be made, however, as part of a periodic review of protection systems even though such studies cannot be used as a general rule in devising a magic formula for computing the cost-per-$1000 actually saved in cash or goods that would otherwise have been lost. Such a review would consider, for example, the savings that could result from the substitution of functionally equivalent electronic or other gear for personnel (the most expensive deterrent) and the feasibility of taking such a step.

Periodic Review

Even after the security plan is formulated, it is essential that the survey process be continued. To be effective a security plan must be dynamic. It must change regularly in various details to accommodate changing circumstances in a given facility. Only regular inspections can provide a basis for the ongoing evaluation of the security status of the company. Exposure and vulnerability change constantly. What may appear to be a minor alteration in operational routines may have a profound effect on the security of the entire facility.

Security Files

The survey and its resultant report are also valuable in the building of security files. From this evaluation emerges a detailed current profile of the firm's regular activities. With such a file, the security department can operate with increased effectiveness, but it should, by inspections and additional surveys, be kept current.

Such a database could be augmented by texts, periodicals, official papers, and articles in the general press related to security matters. Special attention should be paid to subjects of local significance. Although national crime statistics are significant and help to build familiarity with a complex subject, local conditions have more immediate import to the security of the company.

As these files are broadened, they will become increasingly useful to the security operation. Patterns may emerge, seasons may become significant, and/or economic conditions may predict events to be alert to. For example:

- Certain days or seasons may emerge as those on which problems occur.
- Targets for crime may become evident as more data is amassed. This may enable the security director to reassign priorities.
- A profile of the types and incidences of crimes—possibly even of the criminal—may emerge.
- Patterns of crime and their modus operandi on payday or holiday weekends may become evident.
- Criminal assaults on company property may take a definable or predictable shape or description, again enabling the security director to shape countermeasures better.

The careful collection and analysis of data concerning crime in a given facility can be an invaluable tool for the conscientious security officer. It can add an important dimension to the regular reexamination of the status of crime in the company.

Given the present atmosphere of litigation for failure to provide adequate security, the files showing an efficient security operation can be invaluable. On the other hand, poorly kept files can be as great a liability as can well-kept files on a poorly designed security plan or poorly operated security organization.

Insurance

As noted earlier, one method of managing risks is risk transfer. Insurance is an option that is regularly pursued in the area of transfer. More specifically, insurance transfer is a means of protecting or safeguarding against risk or harm to the corporate enterprise. Yet far too many security managers falsely assume that the most effective means of guarding against unforeseen business losses is insurance. Many still attempt to use insurance as a substitute for a comprehensive security program which can lull one into a false sense of security. The fallacy in this attitude is twofold.

> Insurance alone can never be a substitute for a comprehensive security program.

In the first place, almost all casualty insurance companies require that adequate safety and security measures be in place before they will cover a potential risk. To the extent such measures are not in place, underwriters will either refuse to offer the insurance, or offer the insurance only at a very high premium, perhaps with exclusions addressing the very risks which the company seeks to insure. In short, insurance cannot replace a safety and security program, but must work hand-in-glove with one. In the second place, it is virtually impossible to insure against all the losses that could be incurred. Insurance policies contain conditions that must be met before the coverage will come into play; various exclusions that could preclude coverage; deductibles, self-insured retentions and waiting periods that require the insured to assume some portion of the loss before the insurer begins to pay; and, of course, limits on the total amount the insurer will pay. Insurance does not cover all losses; it only helps to transfer the risk. Damages may include loss of company morale, loss of customer confidence or loss of business in a highly competitive marketplace—all are serious, if not fatal, blows to any business's brand value and can never be recompensed.

Clearly insurance can never be a substitute for a security program but rather one of the many tools to ensure a comprehensive program. In many cases, the over-dependence on asset insurance to some degree tends to reduce the interest of the proprietor in driving reasonable security procedures beyond those minimums specified in a policy. An over-dependence on insurance can also serve to reduce a proactive interest the Company may have in capturing or prosecuting perpetrators of crimes. The unintended consequence of this in effect could encourage the proliferation of like criminal acts.

The Value of Insurance as Part of a Total Loss Prevention Program

Insurance is certainly important. It is necessary for any business that wishes to be protected against loss—to transfer the risk—but it must be thought of as an important pillar rather than the principal defense against losses from crime. It is equally important to realize that insurance carriers provide coverage on the basis of formulas that take into account both the likelihood of loss and that the "estimated value" (EV) along with the estimated loss is always less than the total of the premiums paid over time. In essence, insurance is based on mathematical formulas, statistics and probabilities.

Insurance must be viewed as an important pillar in a comprehensive security program, rather than the principal defense.

Types of Insurance

There are many types of insurance that indemnify against loss and are available to a company. For the purposes of this discussion, however, the focus will be on only those types of insurance that typically play a prominent role in loss prevention.

Fidelity Coverage

Commonly referred to as *employee honesty insurance*, this coverage provides payment for losses due to employee acts of dishonesty. Employee dishonesty can come in many shapes and forms, including but not limited to: falsifying expense reports, stealing cash, shipping and billing scams, or adding "ghost employees" to the payroll that don't actually exist, and finally committing cybercrimes like stealing a corporation's or individual's identity. It is important to note that electronic crimes (commonly known as "cybercrimes") perpetrated through the Internet can be covered by new policies or by traditional crime insurance.[3] This is something the security manager needs to consider to ensure appropriate coverage.

An important aspect of fidelity insurance is *blanket bonds.* This type of bond is in general use because it covers categories of employees and thus allows automatic coverage of new employees; the *name* or *position* bonds are also popular because of the lower premium costs. The name bond covers only certain specifically named individuals, while the position bond covers only those persons who hold a specific position of trust within the company. Generally speaking, it would include owners, officers, executives and finance personnel.

This type of coverage is frequently badly underestimated. In effect, the bonding company is guaranteeing the insured that bonded employees will perform in good faith—that is, that they will not commit any dishonest or fraudulent acts against their employer. If any so bonded employees violate this trust, the guarantor—the bonding company—will stand the loss up to the amount insured.

The investigation by the bonding company is valuable in that it provides a further check on the background of employees in sensitive positions in addition to underwriting possible losses resulting from a violation of trust. Any employee with a past criminal history is excluded.

Proper bonding of employees is important to ensure maximum coverage!

Most companies require that employees handling cash or high-value merchandise be insured for employee dishonesty. But too many of these companies go along on a program providing limits far too low to recover any real risk of loss posed by a theft scheme, which often may be perpetrated over a period of months, or even years. In situations where there is no system providing a regular, foolproof audit of cash and valuable merchandise, for example, an employee might steal enormous sums over a period of time even if the daily amount is relatively small. Importantly, under employee dishonesty coverage, all of these sums stolen by the same person are generally aggregated together to constitute one loss, subject to one limit under the crime coverage, and typically only one policy period will apply, even if the theft occurs over more than one year.

Employee theft is a rapidly growing crime today, especially in light of the severe economic downturn. With losses attributable to internal theft estimated in the billions of dollars, which is ultimately passed on to the consumer, it is easy to see why fidelity coverage is thought of as high-priority coverage, especially since it provides particular protection in areas where exposure is generally the greatest. As important as this form of coverage is, it is essential that it be handled properly to provide the full protection of which it is capable.

It is important in cases where employees are discovered to have been involved in theft to report the matter to law enforcement and to the insurer in strict compliance with policy obligations. It is also important not to agree with the employee regarding restitution of the amount stolen without the insurer's approval. Crime policies, like all insurance policies, contain restrictions. Settlement with the employee, absent the insurer's involvement and consent, could impair the insurer's rights to recover from the employee for the losses. If this occurs, the insured might be found in breach of a policy condition necessary for the coverage. In short, it is always best to leave the matter to the insurer.

Surety Coverage

Also called *performance bonds,* this coverage provides protection for failure to live up to contractual obligations. The *penal sum* is an important aspect of the surety bond, essentially designating a specified amount of money which is the maximum amount the surety will be required to pay should the principal default. Surety coverage should be considered for guard and investigative services to guarantee contractual obligations and performance.

Federal Crime Insurance

The Federal Crime Insurance Program provides for federally funded crime insurance at reasonable rates, based on the size and accepted risk of the insured property. This insurance is intended to fill in a gap where private insurers withdrew from the market and refused to write policies, based on the high level of risk exposure.

A substantial number of businesses in this country have had some kind of problem with property insurance. These problems include canceled policies, refusal to issue or renew insurance, prohibitive rates, and limiting coverage to well below the cash value of insured property.

In inner-city locations or in certain types of businesses, policies, when they are issued, substantially limit the insurer's liability—frequently to the point where the policy is virtually useless as support protection. The Federal Crime Insurance Program requires the participation of individual states, provides for federally funded crime insurance at reasonable rates, based on the size and accepted risk of the insured property.

In order to qualify for protection under this program, however, a business must establish certain minimum protective devices and procedures. The business must, in short, recognize that it can get the supportive protection that insurance offers provided it makes at least minimal efforts to protect itself.

The program prescribes locks, safes, alarm systems, and other protective devices, and establishes the kind of protection that various businesses must provide for themselves in order to qualify for this insurance. For example, gun stores, wholesale liquor and fur stores, jewelry firms, and drugstores must all have a central station alarm system; service stations must have a local alarm system; and so on. Small loan and finance companies, theaters, and bars—businesses rated as high risk—are also eligible for insurance under the program.

Not only does it provide for insurance coverage of premises otherwise difficult or impossible to insure adequately or reasonably, but it also focuses attention on the very real need for the insured to take positive steps to provide protection of the premises to prevent loss and to use insurance to defray those losses that do occur only when security measures fail. In short, it takes insurance from the front line of crime prevention—where it clearly cannot perform—and puts it into a backup position where it can.

3-D Coverage

Comprehensive dishonesty, destruction, and disappearance (3-D) coverage is extremely flexible. Policies will vary in coverage and premiums, based on the needs of the firm. Possible areas covered may include burglary, robbery, employee theft, and counterfeit goods. These policies are designed to provide the widest possible coverage in cases of criminal attack of various kinds. The standard form is set up to offer five different kinds of coverage. The insured has the option of selecting any or all of the insuring agreements offered and of specifying the amount of coverage on each one selected. In addition to the coverage options in the standard form, 12 endorsements are also available to the security manager having a need for any or all of them.

The coverage available on the standard form consists of the following:

1. Employee dishonesty coverage
2. Money and securities coverage on the premises
3. Money and securities coverage off the premises
4. Money order and counterfeit paper currency coverage
5. Depositors' forgery coverage

Additional endorsements available include:

1. Incoming check forgery
2. Burglary coverage on merchandise
3. Paymaster robbery coverage on and off premises
4. Paymaster robbery coverage on premises only
5. Broad-form payroll on and off premises
6. Broad-form payroll on premises only
7. Burglary and theft coverage on merchandise
8. Forgery of warehouse receipts
9. Securities of lessees of safe-deposit box coverage
10. Burglary coverage on office equipment
11. Theft coverage on office equipment
12. Credit card forgery

Obviously, the premium on this coverage will vary according to the number of options selected and the amount of coverage desired for each.

Insurance against Loss of Use and Extra Expense Coverage

Most standard policies do not provide loss of use (the business or some sub-operation ceases production, resulting in losses) or extra expense coverage (the business cannot afford to be down and therefore must pay for rental space and the like to continue operations). Since both these matters can represent a very substantial loss to most companies, consideration must be given to expanding the provisions of the coverage to include one or the other. Both of these losses can be covered either by endorsement or by additional policies that will provide that coverage on a broad basis.

Even a small fire in an office may render it inoperable from smoke and water damage or damaged equipment for a substantial period of time. Even though all the damage is covered and will be cared for, the interim period during which revenues may be lost and new facilities are being occupied may be as expensive as the fire itself. In this scenario, loss of use or business interruption coverage would be suitable.

Here, too, there are options. A business interruption policy can be drawn up on a comprehensive basis, which means that it will cover a broad base of situations that might create a stoppage. Such a policy must, of course, be examined for types of incidents specifically excluded from coverage. On the other hand, such a contract might be drawn up in which the incidents covered are specified and perhaps limited to just a few potential hazards.

Basic insurance coverage for a specific loss often fails to provide coverage for business interruption costs.

The amount of coverage and the nature of recovery in business interruption contracts can be complicated. If recovery is on an actual loss basis, a careful audit of actual demonstrable losses must be presented to the insurer in order to collect. If the policy is drawn as a valued loss contract, an accountant must certify the daily amount that would be lost if an interruption were to occur. This amount is entered as part of the contract. The premium and recovery are based on this amount, computed on the specified number of days to be covered for each interruption.

Recent court decisions point out the need to clearly state coverage. After the 2001 terrorist attacks, companies connected to the World Trade Center filed claims on their business interruption insurance policies. Courts found that only those with close and direct causal connections between the firm's loss of business and the attacks were accepted. For example, a hotel claimed to have lost business as a result of grounded flights. This claim was rejected. On the other hand a cleaning business that cleaned for firms in the Trade Center made a claim that was accepted.[4]

Extra expense coverage would be called for in a situation where the operation must immediately be transferred to another location and equipment rented until the damaged facility is back in operation. A good example would be a newspaper where the operation must continue in order to retain readership. If the business ceased operation, even temporarily, subscribers would look elsewhere for their news.

Kidnap and Ransom Insurance

With the alarming increase in international terrorism and the accompanying increase in kidnapping that naturally follows, the demand for insurance for executives and key employees against such incidents is also important. These policies generally cover all costs associated with the successful recovery of a kidnapped executive or key employee including costs of information and loss of ransom money. Lightning kidnaps (also known as ATM kidnaps) are now in fashion, especially in South and Central America, with the crime being primarily economic over principle or social messaging. These kidnaps generally occur within 24 hours or less. It is important to understand if the policy covers these kinds of incidents. Some policies also cover the cost of lawsuits filed against the firm for inadequate protection, especially if the crime is foreseeable, and insufficient efforts to win release.

Such coverage requires that companies institute certain basic security measures:

1. Executives and key employees must maintain secrecy about the existence of coverage.
2. Verifiable education and awareness training including countermeasures and defensive driving may also be called out in the policy.
3. Every reasonable effort must be made to contact the police, FBI, and insurance company before payment is made.
4. Serial numbers on ransom money must be recorded.
5. A plan of action for dealing with kidnapping must be in place.

Fire Insurance
While fire insurance is a must for homeowners, the use of these once popular policies for business purposes has to a great extent been supplanted by broad coverage policies.

Business Property Insurance
These special multi-peril policies (SMP) generally offer coverage against a multitude of losses including crime, property, liability, and machinery. In simple terms, this type of policy may be likened to homeowner's property insurance.

Liability Insurance
With the growth in the number of lawsuits filed against businesses each year for various negligent acts, the popularity of these insurance policies has grown. Coverage may include employer–employee, customer–employer, contracts, and professional services.

Workers' Compensation Insurance
This basic insurance provides for medical costs, lost wages, and rehabilitation of workers injured on the job. There are also death benefits available. In most states, this coverage is required by law.

Portfolio Commercial Crime Policies are effectively replacing many of the individual types of crime insurance.

Portfolio Commercial Crime Coverage
This form of coverage is replacing some of the individual policies listed above. The policy is composed of standard modules and allows for up to 14 endorsements:

1. Employee dishonesty
2. Forgery or alteration
3. Theft, disappearance, and destruction
4. Robbery and safe burglary
5. Premises burglary

6. Computer fraud
7. Extortion
8. Premises theft and robbery outside the premises
9. Lessees of safe deposit boxes
10. Securities deposited with others
11. Liability for guests' property—safe deposit box
12. Liability for guests' property—premises
13. Safe depository liability
14. Safe depository direct loss

Cyber Liability Coverage

There is no insurance coverage that fits neatly within the definition of "cyber liability coverage." Cyber liability coverage can be found in many forms. For certain risks, such as denial of service attacks, the introduction of a virus or trojan horse into a customer's computer network or misappropriation of personal information stored or transmitted electronically, this coverage may be purchased as part of an errors and omissions policy. These risks generally arise in the performance of services, so it is logical that coverage would be available. Standalone policies covering these risks are also available, but seem to be favored by companies strictly in the Internet business (e.g., performing services solely on the Internet).

For other risks, coverage for "cyber threats" may be found as extensions under a property insurance program (e.g., damage to data and programs) or a crime insurance program (e.g., theft of information). The availability of coverage for various types of risks is not always straightforward, and exclusions must be carefully considered in assessing the availability of coverage for any of the risks. Whether the perpetrator or the computer being used is on premises or off premises may make the difference between coverage and no coverage.[5]

Terrorism Risk Insurance Act of 2002

On November 26, 2002, President George W. Bush signed into law the Terrorism Risk Insurance Act. The Act established a federal program to share the burden of commercial property and casualty losses resulting from acts of terrorism with the private sector. The Act was to sunset (i.e., cease to exist) December 31, 2005, but was extended for two years. However, Congress has extended the act again for 15 years, with the introduction of the Terrorism Risk Insurance Revision and Extension Act (TRIREA) of 2007. Specific information regarding this act may be found at www.treasury.gov/trip.[6]

The aftermath of the World Trade Center attacks in 2001 and Hurricane Katrina in 2005 brought to light new concerns over the costs associated with mass destruction and who should pay for rebuilding.

In the business environment, a vast majority of small businesses are choosing to not purchase additional coverage. The Council of Insurance Agents and Brokers report that less than 10 percent of the small business clients represented by their surveyed insurance brokers were interested in the expanded coverage. Two reasons for declining the coverage are: expense, and belief that the business would not be a terrorist target.[7]

The cost varies tremendously by company, ranging from a low of 2 percent of the value of the property being insured to a high of 150 percent. The average cost is 12 percent of the property value.[8]

How Much Insurance?

Since the options are essentially a choice between recovery of the cash value of the property or recovery of its replacement cost, there can be little hesitation in making a decision. Few experts disagree that insuring to the amount of replacement is clearly the wiser course to follow. When property is insured for its cash value only, there will almost inevitably be a loss to the insured unless property values decline enough to make up for the extra costs involved in replacement. The latter might include demolition of the remaining structure in the event of fire, clearing the site for rebuilding, or the declining value of the dollar. History shows us that property values, or more specifically building costs, rarely decline in this manner. Protection should therefore be arranged on the basis of resuming business as it was before the damage took place. This can normally be done only by insuring for the replacement cost of the property.

Cost must be weighed against the risk and its consequences.

Replacement cost coverage is more expensive than is cash value coverage since the insurer must set the premium sufficiently high not only to cover the estimated likelihood of a fire occurring, for example, but also to try to anticipate the rate of increase in the cost of labor and materials in the reconstruction of the building in whole or in part.

Even insuring at replacement cost will not, as a rule, cover the full cost involved in a major disaster. Business interruption, extra expenses, site clearance, intervening passage of new and more exacting building codes—all add to the already inflated costs of replacing the existing structure so that business may resume as before as rapidly as possible.

There are many endorsements available to extend coverage to fully compensate for all expenses involved in replacement. They should all be considered in the light of individual needs. Obviously the greater the coverage, the higher the cost of coverage, but this cost must be weighed against the risk and its consequences.

CASE STUDY

Verne Jacobs, CEO of Superior Corporation, has decided to lease additional office space for Superior's growing workforce while plans are reviewed and underway to add additional buildings to the existing Corporate campus. The leased office space is about 35 miles from the Corporate campus and is located in a commercial office complex. The lease has been signed for two years with the lessee accepting all risk and liability during the terms of the lease. Either party can cancel the lease with 30 days notification.

Leased office space is difficult given the growing number of Companies in the area with similar needs for office space to house their employees some of which are competitors of Superior and who happen to be located in the same building and on some of the same floors with Superior employees. In fact, Superior has temporarily relocated its research and development lab to one of the floors in the leased complex. There is a central cafeteria that services all tenants in the building.

The area surrounding the commercial office complex is economically challenged with petty theft and vandalism on the rise. The local police are aware of the problem and are working to curb the rise in crime. Superior is a major employer in the area who pays Corporate taxes that support local funding of the Police, the Fire Department among other city services.

The commercial office complex is surrounding by a decorative fence and is monitored by CCTV on a 24 hour basis. Security, overall, is minimal with the lease agreement requiring each tenant to provide its own security.

1. What unique challenges are presented by having competitors in and around Superior Corporation's new leased office space?
2. What could Superior Corporation do to mitigate the challenges of having competitors near their location?
3. How could employees of Superior Corporation help to ensure that critical information is not lost to a competitor?

Summary

The area of risk analysis, as with law, can be complicated. Security managers who assume major responsibility for risk management should consider enrolling in courses directed toward risk management, cyber security and insurance. When threats and vulnerabilities are identified and their impact on assets analyzed, the risk to those assets can be properly understood. To overlook the role of this important area of security is to move forward without a carefully developed plan. Dollars may be spent on security measures that have little impact on the actual protection of company assets while other assets remain vulnerable to destruction, theft or other vulnerabilities.

Insurance has become an integral part of the overall loss prevention plan. Security planning in turn has an impact on the cost associated with insurance. The more comprehensive the security plan, the lower the costs.

■ ■ CRITICAL THINKING ■

Why would a security loss prevention manager choose to hire an outside firm to review security operations and make recommendations for changes as well as insurance options rather than conduct the study internally?

Review Questions

1. What is the difference between vulnerability to loss and loss probability?
2. What is meant by criticality of loss?
3. If a security countermeasure costs as much as, or more than, the loss being protected against for a given period, does it follow that the security measure should be discontinued because it is not cost-effective?
4. Why should accounting procedures be a part of a security survey? Why are security files significant in protection planning?
5. Why is it essential to include information systems as part of the security survey?
6. List four typical limited loss-prevention responses.
7. What are the steps involved in a good risk-management program?
8. What is an operational audit?
9. Do you agree with the statement that "insurance must be thought of as a supportive rather than the principal defense against losses due to crime"? Why?
10. Why is it important for the insured to clearly understand the terms describing criminal activity (for example, burglary or robbery)?
11. In terms of property insurance, define direct loss, loss of use, and extra expense losses.
12. What are the differences between specific and comprehensive property insurance policies? If you were a business executive, which would you prefer if you had to give priority to cost-effectiveness?
13. What were the problems that led to the establishment of the Federal Crime Insurance Program? What are some of the crime-prevention measures prescribed by the program?

References

[1] Christopher T. Déjà vu all over again: a mid-year P/C outlook. Risk Manage, July 2007.
[2] http://www.iii.org/facts_statistics/property-casualty-insurance-cycle.html.
[3] Jones S. Cybercrime: covered or not? Insur J, January 2002.
[4] Gosnell S. Disasters and the law. Can Insur, January 2007;112(1):21.
[5] Stein W. In: Interview and written comments. Insurance Law Group; August 2010.
[6] Interim Guidance Concerning Certain Conditions for Federal Payment, Non-U.S. Insurers, and Scope of Insurance coverage in the Terrorism Risk Insurance Act of 2002. Department of the Treasury, downloaded 3/24/2003 (checked for updates, 07/04/2007). <www.treasury.gov.press/releases/reports/interimguidance.htm>.
[7] Most Small Businesses Not Buying Terror Insurance. National Federation of Independent Businesses, downloaded 3/29/2003. <www.nfib.com/cgi-bin/NFIB.fll/public?Advocacy/newsReleaseDisplay>.
[8] Terrorism Insurance Meets Deadline Today. National Federation of Independent Business E-News, downloaded 2/25/2003. <www.NFIB.com>.

8

Interior and Exterior Security Concerns

OBJECTIVES

The study of this chapter will enable you to:

1. Recognize the role of doors and windows as security risks.
2. Identify various types of locks and associated hardware.
3. Have a basic understanding of surveillance equipment and its proper use in security operations.
4. Understand the significance of having security involved in building design and remodeling.
5. Realize the growing importance of "interoperability" among various security devices through the use of information and equipment technology.

Introduction

Besides the clear-cut concerns for the perimeter and exterior security, other security vulnerabilities for the facility might be either a part of the perimeter, part of the interior, or both. This chapter deals with security concerns that could be either interior or exterior problems depending on the type of facility. For example, a freestanding retail outlet store has problems of perimeter security that include windows, doors, and roof. The same store located within the confines of a mall may not be as concerned with the windows and doors or with perimeter defense that must be overcome before the attacker can concentrate on the store, because the mall is a buffer.

On another level, the entire area of physical security thinking is changing as technology continues to develop. Not only are the devices used in physical security more sophisticated, but the ability to integrate various physical security operations is now not only possible, but becoming the standard. That is not to say that the traditional lock and key, guard and camera are no longer in use. But the truth is that in many operations this is old technology that has been replaced by 21st century modern computer-based operations. In fact, network and web-based platforms are allowing companies to develop security systems that at one time were only science fiction dreams.

However, with this new technology come new challenges in the protection of the hardware and software used in these operations. More will be said on protection of the computer and its components in Chapter 17.

Buildings On or As the Perimeter

When the building forms part of the perimeter barrier or when, as in some urban situations, the building walls are the entire perimeter of the facility, it should be viewed in the same light as the rest of the barrier, or it should be evaluated in the same way as the outer structural barrier. It must be evaluated in terms of its strength, and all openings must be properly secured.

In cases of a fence joining the building as a continuation of the perimeter, there should be no more than 2 inches between the fence and the building. Depending on the placement of windows, ledges, or setbacks, it might be wise to double the fence height gradually to the point where it joins the building. In such a case, the higher section of the fence should extend 6 to 8 feet out from the building.

Windows and Doors

Windows and other openings larger than 96 square inches should be protected by grilles, metal bars, or heavy screening when they are less than 18 feet from the ground or when they are less than 14 feet from structures outside the barrier (that is, trees and other buildings). Doors that penetrate the perimeter walls must be of heavy construction and fitted with strong locks.

Since both the law and good sense require that there be adequate emergency exits in the event of fire or other danger, provision must be made for such eventualities. Doors created for emergency purposes only should have exterior hardware removed so that they cannot be opened from the outside. A remotely operated electromagnetic holding device can secure them, or they can be fitted with alarms so that their use from inside will be substantially reduced or eliminated.

Windows

It is axiomatic that windows should be protected. Since the ease with which most windows can be entered makes them ready targets for intruders, they must be viewed as potential weak spots in any building's defenses. Most forced break-ins are through window glass—whether such glass is installed in doors or windows.

In most industrial facilities, windows should be protected with grillwork, heavy screening, or chain-link fencing. In some cases, however, caution dictates that they may be needed as emergency exits beyond strict requirements of fire laws. Or where they might be needed to lead in fire hoses, consideration should be given to hinging and padlocking protective coverings for easy removal.

BURGLARY-RESISTANT GLASS

In applications such as prominent administration and office buildings, where architectural considerations preclude the use of such relatively clumsy installations as mesh or industrial screen, the windows can be immeasurably strengthened by the use of either UL-listed (which means that the material or item so designated has met the standards of Underwriters Laboratories) burglary-resistant glass or one of the brands of UL-listed polycarbonate glazing material. Both of these products are considerably more expensive than is plate glass and are

FIGURE 8-1 Shatter-resistant film. *(Courtesy of the www.ShatterGARD.com Glass Protection Experts.)*

generally used only in those areas where attack can be expected or where a reduction in insurance premium would justify the added expense (see Figure 8-1).

Standard plate glass can be given some measure of resistance if it is covered with a 4- to 6-millimeter cover of mylar. This is a low-cost operation, but the mylar needs to be replaced every 5 years. As little as a 2-millimeter cover of mylar can keep glass from fragmenting. Thus the mylar cover is good protection from flying shards associated with bombings. Since the first bombing of the Trade Center and the bombing of the Federal Building in Oklahoma City, various studies have supported the need to mandate some type of coating for windows in buildings that might be targeted for terrorists' attacks.

As opposed to tempered glass, which is designed to protect people from the danger of flying shards in the event of breakage, UL-listed burglary-resistant glass (frequently referred to as *safety glass*) resists heat, flame, cold, picks, rocks, and most other paraphernalia from the intruder's arsenal. It is a useful security glazing material because it is durable, weathers well, and is noncombustible. On the other hand, it is heavy, difficult to install, and expensive.

The plastic glazing sold under various trade names is optically clear, thin, and easy to install.

Acrylic glazing material that appears as Plexiglas generally does not meet UL standards for burglar-resistant material; it is much stronger than ordinary glass, however, and has many useful applications in window security applications. It is also lighter in weight and cheaper than either safety glass or plastic glazing and at 1¼ inches thick is UL approved as a bullet-resistant barrier.

All these materials have the appearance of ordinary glass. Obviously, any window so hardened against entry must be securely locked from the inside to protect against intrusion from the outside. This implies a strong window frame and supporting construction.

"SMASH AND GRAB" ATTACKS

Burglary-resistant glass is used to a considerable degree in banks and retail stores where there has been a very real need to prevent "smash and grab" raids on window displays and showcases.

It should be noted that UL-listed burglary-resistant glass is a laminate of two sheets of flat glass (usually ³⁄₁₆-inch thick) held together by a ¹⁄₁₆-inch layer of polyvinyl butyl, a soft transparent material. In this thickness, laminated glass is virtually indistinguishable from ordinary glass; hence burglars may try with a hammer or iron bar to break what they suppose to be a plate-glass window. It is only after they have made a few unsuccessful tries that they realize the material is not penetrable.

Even though such attackers may flee empty-handed, the owners in such situations are left with windows with webs of cracks over the surface of the outer layer of glass, making replacement of the entire pane necessary in applications where appearance is important. Insurers in many cases require that laminated glass be clearly identified to discourage what would be a futile but damaging assault by the "smash and grab" attacker.

SCREENING

It can also be important to screen windows. First, this might protect their use as a means by which employees can temporarily dispose of goods for later recovery. The smaller the goods being manufactured or available on the premises, the smaller the mesh in the screen must be to protect against this kind of pilferage.

Second, as was noted earlier, any windows less than 18 feet from the ground or less than 14 feet from trees, poles, or adjoining buildings should receive some protective treatment unless they are well within the perimeter barrier and open directly onto an area outside the building that is particularly well secured.

Doors

Every door, whether exterior or interior, must be carefully examined to determine the degree of security required. Such an examination will also determine the type of construction as well as the locking system to be used on each door.

The required security measures at any specific door will be determined by the operations in progress within the facility or by the value of the assets stored or available in the various areas. The need for adequate security cannot be overemphasized, but it must be provided as part of an overall plan for the safe and efficient conduct of the business.

When this balance is lost, the business may suffer. Either the security function will be downgraded in favor of a more immediate convenience, or the smooth flow of business will be impeded to conform to obtrusive security standards. Either of these conditions is intolerable in any business, and it is a management responsibility to determine the balance required in establishing systems that will recognize and accommodate production and security needs.

Door Construction and Hardware

Doors are frequently much weaker than the surface into which they are positioned. Panels may be thin, easily broken wood or glass. Locks may be old and ineffective. The door frame may be so constructed that a lever or a plastic card can be inserted between the door and jamb to disengage the bolt in the lock. Even with a properly hung door, if the jamb is of soft material (unreinforced aluminum or light pine), it can be peeled or ripped away from the bolt. This technique is referred to as *spreading*. The locking bolt must throw at least an inch into the jamb for security applications. Heavy wood or metal doors with reinforced jambs can go a long way in reducing the potential for spreading.

In some cases, doors are entered by *pulling*, a technique whereby the lock cylinder is ripped from the door and the locking mechanism is operated through the opening left in its face. The installation of a special, hardened steel key cylinder guard can overcome this kind of assault. Or cylinders should be flush or inset to prevent them from being wrenched out or "popped." Figure 8-2 illustrates common techniques of attacking doors and door frames.

Door hinges may also contribute to a door's weakness. Surface-mounted hinges with mounting screws or hinge pins exposed on the exterior side of the door can be removed and entrance gained on the hinge side. To complicate the matter, the door can be replaced on its hinges after the intruder has finished, and in most cases, the intrusion will never be detected. Without any visible sign of forced entry, very few insurance policies would pay off on the stolen merchandise.

To prevent this unhappy chain of events, hinges should be installed with the screws concealed and with the hinge pins either welded or flanged to prevent removal.

Locks and Keys

Attacks against Locks

Although direct forcible assault is the method generally used to gain entry, more highly skilled burglars may concentrate on the locks. This may be their only practical means of ingress if the door and the jamb are well designed in security terms and essentially impervious to forcible attack.

Picking the lock or making a key by impression are the methods generally used in attacks against traditional locking systems. Both require a degree of expertise. In the former method, metal picks are used to align the levers, pins, discs, or tumblers as an authorized key would, thus enabling the lock to operate. Making a key by taking impressions is a technique requiring even greater skill because it is a delicate, painstaking operation requiring repeated trials.

Because both of these techniques are apt to take time, they are customarily used to attack those doors where the intruder may work undisturbed and unobserved for adequate periods of time. The picked lock rarely shows any signs of illegal entry, and often insurance is not collectible.

Locks as Delaying Devices

The best defense against lock picking and making keys by impression is the installation of special pick-resistant, impression-resistant lock cylinders or the use of magnetic cards

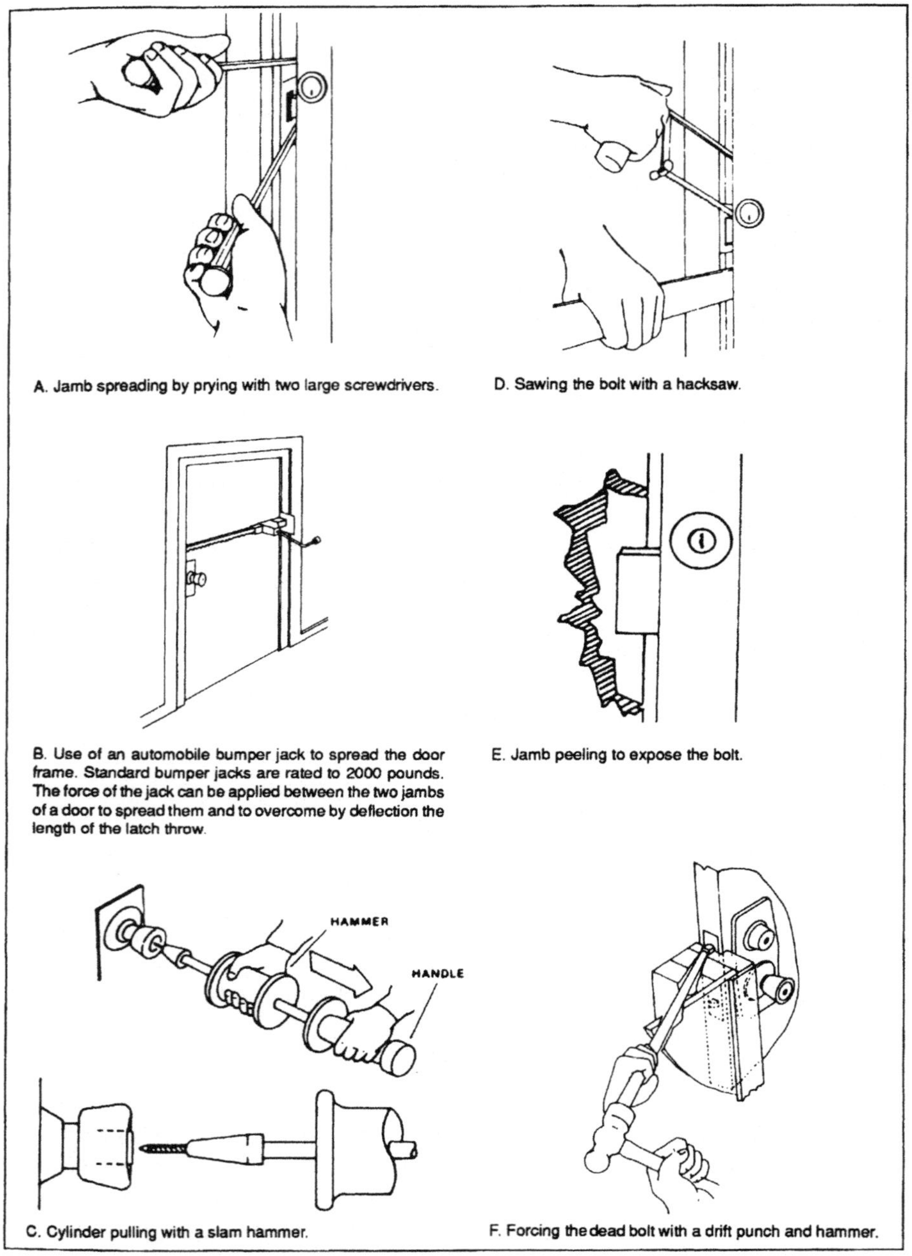

A. Jamb spreading by prying with two large screwdrivers.

D. Sawing the bolt with a hacksaw.

B. Use of an automobile bumper jack to spread the door frame. Standard bumper jacks are rated to 2000 pounds. The force of the jack can be applied between the two jambs of a door to spread them and to overcome by deflection the length of the latch throw.

E. Jamb peeling to expose the bolt.

C. Cylinder pulling with a slam hammer.

F. Forcing the dead bolt with a drift punch and hammer.

FIGURE 8-2 Common attack methods on doors and door frames. *(Reprinted with permission of National Crime Prevention Institute, School of Justice Administration of National Crime Prevention Institute, School of Justice Administration, University of Louisville, from Edgar et al., The Use of Locks [Boston: Butterworth-Heinemann, 1987], pp. 72–76.)*

(as commonly used in the hotel industry and discussed later in this chapter) in place of traditional keys. They are more expensive than standard cylinders but in many applications may well be worth the added cost. Generally speaking, in fact, locks are the cheapest security investment that can be made. Cost cutting in their purchase is usually a poor economy since a lock of poor quality is virtually useless and effectively no lock at all.

The elementary but often overlooked fact concerning locking devices is that in the first place they are simply mechanisms that extend the door or window into the wall that holds them. If, therefore, the wall or the door itself is weak or easily destroyed, the lock cannot be effective.

In the second place, it must be recognized that any lock will eventually yield to an attack. They must be thought of only as delaying devices. But this delay is of primary importance. The longer an intruder is stalled in an exposed position while he or she works at gaining entry, the greater are the chances of discovery. Since many types of locks in general use today provide no appreciable delay to even the unskilled prowler, they have no place in security applications.

Even the highest-quality locking devices are only one part of door and entrance security. Locks, cylinders, door and frame construction, and key control are inseparable elements; all must be equally effective. If any one element is weak, the system breaks down.

All locks are essentially composed of three parts: the operating mechanism, the keying device, and the latch or bolt. Any number of combinations may be involved in any lock, and to understand a lock, one must understand the variety of items that exist in each of the essential parts.

Latches and Bolts

The simplest latch is the spring lock. Its value as a security device is negligible. Spring locks are designed primarily as latching devices to hold a door closed for privacy. Since the latch is spring-loaded and has a tapered side so it will slide smoothly over the strike plate, it can easily be opened with a plastic or celluloid strip or a credit card.

The next type of latch is the dead latch. This latching device combines the advantages of the spring lock with a means of protecting the latch from carding. The dead latch is a simple device that holds the spring latch in position when the door is closed. When the door is unlocked, the device is in an open position to allow the latch to operate as a simple spring latch. But when the door is locked, the strike plate depresses the device so that the spring latch becomes "dead": it will no longer move unless the operating device manipulates it. The basic problem with the dead latch as a security device is that the overall length of the latch is still not long enough to keep the door from being forced open by prying between the door frame and the door.

To overcome the problem of springing doors, a dead bolt lock is needed (see Figure 8-3). The dead bolt gets its name from the fact that it does not have a tapered side and is dead in the door whether it is open or closed. The only way to manipulate a dead bolt is with an operating mechanism (key, electric switch, and so on). Dead bolts are generally long enough to overcome the problem of springing the door. Some intruders have attacked dead bolts with hacksaws, however, and cut their way through the brass alloy. To overcome this problem, the better dead bolts have a case-hardened pin in the center of the dead bolt to frustrate the use of a hacksaw.

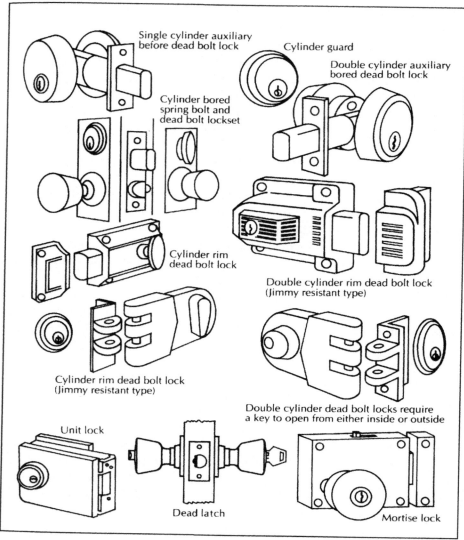

FIGURE 8-3 Common dead bolt locks. The bolt should extend at least 1 inch beyond the door edge. Other locks, however, that use an interlocking principle (for example, jimmy-resistant rim locks) also offer good security. If glass is within 40 inches of the lock, a double cylinder dead bolt (with keys needed for opening both sides) should be installed. This makes it impossible for a criminal to break the glass and reach inside to unlock the door. Be certain to have the key readily available so that fast exits are possible in the event of an emergency.

Keying Devices and Systems

Keying devices (which include lock and key) and the mechanisms they operate are many and varied in usage and style. A brief review of the types of keying and locking devices in general use and of their characteristics follows.

1. *Warded locks* are generally found in pre-World War II construction in which the keyway is open and can be seen through. These are also recognized by the single plate that includes the doorknob and the keyway. The security value of these locks is nil. These locks are only found in older construction and are becoming rare.

2. *Disc tumbler locks,* initially designed for use in the automobile industry, have been replaced in that industry with pin locks, combination and proximity devices (discussed later in this chapter). Because the disc tumbler lock is easy and cheap to manufacture, however, its use has expanded to other areas such as desks, files, and padlocks. The life of these locks is limited because of their soft metal construction. Although these locks provide more security than do warded locks, they cannot be considered very effective. The delay afforded is approximately 3 minutes.

3. *Pin tumbler locks* are in wide use in industry as well as in residences (see Figure 8-4). They can be recognized by the keyway, which is irregular in shape, and the key, which is grooved on both sides. Such locks can be master keyed in a number of ways, a feature that recommends them to a wide variety of industrial applications, although the delay factor is 10 minutes or less.

4. *Lever locks* are difficult to define in terms of security since they vary greatly in effectiveness. The best lever locks are used in safe deposit boxes and are for all practical purposes pickproof. The least of these locks are used in desks, lockers, and cabinets and are generally less secure than are pin tumbler locks. The best of this variety are rarely used in common applications, such as doors, because they are bulky and expensive.

Removable Cores

In facilities that require a number of keys to be issued, the loss or theft of keys is an ever-present possibility. In such situations, it might be well to consider removable cores on all locks. These devices are made to be removed if necessary with a core key, allowing a new core to be inserted. Since the core is the lock, this has the effect of rekeying without the necessity of changing the entire device, as would be the case with fixed-cylinder mechanisms.

Keying Systems

Keys are generally divided into change, submaster, master, and occasionally grand master keys.

1. *The change key:* One key to a single lock within a master-keyed system.

2. *The submaster key:* Will open all the locks within a particular area or grouping in a given facility. In an office, a submaster might open all doors in the accounting department; in an industrial facility, it might open all locks in the loading dock area. Typically, such groupings concern themselves with a common function, or they may simply be located in the same area even if they are not otherwise related.

3. *The master key:* Where two or more submaster systems exist, a master key system is established. Such a key would open any of the systems.

4. *The grand master key:* One that will open everything in a system involving two or more master-key groups. This system is relatively rare but might be used by a multipremise operation in which each location was master keyed, while the grand master would function on any premise.

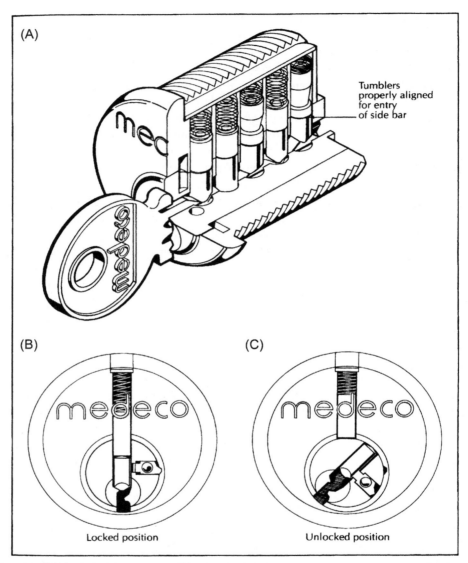

FIGURE 8-4 Pin tumbler lock. **(A)** A cutaway of a pin tumbler lock, showing the springs and tumblers. When the correct key is inserted into the lock, it will align all of the tumblers in a straight line to allow the plug to turn and operate the locking mechanism. **(B)** Locked position. Notice how the spring is forcing the tumbler to project partway into the inner core (plug) of the lock, making it impossible for the plug to rotate. **(C)** Unlocked position. The tumbler is now outside the plug, thereby allowing it to be rotated. *(Courtesy of Medeco Security Locks, Inc.)*

Obviously, master and submaster keys must be treated with the greatest care. If a master key is lost, the entire system is threatened. Rekeying is the only really secure thing that can be considered, but the cost of such an effort can be enormous.

Any master-key system is vulnerable. Beyond the danger of loss of the master itself and the subsequent staggering cost of rekeying—or, even more unfortunate, of the use of such a key by enterprising criminals to loot the facility—there is the problem that it necessarily serves a lesser lock. Locks in such a system are neither pick-resistant nor resistant to making a key by impression.

On the other hand, relative security coupled with convenience may make such a system preferable in some applications but not in others. Only the most careful evaluation of the particular circumstances of a given facility will determine the most efficient and effective keying system.

Rekeying

In any sizable facility, rekeying can be very expensive, but there are methods of lessening the disruption and staggering cost that can be involved in rekeying. Outer or perimeter locks can be changed first, and the old locks can be moved to interior spaces requiring a lower level of security. After an evaluation, a determination of priorities can be made and rekeying can be accomplished over a period of time, rather than requiring one huge capital outlay all at once. Of prime importance is securing keys so that such problems do not arise.

Key Management

Whether traditional keying systems or state-of-the-art digital operations, every effort should be exerted to develop ways whereby keys and access to keying information remain in the hands of security or management personnel. In those cases where this is not possible or practical, there must be a system of inventory and accountability. In any event, keys should be issued only to those demonstrably responsible persons who have compelling need for them. In the case of digital keying systems, management requires understanding the hardware, software, digital security and encryption, as well as a carefully developed program of information/key management. Though possession of keys and access to restricted areas is frequently a status symbol in many companies, management must never allow access on that basis.

Keys should never be issued on a long-term basis to outside janitorial personnel. The high employee turnover rate in this field would suggest that this could be a dangerous practice. Employees of this service should be admitted by guards or other building employees and issued interior keys that they must return before leaving the building.

By the same token, it is bad practice to issue entrance keys to tenants of an office building. If this is done, control of this vital security point is lost. A guard or building employee should control entry and exit before and after regular building hours. If keys must be issued to tenants, however, the lock cylinder in the entrance should be changed every few months and new keys issued to authorized tenants.

The security department must maintain a careful, strictly supervised record of all keys issued. This record should indicate the name and department of the person to whom the key was issued as well as the date of issue. In today's environment records should be kept on keys from creation to disposal.

A key depository for securing keys during nonworking hours should be centrally located, locked, and kept under supervision of security personnel. Keys issued on a daily basis or those issued for a specific, one-time purpose should be accounted for daily. Keys should be counted and signed for by the security supervisor at the beginning of each working day.

When a key is lost, the circumstances should be investigated and set forth in writing. In some instances if the lost key provides access to sensitive areas, locks should be changed. All keys issued should be physically inspected periodically to ensure that they have not been lost, though unreported as such.

Master keys should be kept to a minimum. If possible, submasters should be used, and they should be issued only to a limited list of personnel especially selected by management. Careful records should be kept of such issuance. The list should be reviewed periodically to determine whether all those authorized should continue to hold such keys.

Before a decision can be reached with respect to the master and submaster key systems and how such keys should be issued, there must be a careful survey of existing and proposed security plans, along with a study of current and planned locking devices. Where security plans have been developed with the operational needs of the facility in mind, the composition of the various keying systems can be readily developed.

Locking Schedules

Door-locking schedules and responsibilities must be established and supervised vigorously. The system must be set up in such a way that a procedure for altering the routine to fit immediate needs is possible, but in all respects the schedule, whether the master or the temporary plan, must be adhered to in every detail. A breakdown in such a system, especially in large offices, institutions, or industrial facilities, could represent just the opportunity an alert criminal is waiting for.

A Managed Key Program

There are perhaps three parts to a good managed key program: key creation, key usage and key breach policies.

Key creation should be controlled. Every new key should be recorded in a permanent record with the usual *who, what, where, when, why, and how* information. Key use in digital systems can be tracked. Traditional keys are generally understood in the context of security and are protected by the user. On the other hand many users do not understand the digital key and its vulnerabilities. Given recent technologies, encrypted information contained on key cards might be read by someone in close proximity without any actual physical contact with the key.

Key breaches are also a problem. It is possible to know when a traditional key system is breached or likely to be breached. If a key is lost or stolen, key management can make appropriate decisions regarding the need to rekey. However, when the breach is from lock compromise the attack might go unnoticed and never addressed. In the case of digital keys, the opposite is true. It may not be obvious that someone has stolen digital key information, but in systems that track key usage it is possible to know that someone accessed a restricted area through tracking software. Again, this information is of value only when audited.

Other Operating Mechanisms for Access Control

Besides the traditional key and lock, other mechanisms have been developed for access control purposes. These systems, as noted earlier, have in many instances replaced the traditional metal key. The following are commonly used in security applications.

1. *Combination locks* are difficult to defeat since they cannot be picked and few experts can so manipulate the device as to discover the combination. Most of these locks have three dials that must be aligned in the proper order before the lock will open. Some such locks may have four dials for greater security. Many also have the capability of having the combination changed quickly.

2. *Code-operated locks* are combination locks in which no keys are used. They are opened by pressing a series of numbered buttons in the proper sequence (see Figure 8-5). Some of them are equipped to sound an alarm if the wrong sequence is pressed. The combination of these locks can be changed readily. These are high-security locking devices. Because this type of lock can be compromised by "tailgating" (more than one person entering on an authorized opening), it should never be used as a substitute for a guard or receptionist.

3. *Card-operated locks* (see Figure 8-6) are electrical or, more usually, electromagnetic. Coded cards are about the size of a credit card. These frequently are fitted with a recording device that registers time of use and identity of the user. The cards serving as keys also serve as company identification cards. As with code-operated locks, tailgating can occur with this lock as well. In addition, the readers identify the card, not the individual. The hotel industry

FIGURE 8-5 Code-operated lock. *(Courtesy of KABA, www.kaba-ilco.com.)*

FIGURE 8-6 Card reader. *(Courtesy of FlexIso-MifareCard/Indala.)*

has been making the switch from traditional key systems to the electronic locking systems over the past 10 years. The advantages of these systems are many, including the following:

a. Security staff no longer need to spend hours rotating key cores and keeping detailed logs.
b. Card keys can be programmed to function in a variety of ways.
c. Lost cards can be deactivated in a matter of seconds.
d. The systems often allow hotels to keep track of the time and number of entries at a given site.[1]

There are several types of card-operated systems on the market.

a. *Magnetic coded cards* are of two basic designs. The first contains a flexible magnetic sheet sealed between two sheets of plastic. The second contains a magnetic strip along one edge of the card. The code is created by magnetizing spots on the sheet or strip. The code can be erased if it is exposed to a strong magnetic field. It is possible to duplicate the magnetic pattern and create false cards.
b. *Wiegan Effect cards* rely on short-length magnetic wires embedded within the card. Cards contain up to 26 wire bits, which make millions of code combinations possible. The card is immune to demagnetization and difficult to copy.
c. *Optical coded cards* contain bar codes similar to those found on products in most grocery stores. Early cards used the visible bar codes and were easy to duplicate. Today's product contains bar codes visible only under ultraviolet or infrared light.
d. *Proximity cards* (see Figure 8-7) do not need to be inserted into a reader or scanned. These cards send a code to a receiver via magnetic, optical, or ultrasonic pulses.
e. *Radio Frequency Identification (RFID)* is a form of proximity technology relying on radio frequency identification. The system has a signaling device (badge or tag) and readers. Coupled through an integrated system, data can be recorded and managed by computer systems that track all types of information such as time admitted and number of entrances versus exits. Frost and Sullivan, industry analysts, predicted that 80 percent of the access control market would be RFID-based by 2006. Today RFID badges have become commonplace due to the flexibility in the ability to recode the cards and associated hardware.

BACK   FRONT

FIGURE 8-7 Proximity tags. *(Photo courtesy of HID Corporation.)*

4. *Biometric systems* are designed to recognize biological features of the individual before access is granted. These systems are in fact identity verification systems that use personnel characteristics to verify identity. While these systems bring the James Bond/Jason Borg gadgetry to real systems, they are also currently handicapped with problems relating to the fact that physical characteristics of people do change with physical injuries, stress, and fatigue. There are several types of this state-of-the-art technology.

a. *Fingerprint recognition systems* (see Figure 8-8) optically scan a chosen fingerprint area and compare the scanned area with the file of the person to be admitted.

b. *Signature recognition systems* rely on the fact that no two people write with the same motion or pressure. Although forgers can duplicate the appearance of the signature, the amount of pressure and motions used in creating the signature will differ.

c. *Hand geometry recognition systems* (see Figure 8-9) use the geometry of the hand. The system basically measures finger lengths and compares them with the authorized files.

FIGURE 8-8 Fingerprint recognition system. *(Courtesy of KABA.)*

FIGURE 8-9 Hand geometry recognition. *(Courtesy of Handkey II Ingersoll Rand, www.handreader.com.)*

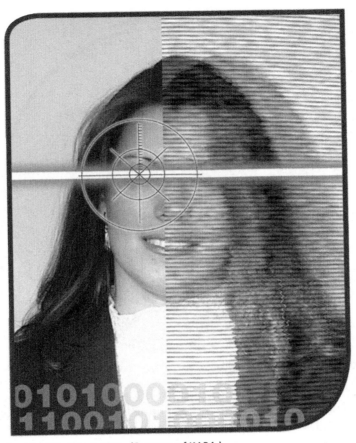

FIGURE 8-10 Eye retina identification system. *(Courtesy of KABA.)*

    d. *Speaker verification systems* use the uniqueness of voice patterns to determine identification and control admittance. The system uses soundproof booths and requires that the person to be identified repeat a simple phrase, usually four words in length.

    e. *Eye retina recognition systems* (see Figure 8-10) analyze the blood vessel pattern in the retina of the eye. These patterns vary widely even between identical twins. The chance of false identification using this system is one in a million.

    f. Facial recognition systems are among the latest entry

5. *Padlocks* are detachable, portable locks that have a shackle adapted to be opened for engagement through a hasp or chain. Padlocks should be hardened and strong enough to resist prying. The shackle should be close enough to the body to prevent the insertion of a tool to force it. No lock that will be used for security purposes should have fewer than five pins in the cylinder. Padlocks can be supplied with a function that prevents the withdrawal of a key until the lock is closed.

FIGURE 8-11 Modern high-security lock with e-key access. *(Photo courtesy of LA GARD, INC, A MASCO Company.)*

It is important to establish a procedure requiring that all padlocks be locked at all times even when they are not securing an area. This will prevent the possibility of the lock being replaced by another to which a thief has the key.

The hardware used in conjunction with the padlock is as important as is the lock itself. It should be of hardened steel, without accessible screws or rivets, and bolted through the door to the inside, preferably through a backing plate. Shackles should be forged of hardened steel, ⅜-inch in both the heel and toe. The bolt ends should be burred.

6. *High-security locks* (see Figure 8-11). Virtually every lock manufacturer makes some kind of special high-security lock that is operated by nonduplicable keys. A reliable locksmith or various manufacturers should be consulted in cases of such need.

7. The *iButton* is an extension of the "smart card" technology being used by the banking industry. The iButton contains a hermetically sealed computer chip in a stainless steel container. The iButton can secure information and can provide information for security.[2]

It appears that as the technology allows, old systems are being supplanted by technology. One example of the ability to use old systems with new technology is the "electronic key." This key looks and is used like a regular metal key, but contains the electronic characteristics of a smart card. The key does not have any cuts: it is an electronic key that can be fitted to existing door hardware.[3]

A 1996 study predicted that the then popular existing magnetic strip cards would be phased out shortly after 2000. The same study predicted that in 2000 there would be 550 million smart cards.[4] In 2010 various versions of smart cards, RFID cards and other similar technologies dominate large firm access control technology. The initial uses were in replacing store cards. However, this high-tech card now is accepted by credit card companies, banks, and others. Over half of all transactions in retailing and banking are now handled by smart card technology. Other uses include access control for physical security, computer information access control, and health care cards. Other uses include access to mobile telephone networks and multipurpose ID cards in colleges, universities, and the workplace.

Locking Devices

In the previous list we considered the types of locks that are generally available. It must be remembered, however, that locks must work in conjunction with other hardware that affects the actual closure. These devices may be fitted with locks of varying degrees of security and may themselves provide security to various levels. In a security locking system, both of these factors must be taken into consideration before determining which system will be most effective for specific needs.

1. *Electromagnetic locking devices* hold doors closed by magnetism. These electrical units consist of an electromagnet and a metal holding plate. When the power is on and the door secured, they will resist a pressure of up to 1,000 pounds. A high frequency of mechanical failures with this type of lock can create problems. Inconvenienced employees will often block the door open or jam the door-bolting mechanism so that the lock no longer operates. Quality equipment, preventive maintenance, frequent inspections, and quick response to problems will minimize these problems.

2. *Double-cylinder locking devices* are installed in doors that must be secured from both sides, requiring a key to open them from either side. Their most common application is in doors with glass panels that might otherwise be broken to allow an intruder to reach in and open the door from the other side. Such devices cannot be used in interior fire stairwell doors since firemen break the glass to unlock the door from the inside in this case.

3. *Emergency exit locking devices* are panic-bar installations allowing exit without use of a key. This device locks the door against entrance. Because such devices frequently provide an alarm feature that sounds when exit is made, they are fitted with a lock that allows exit without setting of the alarm when a key is used.

4. *Recording devices* provide for a printout of door use by time of day and by the key used.

5. *Vertical throw devices* lock into the jamb vertically instead of the usual horizontal bolt. Some versions lock into both jamb and lintel. A variation of this device is the police lock, which consists of a bar angled to a well in the floor. The end of the bar contacting the door is curved so that when it is unlocked it will slide up the door, allowing the door to open. When it is locked, it is secured to the door at one end and set in the floor at the other. A door locked in this manner is virtually impossible to force.

6. *Electric locking devices* are installed in the same manner as are other locks. They are activated remotely by an electric current that releases the strike and thus permits entrance. Many of these devices provide minimal security since the engaging mechanisms frequently offer no security feature not offered by standard hardware. The electric feature provides a convenient method of opening the door; it does not in itself offer locking security. Because such doors are usually intended for remote operation, they should be fitted with a closing device.

7. *Sequence locking devices* are designed to ensure that all doors covered by the system are locked. The doors must be closed and locked in a predetermined order. No door can be locked until its designated predecessor has been locked. Exit is made through the final door in the sequence, and entry can be made only through that same door.

While traditional locking devices and systems continue to dominate many security operations, the use of electronic access controls continues a trend that is projected to only increase. According to the Freedonia Group, this market will continue to increase at a rate of 11.9% through 2012 with expenditures of approximately $6 billion.[5]

Roofs and Common Walls

An important though often overlooked part of the perimeter is the roof of the building. In urban shopping centers or even in small, freestanding commercial situations where the building walls are the perimeter, entry through the roof is common. Entry can be made through skylights or by chopping through the roof—an activity rarely detected by passersby or even by patrols.

Buildings sharing a common wall have also frequently been entered by breaking through the wall from a poorly secured neighboring occupancy. All of these means of entry circumvent normal perimeter alarm systems and can therefore be particularly damaging.

Surveillance Devices

Surveillance of a facility both internally and externally has traditionally been conducted by patrolling security personnel who watch for any signs of criminal activity. If they spot any trouble, they are in a position to take such action as necessary. Patrols cannot be everywhere, however, and with the present emphasis on cost-effectiveness, other methods have been introduced to supplement or replace patrols. A wide variety of surveillance devices, including motion picture cameras, sequence cameras, and closed-circuit television (CCTV) monitors with video and digital cameras, are being used. However, the traditional systems have rapidly been replaced with digital technology.

Effective surveillance systems are expected to produce two possible end results. First, a good system should produce an identifiable image of persons engaging in criminal behavior or violating company policy. Second, the system should also serve as a deterrent. Although there is no way to determine how many attempts are discouraged because of the presence of the system, one definite advantage is that surveillance systems generally mean lower insurance rates.

The major factor limiting the use of surveillance devices is the cost of installation and maintenance. In addition, some companies worry about the possible negative impact of these

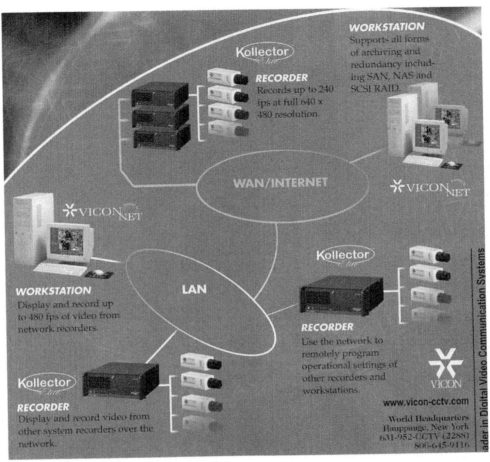

FIGURE 8-12 CCTV systems. *(Courtesy of Vicon Industries.)*

systems on employee morale, although this is becoming increasingly rare as cameras become commonplace not only in the workplace, but in public areas as well.

CCTV systems are the state-of-the-art surveillance devices and in most cases have replaced still and motion picture systems. The CCTV systems coupled with recording (VCR) equipment or computers (digital systems) are exceptionally flexible (see Figure 8-12). The tapes or digital records can be erased and reused, a definite cost savings in comparison with other systems. In today's market digital has rapidly replaced VHS, which is viewed as too limiting—maintenance intense, remote accessibility problems, and difficult to integrate with other systems.[6] Problems associated with VHS are reduced or not present with digital technology.

Current technology developments have produced CCTV that is used for laparoscopic surgery. Improved lens design has also produced cameras that can identify objects within a ¾-square foot from more than 100 miles in space. With the improved technology, the CCTV has become one of the most sought-after systems in the security market.

FIGURE 8-13 The new first-generation smart camera. *(Courtesy of Pelco, www.pelco.com.)*

Technological advances allow CCTV technology to be used where it would not have been effective 10 years ago—in areas such as loading docks and automatic teller machines. The reduced size of the camera also allows for a greater number of applications (see Figure 8-13). Cameras may easily be placed in covert locations such as wall receptacles, clocks, and mannequins. The reduced price and improved reliability of color cameras have enabled banks, retail stores, museums, and others to add color evidence to their security capabilities.

The "starlight" cameras allow for good video reproduction at 0.0001 lux (the amount of light produced by stars on a clear night) compared to the previous 0.1 lux level. CCTV technology adds the ability to use thermal cameras with current systems. These cameras detect and transmit heat images. They will work in the light or in pitch-black environments.

Finally, the ability to network a widespread surveillance system and monitor multiple sites remotely from a central location is a reality (see Figure 8-14). Remote video management systems (RVMS) are now being used to monitor multiple VCRs and video-signaling devices at

FIGURE 8-14 Control console. *(Courtesy of Winsted Corp.)*

thousands of locations.[7] Modern integrated surveillance systems through the use of digital cameras connected worldwide via wide area networks (WANs), local area networks (LANs), and the Internet have truly expanded and revolutionized monitoring capabilities.

The above improvements in CCTV technology have resulted in a move to replace or augment existing security systems with CCTV. As the demand for digital cameras continues to increase, analog CCTV equipment will eventually be phased out of use. This trend is predictable as security operations continue to integrate video, alarm systems, and access control into a seamless operation. Networking is becoming the technology to build complete security systems that allow central stations to monitor operations around the world.

Once a decision has been made to purchase a system, careful planning must precede the purchase. Poor planning generally means wasted funds and a system that does not do the required job. Several questions should be asked before any purchase, among them the following:

- Is the camera to be visible and used as a deterrent to crime or hidden and used in civil or criminal prosecutions? Most businesses would rather prevent a crime than go to the effort and expense of prosecution and therefore prefer visible camera locations. In addition, hidden camera sites cost more, because there is not only the investment in the camera, but also expenses for hiding the camera.
- What effect, if any, will the sun have on the operation of the system? Sunlight is variable in intensity, and good light conditions may deteriorate as the day progresses into dusk. In addition, sunlight can cause glare. A CCTV system may allow for changes in the setting of the recording cameras to help adjust for changes in light intensity.
- Where is the best location for a camera? In banks, for instance, placement might be where customers do not immediately notice the camera. In many cases, this is accomplished by placing the camera over the exits. This permits narrow-angle coverage of an area where the subject must approach the camera directly at a time when he may be comparatively off-guard (for example, you might catch a bank robber removing his disguise). In these cases, the teller does not have to signal the camera until the robber is on the way out of the bank.

When the camera is placed to photograph the teller/cashier area, attempting to trigger the camera manually may endanger the employee.

- Should the placement of the camera be high? High placement is not as efficient as it is often thought to be. Persons photographed from high locations may not be recognizable. A good spot is just high enough to see over obstacles and to protect the camera from curious observers.
- What type of lighting is in use? Sodium systems produce poor color accuracy. Ideally the camera system and lighting plan should be coordinated to allow for the most accurate recording of details.

A site survey is essential to effective planning. This survey is generally presented in the form of a diagram with the areas to be protected drawn to scale. The diagram should include blind spots, areas of high loss potential, exits, windows, cash registers, electrical outlets, and other data significant to the site. Lighting requirements can be determined by using an illumination meter. Record the information on the diagram, and measure illumination at both the brightest and the darkest times of the day. Levels of light should generally not be below 20 to 40 or above 250 foot-candles (fc). If the level falls below 20 fc, you will need additional lighting or other camera equipment; if it rises above 250 fc, you will need special filters. Study the traffic flow to discover the greatest usage.

The future might increase the role of surveillance into the area of decision making. Manufacturers are developing cameras that not only watch and record, but also interpret what they see. There are hints that cameras may soon be able to detect unattended baggage at airports, guess a person's weight or analyze the way you walk.[8] The European Union is funding research in this area to include cameras and robots. According to Richard Bowden, University of Surrey professor, "if you have a robot with a camera that looks down a road and it knows it is normal behavior for people to just walk along, then it will know that if somebody shimmies up a drainpipe, it is something the system has not seen before."[9]

Cameras systems are becoming an increasing part of a company's and community's program of protection. A Philadelphia neighborhood recently installed a $120,000 high-tech camera system that can detect suspicious activity, photograph faces and license plates from three blocks away and eventually send real-time video straight to a police squad car.[10]

In Baltimore where 350 cameras were installed in 2005, violent crime in the areas with cameras has fallen 15%.[11]

Old Construction

Older buildings—particularly, though certainly not exclusively, office buildings—present a host of different and difficult security problems. Exterior fire escapes, old and frequently badly worn locks, common walls, roof access from neighboring buildings, unused and forgotten connecting doors—all increase the exposure to burglary.

It is vital that all such openings are surveyed and plans made for securing them. Those windows not designated as emergency exits must be barred or screened. Where windows lead to a fire escape or are accessible to adjacent fire escapes, their essential security must be accomplished within the regulations of the local fire codes. Fire safety must be a primary consideration. In cases where prudence or the law (or both) dictate that locks would be a hazard to safety, windows should be alarmed and the interior areas to which these windows provide access must be further secured. Here, security can be likened to any army retreating to secondary or tertiary lines of defense to establish a strong and defensible position.

It is also well to consider the danger of attack from neighboring occupancies in shared space where entry might be made from a low-risk, badly secured premise into a higher-risk area that might otherwise be well protected against a more direct attack.

New Construction

Modern urban buildings, though security conscious in varying degrees, present their own problems. Most interior construction is standardized. Fire and building codes are such that corridor doors can resist most attacks if the hardware is adequate. Corridor ceilings are fixed, and entrances to individual offices usually offer a fairly high degree of security.

On the other hand, modern construction creates offices that are essentially open-top boxes. They have solid exterior walls (though interior walls are frequently plasterboard) and a concrete floor. But nothing of any security value protects the top. The ceiling is simply a layer of acoustical tiles lying loose on runners suspended between partition walls. In the space above these tiles—between them and the concrete slab above—are vital air conditioning ducts and wiring for power and telephones.

In effect, any given floor of a building has a crawl space that runs from exterior wall to exterior wall. This may not be literally so in every case, but the net result stands. It means that virtually every room and every office is accessible through this space. Once this crawl space is reached from any occupancy, the remaining offices on that floor are accessible.

Extending dividing walls up to the next floor will not solve the problem, because this drywall construction is easily broken through and, in any case, it must be breached to allow passage of all utilities. Alarms of various kinds, which are discussed later in this book, are recommended to overcome this problem.

Security at the Building Design Stage

Once a building has been constructed, the damage has been done. Security weaknesses begin to manifest themselves, but it is far too expensive to make basic structural changes to correct them. Guard services and protective devices that might not otherwise have been necessary must be instituted. In any event, there will be some considerable expense for protection that could easily have been incorporated into the design of the building before construction began. This kind of oversight can be very expensive indeed.

Unfortunately, we have not yet arrived at the point where the need for security from criminal acts is as automatic a consideration as is the need for efficiency or profits. Architects' interest in design for protection of buildings and grounds is usually minimal. They traditionally leave such demands to their clients, who usually are unaware of the availability of protective hardware and who are rarely competent to deal with the problems of protective design.

This situation may be changing following the growing apprehension of possible terrorist attacks. Events such as the destruction of the World Trade Center, as well as attacks on buildings in many other countries, may prompt designers to consider the incorporation of security in their designs. The construction industry, in response to issues raised in the review of the failure of the twin towers of the World Trade Center, has revised its MasterFormat system to include more security concerns. The MasterFormat system is the work of the Construction Specifications Institute.

In addition, the growing awareness of potential crime problems has directed more attention toward the important role that building design can play in security. There have been some efforts on the part of the federal government to accentuate the architect's role in security.

Under the umbrella of environmental security, concepts of crime prevention through environmental design (CPTED) have received added attention in recent years. Early work in this field concentrated on residential security, particularly in public housing, with Oscar Newman's major study of "defensible space" being a pioneering work.[12]

This approach to crime prevention through environmental design has important implications for private security. It seeks to bring together many disciplines—among them urban planning, architectural design, public law enforcement, and private security—to create an improved quality of urban life through crime prevention. And in particular it encourages awareness of crime-prevention techniques through physical design.

Security Principles in Design

Certain principles should always be considered in planning any building. Without them, it can be dangerously vulnerable. Some areas of consideration are listed below.

1. The number of perimeter and building openings should be kept to a minimum consistent with safety codes.
2. Perimeter protection should be planned as part of the overall design.
3. Exterior windows, if they are less than 14 feet above ground level, should be constructed of glass brick, laminated glass, or plastic materials, or they should be shielded with heavy screening or steel grilles.
4. Points of possible access or escape that breach the exterior of the building or the perimeter protection should be protected. Points to be considered are skylights, air-conditioning vents, sewer ducts, manholes, or any opening larger than 96 square inches.
5. High-quality locks tied to smart card technology should be employed on all exterior and restricted area doors for protection and quick-change capability in the event of cardkey loss.
6. Protective lighting should be installed.
7. Shipping and receiving bays should be widely separated from each other.
8. Exterior doors intended for emergency use only should be fitted with alarms.
9. Exterior service doors should lead directly into the service area so that nonemployee traffic is restricted in its movement.

10. Dock areas should be designed so that drivers can report to shipping or receiving clerks without moving through storage areas.
11. Employment offices should be located so that applicants either enter directly from outside or move through as little of the building as possible.
12. Employee entrances should be located directly off the gate to the parking lot.
13. Employee locker rooms should be located by employee entrance and exit doors.
14. Doors in remote areas should be fitted with alarms.

Of growing interest is the need for some type of standardization of products in both the surveillance and access control areas. The need for interoperability between network video products and access control drives the desire for hardware and software standardization. ONVIF reports that standardization would benefit system integrators, manufacturers and, of course, end users.[13]

Summary

While the basics of interior and exterior security theory remain constant, the tools used to establish the systems have improved. Systems that integrate all aspects of security are becoming commonplace. Smart cameras work with monitors, switchers, recorders, access control devices, and computers to allow operators to efficiently monitor and control a multitude of locations from one central station operation. Biometric technology implementation is growing at a rapid pace and should continue to be an area of advancement in the security industry. Geoff Kohl, editor, SecurityInfoWatch.com noted in 2007 that "our industry (security) is still heavily focused on gates, fences, analog cameras, guards and old reed-style contacts for intrusion detection. While... mag strip cards and security fencing may still define much of commercial security, it's rapidly moving beyond that. The industry as a whole is paying a lot more attention to IT."[14] Along with the interest in technology is the desire to integrate security systems into one interactive operation that includes: access control, surveillance, assets tracking and employee monitoring. And along with the security industry's interest will be increased interest from the criminal element. According to Ray Bernard, "The security industry has changed. It will be the rule that hacker's conferences will include sessions on how to hack physical security systems."[15]

CASE STUDY

You are the security director for IBID International, a large manufacturer. You have decided to conduct a security survey of the company's administrative building. You have received the security survey report form (see Appendix C) and a blueprint of the building (see Figure 8-15). Your approved recommendation will be presented to senior management for final approval and funding. Fill out the survey form. As you do so, look over the blueprint and note directly on your blueprint what you would do and where you would make security improvements in the building and its interior.

You should succinctly, clearly, and logically draw together your findings on the physical security survey. This narrative should briefly provide an overview of each problem and its recommended solution or correction.

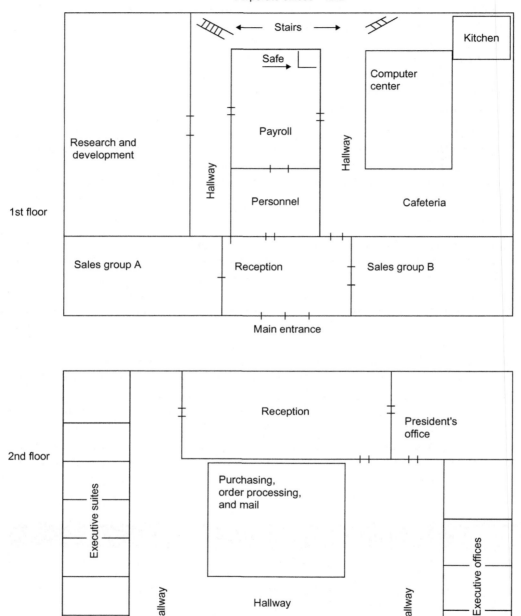

FIGURE 8-15 Floor plan.

Review Questions

1. What factors need to be considered when you are purchasing and installing locking devices for security purposes?
2. Describe the basic principles of an effective key control plan.
3. What are the two end results of an effective surveillance system?

References

[1] Beaudry M. Locking in hotel security. Secur Manage November 1996;37.

[2] Access and beyond: read/write tokens open new possibilities. Security December 1996;23.

[3] Electronic keys: hardware feel, high-tech benefits. Security August 1996;19.

[4] Market for smart cards to increase 8-fold by 2000. Security January 1996;15–6.

[5] Kosk N. Access control: leveraging the legacy. Access Control Trends Technol May/June 2010;p. S-8.

[6] Security surveillance and monitoring systems to 2000. April 1996, <www.freedoniagroup.com>.

[7] Messenbrink J. The Digital rEVOLUTION. Security, downloaded 1/14/2003, <www.securitymagazine. com/security/cda/articleinformation/coverstory/bnpcoverstoryitem/0,5409,82693,00.html>.

[8] <www.boston.com/business/technology/articles/2007/02/26/surveillance_cameras_latest_job_interpret_ the_threats_they_see/>.

[9] Security management daily, February 28, 2007 excerpted from Engineer (02/26/2007).

[10] Security management daily, February 28, 2007 excerpted from Philadelphia Inquirer (02/26/2007).

[11] Security management daily, February 28, 2007 excerpted from Columbus Dispatch (02/25/2007).

[12] Newman O. Defensible space: crime prevention through urban design. New York: Macmillan; 1973.

[13] ONVIF highlights in store. Access control trends and technology, May/June 2010. p. S-10.

[14] Kohl G. The week that was: a recap – February 3–9, 2007, listserve e-mail to rjfish@macomb.com, 2/09/2007.

[15] Security Technology Executive, May 10, 2010. p. 15.

9

The Outer Defenses
Building and Perimeter Protection

OBJECTIVES

The study of this chapter will enable you to:

1. Recognize the basic tools of perimeter security.
2. Discuss the various types of lighting fixtures and systems.
3. Know the specifics in designing various types of barrier protection.

Introduction

Delay and deny are two of the major tenets of good security system design, beginning at a planned outer perimeter premises barrier. These premises should then be further protected from criminal attack by denying ready access to interior spaces in the event that a determined intruder surmounts exterior controls. This must be the first place to start in security planning.

This basic concept is being applied to facilities such as water treatment plants and electric substations, where in the past security was rarely considered. Basic security protection is being implemented by these operations as a result of federal guidelines established following the events of 9–11. For additional information on these guidelines see Chapter 5.

Each security program element should be interconnected into an integrated whole program. Each element must grow out of the specific needs dictated by the circumstances affecting the facility to be protected. The first and basic defense is still the physical protection of the facility. While planning this defense is neither difficult nor complicated, it does require meticulous attention to detail.

The development of computer protection programs, anti-embezzlement systems, or even the establishment of shipping and receiving safeguards generally requires some particular sophistication and expertise. The implementation of an effective program of physical security most often starts with the application of common sense and a lot of legwork expended in the inspection of the facility and surrounding area.

Physical security concerns itself with those means by which an organization protects its facilities against intrusion, theft, vandalism, sabotage, unauthorized entry, fires, accidents, and natural disasters. And in this context, a *facility* might be a plant, building, office, institution, or any commercial or industrial structure or complex with all the attendant structures and functions that comprise an integrated operation. An international manufacturing operation, for example, might have many facilities within its total organization.

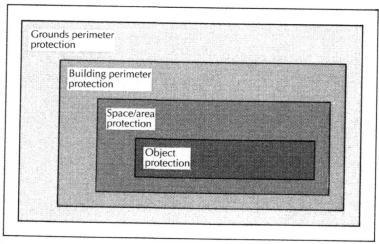

FIGURE 9-1 The four lines of protection. *(From Don T. Cherry,* Total Facility Control *[Boston: Butterworth-Heinemann, 1986], p. 100.)*

With the ever-increasing sophistication of criminals, it is evident that the use of the burglar alarm and padlocks alone may not be enough to constitute a good security program. Good systems combine controlled access with advanced security technologies such as intrusion detection sensors, video monitoring, and perhaps smart cards and biometrics. When these security elements are integrated through physical security information management (PSIM) platforms along with sound security policies, you have an excellent beginning to a total asset protection program.

Physical security planning includes protection of (1) the grounds around the building, (2) the building's perimeter surfaces, (3) the building's interior, and (4) its contents. Figure 9-1 illustrates these four lines of protection.

Barriers, Fences, and Walls

A facility's perimeter will usually be determined by the function and location of the facility itself. An urban office building or retail enterprise will frequently occupy all the real estate where it is located. In such a case, the perimeter may well be the walls of the building itself. Most industrial operations, however, require yard space and warehousing even in urban areas. In that case, the perimeter is the boundary of the property owned by the company. But in either case, the defense begins at the perimeter—the first line that must be crossed by an intruder.

Barriers

Natural and structural barriers are the elements by which boundaries are defined and penetration is delayed or deterred. Natural barriers comprise the topographical features that assist in impeding or denying access to an area. They may consist of rivers, cliffs, boulders, canyons, dense growth, or any other terrain or feature that is difficult to traverse. Structural barriers are permanent or temporary devices such as fences, walls, grilles, doors, roadblocks, screens, or any other construction that will serve as a deterrent to unauthorized entry.

It is important to remember that structural barriers rarely if ever prevent penetration. Fences can be climbed, walls can be scaled, and locked doors and grilled windows can eventually be bypassed by a resolute assault.

The same is generally true of natural barriers. They almost never prevent a determined and resourceful criminal from intrusion. Ultimately all such barriers must be supported by additional security layers. Structural barriers of some kind should further strengthen most natural barriers. It is a mistake to suppose that a high, steep cliff, for example, is by itself protection against unauthorized entry.

Fences

The most common type of structural barrier, familiar to most, is the fence. The most common type of fencing normally used for the protection of a facility is chain link. Barbed wire and concertina barbed wire are often added to the top of the fence to deter attempts to climb the barrier. In situations where aesthetics is the driving force, there may be fencing that is not visible. In this situation the barrier is not physical, but rather it is composed of some type of sensing system, and the governing physical security design principle shifts from "deter or delay" to "rapid detection."

Chain Link

Chain-link fencing should meet the specifications developed by the U.S. Department of Defense in order to be fully effective. (See Table 9-1 for common characteristics of chain-link fences.) It should be constructed of a 9-gauge or heavier wire with twisted and barbed selvage top and bottom. The fence itself should be at least 6 feet tall and should begin no more than 2 inches from the ground. The bottom of the fence can be stabilized against crawling under or

Table 9-1 Common Chain-Link Fence Characteristics

Characteristic	Option
Gauge	#9 (3.8 mm), #11 (3.0 mm)
Mesh	2 in. (50 mm), 1.6 in. (40 mm), 2.4 in. (60 mm)
Coating	vinyl, galvanized
Tension wires	wire, rail, cable (attached at top or bottom)
Support posts	metal posts (see Federal Specifications RR-F-191H/GEN and RR-F-191/33)
Height	6 ft. (1.8 meters), 7 ft. (2.1 meters), 8 ft. (2.4 meters)
Fabric tie-downs	buried, encased in concrete, staked
Pole reinforcement	buried, encased in concrete
Gate opening	swing, slide, lift, turnstile

Source: Gary R. Cook, "The Facts on the Fence," *Security Management* (June 1990): 86.

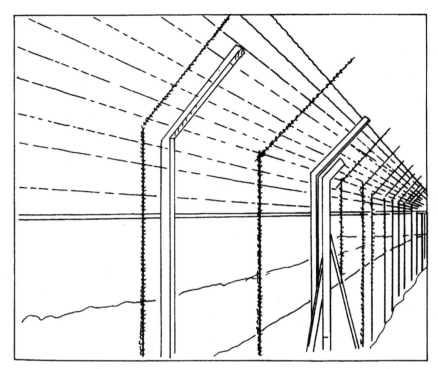

FIGURE 9-2 D.T.R. taut-wire intrusion detection systems—the solution for airport security and high-risk facilities. *(Courtesy of Safeguards Technology, Inc.)*

lifting by tying it to rigid metal poles or concrete sills. The sills are usually precast with AWF #9 wires for ties. If the soil is sandy or subject to erosion, the bottom edge of the fence should be installed below ground level. The fence should be stretched and fastened to rigid metal posts set in concrete with additional bracing as necessary at corners and gate openings. Mesh openings should be no more than 2 inches square. In addition, the fence should be augmented by a top guard or overhang of three strands of stretched barbed wire angled at 45 degrees away from the protected property (see Figure 9-2). This overhang should extend out and up far enough to increase the height of the fence by at least 1 foot to an overall height of 7 feet or more.

To protect the fence from washouts or channeling under it, culverts or troughs should be provided at natural drainage points. If any of these drainage openings are larger than 96 square inches, they too should be provided with physical barriers that will protect the perimeter without, however, impeding the drainage.

High-strength welded metal or wire mesh fencing solutions are common, although more expensive, alternatives to chain-link fencing. They are generally made of heavier gauge metal and therefore offer greater ability to resist cutting. The mesh spacing may sometimes be smaller than chain link, in which case they become more difficult to climb.

If buildings, trees, hillocks, or other vertical features are within 10 feet of the fence, it should be heightened or protected with a Y-shaped top guard.

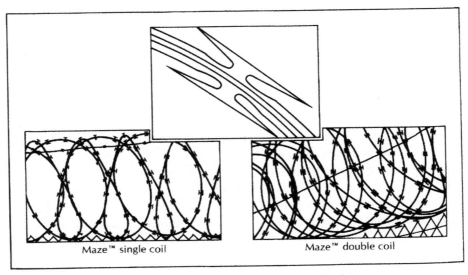

Maze™ single coil   Maze™ double coil

FIGURE 9-3 Concertina wire. *(Courtesy of American Security Fence Corporation.)*

Another extension of chain-link technology has been developed by Amico (Alabama Metal Industries Corporation). The fence, similar in appearance to traditional chain-link fencing, is actually a solid sheet of steel with holes cut in it. The construction makes this fence a less attractive target for many thieves. For more information see www.amico-online.com.

Some security plans use the standard chain-link fence and incorporate monitoring technology, putting the monitoring cables into the fence fabric. This allows early detection of climbing or cutting of the fence fabric. The fence is not noticeably different from standard chain-link fencing except in its built-in telemetry.[1]

Barbed Wire

When a fence consists of barbed wire, a 13.5-gauge, twisted double strand with 4-point barbs 4 inches apart is generally used. These fences, like chain-link fences, should also be at least 6 feet high and they should in addition carry a top guard. Posts should be metal and spaced no more than 6 feet apart. Vertical distance between strands should be no more than 6 inches, preferably less.

Concertina Wire

Concertina wire is a coil of steel wire clipped together at intervals to form a cylinder (see Figure 9-3). When it is opened, it forms a barrier 50 feet long and 24, 30, 36, 40, or 60 inches high. Developed by the military for rapid laying, it can be used in multiple coils. It can be used either with one roll atop another or in a pyramid with two rolls along the bottom and one on the top. Ends should be fastened together, and the base wires should be staked to the ground. Concertina wire is probably the most difficult fence to penetrate, but it is unsightly and is rarely used except in a temporary application, although in recent years, concertina wire has replaced traditional barbed wire as fence "top guards."

With the exception of concertina wire, most fencing is largely a psychological deterrent and a boundary marker rather than a barrier. In most cases such fences can be rather easily penetrated unless added security measures are taken to enhance the security of the fence. Fences may deter the undetermined, but they will only delay the determined.

Walls

In some instances, masonry, stone, brick, or block walls may be used to form all or part of the perimeter barrier. Such walls may be constructed for aesthetic or strength reasons to replace the less decorative barriers within that part of the facility.

In those areas where masonry walls are used, they should be at least 7 feet high with a top guard of three or four strands of barbed wire, as in the case of chain-link fences.

Since concealment of the inside activity must be paid for by also cutting off the view of any activity outside the wall, extra efforts must be made to prevent scaling the wall. Ideally, the perimeter line should also be staggered in a way that permits observation of the area in front of the wall from a position or positions inside the perimeter. (Think Great Wall of China.)

Gates and Other Barrier Breaches

Every opening in the perimeter barrier is a potential security hazard. The more gates, the more security personnel or surveillance devices must be deployed to monitor the traffic through them. Obviously, these openings must be kept to the minimum number needed to support the workflow of the organization and to remain in compliance with any prevailing regulations or codes.

During shift changes, there may be more gates open and in use than at other times of the day when the smallest practical number of such gates are in operation. Certainly there must be enough gates in use at any time to facilitate the efficient movement of necessary traffic. The number can only be determined by a careful analysis of needs at various times of day, but every effort must be made to reduce the number of operating gates to the minimum, balancing safety concerns with business efficiencies.

If some gates are not necessary to the operation of the facility or if changing traffic patterns can eliminate them, the openings should be sealed off and retired from use. Barriers can also include walk-in entrances where vehicle access must be restricted (see Figure 9-4).

Padlocking

Gates used only at peak periods or emergency gates should be padlocked or otherwise secured. If it is possible, the lock used should be distinctive and immediately identifiable. It is common for thieves to cut off the plant padlock and substitute their own padlock so they can work without alarming a passing patrol that might have spotted an otherwise missing lock. A lock of distinctive color or design could compromise this ploy.

It is important that all locked gates be checked frequently. This is especially important where, as is usually the case, these gates are out of the current traffic pattern and are remote from the general activity of the facility.

FIGURE 9-4 Raised stainless steel bollards protect access to the U.S. Department of State headquarters. *(Courtesy of Delta Scientific, Inc.)*

Personnel and Vehicle Gates

Personnel gates are usually from 4 to 7 feet wide to permit single-line entry or exit. It is important that they not be so wide that control of personnel is lost. Vehicular gates, on the other hand, must be wide enough to handle the type of traffic typical of the facility. They may handle two-way traffic, or if the need for control is particularly pressing, they may be limited to one-way traffic at any given time. Good security planning must account for the types of vehicles that will be using the gate. Gates needed to accommodate large cargo trucks will be different from those needed for small passenger cars or motorcycle only traffic.

A drop arm or railroad crossing type of barrier is normally used to cut off traffic in either direction when the need arises. The gate itself might be single or double swing, rolling or overhead. It could be a manual or an electrical operation. Railroad gates should be secured in the same manner as other gates on the perimeter except during those times when cars are being driven through them. At these times, the operation should be under inspection by a security guard.

Another effective traffic direction control device is the spiked security track that allows for one-way traffic to egress the facility, but is designed to puncture tires of vehicles attempting to travel into the facility (unless the spikes are retracted by security personnel).

Miscellaneous Openings

Virtually every facility has a number of miscellaneous openings that penetrate the perimeter. All too frequently these are overlooked in security planning, but they must be taken into account because they are frequently the most effective ways of gaining entrance into the facility without being observed.

These openings or barrier breaches consist of sewers, culverts, drainpipes, utility tunnels, exhaust conduits, air intake pipes, maintenance hole covers, coal chutes, and sidewalk elevators. All must be accounted for in the security plan.

Bars, grillwork, barbed wire, or doors with adequate locking must protect any one of these openings having a cross-section area of 96 square inches or more. Sidewalk elevators and maintenance hole covers must be secured from below to prevent their unauthorized use. Storm sewers must be fitted with deterrents that can be removed for inspection of the sewer after a rain.

Barrier Protection

In order for the barrier to be most effective in preventing intrusion, it must be patrolled and inspected regularly. Any fence or wall can be scaled, and unless these barriers are kept under observation, they tend to neutralize the security effectiveness of the structure. Such observation can be in person, through video surveillance, or through other perimeter incursion detection sensors.

Clear Zone

A clear zone should be maintained on both sides of the barrier to make any approach to the barrier from the outside or any movement from the barrier to areas inside the perimeter immediately visible. Anything outside the barrier, such as refuse piles, weed patches, heavy undergrowth, or anything else that might conceal someone's approach, should be eliminated. Inside the perimeter, everything should be cleared away from the barrier to create as wide a clear zone as possible. Where possible, methods should be deployed to limit the ability to drive vehicles up to the fence, stopping them from being used as a platform to aid an intruder in scaling the fence.

Unfortunately it is frequently impossible to achieve an uninterrupted clear zone. Most perimeter barriers are indeed on the perimeter of the property line, which means that there is no opportunity to control the area outside the barrier. The size of the facility and the amount of space needed for its operation will determine how space can be given up to the creation of a clear zone inside the barrier. It is important, however, to create some kind of clear zone, however small it might be. In situations where the clear zone is necessarily so small as to endanger the effectiveness of the barrier, thought should be given to increasing the barrier height and strength in critical areas. In such areas, the installation of intrusion detection devices to give due and timely warning of an intrusion attempt to an alert guard force may also be warranted.

Inspection

Having established the perimeter defense and the clear zones to the maximum that is possible and practical, it is essential that a regular inspection routine be set up. Gates should be examined carefully to determine whether locks or hinges have been tampered with; fence lines should be observed for any signs of forced entry or tunneling; walls should be checked for marks that might indicate they have been scaled or that such an attempt has been made; top guards must be examined for their effectiveness; miscellaneous service penetrations must be examined for any signs of attack; brush, trees and branches, and weeds must be cleared away. Erosion areas must be filled in; and any potential scaling devices such as ladders, ropes, oil drums, or stacks of pallets must be cleared out of the area. Any condition that could even in the smallest degree compromise the integrity of the perimeter must be both reported and corrected. Such an inspection should be undertaken no less than weekly and possibly more often if conditions so indicate.

Hydraulic Defenses

Hydraulic defenses (barricades) are used to protect gates. These devices are designed to deploy in 1 to 3 seconds and can stop a 15,000-pound vehicle moving at 50 mph. In most cases the devices are installed in the roadway, where they are unobtrusive except for the outline of the top of the barrier, which is level with the surface into which it is installed.

Fence Protection Devices

Over the past decades, various devices have been designed to protect the integrity of fences. The earliest systems used electrification. Since those early days, the introduction of sensors to alert security staff of the presence of intruders has supplemented the use of fencing and in some cases replaced it. The most commonly used external intruder detection sensors are described below.

Fluid Pressure

In this system, a fluid-pressure sensor (a small diameter tube sealed at one end and filled with fluid) is placed in the barrier. If a load is applied to the barrier, the tubes compress, placing force on the monitored fluid. This is useful in detection of persons crossing open ground and is commonly used in military installations and tank farms.

Electromagnetic Cable

This sensing device (see Figure 9-5) operates on the principle of electromagnetic capacitance. In simple terms, the cable creates an electromagnetic field that is constant in output. If the field is interrupted by cutting or pressure on the cable, it can be sensed and reported. The cable is normally mounted on inner chain-link or mesh fences.

Buried Cable Detection

Various technologies may be used for this type of sensing system. Generally a cable is buried along the perimeter path. When an intruder crosses or walks over the buried cable the system senses a disturbance and sends an alarm condition.

FIGURE 9-5 A sensing device. *(Courtesy of Southwest Microwave, Inc.)*

Fiber-optic Cable

A beam of pulsed light is transmitted through the cable, and this is sensed at the other end. If the cable is cut or interfered with, the pulsing stops and there are changes in amplitude. The application of the cable is similar to the electromagnetic cable. Some manufacturers of high-security fences, however, have incorporated the fiber-optic cable into the hollow strand of their normal fence material, making it impossible to detect. Fiber-optic cable is finding many uses as a sensor. The sensing cable may be attached to a fence, buried alongside a pipeline, or used in conjunction with power or communications lines. The use of fiber-optic cable allows for the monitoring of miles of fence, pipe or transmission lines with no electronics or power in the field.

Capacitive Field Effect

This sensor (see Figure 9-6) operates on the same principles as do the electromagnetic systems but uses electrical rather than magnetic fields. These systems are extremely sensitive and can be affected by snow and ice. Weeds and paper debris contacting the sensor wire will also give rise to false alarms.

Active Infrared System (AIRS)

While more will be said about this system in Chapter 10, the basic premise behind it is a beam of infrared light sent to a sensor. If the beam is interrupted, the indicates an alarm condition. These systems come in a variety of configurations from a single beam (transmitter and receiver) to a tower of multiple beams with crossing patterns.

External Microwave

Whether used internally or externally, this system relies on sending microwaves and on the Doppler effect. When the sensor receives an unfamiliar return of waves, an alarm condition

FIGURE 9-6 Capacitive field effect. *(Courtesy of Southwest Microwave, Inc.)*

is noted. Most external systems, however, rely on the transmission of a microwave beam to a receiver. If the beam is interrupted or changed, the system goes into alarm. More will be said about microwave sensors used internally in Chapter 10.

Taut Wire

This sensor, which is placed on the wires of a fence, relies on a pendulum that is in the off position until tension on the taut wire forces the pendulum to swing into the on position, tripping the alarm.[2]

Vibration Sensors

This sensor technology can be used on multiple fence types and materials. Vibration sensors are attached to the fencing material, spaced out along the entire fence line at specific intervals. They detect particular amounts and types of vibrations that indicate a possible climbing or cutting activity.

Inside the Perimeter

Unroofed or outdoor areas within the fenced perimeter must be considered a second line of defense. These areas can usually be observed from the outside so that targets can be selected before an assault is made.

In an area where materials and equipment are stored in a helter-skelter manner, it is difficult for guards to determine if anything has been disturbed or taken. On the other hand, neat, uniform, and symmetrical storage can be readily observed, and any disarray can be detected at a glance.

Discarded machinery, scrap lumber, and junk of all kinds haphazardly thrown about the area create safety hazards and provide cover for any intruder. Such conditions must never be permitted to develop. Efficient housekeeping is basic security.

Parking

The parking of privately owned vehicles within the perimeter barrier creates a serious security risk. Facilities that cannot or will not establish parking lots outside the perimeter barrier are almost invariably plagued by a higher incidence of pilferage because of the ease with which employees can conceal goods in their cars at any point during the day. In addition, with today's heightened security concerns over terrorism, allowing unscreened cars into the perimeter introduces yet another risk to the organization.

In cases where the perimeter barrier encompasses the employee and visitor parking areas, additional fencing should be constructed to cordon off the parking areas. Appropriate guarded pedestrian gates or turnstiles must, of course, be installed to accommodate and monitor the movement of employees to and from their cars.

The fenced parking lot should be patrolled to protect against car thieves and vandals. An unprotected parking area in a crime-ridden neighborhood can create feelings of insecurity that quickly damage organizational morale. According to Liability Consultant, Inc., the greatest number of negligent security lawsuits filed between 1992 and 2002 were the result of poor parking lot security.[3]

Company cars and trucks—especially loaded or partially loaded vehicles—should be parked within the fenced perimeter for added security. This parking area should be well lighted and regularly patrolled or kept under constant surveillance.

Loaded or partially loaded trucks and trailers should be sealed or padlocked and should be parked close enough together and close enough to a building wall (or even back to back) so that neither their side doors nor rear doors can be opened without actually moving the vehicles. Valuable materials should be removed from the vehicle each night or at a minimum they should not be visible from outside of the vehicle.

Surveillance

The entire outside area within the security perimeter barrier must be kept under surveillance at all times, particularly at night. Goods stored in this area are particularly vulnerable to theft or pilferage. This is the area most likely to attract the thief's first attention. With planning and study, in cooperation with production or operations personnel, it should be possible to lay out the yard area so that there are long, uninterrupted sight lines that permit inspection of the entire area with a minimum of movement. (Surveillance is discussed in detail in Chapter 8.)

Lighting

Depending on the nature of the facility, protective lighting will be designed either to emphasize the illumination of the perimeter barrier and the outside approaches to it or to concentrate on the area and the buildings within the perimeter. In either case, it must produce sufficient light to create a psychological deterrent to intrusion in addition to making detection virtually certain in the event an entry is made. The Illuminating Engineering Society of North America (IESNA) in its *Guideline for Security Light for People, Property, and Public Spaces* (2003) recommends a minimum of 3 footcandles in all parking facilities.[4] A footcandle is a unit for measuring the intensity of illumination equal to 1 lumen per square foot— that is, the amount of light a single candle provides over 1 square foot (see Figure 9-7).

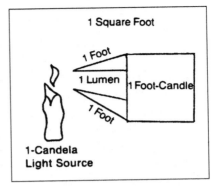

FIGURE 9-7 Footcandle. *(From Richard Gigliotti and Ronald Jason,* Security Design for Maximum Protection *[Boston: Butterworth-Heinemann, 1984].)*

While light must provide a specified level of illumination, creating glare to deter intruders, it must also avoid casting annoying or dangerous light into neighboring areas. This is particularly important where the facility abuts streets, highways, or navigable waterways.

The system must be reliable and designed with overlapping illumination to avoid creating unprotected areas in the event of individual light failures. It must be easy to maintain and service, and it must be secured against attack. Poles should be within the barrier, power lines should be buried, and the switch box or boxes should be secure.

There should be a backup power supply in the event of power failure. Supplementary lighting, including searchlights and portable lights, might also be a part of the system. These lights are provided for special or emergency situations, and although they should not be used with any regularity, they must be available to the security force.

The system could be operated automatically by a photoelectric cell that responds to the amount of light to which it is subjected. Such an arrangement allows for lights to be turned on at dusk and extinguished at daylight. This can be set up to activate individual lamps or to turn on a lighting zone, or the entire system at once.

Other controls are timed, which simply means that lights are switched on and off by a clock schedule. Such a system must be adjusted regularly to coincide with the changing hours of sunset and sunrise. The lights may also be operated manually.

Types of Lighting

Lamps used in protective lighting are either incandescent, fluorescent, metal halide, mercury vapor, quartz, or high- or low-pressure sodium. Each type has special characteristics suitable for specific assignments. Each lamp type offers differing effects on natural colors. The effect of the light on the color appearance of objects is measured by the color-rendering index (CRI). Light with a low CRI makes colors appear less normal. The higher the CRI, the more the object appears normal. Sodium light has a low CRI, whereas metal halide has a high CRI.[5]

Incandescent

These are common lightbulbs of the type found in the home. They have the advantage of providing instant illumination when the switch is thrown and are thus the most commonly used in protective lighting systems. Some incandescents are manufactured with interior coatings that reflect the light and with a built-in lens to focus or diffuse the light. Regular high-wattage incandescents can be enclosed in a fixture that will give much the same result. Because of their inefficient power-to-effective-light ratio, recent federal law will soon make the purchase of incandescent bulbs more difficult, driving the consumer to other forms of lighting.

Fluorescent

These are generally of a mercury vapor type and are highly efficient, giving off approximately 62 lumens per watt. Most fluorescent lamps are temperature sensitive and thus have limited value for outdoor use in colder climates. In addition, the common flickering effect created by these lamps can have a disorienting effect on both security personnel and intruders. Fluorescent lamps sometimes interfere with radio, audio and other data transmissions.

Mercury Vapor Lamps

These common security lamps give out a strong light with a bluish cast. They are more efficient than are incandescents and have a considerably longer lamp life. In general, these lamps can tolerate power dips of up to 50 percent. The time for these lights to reach full luminance after start up can be considerable.

Metal Halide

These lamps are also very tolerant of power dips. As with mercury vapor lamps, the start-up time is long. A power outage as short as 1/20th of a second may be enough to knock this lamp off-line.

Sodium Vapor Lights

Sodium lamps give out a soft yellow light and are even more efficient than are mercury vapor lamps. They are widely used in areas where fog is a frequent problem because yellow penetrates the mist more readily than does white light. They are frequently found on highways and bridges.[6]

Quartz Lamps

These lamps emit a very bright white light and snap on almost as rapidly as do incandescent bulbs. They are frequently used at very high wattage—1,500 to 2,000 watts is not uncommon in protective systems—and are excellent for use along the perimeter barrier and in troublesome areas (see Table 9-2).

Table 9-2   Types of Luminaire Lamps Found in a Maximum-Security Environment

Lamp Type	Mean Lumens per Watt	Start	Restrike	Nominal Life of Lamp (hrs.)	Percent Lumen Maintenance at Rated Life	Color Discrimination
Incandescent	4 (21) 22*	Instant	Instant	750–1,000	85–90	Excellent
Fluorescent	35 (62) 100	Rapid	Rapid/Instant†	7,500–10,000	70–90	Excellent
Metal halide	68 (80) 100	3–5 min.	10–20 min.	10,000–15,000	65–75	Excellent
Mercury vapor	20 (48) 63	3–7 min.	3–6 min.	16,000–24,000	50–75	Good
Sodium	92 (127) 140	4–7 min.	Instant‡	16,000–24,000	75–85	Fair
Xenon ARC	__	Rapid/Instant	Instant	1,500	__	Excellent

*4 (21) 22:4 = minimum mean; (21)=nominal rating for most protective lighting applications; 22 = maximum mean.
†Low-temperature ballast must be considered.
‡Instant for most lamps if less than 1 minute of power interruption but a reduced lumen output.
__Used for searchlights only.
Source: Richard Gigliotti and Ronald Jason, *Security Design for Maximum Protection* (Boston: Butterworth-Heinemann, 1984), p. 138.

Light-Emitting Diodes (LEDs)

These lights are very small and energy efficient. LEDs provide a sharper image than their incandescent predecessors. They have a very long useful life but are more expensive. LED technology has advanced greatly in the past few years and is now becoming more commonplace in high-efficiency applications.

Electroluminescent

These lights are similar to their fluorescent cousin. However, they do not contain mercury and are more compact.[7]

Types of Equipment

No one type of lighting is applicable to every protective lighting situation.

Amid the great profusion of equipment in the market, there are four basic light types that are in general use in security applications: floodlights, searchlights, Fresnel lenses, and streetlights (see Figure 9-8). (The first three of these might in the strictest sense be considered as a single type because they are all basically reflection units in which a parabolic mirror directs the light in various ways. We will, however, deal with them separately.)

Streetlights

Streetlights are pendant lighting units that are built as either symmetrical or asymmetrical. The symmetrical units distribute light evenly. These units are used where a large area is to be lighted without the need for highlighting particular spots. They are normally centrally located in the area to be illuminated.

Asymmetrical units direct the light by reflection in the direction where light is required. They are used in situations where the lamp must be placed some distance from the target area. Because these are not highly focused units, they do not create a glare problem.

Luminaire		Photometric designation	Typical distribution characteristics
Streetlight		Medium wide asymmetric	 70° to 75°
Fresnel lens		Asymmetric, glare protection	
Searchlight	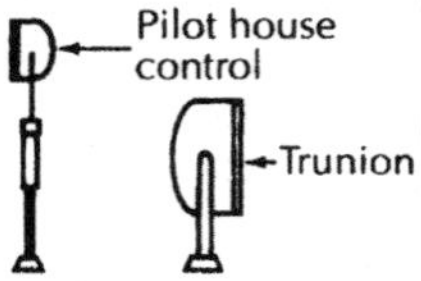 Pilot house control / Trunion	Extremely narrow beam	 Less than 10°
Floodlight		Medium to wide beam	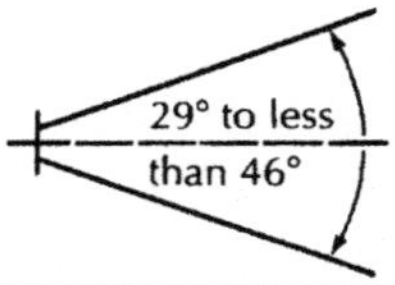 29° to less than 46°

FIGURE 9-8 Typical equipment for protective lighting.

Streetlights are rated by wattage or even more frequently by lumens and in protective lighting applications may vary from 4,000 to 10,000 lumens, depending on their use.

Floodlights

Floodlights are fabricated to form a beam so that light can be concentrated and directed to specific areas. They can create considerable glare.

Although many floodlights specify beam width in degrees, they are generally referred to as wide, medium, or narrow, and the lamp is described in wattage. Lamps run from 300 to 1,000 watts in most protective applications. However, there is a wide latitude in this, and the choice of one will depend on a study of its mission.

Fresnel Lenses

Fresnel lenses are wide-beam units, primarily used to extend the illumination in long, horizontal strips to protect the approaches to the perimeter barrier. Unlike floodlights and searchlights, which project a focused round beam, Fresnel lenses project a narrow, horizontal beam that is approximately 180° in the horizontal plane and from 18° to 30° in the vertical plane.

These units are especially good for creating a glare for the intruder while the facility remains in comparative darkness. They are normally equipped with a 300- to 500-watt lamp.

Searchlights

Searchlights are highly focused incandescent lamps that are used to pinpoint potential trouble spots. They can be directed to any location inside or outside the property, and although they can be automated, they are normally controlled manually.

They are rated according to wattage, which may range from 250 to 3,000 watts, and to the diameter of the reflector, which may range from 6 inches to 2 feet (the average is around 18 inches). The beam width is from 3° to 10°, although this may vary in adjustable or focusing models.

Maintenance

As with every other element of a security system, electrical circuits and fixtures must be inspected regularly to replace worn parts, verify connections, repair worn insulation, check for corrosion in weatherproof fixtures, and clean reflecting surfaces and lenses. Lamps should be logged as to their operational hours and replaced at between 80 and 90 percent of their rated life.

Perimeter Lighting

Every effort should be made to locate lighting units far enough inside the fence and high enough to illuminate areas both inside and outside the boundary. The farther outside the boundary the lighted areas extend, the more readily guards will be able to detect the approach of an intruder.

Light should be directed down and away from the protected area. The location of light units should be such that they avoid throwing a glare into the eyes of the guard, do not create shadow areas, and create instead a glare problem for anyone approaching the boundary.

Fixtures used in barrier and approach lighting should be located inside the barrier. As a rule of thumb, they should be around 30 feet within the perimeter, spaced 150 feet apart, and about 30 feet high. These figures are, of course, approximations and will not apply to every installation. Local conditions will always dictate placement.

Floodlights or Fresnel lenses are indicated in illuminating isolated or semi-isolated fence boundaries where some glare is called for. In either case, it is important to light from 20 feet inside the fence to as far into the approach as is practical. In the case of the isolated fence, this could be as much as 250 feet. Semi-isolated and non-isolated fence lines cannot be lighted as far into the approach because such lighting is restricted by streets, highways, and other occupancies. Because glare cannot be employed in illuminating a non-isolated fence line, streetlights are recommended.

Where a building of the facility is near the perimeter or is itself part of the perimeter, lights can be mounted directly on it. Doorways of such buildings should be individually lighted to eliminate shadows cast by other illumination.

In areas where the property line is on a body of water, lighting should be designed to eliminate shaded areas on or near the water or along the shoreline. This is especially true for piers and docks, where both land and water approaches must be either lighted or capable of being lighted on demand. Before finalizing any plans for protective lighting in the vicinity of navigable waters, however, the U.S. Coast Guard must be consulted.

Gates and Thoroughfares

It is important that the lighting at all gates and along all interior thoroughfares be sufficient for the operation of the facility.

Because both pedestrian and vehicular gates are normally staffed by guards inspecting credentials as well as checking for contraband or stolen property, it is critical that the areas be lighted to at least 2 foot-candles. Pedestrian gates should be lighted to about 25 feet on either side of the gate if possible, and the range for vehicular gates should be twice that distance. Street lighting is recommended in these applications, but floodlights can also be used if glare is strictly controlled.

Thoroughfares used for pedestrians, vehicles, or forklifts should be lighted to 0.10 fc for security purposes. Much more light may be required for operational efficiency, but this level should be maintained as a minimum, no matter what the conditions of traffic may be.

Other Areas

Open or unroofed areas within the perimeter, but not directly connected to it, require an overall intensity of illumination of about 0.05 fc (up to 0.10 fc in areas of higher sensitivity). These areas, when they are nonoperational, are usually used for material storage or for parking. Particularly vulnerable installations in the area should not, according to many experts, be lighted at all, but the approaches to them should be well lighted for at least 20 feet to aid in the observation of any movement.

Searchlights may be indicated in some facilities, especially in remote mountainous areas or in waterfront locations where small boats could readily approach the facility.

General

A well-thought-out plan of lighting along the security barrier and the approaches to it; an adequate overall level of light in storage, parking, and other nonoperational areas within the perimeter; and reasonable lighting along all thoroughfares are essential to any basic security program. The lighting required in operational areas will usually be much higher than the minimums required for security and will, therefore, serve a security purpose as well.

Can better lighting be sold to security management on the basis of cost-effectiveness? High-pressure sodium lighting uses about 50 percent less energy to produce the same light as older, incandescent streetlights. Sodium systems have a lumens-per-watt efficiency five or six times that of incandescent lighting and produce 106 percent more light than do the most common mercury streetlights, but use about 14 percent less electricity. Table 9-3 compares wattage ranges, lumens, and rated life for six basic lamp families.

Table 9-3   Six Basic Lamp Families

Type of Lamp	Wattage Range	Initial Lumens[†] per Watt Including Ballast Losses	Average Rated Life (Hours)
Sodium	35–1,000	15–130	7,500–24,000[*]
Metal halide	70–2,000	69–115	5,000–20,000
Mercury vapor			
Standard	40–1,000	24–60	12,000–24,000[*]
Self-ballasted	160–1,250	14–25	12,000–20,000
Fluorescent	4–215	14–95	6,000–20,000[*]
Incandescent	15–1,500	8–24	750–3,500

[*]Data are based on the more commonly used lamps and are provided for comparison purposes only. Actual results to be derived depend on factors unique to the specific products and installation involved. Consult manufacturers for guidance.

[†]Lumens (of light output) per watt (of power input) is a common measure of lamp efficiency. Initial lumens-per-watt data are based on the light output of lamps when new. The light output of most lamps declines with use. The actual efficiency to be derived from a lamp depends on factors unique to an installation. The actual efficiency of a lighting system depends on far more than the efficiency of lamps or lamps/ballasts alone. More than efficiency should be considered when evaluating a lighting system.

Source: John P. Bachner, "The Myths and Realities Behind Security Lighting," *Security Management* (August 1990): 109.

Planning Security

No business exists without a security problem of some kind, and no building housing a business is without security risk. Yet few such buildings are ever designed with much thought given to the steps that must eventually be taken to protect them from criminal assault.

A building must be many things in order for it to satisfy its occupant. It must be functional and efficient, achieve certain aesthetic standards, be properly located and accessible to the markets served by the occupant, and provide security from interference, interruption, and attack. Most of these elements are provided by the architect, but all too frequently the important element of security is overlooked.

Good security requires thought and planning that results in a carefully integrated system. Most security problems arise, in large part, because no one has anticipated and planned for them.

This is especially true where a company building is concerned. Only now are many architecture credentialing programs providing any training on security matters. Therefore, many building designs do little to deter burglars or vandals. When building owners give insufficient consideration to security during the planning stages, buildings are erected that may inadvertently provide opportunities for crime.

Summary

Physical security devices are the most commonly thought of security measures. The discussion in this chapter covered security measures that are as old as security, as well as new technologies that are improving these basic concepts. Clear zones, moats, and fences still exist, but are now augmented with state-of-the-art sensors and computerized monitoring stations. The next two chapters focus on specific basic security measures that protect the interior as well as the exterior of facilities.

These basic tools are the fundamentals of which all security professionals should have at least conversational knowledge.

CASE STUDY

You are the security director for IBID International, a large manufacturer. You have assigned one of your staff to conduct a security survey of the company's exterior grounds. You have received the report and now must approve or disapprove each recommendation. Your approved recommendations will be presented to senior management for final approval and funding.

CRITICAL CONSIDERATIONS

1. Senior management requires all expenditures over $10,000 to go to the capital expenditure committee (a 6-month process).
2. Senior management is concerned about your large head count.
3. Your personal bias is for security operations to be "low profile."
4. Your plant is in a low-threat environment.
5. If approved, expensive items can be costed out over six years according to company policy.

PHYSICAL SURVEY REPORT

1. There are no perimeter lights around company property. The city provides streetlights, causing deep shadow along the perimeter.
   RECOMMENDATION: Initiate a major perimeter lighting program involving placement of mercury vapor lights in such a manner as to effectively illuminate the property perimeter. COST: $45,000
   APPROVE   DISAPPROVE
2. The property presently lacks any perimeter fence.
   RECOMMENDATION: Install 7-foot high steel chain link fencing topped with triple-strand barbed wire. This will prevent access to all but authorized personnel, visitors, etc. COST: $800,000
   APPROVE   DISAPPROVE
3. At present, employees and visitors park in three separate unguarded, unlighted and open parking lots.
   RECOMMENDATION: To insure protection of visitors and personnel and their vehicles, the following recommendations are made:
   - Establish regular security patrols of parking lots.   COST: $0
     (change in patrol patterns)
     APPROVE   DISAPPROVE
   - Light parking lots to a level where vehicles can be observed clearly
     24 hours a day.   COST: $15,000
     APPROVE   DISAPPROVE
   - Use closed circuit cameras to monitor parking areas.   COST: $7,000
     APPROVE   DISAPPROVE
   - Place an emergency phone in each parking lot.   COST: $ 400
     APPROVE   DISAPPROVE

- Establish a vehicle sticker program to aid in identifying authorized vehicles.   COST: $1200
  APPROVE   DISAPPROVE
- Combine all parking areas into one large lot for better overall protection.
                                                 COST: $125,000
  APPROVE   DISAPPROVE
- Fence in the two existing parking areas for employees using electronic gates operated by an electronic pass system.   COST $62,000
  APPROVE   DISAPPROVE
- Purchase a scooter-type vehicle for parking lot patrols.   COST $3,650
  APPROVE   DISAPPROVE

4. There is no access control at any of the four points of vehicle entry and exit.
   RECOMMENDATION: Establish all-weather entry control points staffed 24 hours a day to control access on and off the plant property.
   - Five all-weather guard huts @ $10,000 each.   COST: $50,000
     APPROVE   DISAPPROVE
   - Sixteen new positions to staff all entry points.   COST: $200,000
     APPROVE   DISAPPROVE

NARRATIVE: You need to succinctly, clearly and logically explain the rationale for each decision. The narrative should strive to briefly present each problem, and why the suggestions are being approved or disapproved.

Review Questions

1. What four lines of protection should be included in physical security planning?
2. How can the various openings in a perimeter be effectively protected and secured?
3. Why should parking not be allowed inside the controlled perimeter?
4. Discuss the different security applications for the various types of lighting equipment.
5. What considerations must be taken into account when installing a security lighting system?

References

[1] Security and the chain link fence. Security December 1996:91.

[2] Cumming N. In: Security: the comprehensive guide to equipment selection and installation. London: Architectural Press; 1987. pp. 70–109.

[3] Anderson T. Laying down the law: a review of trends in liability lawsuits. Secur Manage October 2002:43–51.

[4] Illuminating Engineering Society of North America, Guideline for Light for People, Property, and Public Spaces, Security Lighting Committee, 2003.

[5] Kangas S. Lighting the way to better business. Secur Manage September 1996:85.

[6] Ibid., p. 84.

[7] Ibid., pp. 85–6.

10

The Inner Defenses
Intrusion and Access Control

OBJECTIVES

The study of this chapter will enable you to:

1. Identify the security measures aimed at protection of assets within a physical structure.
2. Evaluate various types of identification systems.
3. Know how to evaluate various types of safes, vaults and files.
4. Discuss various alarms and alarm systems.

Introduction

Once the facility's perimeter is secured, the next step in physical security planning is to minimize or control access to the facility's or the building's interior. The extent of this control will depend on the nature and function of the facility: The controls must not interfere with the facility's operation. It is theoretically possible to seal off access to a given operation completely, but it would be difficult to imagine how useful the operation would be in such an atmosphere.

Certainly no commercial establishment can be open for business while it is closed to the public. A steady stream of outsiders, from customers to service personnel, is essential to its economic health. In such cases, the security problem is to control this traffic without interfering with the function of the business being protected. Isolated manufacturing facilities must also provide for the traffic created by the delivery of raw materials, the shipping of fabricated goods, the provision of services, and of course the labor force, which may be operating in several shifts.

All such traffic tends to compromise the physical security of the facility. But security must be provided, and it must be provided appropriately for the operation of the facility being served.

Within any building, no matter whether it is located inside a perimeter barrier or is a part of the perimeter wall, it is necessary to consider the need to protect against the internal thief as well as the potential intruder. Whereas the boundary fence is primarily designed to keep out unwanted visitors (not altogether forgetting its function in the control of movement of authorized personnel), interior security must provide some protection against the free movement of employees and others bent on pilferage as well as establishing a second line of defense against the unannounced intruder.

Since every building is used differently and has its own unique traffic composition and flow, each building presents a different security problem. Each must be examined and analyzed in great detail before an effective security program can be developed.

It cannot be overemphasized that such a program must be implemented without in any way interfering with the orderly and efficient operation of the facility to be protected. It must not be obtrusive, and yet it must provide a predetermined level of protection against criminal attack from outside or inside.

The first points of examination must be the doors and windows of buildings within the perimeter. These must be considered in terms of effectiveness no matter whether the building walls form a part of or constitute in themselves the perimeter barrier (as we have already discussed in Chapter 8) or they are a true second defense line where the building under examination is completely within the protection of a barrier.

Doors to Sensitive Areas

Doors to equipment rooms, computer installations, research and development, and other sensitive areas should be equipped with automatic door-closing devices and fitted with strong dead bolts and heavy latches.

In cases where an area is under heavy security but has any degree of traffic, it might be well to consider the installation of an electric strike to secure the operation and control the traffic. This kind of unit is a locking device controlled remotely by a security person, permitting entry of a recognized, authorized person only when a button is pressed to release the lock. Because it requires someone on hand at all times for its operation, this system can be expensive. It must be examined with the cost-versus-security cost equation in mind. Continuing development in biotechnology and related computer technology is providing other options that will be discussed later in this chapter.

Supply room and tool room doors should be secured whenever those rooms are not actually in use. Even when they are in use, entrance into these areas must be restricted. The usual construction of such restraints consists of either a "dutch" door, in which the bottom half is secured, or a counter that can be closed off by heavy screening, chain-link fencing material, or reinforced shutters.

Special care should be taken in the storage of small items of value. Such merchandise or material is highly subject to theft by virtue of its value for resale or personal use combined with the ease with which it can be stolen. Although such items may be stored in a facility of any construction capable of providing security, it has been the experience of many firms that uniformly stacked rows, piles, or pallets of such items within a cage-type construction that provides instant eyeball inventory is the best protection. Such precautions will vary from business to business, but they must be carefully systematized to control this potentially troublesome area of loss.

Office Area Doors

Doors between production and office areas or between heavily trafficked areas and office spaces must be examined for the likelihood of their use for criminal purposes. Their construction and locking hardware will be determined by such a survey. In most cases, these passages will be minimum-security areas during regular working hours because there is usually a need for movement between these areas. When there is little or no use of the office area, these doors should be secured.

Traffic Patterns

Doors must be analyzed for their function when laying out the security plan. In some cases, they may serve a dual purpose as, for example, fire doors, which are designed to close automatically in the event of a fire. These doors, which may remain open at the discretion of management, must be fitted to form an effective and automatic barrier to the spread of fire. They may be desirable when fire doors separate a production area from a warehouse or storage area. During those times when the production area is in operation but the warehouse is not, such fire doors can perform a security function by remaining closed.

In other cases, doors must be examined in an effort to establish a schedule for their use. Employee entrances that are the authorized points of passage for all employees may be staffed by security personnel, depending on whether the control point is established there or farther out on the perimeter. These doors could be secured once the employees have entered, thus denying entrance to unauthorized visitors as well as preventing any employees from wandering out to the fence or the parking lot or any other location where they might cache contraband for later pickup or transport.

It is axiomatic, however, that any door used as an entrance will in a time of emergency be used as an exit by some employees. This is true in apartments and office buildings and even in industrial facilities where the employees are thoroughly familiar with the premises. No matter what or how many designated emergency exits or procedures there may be, some individuals in a time of tension or near panic will seek out the door with which they are most familiar. The entrance then must always be considered an emergency exit, and it should be equipped with panic hardware. To protect against surreptitious use, it should also be fitted with at least a local alarm.

The same, of course, is true of the designated emergency exits. These doors should, in addition, be stripped of all exterior hardware since they are not intended for operational use at any time.

Personnel doors leading to and from the dock area must be carefully controlled and supervised at all times when the dock is in use. These and dock doors must be secured when the area is no longer operational.

Fire doors in office buildings should be fitted with alarms to prevent surreptitious use and access to the interior. Use of public stairwells should be prohibited or discouraged unless doors from them open into reception areas.

Traffic Control

Controlling traffic in and out and within a facility is essential to the facility's security program. Perimeter barriers, locked doors, and screened windows prevent or deter the entry of unauthorized visitors. But because some traffic is essential to every operation no matter how highly classified it may be, provision must be made for the control of this movement.

Specific solutions will depend on the nature of the business. Obviously retail establishments, which encourage high-volume traffic and regularly handle a great deal of merchandise both in and out, have a problem of a different dimension from that of the industrial operation working on a highly classified government project. Both, however, must work from the same general principles toward providing the greatest possible security within the efficient and effective operation of the job at hand.

Controlling traffic includes the identification of employees and visitors and directing or limiting their movements and the control of all incoming and outgoing packages and of trucks and private cars.

Visitors

All visitors to any facility should be required to identify themselves. When they are allowed to enter after they have established themselves as being on an authorized call, they should be limited to predetermined, unrestricted areas. The obvious exception is in firms where the public has free access to the facility, for example, retail stores.

If possible, sales, service, and trade personnel should receive clearance in advance on making an appointment with the person responsible for their being there. Although this is not always possible, most businesses deal with such visitors on an appointment basis and a system of notifying the security personnel can be established in a majority of cases.

Businesses regularly called on unannounced—by salespeople or other tradespeople—should set aside a waiting room that can be reached without passing through sensitive areas.

In some cases, it may be advisable to issue passes that clearly designate them as visitors. If they will be escorted to and from their destination, a pass system is probably unnecessary.

Ideally, all traffic patterns involving visitors should be short, physically confined to keep them from straying, and capable of being observed at all points along the route. In spread-out industrial facilities, they should take the shortest, most direct route that will not pass through restricted, sensitive, or dangerous areas and will pass from one reception area to another.

To achieve security objectives without alienating visitors or in any way interfering with the operation of the business, any effective control system must be simple and understandable (see Figure 10-1). It must incorporate certain specific elements in order to accomplish its aims. It must limit entry to those people who are authorized to be there and be able to

FIGURE 10-1 Visitor/contractor ID. (*Courtesy of Temp Badge.*)

identify such people. It must have a procedure by which persons may be identified as being authorized to be in certain areas and must prevent theft, pilferage, or damage to the assets of the installation. It must also prevent injury to the visitor.

Employee Identification

Small industrial facilities and most offices find that personal identification of employees by guards or receptionists is adequate protection against intruders entering under the guise of employees. In organizations of more than 50 employees per shift or in high-turnover businesses, this type of identification is inadequate. The opportunity for error is simply too great. In 2004, the White House noted that there were "wide variations in the quality and security of forms of identification used to gain access to secure Federal and other facilities where there is a potential for terrorist attacks."[1] Breaches of security at the White House are rare but not impossible. In November 2009, Tareq and Michaele Salahi were admitted to the White House state dinner for the Indian Prime Minister. They were not on the guest list.[2]

With the continually changing technology, James Bond type access controls are becoming common. Facial imaging along with already in use biometric identity and access control technology will make it unlikely that access to secured areas will be gained by another, other than the person intended. The International Biometrics Group projects that facial recognition will represent more than $1.4 billion or 19 percent of the $7.4 billion non-AFIS market by 2012. Fingerprint matching is predicted to be 25%.[3]

The most practical and generally accepted system is the use of badges or identification cards (see Figure 10-2). Generally speaking, this system should designate when, where, how, and to whom passes should be displayed; what is to be done in case of loss of the pass; procedures for retrieving badges from terminating employees; and a system for cancellation and reissue of all passes, either as a security review or when a significant and specific number of badges have been reported lost or stolen.

To be effective, badges must be tamper-resistant, which means that they should be printed or embossed on a distinctive stock that is worked with a series of designs difficult to reproduce. The use of holographic designs embedded in the badge has made duplication more difficult. They should contain a clear and recent photograph of the bearer, preferably in color. The photograph should be at least 1 inch square and should be updated every 2 or 3 years or when there is any significant change in facial appearance, such as the growing or removal of a beard or mustache. It should, in addition, contain vital statistics such as date of birth, height, weight, color of hair and eyes, sex, and both thumbprints. It should be laminated and of sturdy construction. The recent increased use of smart card ids, which contain all the above information in electronic format, has also reduced

FIGURE 10-2 Employee identification card.
(Courtesy of Temp Badge.)

the ability to counterfeit ids in operation using this type of technology. In cases where there are areas set off or restricted to general employee traffic, it might be color-coded to indicate those areas to which the bearer has authorized access.

The latest entry into badge protection is holography. The introduction of holography into badge control systems reduces the chance of counterfeiting cards. Holograms incorporated with other technology advances from the past 15 years have produced an identification card system that is almost forgery-proof while incorporating computer accountability.

If a badge system is established, it will only be as effective as its enforcement. Facility guards are responsible for seeing that the system is adhered to, but they must have the cooperation of the majority of the employees and the full support of management. If the system is simply a pro forma exercise, it becomes a useless annoyance and would better be dispensed with.

Under the Department of Homeland Security Presidential Directive/Hspd-12, issued August, 2004, the Federal government established a mandatory, government-wide standard for secure and reliable forms of identification issued by the Federal government to employees and government contractors and their employees.[4]

Pass Systems

As we have just noted, all employees entering or leaving a facility or area should be identified and their authorizations to be there should be checked. This can be achieved through one of many badge systems. Three possible systems are as follows:

- *The single-pass system*, in which a badge or pass coded for authorization to enter specific areas is issued to an employee, who keeps it until the authorization is changed or until leaving the company.
- *The pass-exchange system*, in which an employee entering a controlled area exchanges one color-coded pass for another that carries a different color code specifying the limitations of the authorization. On leaving, the employee surrenders the controlled-area pass for the basic authorization identification pass. (In this system, the second pass never leaves the controlled area, thus reducing the possibility of switching, forging, or altering.)
- *The multiple-pass system* is essentially the same as the exchange system, but it provides an extra measure of security by requiring that exchanges take place at the entrance to each restricted area within the controlled area.

Package Control

Every facility must establish a system for the control of packages entering or leaving the premises. However desirable it might seem, it is simply unrealistic to support a blanket rule forbidding packages either in or out as a workable procedure. Such a rule would be damaging to employee morale and in many cases would actually work against the efficient operation of the facility. Therefore, since the transporting of packages through the portals is a fact of life, they must be dealt with in order to prevent theft, misappropriation of company property, and concealment of dangerous materials (i.e., biological or chemical agents or bombs).

If it is deemed necessary, the types of items that may be brought in or taken out may be limited. If such is the case, the fact must be publicized and clearly understood by everyone.

Packages brought in should be checked for content. If possible, where they are not to be used during work, they should be checked with the guard to be picked up at the end of the day. In most cases, spot-checking will suffice.

Whatever the policy concerning packages—whether they are to be checked or inspected—that policy must be widely publicized in advance. This is to avoid the appearance of discrimination against those whose packages are opened and examined or those that are denied entrance in conformity with company policy.

Files, Safes, and Vaults

The final line of defense at any facility is in the high-security storage areas where papers, records, plans, cashable instruments, precious metals, or other especially valuable assets are protected. These security containers will be of a size and quantity that the nature of the business dictates.

Every facility will have its own particular needs, but certain general observations apply. The choice of the proper security container for specific applications is influenced largely by the value and the vulnerability of the items to be stored in them. Irreplaceable papers or original documents may not have any intrinsic or marketable value so they may not be a likely target for a thief, but because they do have great value to the owners, they must be protected against fire. On the other hand, uncut precious stones or even recorded negotiable papers that can be replaced may not be in danger from fire, but they would surely be attractive to a thief. They must therefore be protected against theft.

In protecting property, it is essential to recognize that, generally speaking, protective containers are designed to secure against either burglary or fire. Each type of equipment has a specialized function and provides only minimal protection against the other risk. There are containers designed with a burglary-resistant chest within a fire-resistant container that are useful in many instances, but these too must be evaluated in terms of the mission.

Whatever the equipment, the staff must be educated and reminded of the different roles played by the two types of containers. It is all too common for company personnel to assume that a fire-resistant safe is also burglary-resistant and vice versa.

Files

Burglary-resistant files are secure against most surreptitious attacks. On the other hand, they can be pried open in less than half an hour if the burglar is permitted to work undisturbed and is not concerned with the noise created in the operation. Such files are suitable for nonnegotiable papers or even proprietary information since these items are normally only targeted by surreptitious assault.

Filing cabinets with a fire rating of 1 hour and further fitted with a combination lock will probably be suitable for all uses but the storage of government-classified documents.[5]

CASE STUDY

Assets Corporation has accepted a contract to assist a small start-up firm, Memory Backup, in creating a security program to protect its new automated memory stick technology. Assets's owner, James Fisk, has assigned you the task of evaluating the need for safe storage. The following information is provided:

> The company keeps all its records on its own memory sticks.
> There is very little cash or other negotiables kept on the premises.
> The information regarding the development and production of the memory stick is also recorded on a memory stick.
> Threats include disasters, both natural and manmade, and theft, both internal and external.
> The firm cannot afford to lose the materials related to the manufacture of the stick to competitors

  What type of storage container would you tell Mr. Fisk to recommend in his security plan for Memory Backup?

Safes

Safes are expensive, but if they are selected wisely, they can be very important investments in security. Emphatically, safes are not simply safes. They are each designed to perform a particular job to provide a particular level of protection. The two types of safes of most interest to the security professional are the record safe (fire-resistant) and the money safe (burglary-resistant). To use fire-resistant safes for the storage of valuables subject to theft—an all too common practice—is to invite disaster. At the same time, it would be equally careless to use a burglary-resistant safe for the storage of valuable papers or records because, if a fire were to occur, the contents of such a safe would be reduced to ashes.

  Safes are rated to describe the degree of protection they afford. Naturally, the more protection provided, the more expensive the safe will be. In selecting the best one for the requirements of the facility, a number of questions must be considered: How great is the threat of fire or burglary? What is the value of the safe's contents? How much protection time is required in the event of a fire or of a burglary attempt? Only after these questions have been answered can a reasonable, permissible capital outlay for their protection be determined.

Record Safes

Fire-resistant containers are classified according to the maximum interior temperature permitted after exposure to heat for varying periods of time. A record safe with an Underwriters Laboratories (UL) rating of 350-4 (formerly designated "A") can withstand exterior temperatures building to 2,000°F for 4 hours without permitting the interior temperature to rise above 350°F.

  The UL tests that result in the classifications are conducted to simulate a major fire with its gradual buildup of heat to 2,000°F, including circumstances where the safe might fall several stories through the fire-damaged building. In addition, an explosion test simulates a cold safe dropping into a fire that has already reached 2,000°F.

The actual procedure for the 350-4 rating involves the safe staying 4 hours in a furnace temperature that reaches 2,000°F. The furnace is turned off after 4 hours but the safe remains inside until it is cool. The interior temperature must remain below 350°F during heating and cooling-off periods. This interior temperature is determined by sensors sealed inside the safe in six specified locations to provide a continuous record of the temperatures during the test. Papers are also placed in the safe to simulate records. The explosion impact test is conducted with another safe of the same model that is placed for ½ hour in a furnace preheated to 2,000°F. If no explosion occurs, the furnace is set at 1,550°F and raised to 1,700°F over a ½ hour period. After this hour in the explosion test, the safe is removed and dropped 30 feet onto rubble. The safe is then returned to the furnace and reheated for 1 hour at 1,700°F. The furnace and safe then are allowed to cool, after which the papers inside must be legible and not charred.

Computer media storage classifications are for containers that do not allow the internal temperature to go above 150°F. This is critical because computer media begin to distort at 150°F and diskettes at 125°F.[6]

Fire resistant safes are rated based on: the type of material being protected, the length of time it is protected without damage, and whether the container can withstand an impact. UL has three basic classes:

350—protect paper products
150—protect magnetic tapes
125—protect flexible computer disks[7]

Insulated vault-door classifications are much the same as they are for safes except that the vault doors are not subjected to explosion/impact tests.

In some businesses, a fire-resistant safe with a burglary-resistant safe welded inside may serve as a double protection for different kinds of assets (see Figure 10-3), but in no event must the purposes of these two kinds of safes be confused if there is one of each on the premises. Most record safes have combination locks, relocking devices, and hardened steel lock plates to provide a measure of burglar resistance. It must be reemphasized that record safes are designed to protect documents and other similar flammables against destruction by fire. They provide only slight deterrence to the attack of even unskilled burglars. Similarly, the resistance provided by burglar-resistant safes is powerless to protect contents in a fire of any significance.

With the increased use of electronic media in place of paper records most safe manufacturers offer some type of media safe, designed with the specific purpose of allowing for protection of electronic data. There are USB desktop drives that are now fire and water resistant. These drives come with many different options regarding amount of storage, temperature rating and length of time.[8]

Money Safes

Burglary-resistant safes (see Figure 10-4) are nothing more than very heavy metal boxes without wheels, which offer varying degrees of protection against many forms of attack. A safe with a UL rating of TL-15, for instance, weighs at least 750 pounds, and its front face can resist attack

FIGURE 10-3 Fire-resistant safe. *(Courtesy of Diebold, Incorporated.)*

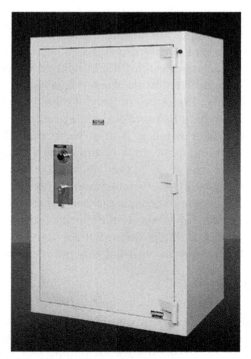

FIGURE 10-4 Burglary-resistant safe. *(Courtesy of Diebold, Incorporated.)*

by common hand and electric tools for at least 15 minutes. Other safes will resist not only attack with tools but also attack with torches and explosives.

The classification system uses the following: TL refers to hand tools and drills, TRTL adds torches, and TXTX adds explosives. The number following the letters indicates the minimum amount of time that the safe will stand attack, in minutes.[9] Because burglary-resistant safes have a limited holding capacity, it is always advisable to study the volume of the items to be secured. If the volume is sufficiently large, it might be advisable to consider the installation of a burglary-resistant vault, which, although considerably more expensive, can have an enormous holding capacity.

Securing the Safe

Whatever safe is selected must be securely fastened to the structure of its surroundings. Police reports are filled with cases where unattached safes, some as heavy as a ton, have been stolen in their entirety—safe and contents—to be worked on in uninterrupted concentration.

A convicted criminal told investigators how he and an accomplice had watched a supermarket to determine the cash flow and the manager's banking habits. They noted that he accumulated cash in a small, wheeled safe until Saturday morning when he banked it. Presumably he felt secure in this practice since he lived in an apartment above the store and perhaps felt that he was very much on top of the situation in every way. One Friday night, the thief and his friend rolled the safe into their station wagon. They pried it open at their leisure to get the $15,000 inside.

Pleased with their success, the thieves were even more pleased when they found that the manager replaced the stolen safe with one exactly like it and continued with the same banking routine. Two weeks later, one man went back alone and picked up another $12,000 in exactly the same way as before.

It has become a common practice to install the safe in a concrete floor, where it offers great resistance to attack. In this kind of installation only the door and its combination are exposed. Because the door is the strongest part of a modern safe, the chances of successful robbery are considerably reduced.

Vaults

Vaults are essentially enlarged safes. As such, they are subject to the same kinds of attack and fall under the same basic principles of protection as safes.

Because it would be prohibitively expensive to build a vault out of shaped and welded steel and special alloys, the construction, except for the door, is usually of high-quality, reinforced concrete. There are many ways in which such a vault can be constructed, but however this is done, it will always be extremely heavy and at best a difficult architectural problem.

Typically vaults are situated at or below ground level so they do not add to the stresses of the structure housing them. If a vault must be built on the upper stories of a building, independent members that do not provide support for other parts of the building must support it. It must also be strong enough to withstand the weight imposed on it if the building should collapse from under it as the result of fire or explosion.

The doors of such vaults are normally 6 inches thick, and they may be as much as 24 inches thick in the largest installations. Because these doors present a formidable obstacle to any criminal, an attack will usually be directed at the walls, ceiling, or floor, which must for that reason match the strength of the door. As a rule, these surfaces should be twice as thick as the door and never less than 12 inches thick.

If it is at all possible, a vault should be surrounded by narrow corridors that will permit inspection of the exterior but that will be sufficiently confined to discourage the use of heavy drilling or cutting equipment by attackers. It is important that there be no power outlets anywhere in the vicinity of the vault; such outlets could provide criminals with energy to drive their tools.

Container Protection

Because no container can resist assault indefinitely, it must be supported by alarm systems and frequent inspections. Capacitance and vibration alarms are the types most generally used to protect safes and file cabinets. Ideally any container should be inspected at least once within the period of its rated resistance. Closed-circuit television (CCTV) surveillance can, of course, provide constant inspection and, if the expense is warranted, is highly recommended.

By the same token, safes have a greater degree of security if they are well lighted and located where they can be seen readily. Any safe located where it can be seen from a well-policed street will be much less likely to be attacked than one that sits in a darkened back office on an upper floor.[10]

Continuing Evaluation

Security containers are the last line of defense, but in many situations, they should be the first choice in establishing a sound security system. The containers must be selected with care after an exhaustive evaluation of the needs of the facility under examination. They must also be reviewed regularly for their suitability to the job they are to perform.

Just as the safe manufacturers are continually improving the design, construction, and materials used in safes, so is the criminal world improving its technology and techniques of successful attack. Because of the considerable capital outlay involved in providing the firm with adequate security containers, many businesspeople are reluctant to entertain the notion that these containers may someday become outmoded—not because they wear out or cease to function, but rather because new tools and techniques have nullified their effectiveness. In 1990 a series of attacks on financial institutions in a major West Coast city, in which the burglars used drainage tunnels to enter vaults from beneath the facilities, pointed out that vaults are not impregnable.

In selecting security containers, it is important that the equipment conform to the needs of the risk, that it be regularly reevaluated, and that, if necessary, it be brought up to date, however unwelcome the additional outlay may be.

Inspections

In spite of all defensive devices, the possibility of an intrusion always exists. The highest fence can be scaled, and the stoutest lock can be compromised. A knowledgeable professional can contravene even highly sophisticated alarm systems. The most efficient system of physical protection can eventually be foiled.

It is necessary, therefore, to support each element of the system continually with another element—remember the concept of defense in depth. The ultimate backup surveillance must never let down.

Guard Patrols

Visual inspections by irregular patrols through office spaces or an industrial complex, or constant CCTV surveillance of these same areas, are vital to the success of the security program.

It is equally important to sweep the facility after closing time. "Hide-ins" are common in offices or retail establishments. These are thieves who conceal themselves in a closet or utility room and wait for the establishment to close and for everybody to go home. After hide-ins take what they are looking for, the only challenge is to break out. The chances of catching such thieves in a premise protected only by perimeter alarms are remote indeed. They must be picked up on the sweep when guards go through the entire facility from top to bottom or from east to west to see that everyone required to do so has left.

Specific duties of guards on patrol are discussed elsewhere, but in general, it should be noted that patrols should be made at least once each hour, more often if the area and the size of the guard force permit.

Particular attention must be paid to any signs of tampering with locks, gates, fences, doors, or windows. The presence of piles of rubbish or materials should be noted for the possibility of concealment—particularly if they are near the perimeter barrier or in the vicinity of storage areas.

In the patrol of office buildings, it is wise to stop occasionally for a long enough period of time to listen for any sounds that might indicate the presence of an intruder.

It is equally important that patrols in any facility be alert to any condition that might prove hazardous. These might be anything from an oil slick in a typically trafficked area to a heater left on and unattended. Those conditions presenting an immediate danger must be corrected immediately; others must be reported for correction. All of them must be noted in the log and on the appropriate form.

Alarms

In order to balance the cost factors in the consideration of any security system, it is necessary to evaluate the security needs and then determine how that security can, or more importantly should, be provided. Because the employment of security personnel can be costly, methods must be sought to improve their efficient use and to extend the coverage they can reasonably provide.

Protection provided by physical barriers is usually the first area to be stretched to its optimum point before looking for other protective devices. Fences, locks, grilles, vaults, safes, and similar means of preventing entry or unauthorized use are employed to their fullest capacity.

Since such methods can only delay intrusion rather than prevent it, security personnel are engaged to inspect the premises thoroughly and frequently enough to interrupt or prevent intrusion within the time span of the deterrent capability of the physical barriers. In order to further protect against entry should both barrier and guard be circumvented, alarm systems are frequently employed.

Such systems permit more economical use of security personnel, and they may also substitute for costly construction of barriers. They do not act as substitutes for barriers as such. But they can support barriers of lesser impregnability and expense, and they can warn of movement in areas where barriers are impractical, undesirable, or impossible.

The latest changes in this area add video clip confirmation of what is detected. This live streaming and recorded information make it possible to determine the priority of response and improve the probability of apprehension.

In determining whether a facility actually needs an alarm system, a review and evaluation of past experience of robbery, burglary, or other crimes involving unauthorized entry should be part of the survey preparatory to the formulation of the ultimate security plan. Such experience, viewed in relation to national figures and the experience of neighboring occupancies and businesses of like operation, may well serve as a guide for determining the need for alarms.

Kinds of Alarm Protection

There are three basic types of alarm systems providing protection for a security system:

1. *Intrusion alarms* signal the entry of persons into a facility or an area while the system is in operation.
2. *Fire alarms* operate in a number of ways to warn of fire dangers in various stages of development of a fire or respond protectively by announcing the flow of water in a sprinkler system, indicating either that the sprinklers have been activated by the heat of a fire or that they are malfunctioning. (Fire alarms will be discussed in detail in the following chapter.)
3. *Special-use alarms* warn of a process reaching a dangerous temperature (either too high or too low), of the presence of toxic fumes, or that a machine is running too fast. Although such alarms are not, strictly speaking, security devices, they may require the immediate reaction of security personnel for remedial action, and thus deserve mention at this point.

Alarms do not, in most cases, initiate any counteraction. They serve only to alert the world at large or, more usually, specific reactive forces to the fact that a condition exists for which the facility was fitted with alarms. However, many modern alarms are integrated with computer systems that can and do take predetermined actions in response to an alarm condition.

Alarm systems are of many types, but all have three common elements:

1. *An alarm sensor*: A device that is designed to respond to a certain change in conditions, such as the opening of a door, movement within a room, or rapid rise in heat.
2. *A circuit or sending device*: A device that sends a signal about whatever is sensed to some other location. This may be done via an electrical circuit that transmits the alarm signal over a telephone, fiber optic lines, or through air waves.

3. *An enunciator or sounding device*: A device that alerts someone that the sensor has detected a change in conditions. The device may be a light, a bell, a horn, a self-dialing phone, or a punch tape.

The questions that must be answered in setting up traditional simple alarm systems are:

1. Who can respond to an alarm fastest and most effectively?
2. What are the costs of such response as opposed to response of somewhat lesser efficiency?
3. What is the comparable predicted loss factor between these alternatives?

In modern integrated systems, the computer makes predetermined choices that include notification of appropriate personnel and perhaps activation of cameras and recording devices.

Alarm Sensors

The selection of the sensor or triggering device is dependent on many factors. The object, space, or perimeter to be protected is the first consideration. Beyond that, the incidence of outside noise, movement, or interference must be considered before deciding on the type of sensor that will do the best job. A brief examination of the kinds of devices available will serve as an introduction to a further study of this field.

Electromechanical Devices

These are the simplest alarm devices used. They are nothing more than switches that are turned on by some change in their status. For example, an electromechanical device in a door or a window, their most common application, is held in the open, or noncontact, position when the door or window is closed (see Figure 10-5). Opening either of these entrances breaks the magnetic contact, engaging the device and thus activating the alarm.

Such devices operate on the principle of breaking the circuit. Since these devices are simply switches in a circuit, they are normally used to cover several windows and doors in a room or along a corridor. Opening any of these entrances opens the circuit and activates the alarm. They are easy to circumvent in most installations by jumping the circuit, or tying back the plungers with string or rubber bands can defeat them from within.

Pressure Devices

These are also switches, activated by pressure applied to them. This same principle is in regular use in buildings with automatic door openers. In security applications, they are usually in the form of mats. These are sometimes concealed under carpeting, or, when they logically fit the existing decor, are placed in a strategic spot without concealment. Wires leading to them are naturally hidden in some way.

Photoelectric Devices

These use a beam of light, transmitted from as much as 500 feet away, to a receiver. As long as this beam is directed into the receiver, the circuit is inactive. As soon as this contact is broken, however briefly, the alarm is activated. These devices are also used as door openers.

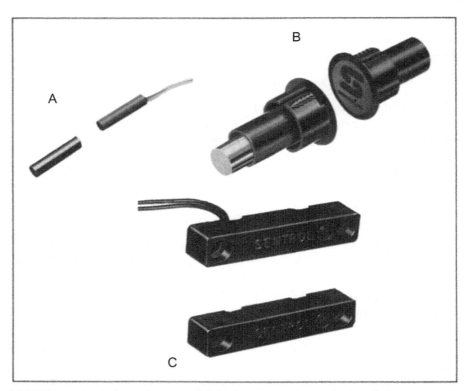

FIGURE 10-5 (A) Recessed switches and magnets help make a neat and attractive installation. Because they are concealed, they are more tamper-resistant. **(B)** Larger switches are sometimes easier to install. Since they are recess mounted, they are hidden from view, and their size is not noticeable. **(C)** Surface-mounted switches, although visible, are the least expensive and easiest to install. *(Photo courtesy of Ademco.)*

In security applications, the beam is modulated so that a flashlight or some other light source cannot circumvent the device, as can be done in non-security applications. For greatest security, ultraviolet or infrared light is used—although an experienced intruder can spot even these unless an electronic flicker device is incorporated into the device. Obviously the device must be undetectable because, once the beam is located, it is an easy matter to step over or crawl under it.

In some applications, a single transmitter and receiver installation can be used—even when they are not in a line of sight—by a mirror system reflecting the transmitted beam around corners or to different levels. Such a system is difficult to maintain, however, since the slightest movement of any of the mirrors will disturb the alignment, and the system will cease operating.

Motion Detection Alarms

These operate by radio frequency or ultrasonic wave transmission (see Figure 10-6). The radio frequency (or microwave) motion detector transmits waves throughout the protected area from a transmitting to a receiving antenna. The receiving antenna is set or adjusted to a specific level of emission. Any disturbance of this level by absorption or alteration of the wave pattern will activate the alarm.

FIGURE 10-6 Motion detector alarms. *(From Robert Barnard,* Intrusion Detection Systems *[Boston: Butterworth-Heinemann, 1981], p. 125.)*

The false-alarm rate with this device can be high, because the radio waves will penetrate the walls and respond to motion outside the designated area unless the walls are shielded. Some such devices on the market permit an adjustment whereby the emissions can be tuned in such a way to cover only a single area without leaking into outside areas, but these require considerable skill to tune them properly.

The ultrasonic motion detector operates in much the same way, as does the radio frequency unit, except that it consists of a transceiver that both transmits and receives ultrasonic waves. One of these units can be used to cover an area. Or they may be used in multiples where such use is indicated (see Figure 10-7). They can be adjusted to cover a single, limited area or broadened to provide area protection.

The alarm is activated when any motion disturbs the pattern of the sound waves. Some units come with special circuits that distinguish between inconsequential movement (such as flying moths or moving drapes) and an intruder.

Ultrasonic waves do not penetrate walls and are therefore unaffected by outside movement. They are not affected by audible noise in itself, but such noises can sometimes disturb the wave pattern of the protective ultrasonic transmission and create false alarms.

Passive infrared motion detectors do not transmit a signal for an intruder to disturb. Rather, moving infrared radiation is detected against the radiation environment of the room (see Figure 10-8). This detector is designed to sense the radiation from a human body. Sunlight, auto headlights, heaters, and air-conditioning units can trigger false alarms.

Dual-tech motion sensors combine the traits of passive infrared detectors with either microwave or ultrasonic technology.

Capacitance Alarm Systems

Also referred to as *proximity alarms*, capacitance alarm systems are used to protect metal containers of all kinds. This alarm's most common use is to protect a high-security storage

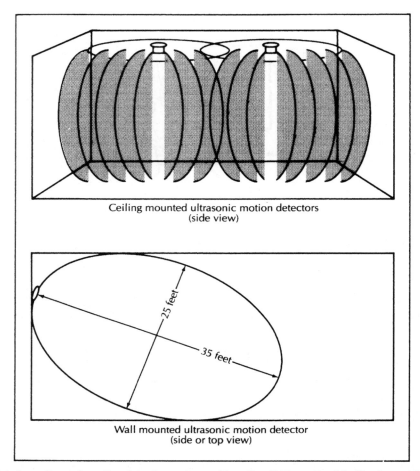

Ceiling mounted ultrasonic motion detectors
(side view)

Wall mounted ultrasonic motion detector
(side or top view)

FIGURE 10-7 Indoor ultrasonic motion detection patterns. *(From Don T. Cherry,* Total Facility Control *[Boston: Butterworth-Heinemann, 1986], p. 134.)*

area within a fenced enclosure. To set the system in operation, an ungrounded metal object (such as the safe, file, or fence mentioned above) is wired to two oscillator circuits that are set in balance. An electromagnetic field is thus created around the object to be protected. Whenever this field is entered, the circuits are thrown out of balance, and the alarm is initiated. The electromagnetic field may project several feet from the object, but it can be adjusted to operate only a few inches from the object when traffic in the vicinity of the object is such that false alarms would be triggered if the field extended too far.

Sonic Alarm Systems

Known variously as *noise detection, sound,* or *audio alarms,* these operate on the principle that an intruder will make enough noise in a protected area to be picked up by microphones that

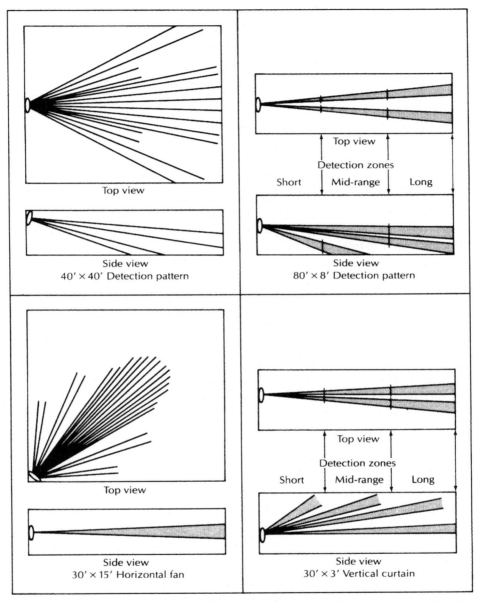

FIGURE 10-8 Indoor infrared motion-detection patterns. *(From Don T. Cherry,* Total Facility Control *[Boston: Butterworth-Heinemann, 1986], p. 135.)*

will in turn activate an alarm. This type of system has a wide variety of uses, limited only by the problems of ambient noise levels in a given area. The system consists simply of a microphone set in the protected area that is connected to an alarm signal and receiver. When a noise activates the alarm, a monitoring guard turns on the receiver and listens in on the prowler.

236 INTRODUCTION TO SECURITY

Such a system must be carefully adjusted to avoid setting off the alarm at every noise. Usually adjustment is set to sound the alarm at sounds above the general level common to the protected area. The system is not useful in areas where background noise levels are so high that they will drown out the anticipated sounds of surreptitious entry. This device may also come with a sound discriminator that evaluates sounds to eliminate false alarms.

Vibration Detectors

Vibration detectors provide a high level of protection against attack in specific areas or on specific objects. In this system, a specialized type of contact microphone is attached to objects such as works of art, safes, or files, or to surfaces such as walls or ceilings. Any attack on, or movement of, these objects or surfaces causes some vibration. This vibration is picked up by microphones, which in turn activate alarms. These units may be adjusted for sensitivity, which will be set according to their application and environment. Here again, discriminator units are available to screen out harmless vibrations. These units are very useful in specific applications because their false alarm rates are very low.

Certain additional alarm devices are currently in use as perimeter protection, as discussed in Chapter 9.

Some of the alarm sensors discussed in the preceding paragraphs can be defined as providing point protection. Electromechanical devices, capacitance alarms, and pressure devices, for instance, will be activated only when a specific area (point of entry) has been crossed, a door or window has been opened, or an object has been moved. On the other hand, ultrasonic, microwave, and passive infrared alarm systems protect large spaces, even entire rooms, against intrusion. These "volumetric" alarms, however, can be triggered by any number of environmental factors normally present in the protected space. Each alarm system has different strengths and weaknesses, and to assure effective performance, the total environment should be analyzed before selecting the specific system. Table 10-1 briefly summarizes the factors that affect these systems.

Alarm Monitoring Systems

Monitoring systems currently available are:

1. *The central station*: This is a facility set up to monitor alarms indicating fire, intrusion, and problems in industrial processes. Such facilities are set up for a number of clients; all are serviced simultaneously. ADT is a familiar name in this business. On the sounding of an alarm, a team of security officers may be dispatched to the scene and the local police or fire department may be notified. Depending on the nature of the alarm, on-duty plant or office protection is notified as well. Such a service is as effective as its response time, its alertness to alarms, and the thoroughness of inspection of premises fitted with alarms.
2. *Proprietary system*: This functions in the same way as does a central station system except that it is owned and operated by the company (for example, Deere and Company) rather than a contractor (for example, ADT), and is located on company property. Response to all alarms is generally by the facility's own security or fire personnel.

Table 10-1   What Detector to Select: Space Protection Guide

Environmental and Other Variables	Ultrasonic	Passive Infrared	Microwave
Vibration	No problem with balanced processing, some problem with unbalanced	Very few problems	Can be a major problem
Effect of temperature change on range	A little	A lot	None
Effect of humidity change on range	Some	None	None
Reflection of area of coverage by large metal objects	Very little	None unless metal is highly polished	Can be a major problem
Reduction of range by drapes, carpets	Some	None	None
Sensitivity to movement of overhead doors	Needs careful placement	Very few problems	Can be a major problem
Sensitivity to small animals	Problem if animals close	Problem if animals close but can be aimed so beams are well above floor	Problem if animals close
Water movement in plastic storm drain pipes	No problem	No problem	Can be problem if very close
Water noise from faulty valves	Can be a problem, very rare	No problem	No problem
Movement through thin walls or glass	No problem	No problem	Needs careful placement
Drafts, air movement	Needs careful placement	No problem	No problem
Sun, moving headlights through windows	No problem	Needs careful placement	No problem
Ultrasonic noise	Bells, hissing, some inaudible noises can cause problems	No problem	No problem
Heaters	Problem only in extreme cases	Needs careful placement	No problem
Moving machinery, fan blades	Needs careful placement	Very little problem	Needs careful placement
Radio interference, AC line transients	Can be problem in severe cases	Can be problem in severe cases	Can be problem in severe cases
Piping of detection field to unexpected areas by AC ducting	No problem	No problem	Occasional problem where beam is directed at duct outlet
Radar interference	Very few problems	Very few problems	Can be problem when radar is close and sensor pointed at it
Cost per square foot large open areas	In between	Most expensive	Least expensive
Cost per square foot divided areas, multiple rooms	Least expensive	Most expensive	In between
Range adjustment required	Yes	No	Yes
Current consumption (size of battery required for extended standby power)	In between	Smallest	Largest
Interference between two or more sensors	Must be crystal controlled and/or synchronized	No problem	Must be different frequencies

This list is intended as a guide only and does not represent absolutes but suggests areas for consideration.
Source: Aritech Corp.

3. *Local alarm system*: In this case, the sensor activates a circuit that in turn activates a horn, a siren, or even a flashing light located in the immediate vicinity of the area fitted with alarms. Only guards within hearing distance can respond to such alarms, so their use is restricted to situations in which guards are so located that their immediate response is assured. Such systems are most useful for fire alarm systems because they can alert personnel to evacuate the endangered area. In such cases, the system can also be connected to local fire departments to serve the dual purpose of alerting personnel and the company fire brigade to the danger as well as calling for assistance from public fire-fighting forces.

4. *Auxiliary system*: In this system, installation circuits are connected to local police or fire departments or 911 centers by leased telephone lines or in very modern designs wireless systems. The dual responsibility for circuits and the high incidence of false alarms have made this system unpopular with public fire and police personnel. In a growing number of cities, such installations are no longer permitted as a matter of public policy.

5. *Local alarm-by-chance system*: This is a local alarm system in which a bell or siren is sounded with no predictable response. These systems are used in residences or small retail establishments that cannot afford a response system. The hope is that a neighbor or a passing patrol car will react to the alarm and call for police assistance, but such a call is purely a matter of chance.

6. *Dial alarm system*: This system is set to dial a predetermined telephone number or numbers when the alarm is activated. The number(s) selected might be the police, the subscriber's home number, or both. When the phone is answered, a recording states that an intrusion is in progress at the location fitted with alarms. This system is relatively inexpensive to install and operate, but because it is dependent on general phone circuits, it could fail if the line(s) being called were busy or if the phone connections were cut.

Cost Considerations

The costs involved in setting up even a fairly simple alarm system can be substantial, and the great part of this outlay is not recoverable should the system prove inadequate or unwarranted. Its installation should be predicated on exposure, concomitant need, and on the manner of its integration into the existing, or planned, security program. Even companies that offer low cost or free installation of home and small business systems require a monthly contract to monitor the newly installed alarms. Nothing is totally free.

This last point is important to consider because the effectiveness of any alarm procedure lies in the response it commands. As elementary as it may sound, it is worth repeating that most alarms take no action; they only notify that action should be taken. There must therefore be some entity near at hand that can take that action. Too often, otherwise effective alarm systems are set up without adequate supportive or responding personnel. This is at best wasteful and at worst dangerous. The best systems are integrated with computers so that the computer can initiate action while waiting for human response.

In many instances, alarm installations are made in order to reduce the size of the guard force. This is sometimes possible. At least if the current guard force cannot be reduced in number, those additional guards needed to cover areas now fitted with alarms will no longer be required.

Good business practice demands that the expense of alarm installations be undertaken only after a carefully considered cost and effectiveness analysis of all the elements. If the existing security personnel can cover the security requirements of the facility, no alarm system is needed. If they can cover the ground but not in a way that will satisfy security standards, more guards are needed, or the existing force must be augmented by an alarm system to extend their coverage and their protective ability. The costs and effectiveness must be studied together with an eye toward the efficient achievement of stated objectives.

Summary

Almost all sources of the security equipment industry predict growth, indicating that there is a growing use of technology and equipment to supplement security staffs. Much new technology

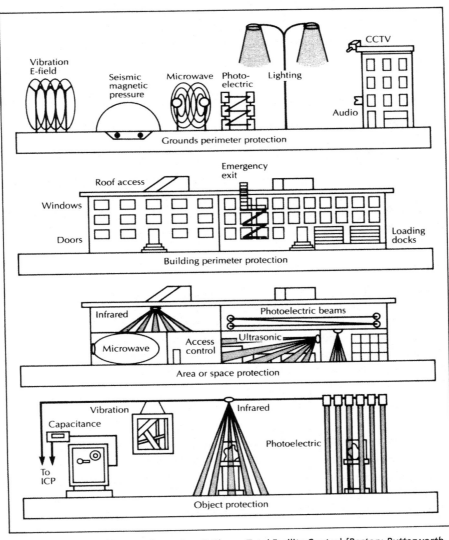

FIGURE 10-9 In-depth physical security. *(From Don T. Cherry,* Total Facility Control *[Boston: Butterworth-Heinemann, 1986], p. 148.)*

and equipment is available to the security manager today. Some of it is useful, some not, but none of it is better than the use to which it is put or the system into which it is integrated. No equipment can stand on its own. It can be used only if it is employed properly, fully, and effectively, and to be effective, all the components of the security system must work together. Figure 10-9 shows what security devices and types of barrier protection are commonly used for each line of defense.

A reasonable level of security cannot necessarily be assured with a single line of defense. Depending on the level of security desired, several layers of protection in an integrated system may be required.

■ ■ CRITICAL THINKING ■

Knowing about the various security techniques that are available for the protection of assets, is it possible to develop a virtually attack-proof security system? Is this even desirable? What are the potential problems, if any?

Review Questions

1. What are the elements necessary for an effective visitor access control system?
2. Explain the characteristics of each of the common types of alarm systems.
3. Give an example of a situation in which a motion-detection alarm might be deployed effectively. Under what circumstances would an ultrasonic system be chosen over a radio frequency system?
4. Describe how passive infrared detectors operate. What environmental variables must be considered to assure proper operation?

References

[1] Homeland Security Presidential Directive/Hspd-12, Office of the Press Secretary, White House, August 27, 2004, <www.whitehouse.gov/news/releases/2004/08>.

[2] Benson P. Secret Service report details 91 breaches, CNN Politics, <http://articles.cnn.com/2009-12-07/politics/secret.service.breaches>, downloaded 9/19/2010.

[3] Face in Hand; Biometrics Company Bioscrypt to Merge with A4Vision Facial Recognition Company Security Magazine, e-mail list serve January 23, 2007.

[4] Homeland Security Presidential Directive/Hspd-12,

[5] Diebold Direct Security Catalog. Canton, OH: Diebold; 1990. p. 10.

[6] Ibid.

[7] <www.vaultandsafe.com/safe_ratings_classifications.shtml>, downloaded 9/24/2010.

[8] <www.klsecurity.com/fireproof_safe.htm>, downloaded 9/24/2010.

[9] <www.vaultandsafe.com/safe_ratings_classifications.shtml>, downloaded 9/24/2010.

[10] Jeffery CR, Hunter RD, Griswold J. Crime prevention and computer analysis of convenience store robberies in Tallahassee, Florida. Florida Police J 1987:65–9.

11

Contingency Planning, Fire Protection, Emergency Response and Safety

OBJECTIVES

The study of this chapter will enable you to:

1. Discuss the role of contingency planning in emergency response, crisis management and business recovery and resumption.
2. Identify the basics of fire science.
3. Consider basic fire-fighting devices and techniques.
4. Discuss why safety is an aspect of many security programs.

Introduction

No facility protection program is complete without clear, well-defined policies and programs confronting the possible threat of fire or any other natural or human-made disaster. While planning for such contingencies is the responsibility of top management, in most situations the task of carrying out emergency response falls specifically on security. This is primarily due to the essence of the security mission—that is, to observe and report. In the best of all possible worlds the responsibility is a shared responsibility between security, fire, and safety departments. Regardless of the functional placement of responsibility, security, fire, and safety personnel must work together when they are confronted with preparing for and responding to disasters.

According to a 2006 IOMA Safety & Security Reports briefing, 39% of U.S. companies lack a basic crisis plan and 56% have not conducted crisis drills or simulation in the last year. Under an amendment to a 9/11 bill passed by the House, the Department of Homeland Security and the American National Standards Institute established a set of "best practices" for disaster preparedness. This includes a certification process to verify compliance.[1]

According to Dennis F. Sigwart, current and future security professionals should be aware of the absolute necessity of disaster planning and preparedness as a viable component of the many facets (fire, earthquake, explosions, flooding, and so forth) of which they will have to perform as a practitioner. Those assigned disaster preparedness tasks must continually play the "what happens if" game.[2]

Fire, safety, and emergency (contingency) planning is designed to anticipate what might happen to endanger people, physical assets, and information (thus causing damage and interruption to normal business) and to take the necessary preventive measures and make provision—through the use of appropriate hardware and/or personnel response—for prompt and effective action when an emergency does occur. While the emphasis in this chapter (as in most actual practice) is on physical safeguards, it is important to emphasize the human aspect of fire, safety and emergency protection. Disastrous losses often occur not from the failure or absence of physical safeguards, but from human error—the failure to close a fire door, to maintain existing protection systems in good working condition, to inspect or to report hazards, and, at the management level, to ensure through continuous employee education and training that the organization remains prepared at any time for any emergency. The Occupational Safety and Health Administration (OSHA), National Fire Protection Association (NFPA), and Life Safety Codes dictate certain safety requirements for all businesses. In addition, the NFPA, Factory Mutual, and Underwriters Laboratories (UL) have established standards for fire and safety that have been adopted by many state and local governments. These standards have been important in helping various insurance companies establish their rating systems.

Contingency Planning

The Association of Contingency Planners (ACP), which is an association dedicated to the evolution of business continuity, describes contingency planning in the following way: "Business continuity planning integrates knowledge from related disciplines such as information technology, emergency response, and crisis communications to create a strategy that ensures a business will remain resilient in the face of adversity."[3]

The purpose of contingency planning is simple. Essentially, contingency planners work to prepare their business, organization or institution to be better able to mitigate any disruption to normal business activities. As an example, if a natural occurrence (e.g., hurricane, fire, or earthquake) disrupts normal business activities, having plans in place for responding to and recovering from such an occurrence will allow for a faster resumption of business, thus reducing the amount of time the business is disrupted.

For our purposes, we will discuss contingency planning in the construct of four major components: emergency response, crisis management, business recovery, and business resumption. The fundamental elements of each component and the need for an effective integrated contingency planning process will be addressed. Furthermore, categories and types of crises, along with basic preparation and awareness needs, will be discussed. The reader will note that emergency response, crisis management, business recovery, and business resumption processes have much in common (for example, communications requirements); however, each is handled as a standalone process.

Security and the Contingency Planning Process

The traditional role of security in the contingency planning process has been to develop emergency evacuation plans for the business and to respond to emergency or crisis situations. Acting as the eyes and ears for an organization, business, or facility and maintaining a 24-hour day,

7-day a week presence, the security organization is best positioned to respond to an emergency and manage a crisis. As crises escalate, they are best managed by a multi-disciplined team.

Because of the ever-ready posture of many security organizations and the increased emphasis on emergency preparedness and contingency planning following the tragic events of September 11, 2001, in New York City and Washington D.C., many security departments have expanded their contingency planning capabilities to include the following components: emergency response, crisis management, business recovery, and business resumption.

Depending upon the scope of the effort, a contingency planning program can take into consideration many activities, events, conditions and processes. Depending upon the size and complexity of a business the process of contingency planning can be quite extensive. Planning for a contingency generally means assessing and understanding all aspects of the business, particularly the business critical processes and supporting information systems. To do this effectively requires the participation of many people from different disciplines, including management, employees, suppliers, and sometimes even customers. It may also include representatives from external organizations such as representatives of an insurance underwriter or the local fire departments.

Having so many people involved and from the many different functional disciplines calls for establishing common parameters. To be effective, everyone involved must have a common understanding of the elements and objectives of the contingency planning program and all must have a common understanding of the process. The first consideration in establishing common parameters is to develop a set of common definitions of terms. When discussing any aspect of contingency planning it is essential that all parties have a common understanding of what is being discussed. Just what does someone mean when he or she refers to the crisis management, business recovery or any other elements of the contingency planning process?

Below are a set of contingency planning terms defined in such a way as to be useful for any organization in establishing a common baseline, points of reference and common jargon for the contingency planning process. Definition of terms must be part of the organization's formal or institutionalized contingency planning process to ensure continuity of planning and success in achieving common preparedness objectives.

- *Business continuity*: Minimizing business interruption or disruption caused by different contingencies—that is, keeping the business going. Business continuity plans encompass actions related to how an organization prepares for, manages, recovers, and ultimately resumes business after a disruption.
- *Business recovery*: Refers to the short-term (less than 60 days) restoration activities that return the business to a minimum acceptable level of operation or production following a work disruption. Commonly used interchangeably with the term *disaster recovery*.
- *Business resumption*: The long-term (more than 60 days) process of restoration activities after an emergency or disaster that return the organization to its pre-event condition. (Keep in mind that restoration to the exact pre-event condition may not be necessary or even desirable. However, making this determination may not be possible without proper planning or going through the actual resumption process.)

- *Contingency*: An event that is possible but uncertain in terms of its occurrence, or that is likely to happen as an adjunct to other events. Contingencies interrupt normal business activities. In some cases the disruption is minor, while in other situations the disruption can be catastrophic.
- *Contingency planning*: The process of planning for response, recovery and resumption activities for the infrastructure, critical processes and other elements of an organization based upon encountering different contingencies.
- *Crisis management*: The process of managing events of a crisis to a condition of stability. This task is best accomplished by an integrated process team (IPT) made up of members from different disciplines throughout the organization. When formed, the team becomes the organization's crisis management team (CMT) and serves as the site, business or organization's deliberative body for emergency response and crisis management planning and implementation.
- *Critical processes*: Activities performed by functions, departments or elements within a business or organization that, if significantly disrupted due to an incident, emergency or disaster, would have an adverse impact on organizational operations, revenue generation ability, production and/or distribution schedules, contractual commitments or legal obligations.
- *Emergency response*: The act of reporting and responding to any emergency or major disruption of the business organization's operations.

Contingency Planning Program

The purpose for contingency planning is to better enable a business or organization to mitigate disruption to the enterprise. Should disruptions occur, and they do all too often, the enterprise must be able to resume normal business activities as quickly as possible. The inability to restore normal operations will have an adverse economic impact on the enterprise. The extent of the impact will correspond to the extent of the disruption or damage. If the damage is severe and the mitigation of such damage has not been properly planned for, the effect could be catastrophic, even to the extent of failure of the business.

Essentially, contingencies fall into three categories:

1. Those that impact the business infrastructure (fire, severe weather, earthquakes: see the definition of hazards further in this section) causing physical damage.
2. Those that impact people, such as accidents, seasonal illnesses (influenza), epidemics or pandemics causing harm to employees, rendering them unavailable to work.
3. Those that impact the reputation of the business (such as a product defect leading to a recall), causing resources to be diverted from normal operations to recovery and/or restoration. Each contingency has the potential to disrupt normal business operations to some degree. A minor building fire may disrupt operations in a limited way for only a couple of days, whereas a major fire may destroy an entire factory, completely stopping operations for an extended period.

Contingency planning is a continuous process. It is not something that can be done once and put away only to be retrieved when needed. It is a continuous process requiring periodic updates and revisions as appropriate to, and consistent with, changing business conditions. It also involves implementing and maintaining awareness and training elements. Those personnel with contingency planning responsibilities require periodic familiarization with plans and processes and training on new techniques and methods. The process of contingency planning should be designed to achieve the following:

- Secure and protect people. In the event of a crisis, people must be protected (employees, visitors, customers and suppliers).
- Secure the continuity of the core elements of the business (the infrastructure and critical processes) and minimize disruptions to the business.
- Secure and protect all information systems that include or affect supplier connections and customer relationships.

Throughout the remaining sections of this chapter, elements of the contingency planning process and program (Figure 11-1) are presented and explained.

Contingency Plans

Contingency plans formally establish the processes and procedures to protect employees, core business elements, critical processes, information systems and the environment in the event of an emergency, business disruption or disaster. These plans should be developed and designed to consider specific categories and types of emergencies and disasters and address the mitigation, preparedness, and response actions to be taken by employees, management and the organizations charged with specific response and recovery tasks. These plans should contain basic guidance, direction, responsibilities, and administrative information and must include the following elements:

- *Assumptions*: Basic assumptions need to be developed in order to establish contingency planning ground rules. As a baseline for planning, it is best to use several possible "worst-case" scenarios relative to time of event, type of event, available resources, building/facility occupancy, evacuation of personnel, personnel stranded on site, and environmental factors such as weather conditions and temperature. Furthermore, consideration should be given to establishing response parameters for emergency events. Define (for your enterprise) what constitutes a minor emergency, a major emergency and a disaster.
- *Risk assessment and vulnerability analysis*: Identify known and apparent vulnerabilities and risks associated with the type of business and geographic location of the enterprise. An assessment of risk and vulnerabilities should be made prior to developing or upgrading contingency plans. All planning will be accomplished in accordance with a thorough understanding of actual and potential risks and vulnerabilities. For example, in a petroleum refining facility, contingency plans for petroleum spillage, contamination, and fires must be considered. Furthermore, if located in an earthquake zone, planning must address

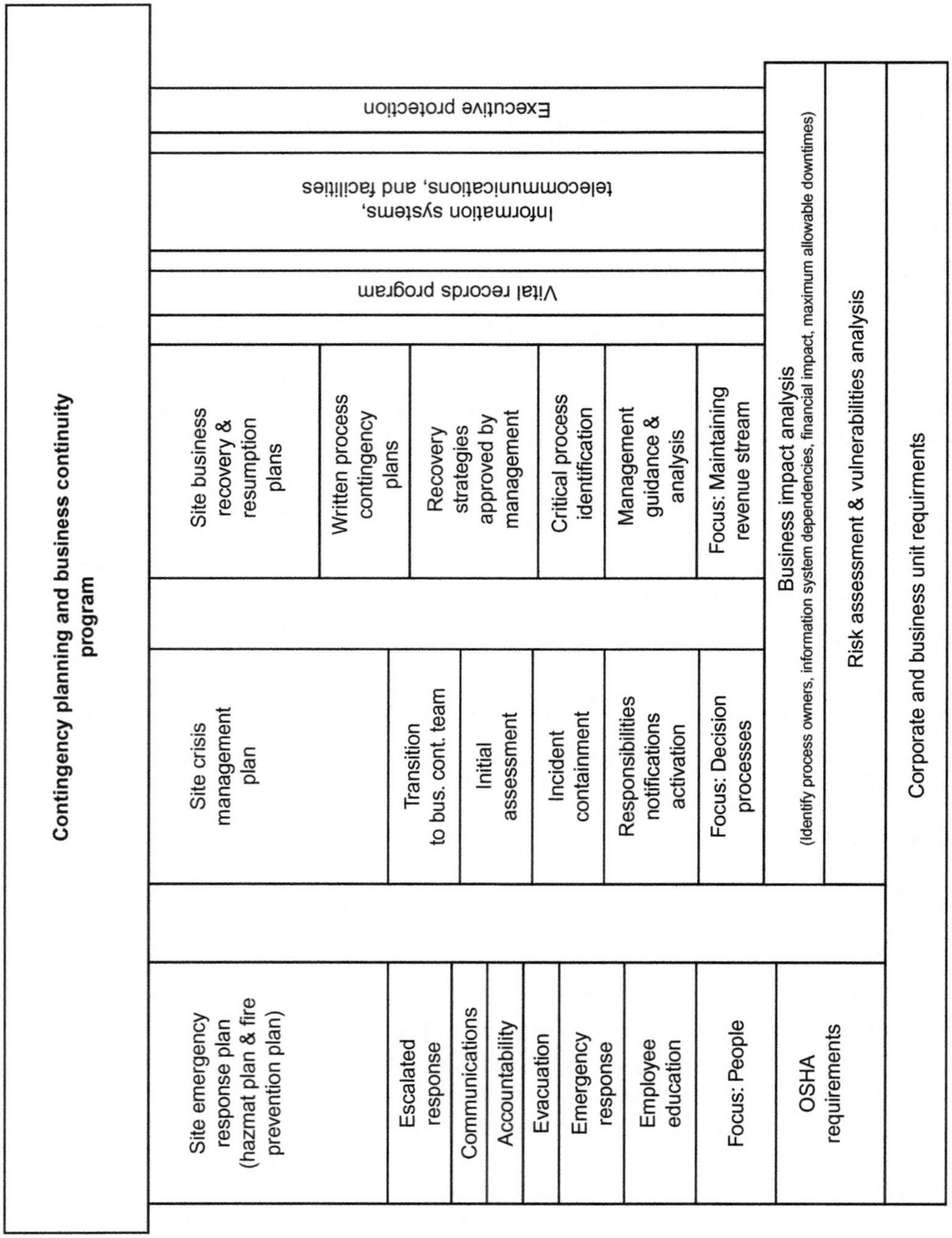

FIGURE 11-1 Elements of a business continuity planning program.

associated hazards. The risk assessment and vulnerability analysis must also include an assessment of enterprise-critical relationships. That means involving suppliers and customers in the contingency planning process. If a critical supplier or many key suppliers are not also prepared for various potential contingencies, their inability to recover will adversely impact your enterprise.

- *Types of hazards*: Planning for each and every type of hazard is not practical, nor desirable. Grouping them into similar or like categories will allow for planning to address categories of hazards. Since many hazards have similar consequences and result in like damages, it is best to plan for them in categories. The following is a list of common hazards: Medical Emergencies; Fires; Bomb Threats; High Winds; Power Interruptions; Floods; Hurricanes; Snow/Ice Storms/Blizzards; Hazardous Materials Issues; Aircraft Crashes; Civil Disorders; Earthquakes; Terrorist Threats/Activities; Workplace Violence; Explosions; and Tornados.

- *Critical process identification*: Critical processes must be ranked in accordance of criticality and importance to the productivity and survivability of the enterprise. Process of recovery must be focused on those critical processes that, when resumed, will restore operations to a minimal acceptable level. In essence, these processes are identified to be the first processes restored in the event of a major interruption to business operations. Failure to restore them presents the greatest possibility of damage or loss to the enterprise and could lead to the loss of a competitive edge, market share or even the viability of the enterprise.

- *Business impact analysis*: A business impact analysis must be accomplished to accurately determine the financial and operational impact that could result from an interruption of enterprise operations. Moreover, all critical interdependencies, those processes or activities critical processes are dependent upon, must be assessed to determine the extent to which they must be part of the contingency planning process.

- *Emergency response*: All participants in the emergency response process, particularly emergency responders, must understand their role. Expectations and responsibilities of emergency response personnel must be well defined and documented. Guidance for all employees on how to react in the event of an emergency and what their individual and collective responsibilities are must be documented and shared. Organizational responsibilities must also be established, to include the development of department-level emergency plans, generally for mid-size and large organizations. Events such as building evacuation and roll-call assembly need to be well defined so, in the event of an actual emergency, there is no confusion or uncertainty as to what must be accomplished.

- *Incident management and crisis management*: As an incident escalates, a crisis management team (CMT) should assume responsibility for managing the crisis. How this process works and who has what responsibilities must be clearly stated in the contingency plans. In the event of an actual emergency, some unauthorized people will attempt to manage the incident or participate in crisis management; however, they should not have any role in this process unless they were previously identified and trained as part of the crisis management team. Without established and well-defined incident management protocols and procedures, chaos is likely to erupt. Essentially, incident management and crisis management personnel must be trained and must understand their responsibilities.

Where practical, back-up supporting personnel should be identified and trained in the event that primary personnel are not available.

- *Incident/event analysis*: After an event occurs and the situation is stable, an analysis of what occurred and why should be conducted to determine the immediate extent of damage and the potential for subsequent additional damage.
- *Business resumption planning*: The process of planning to facilitate recovery of designated critical processes and the resumption of business in the event of an interruption to the business should be performed in two parts. The first part focuses on business recovery in the short term while the other part focuses on business restoration in the long term. This process will also include establishment of priorities for restoration of critical processes, infrastructure and information systems.
- *Post-event evaluation*: An assessment of preceding events to determine what went well, what did not go so well, and what improvements to existing plans need to be made must also be part of the process. Learning from real events is an unfortunate opportunity. There is no better way to learn how to handle an emergency than to actually handle one.

Emergency Response

When an emergency occurs, and unfortunately emergencies occur at even the most prepared businesses, being able to effectively respond is critical. The type and nature of emergencies that can occur vary widely. From a medical emergency in which an employee becomes injured or sick, to a natural or manmade disaster causing extensive damage to buildings and equipment, being prepared to respond will usually lessen the damage or impact of the event. Preparedness takes many forms. Being prepared to respond to a medical emergency is different from being prepared to respond to a natural disaster. The medical emergency may only require applying first aid to a victim or it may require the assistance and services of medical professionals. A natural disaster may require support from emergency medical services along with law enforcement, fire departments, search and rescue operations and hazardous material crews.

When planning for emergencies, types of emergencies should be grouped into like categories so that planning is accomplished for only categories of emergencies, as opposed to each and every possible emergency occurrence. This strategy recognizes the similarities of different types of emergency and is efficient in terms of creating fewer, but flexible, actual plans.

The purpose of preparing an emergency response plan is to document the planning accomplished in preparation for an emergency. This documentation provides the ground rules for emergency response activities. It also provides a reference for all who need to know how the process works. The plan will identify general and specific responsibilities for emergency response personnel and for all employees, both management and non-management. Having a plan in place will assist emergency response personnel in their effort to return the business to normal operations.

- *Reporting emergencies*: Employees must know how, and to whom, emergencies should be reported. If handling an emergency is beyond the internal capability of an organization, additional external assistance can be sought. For example, a seriously ill employee may require immediate medical attention. If paramedic capabilities exist within the company then the in-house paramedic should be the first respondent. If the situation calls for more sophisticated expertise and capabilities, external emergency medical services can be called for.

- *Communications and warning systems during an emergency*:
  - *Fire alarm systems*: These systems are generally the most widely used. Linked to a variety of sensor detectors and manual pull stations, fire alarms do just that: sound an alarm. These systems are sufficiently unique in sound and volume as to clearly indicate the need for building and facility evacuation. Employees must be conditioned to respond immediately.
  - *Public address systems*: These systems can be used to augment the fire alarm system. Announcements can be made alerting employees to the danger of fire. Announcements alerting employees to other types of dangers can also be made. Public address systems are particularly useful during emergencies when a building or facility evacuation is just the opposite of what is needed. For example, in the event of a chemical discharge or other environmental hazard, it may be necessary to keep people inside the facility and shut down all air movement systems, thus preventing employees from exposure to hazardous airborne substances. Since employees are conditioned to evacuate a building or facility when a fire alarm is sounded, they can be conditioned to wait and listen for specific instructions provided over a public address system.
  - *Floor wardens*: The use of employees to augment the emergency notification system has much value. Specially selected and trained employees can be given responsibility to act during an emergency to spread the word to evacuate a building or facility during an emergency. Assigning each a specific area of responsibility (or floor, hence the term floor warden) ensures complete coverage of the building or facility. Communications between floor wardens and emergency response personnel or a security emergency operations center can be easily established. Floor wardens can be alerted by pager, cell-phone or other means (including a variety of wireless devices) in the event of an emergency and be instructed to react to the specific situation. Floor wardens can and should be empowered and trained to react on their own in the event they recognize danger. Authority should be provided to floor wardens to evacuate a building or facility based upon their judgment and assessment of an emergency situation. In the event of a complete communications failure, it may be necessary to empower them to dispatch people to a safe environment.
  - *Response to emergencies:* Since security officers are generally located throughout the facility, they are usually the first to respond. Here, the officer can assess the situation and make a determination if additional assistance is necessary. In some cases, they may not be able to make an assessment and may require support from others. For example, in the event of a hazardous chemical spill, it will be necessary to have an expert in environmental and safety issues on the scene to make the assessment. It may even be necessary for a hazardous materials (HAZMAT) crew to respond to handle the event. Clean up of a chemical spill should only be done by skilled and certified personnel.

Clearly, defining who has what response capabilities and responsibilities will impact the effectiveness of any response.

- *Department-specific emergency plans*: It is best to have one emergency response plan for each company facility. These plans should be incorporated into a master plan and provide a common framework for all sub-elements of the plan. A key sub-element of an emergency plan is the individual departments' emergency plans. Those plans must specifically identify the following information:
  - Common and unique responsibilities in the event of an emergency to include:
    - A roster of department employees
    - Emergency contact/notification roster (not all emergencies occur during working hours so it may be necessary to reach people at home)
  - Identify floor wardens
  - Evacuation routes, procedures and assembly areas
  - Roll-call instructions
  - Procedures for evacuation of people requiring assistance
  - People identified as members of a search-and-rescue team
  - Additional manager- or employee-specific responsibilities
- *Incident management*: Personnel trained in handling emergencies should manage the incident at the scene. If the incident escalates to a crisis, a company crisis management team should be convened to manage the crisis. The senior emergency response person, when at the scene, should manage the incident with the assistance of specialists as appropriate.
- *Evacuation and assembly*: A critical objective during any emergency is employee safety. In the event it is necessary to evacuate a building or facility, having an established and orderly process is essential. Once a warning system sounds the notice to evacuate, employees must be aware of pre-established procedures for quick evacuation, including primary and alternate evacuation routes and where they should assemble. Maps or diagrams with this information should be included in the plan and posted throughout the work area. A floor warden or an employee with the assignment to facilitate evacuation should make a sweep of the area prior to their own evacuation to ensure all personnel have exited the building or facility. Once in the predetermined assembly area, a roll call must be taken. Primary, secondary and tertiary responsibilities should be assigned to ensure someone is available to take roll call and report the results to security. If someone did not evacuate the facility, a search-and-rescue team or other emergency personnel may be required to re-enter the facility and provide assistance.
- *Emergency evacuation drills*: The efficient and complete evacuation of personnel from a building or facility in the event of an emergency is such an important event that periodic drills should be conducted to reinforce the process and its importance. At least annually, each building or facility should undergo an evacuation drill where employees respond to a warning and completely evacuate the building or facility. A roll call should be conducted and results reported to security and senior management.
- *Search and rescue*: In the event of serious damage such as a fire or collapse of building, it may be necessary to search for persons not accounted for. Search and rescue is the

responsibility of responding emergency personnel who have proper protective equipment. Persons not trained in search-and-rescue techniques or who do not have proper equipment should not enter hazardous areas and conduct searches.

- *Return to work*: The process for returning to work after a crisis should also be included in the emergency plan. After any incident where employees are required to leave their work area and evacuate a building or facility, a process for having them return to work is necessary. For example, in the event of a false fire alarm where employees have evacuated a building, a means of communicating to them an all-clear, safe to return to work signal, is needed. This can be accomplished in many ways. Public address announcements can be made or plant protection personnel can go to assembly areas, directing employees to return to work. As appropriate, other methods may also be employed. In the event there is actual damage and employees cannot return to work, a process should be established identifying who makes the decision to send employees home, as well as how that is communicated to them and how they are kept apprised of event updates. For example, if a building was severely damaged due to fire and cannot be occupied for several days, posting daily direction and guidance for employees on the company website or on an emergency toll-free phone line will allow employees to call each day for specific instructions. For this to be effective, employees must know this process, must know the phone number to call or website to access and, as with all other processes, this one must be updated regularly.
- *After action*: When any incident occurs that necessitates evacuation or results in injuries, major damage, or presents the possibility of major business interruption, an after action report must be prepared. The primary focus is twofold:
  - Document the events, circumstances and chronology.
  - Prepare a lessons-learned review. Include key personnel involved in responding to and managing the emergency so as to assess what occurred and how it could have been better handled.

Crisis Management

Emergencies, contingencies, business interruptions and other unplanned events happen. Sometimes the event itself is a crisis, such as a fire burning a building or facility. In other cases, an incident not responded to or managed properly at the scene may turn into a crisis. For example, failing to respond promptly to that small fire may allow for it to turn into a large fire.

Crisis management is the process of managing events of a crisis to a condition of stability. Crisis management is *not* incident management. Emergency response personnel at the scene of an incident manage the incident. If the incident escalates, becoming a crisis, it is then necessary to have a different group take charge. Ideally, a crisis management team, or CMT, consisting of experienced personnel from multiple disciplines, would come together to manage the incidents that develop beyond the capability and decision authority of emergency response personnel. Essentially, the CMT manages the crisis to closure.

After emergency response planning, crisis management planning is the next step in the continuum of the contingency planning process. A crisis management plan should address the following activities and concerns:

- *Crisis management teams*: Managing a crisis can't be left to emergency personnel only. When an incident escalates into a crisis, the situation becomes more complex, affecting different aspects of the business if not the entire business and requiring different skills to manage it. Employees with a broad understanding of the enterprise and its mission, goals and objectives are much better suited to manage a crisis than those with a more narrow perspective of the business. Ideally, a crisis management team is like an integrated process team. Skilled professionals representing different disciplines come together on a short-term basis to work on a specific issue or tasking. In the case of crisis management teams, the task is to serve as a deliberative body to plan and prepare for a crisis and, when a crisis occurs, manage that crisis so as to mitigate damage or its impact. Crisis management teams should include members with expertise in the following areas: security, human resources, site management, safety and environmental and safety services, business management and communications.
- *Disaster operations*: In the event of a crisis or disaster, it is to be expected that some personnel may not be able to immediately leave the site. For example, following an earthquake the surrounding area may not be safe for travel. Employees may have no choice but to seek shelter at the workplace for hours or days. Furthermore, emergency personnel may be needed on site for an extended period to assist in recovery operations. Being prepared to deal with this or similar scenarios is essential. Preparation will include ensuring sufficient supplies are on hand to meet the needs of a reasonable number of stranded or support personnel. It is necessary to ensure that sufficient food, water, medical supplies and emergency sanitation and shelter facilities are available. All of these items can be acquired and placed in a long-term storage condition, providing they are regularly checked for serviceability, spoilage and maintained within the expected shelf life. During a crisis, much uncertainty exists. Consequently, it will be necessary to communicate to employees, keeping them as up to date as possible about the situation and events and providing guidance concerning their safety and work expectations. During a crisis, employees are naturally anxious. Prompt and clear communications can help reduce this anxiety and keep employees informed. Communication may need to extend beyond the duration of a crisis into an undefined subsequent period. Using the previously referred to emergency contact and notification number, or company Web site, can be very effective. Messages can be updated regularly as needed so the information is current. Also, information broadcast on local news radio stations can reach a large population of employees. At the point in time where an incident escalates into a crisis, the crisis management teams become involved, managing the crisis to closure. At some point during a crisis, a de-escalation of events will occur and eventually the crisis will terminate. If the impact or damage from the crisis is significant, the crisis management team will commence with restoration activities. These activities may be led by the crisis management team or passed on to a business continuity team. How this can work will be discussed further in the next section.

- *Media relations*: During a crisis, it is possible that the local, national or even international media will become interested in events. For example, large industrial fires always draw the attention of local media. Natural disasters also draw much media attention. Even isolated events such as incidents of workplace violence can draw significant media attention. It is therefore important to have a media relations plan. The company media representative should be part of the company crisis management team. Since there is always a degree of unpredictability during a crisis, it is best that all crisis management team members understand how to deal with the media and be prepared should they be thrust into such a situation.

- *Damage assessment*: During a crisis, emergency personnel will make ongoing damage assessments, reporting status back to the crisis management team. These assessments are useful in determining actions to be taken next. However, these assessments are situational and due to the circumstances and nature of a crisis, do not have the luxury of thoroughness. The true extent of damage is not determined until after the crisis has terminated and a complete building, facility or site assessment can be made. Immediately following a crisis, a damage assessment for infrastructure safety and functionality must be made. Without this, a return-to-work decision cannot be made. The damage assessment is also the starting point for all restoration and resumption activities.

- *Business continuity team*: Earlier reference was made to the transition of responsibility from a crisis management team to a business continuity team. This is an important step in the effort to resume business. While the crisis management team's focus is on managing through the crisis, the business continuity team's focus is recovery and resumption. The role of a business continuity team will be discussed further in the next section.

- *After action/post event assessments*: After every crisis, an assessment of what occurred should be conducted. The chronology and circumstances of the event will be recorded. The crisis management team will review what went well and what did not. Performance to plan will be reviewed and a lessons-learned document will be created for all team members and supporting personnel to review and hopefully learn from.

Business Continuity

Earlier in this chapter, we defined business continuity as the effort to minimize business interruption or disruption caused by different contingencies. When contingencies occur, business recovery and resumption needs to happen as rapidly as possible. In essence, business must continue. Business disruptions can be costly and even catastrophic. Customers, shareholders and stakeholders demand the business remain viable. Preparation to deal with contingencies is a critical component of keeping the business going and maintaining the viability of the enterprise.

Business continuity is a two-stage process. Business recovery is the first stage. Business resumption is the second. The recovery effort is the process of getting the business up and running again but only in a minimal acceptable condition. It is not a recovery to a pre-event condition, but rather a recovery to produce product, make deliveries to customers and accomplish the basic activities to keep the business going.

The business resumption stage is the effort to recover from a contingency and resume business in a pre-event condition. This is not to say that all critical processes and other processes will be exactly the same as they were pre-event. Resumption planning may call for new or modified processes. The intent is to resume business operations to a level similar to the pre-event operations level, but not necessarily exactly the same.

A business continuity team should be established to provide oversight of the development of business resumption plans. Representation from each of the major business functions should be part of this team. Manufacturing, business management, finance, engineering, information technology, human resources, legal and others major areas and disciplines within the business, depending upon the nature of the business, need to participate. Business resumption teams lead the effort and planning process to ensure the business is prepared to recover from contingencies and resume full business operations. In some cases it may be necessary to have a major supplier or customer participate as a member of this team. Business recovery and resumption planning have common elements. The difference is the stage of recovery and the time necessary to get there. Following are common elements of the processes for business recovery and resumption:

- *Business impact analysis of critical processes and information systems*: The most fundamental aspect of recovery and resumption planning is conducting a business impact analysis of critical processes. Critical processes must first be identified. Knowing what they are and having the business continuity team agree to their criticality will allow for proper planning and prioritization. Failure to properly identify critical processes may lead to wasted time, effort and money. Even worse, non-critical processes may be given priority over critical processes, leading to further delays in recovery and causing the unnecessary expenditure of resources. It is not uncommon for organizations to identify their processes as critical, where upon further examination they are determined not to be critical. Process owners have a tendency to believe all of their processes are critical. This is precisely why it is necessary to have the business continuity team make this assessment. When developing recovery and resumption plans, the following areas must be considered and addressed:
  - *Define critical processes*: Each major business area, function and discipline should provide to the business continuity team a listing of all critical processes. The business continuity team should then review these processes for criticality and prioritize them, creating an official critical process list. Planning for recovery of the critical processes is the primary concern. Non-critical processes should be recovered and resumed after the critical processes. Resource and time limitations do not allow for resumption of all processes at the same time. Processes critical to the business must have top priority. Any processes determined not to be critical should be planned for during the later stages of the resumption effort.
  - *Critical process interdependencies*: As part of the critical process assessment, particular emphasis must be placed on information systems and process interdependencies. For example, an information system in and of itself may not be determined by its process owner to be critical. However, if it supports a critical process and that critical process

can't be completely restored without the information system, then that information system itself becomes critical. Examining processes as part of a system is essential in the assessment of criticality. Interdependencies need to be identified in order to properly assess criticality to the business. Other interdependencies may exist in the form of relationships with organizations outside of the enterprise. These too must be considered. Different methodologies can be used to estimate potential impact a contingency or disaster may have on a critical process. When considering the criticality of a process, the financial effect, operational effect and any less tangible or quantifiable concerns, such as customer satisfaction, must be addressed.

- *Resources*: Critical process recovery requires an assessment of resources. Planning for process restoration means considering what resources may no longer be available and will need to be acquired or obtained to get the critical process up and functional again. What type of facilities will be needed and where? Will additional hardware, software or equipment be required? Will people capable of managing and working the processes be available? Will there be effective means of communications? If not, what must be done to provide a minimum capability of communications until full communications can be restored? These are some of the resource issues and questions the team must grapple with.

- *Mitigation strategies:* For those processes identified as critical, pre-event actions can be taken to help mitigate the impact, both operationally and financially, of interruptions to the business. When developing contingency plans for critical processes, strategies will become apparent that may be implemented prior to an event that will lessen the impact of an event if and when it occurs. A cost/benefit analysis may be required to assess the feasibility of implementing a pre-event action and if the analysis shows it to be an effective action, it should be taken. For example, an old building not built to current building codes may be vulnerable to damage from an earthquake. If that building supports a critical process, it may be more cost effective to retrofit the building with the necessary structural supports and bring it into compliance with current standards than to risk severe damage in the event of an earthquake, rendering a critical process inoperative.

- *Vital records*: The ability to recover vital records is critical to the recovery and restoration process. Having a vital records protection and management program will enable the recovery of essential information during a contingency.

- *Customers and suppliers*: The importance of considering input, participation, and impact to customers and suppliers cannot be overstated. Any business continuity planning must take into consideration customer and supplier relationships. Moreover, it is important to work with your suppliers and providers of goods and services to ensure they too have contingency plans in place. In the event a supplier supports one or more of your critical processes, a disruption to their business will impact your business operations.

- *Communications*: Communicating during the recovery and resumption process can be just as important as communications during other phases of a contingency. Employees who may have been affected by the events of a crisis or disaster need to be kept abreast of developments affecting them and their employment. Customers and suppliers need to understand the progress made toward resumption of business, as it may have a serious impact on their operations. Even the external worlds of stakeholders and shareholders have an interest in these events.

- *Lessons learned*: There is an old adage that lightning doesn't strike twice in the same place. If only that were certain and true, and applicable to the critical processes of a business. However, it is not. Therefore, much can be learned from each phase of managing and recovering from a contingency. Documenting the process of recovery and restoration will help in identifying the things learned, both good and bad, and will go a long way toward helping to deal with other crises when they occur.

Business Recovery

The previous section addressed areas and issues common to resumption and recovery aspects of the total contingency planning process. This section will discuss areas specific to recovery and the short-term process of resuming normal business operations.

Recovery plans focus on getting the business up and running—in essence, the actions that need to be taken within the first 30 to 60 days to restore critical processes and resume operations. These should be the most critical processes focused on infrastructure, product delivery and keeping damage or loss to an absolute minimum. As difficult as it may be, people need to be part of this equation. For example, should a natural disaster occur, causing severe damage to a building or facility, there is a good chance that some key employees may have experienced something similar. Some may be preoccupied with their own issues of recovery and restoration and may not be able to support the company. Generally, you can expect this to be limited to a few, but it could be a critical few. Part of the critical process planning should take this into consideration and identify alternatives.

Vital records recovery is very much part of the recovery process. Being able to access off-site records storage, hard copy and electronic, is critical to expediently moving this process forward. Many companies use outsource providers to handle, store and retrieve their vital records. This process allows for separate storage, away from company facilities, and reduces the possibility of damage or destruction to these records. There are many capable and reliable companies throughout the world who perform vital records handling, storage and recovery.

Business Resumption

Issues and areas of focus and concern that are common with recovery and resumption were addressed earlier. This section discusses areas specific to resumption and the long-term process of resuming normal business. Long-term priorities are addressed in business resumption plans with the intention of restoring operations to a pre-event condition. Restoration to a pre-event condition does not necessarily mean that all is the same or equal to the conditions prior to contingency occurrence, crisis or disaster. During the process of recovery and restoration it may be learned or discovered that the implementation of a critical process or other processes can be accomplished differently, in the sense that improvements can make the process more efficient and more cost effective. Consequently, changes can and should be made. Furthermore, it may be learned that some processes can be eliminated altogether. Recovery and resumption in many ways are similar to a re-engineering process. Process owners are usually the best source for ideas and as they participate in resumption they may develop new approaches and methods to implement and execute their process.

If the process is simple, changes can be implemented quickly with little or no additional review from management or the business continuity team. If the process is complex, affecting or dependent on other processes, a cost-benefit analysis is warranted to accurately assess the impact of any proposed changes.

Within this chapter the authors have attempted to provide the reader with a framework for understanding the complexities of contingency planning and the development of contingency plans. A particular point we attempt to make lies with the importance of planning for categories of contingencies. It is a daunting task to attempt to plan for each and every possible contingency. However, contingencies can be grouped into categories and planned for accordingly. This allows for consistency in preparedness and best utilization of resources. Types of contingencies develop and change over time as societies and organizations change and progress. Prior to the 20th century, nuclear contamination was not a concern, but today countries with nuclear power generation capabilities have in place extensive contingency plans that are regularly tested. More common hazards such as severe weather and other natural events have caused enough damage to drive organizations to better preparedness. State and local governments along with private enterprises in states like California and Mississippi spend large sums of money to prepare to mitigate the effects of earthquakes and hurricanes.

Pandemics

Furthermore, not-so-common hazards drive governments and private organizations to take mitigating measures. Pandemic preparedness continues to receive much attention as the H5N1 Avian Flu and the H1N1 Swine Flu viruses remain active in various parts of the world, with the H5N1 being active mostly in Asia.[4] Pandemics are not new, having been with us since humankind's earliest time. They don't occur frequently but when they do, the effects can be devastating. The last devastating pandemic occurred in 1918, when the Spanish flu affected more than 30% of the population, killing between 50 and 100 million people worldwide and disrupting the normal lives of societies around the globe.[5]

Planning for a pandemic requires an emphasis on people. The focus is on planning to keep employees, and their families, healthy and in the workplace where they can be productive. Pandemics affect people, not infrastructure, although without people operating an infrastructure is at best difficult, and may be nearly impossible. Consider running the air transportation infrastructure without people. With a 30% reduction in the number of air traffic controllers, pilots and maintenance personnel, would this system work effectively, or would it even work at all? How would your business be affected if air transportation was limited or shut down for operating for 30 days?

The Center for Disease Control and Prevention (CDC) has created a Pandemic Severity Index to assist local and state governments in assessing the severity of a viral outbreak.

The level will help officials determine the extent of school closure, quarantines and work-from-home assignments.

- Category 1 involves less than 90,000 deaths and would not require school closures.
- Category 2 and 3 would recommend school closures and limiting personal contact for up to one month.
- Category 4 or 5 would potentially involve over 1.8 million deaths, school closures of up to 3 months and limits on public events.[6]

Fire Prevention and Protection

Although all industry-specific vulnerabilities should be considered in contingency planning, the threat of fire is universal. Because it is also one of the most damaging and demoralizing hazards, fire prevention and control must be a major part of any comprehensive loss-prevention program. The following materials are designed to provide an overview of this important area. For a complete discussion on fire seek out professional literature for each topic, such as arson, fire suppression, or fire prevention.

For someone who has administrative oversight of fire issues, it is important to note that any defense against fire must be viewed in two parts. First, *fire prevention*, which is usually the major preoccupation of most businesses, embodies the control of the sources of heat and the elimination or isolation of the more obviously dangerous fuels. This commendable effort to prevent fire must not, however, be undertaken at the expense of an equal effort for the second part of defense, *fire protection*.

Fire protection includes not only the equipment to control or extinguish fire, but also those devices that will reduce the effect of fire in relationship to the building, its contents, and particularly its occupants in the event of fire. Fire doors, firewalls, smokeproof towers, fireproof safes, nonflammable rugs and furnishings, fire detector and signalling systems—all are fire protection components that are essential to any fire safety program.

Security Personnel

One of the key elements in fire prevention and protection is security personnel. While a smoke detector senses smoke and a heat detector senses heat, the human brain senses much more. The trained security officer can think, solve problems, and sense what detectors cannot. We can often sense when someone is having a bad day. Likewise, we can also sense when the building just "doesn't feel right." There are no detectors to tell you that the exits are blocked, or the exit doors are not opening properly.

Vulnerability to Fire

There are no fireproof buildings, although frequently the term is misapplied. However, there are fire-*resistant* buildings, meaning one that will not collapse quickly under fire conditions and that does not readily add fuel to the fire. But combustible materials inside a fire-resistant building, such as furnishings, paneling, stored flammable materials, and so on, can make ovens out of these buildings, generating heat of sufficient intensity to destroy everything inside. Eventually such heat can even soften the structural steel to such an extent that part or all of a building may collapse. By this time, however, the collapse of the building may endanger only outside elements because many things inside, with the possible exception of certain fire-resistant containers and other metal or fire-resistant items and their contents, may already have been destroyed.

Heat was a major factor in the collapse of the World Trade Center. The heat generated by the burning jet fuel resulted in expansion and thus weakening of the steel superstructure. Coupled with the intense pressure of the burn, the building began to collapse. The weight of the collapsing structure created additional weight on the lower floors, eventually resulting in the collapse of the entire structure.

The particular danger of this situation is that while wood frame construction is recognized for the fire hazard it represents, many otherwise knowledgeable people are oblivious to the potential dangers from fire in steel and concrete construction. We have a normal tendency to think that because something does not burn, it is safer. The normal reaction would be that steel is better in a fire situation than wood. This is not always the case.

We cannot be blind to the fact that steel has its shortfalls. Though wood will ignite between 400 to 600 degrees, steel will start losing strength at 600 degrees and loses 40 percent of its strength at 1100 degrees, well within what common fires develop. While wood will burn, the shear mass of wood needed to match the strength of steel in any given construction project may mean that the structure will withstand the fire longer.[7]

Fireload

The degree of fire exposure in any fire-resistant building is dependent on its fire load—the amount of combustible materials that occupies its interior spaces. Fire load is often misunderstood when we look at different occupancies. We tend to look at some businesses as more hazardous, due to their operation, rather than taking into account their fire load. There may be more of a concern for a processing plant that fabricates steel products because of the intense heat present in the fabrication process than a doctor's office or hospital. A factory atmosphere catches our interest due to electrical equipment, machinery, and stock. But what is the true fire load? A doctor's office could warehouse many years of patient records, files, and x-rays, that could create a greater fire load than a factory. A hospital with its clean and safe environment needs to have supplies and replacement equipment. There could be storerooms of additional beds, furniture, and records. An in-house laundry department will add fire load with the bedding and gowns that workers and patients use. We need to look past the operation to see what will burn.

In the case of multiple occupancies where general businesses, retail, and residential units are under one roof, as in a large office building, no one office manager can have control over the building's fire load. In such an environment, new furniture, decorative pieces, drapes, carpeting, unprotected insulated cables, or even volatile fluids for cleaning or lubricating are moved through the building every day. The building fire load may continue to increase without much thought from most of the occupants.

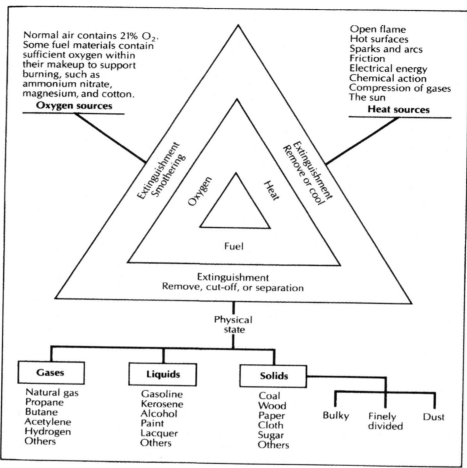

FIGURE 11-2 Fire triangle.

The Nature of Fire

The classic triangle frequently referred to in describing the nature of fire consists of heat, fuel, and oxygen. This triangle has been augmented by the fire tetrahedron theory, which adds a fourth element—chemical reaction—pyrolysis or vaporization. *Pyrolysis* is the decomposition of solids to the point where they give off enough flammable vapors and gases to form an ignitable mixture. In liquid fuels the process is called *vaporization*. Flammable gases require no pyrolysis because they are already in a form capable of combining with oxygen. If all four components exist, normally there will be a fire; remove or reduce any one, and the fire will be reduced or extinguished (see Figure 11-2).[8]

Fuel and oxygen are always present. It would be difficult to imagine any facility that had no combustible items exposed, and air most certainly will be present. Only sufficient heat and the chemical breakdown associated with it are missing, and these factors can be readily supplied by a careless cigarette or faulty wiring, two of the most common causative factors. There are, in fact, an almost infinite number of heat sources that can complete the deadly tetrahedron and start a fire raging in virtually any facility.

Every fire prevention program begins with the education of staff and visitors, because it is nearly impossible to change existing fire-load problems overnight. Yet every fire prevention program must also work to control the amount and nature of the fire load or fuel, and institute programs to prevent the occurrence of any heat buildup, whether from careless smoking or from sparks from a welding torch. This is actually a two-prong approach. First, control ignition sources. Second, set fire loads at appropriate working levels.

By-products of Fire

Contrary to popular opinion, flame or visible fire is rarely the killer in deaths from fire. Death is usually caused by smoke or heat or from toxic gas, explosion, or panic. Several such by-products accompany every fire; all must be considered when defenses are being planned.

Smoke will blind and asphyxiate in an astonishingly short time. Tests have been conducted in which smoke in a corridor reduced the visibility to zero in 2 minutes from the time of ignition. A stairway 2 feet from a subject in the test was totally obscured.

Fire gases, composed mostly of carbon dioxide and carbon monoxide, make up a large part of smoke. Still, most persons have a relatively lax attitude about smoke in comparison to fire. We need to understand that smoke is actually unburned fuel and gases that are still flammable. While most people would not walk into a kitchen when gas is coming out of the four burners of the stove, many people often fail to think of smoke in the same terms. Fire fighters are taking a much harder look at those burning gases in how they add to the fire and fire extension.

Heat plays an important part in the destructive capabilities of fire and its spread. A small fire that does not consume large amounts of fuel can produce large amounts of destructive heat. A fire that may just consume a sofa or chair can produce heat damage many times greater than the cost of the fire-involved item. Heat is also the main means by which fire travels. Flames do not cause a fire to move across an area, room, or building. It is the heat associated with the flames that leads to fire spread. Furthermore, heat can travel great distances and heights without flames, to start new points of ignition.

Ignition temperature is the temperature where a solid fuel will ignite without direct flame contact. Most materials we use in our everyday lives have an ignition temperature somewhere between 400 and 600 degrees, well below the temperatures developed in a fire. If one of these fuels is put into an oven that has no flame, when the fuel reaches its ignition temperature, it will ignite. No flame is required to start combustion. The point is that it does not matter where the fire is, the important question is, where is the heat going? For example, in a high rise building, a fire that starts in a basement where someone has left the stairwell doors open allows heat from the fire to travel to the upper floors of the building. Once the heat reaches that 400 to 600 degree range on those upper floors, fire will develop. It does not matter that the original fire is located 300 feet below the 30th floor. Fires will start!

Kind of fire	Approved type of extinguisher							
Decide the class of fire you are fighting …	Foam Solution of aluminium sulphate and bicarbonate of soda	Carbon Dioxide Carbon Dioxide gas under pressure	Pump tank Plain water	Gas cartridge Water expelled by carbon dioxide gas	Multi-purpose dry chemical	Ordinary dry chemical	Dry powder	Wet chemical
A **Class A fires** Ordinary combustibles • Wood • Paper • Cloth etc.	●		●	●	●			
B **Class B fires** Flammable liquids, grease • Gasoline • Paints • Oils etc.	●	●			●	●		
C **Class C fires** Electrical equipment • Motors • Switches etc.		●			●	●		
D **Class D fires** Combustible metals • Magnesium • Sodium • Potassium etc.							●	
K **Class K fires** • Cooking oils • Fats								●

How to operate

Foam: Don't spray into the burning liquid. Allow foam to fall lightly on fire.

Carbon Dioxide: Direct discharge as close to fire as possible, first at edge of flames and gradually forward and upward.

Pump tank: Place foot on footrest and direct stream at base of flames.

Dry chemical: Direct at the base of the flames. In the case of class A fires, follow up by directing the dry chemicals at the material that is burning.

Wet chemical: Direct at the base of the flames.

FIGURE 11-3 Use of fire extinguishers.

Fire evolves through three stages: incipient, free burning, and smoldering. The incipient stage is the moment the fire begins, when fuel, air, and an ignition source all come together at the right rate, for the correct period of time. Picture the instant a match falls in a trash can. There is little heat and little smoke; you can stand right in front of the trash can with no real concern. In the free burning stage, the fire is doing just what it wants, as it has plenty of fuel and air. However, as flames continue to develop with heat increasing, being close to the fire is almost impossible. In the smoldering stage air has been reduced or fuel is dwindling to the point where there is no visible flame, just an entire area filled with tremendous heat.

The plan of how to "fight" a given fire for extinguishment is determined by the stage that the fire is in. Another determining factor is whether the fire is confined or unconfined.

Most fires in a commercial setting are of a confined nature. An unconfined fire is similar to a campfire. The fire is doing just what it wants; the heat is going up, the smoke is going up, and it is doing what it naturally does. We could walk up to that fire and warm our hands or put the fire out by simply pouring a bucket of water on it. However, put that same fire in a room, confine it, and the heat and smoke still go up until they hit the ceiling, and then they to start to move laterally. The smoke and heat will not allow direct access to the fire anymore. This is why the fire stage and whether the fire is confined or unconfined are so important in determining how a fire is fought and the amount of extinguishment time involved.

Classes of Fire

All fires are classified into one of five groups. It is important that these groups and their designations be widely known because the use of various kinds of extinguishers is dependent on the type of fire to be fought (see Figure 11-3).

- *Class A*. Fires of ordinary combustible materials, such as wastepaper, rags, drapes, and furniture. These fires are most effectively extinguished by water or water fog. It is important to cool the entire mass of burning material to below the ignition point to prevent rekindling.
- *Class B*. Liquid fuel fires such as gasoline, grease, oil, or volatile fluids, with the exception of some cooking oils. In this type of fire an oxygen displacement effect such as carbon dioxide (CO_2) is used, or other fire extinguishing agents. A stream of water on such fires would simply serve to spread the substances, with disastrous results. Water fog, however, is excellent because it cools without spreading the fuel if applied by properly trained personnel.
- *Class C*. Fires involving live electrical equipment such as transformers, generators, or electric motors. The extinguishing agent is nonconductive to reduce the danger of electrocution to the firefighter. Electrical power should be disconnected before beginning extinguishing efforts, because you are basically extinguishing the Class A or B fuel around the energized equipment.
- *Class D*. Fires involving certain combustible metals such as magnesium, sodium, and potassium. Dry powder is usually the most, and in some cases the only, effective extinguishing agent. Because these fires can only occur where such combustible metals are in use or production, they are fortunately rare.

- *Class K.* In recent years studies have found that some cooking oils produce too much heat to be controlled and extinguished by the traditional Class B extinguishing agents. Class K fires and extinguishers deal with cooking oil fires.

Extinguishers

The security department must evaluate the fire risk for each facility or department and determine the types of fires most likely to occur. Although the potential for all types of fires exists and should be planned for, certain production areas are more likely to have a specific type of fire than are others. This condition should be considered when assigning extinguishers to the department or facility. Every operation is potentially subject to Class A and C fires, and most are also threatened by Class B fires to some degree.

Having made such a determination, security must then select the types of fire extinguishers most likely to be useful. The choice of extinguisher is not difficult, but it can only be made after the nature of the risks is determined. Extinguisher manufacturers can supply all pertinent data on the equipment they supply, but the types in general use should be known. It is important to know, for example, that over the years the soda/acid and carbon tetrachloride extinguishers have been prohibited. They are no longer manufactured. An extinguisher that must be inverted to be activated is no longer legal.

There are a number of considerations when choosing fire extinguishers to be used in a given facility. First, what type of fire do you anticipate, as stated above? Second, how compatible is the fire extinguisher with the environment and personnel who are going to use it? Is the fire extinguishing agent inside the fire extinguisher going to do more harm than the fire? Are personnel of a size and stature to handle such an extinguisher?

Fire extinguishers must be matched up with the type of fire they will be effective on. Again, all extinguishers have their good and bad points to consider:

- *Dry chemical.* These are designed for Class B and C fires.
- *Multi-purpose dry chemical.* These extinguishers are designed for use on Class A, B, and C fires.
- *Dry powder.* This is used on Class D fires. It smothers and coats.
- *Foam extinguishers.* These are effective for Class A and B fires where blanketing is desirable.
- *CO_2.* Used on Class B or C fires, CO_2 has no lasting cooling effect due to the fact that the gas dissipates so quickly, making flashbacks of fires a concern.
- *Wet agents.* New clean water-based wet agents have been developed for different classes of fires. They are cleaner then the powders and more desirable in some occupancies. It is necessary to match up the proper extinguisher with the type of protection desired.

Halogenated agents, though no longer in production, do have substitutes taking their place. Research the proper substitution agent to meet protection needs.

After extinguishers have been installed, a regular program of inspection and maintenance must be established. A good policy is for security personnel to check all devices visually once a month and to have the extinguisher service company inspect them twice a year. In this process, the service person should retag and if necessary recharge the extinguishers and replace

Fire Extinguisher Checklist

No.	Location	Description and size	Fully charged	Operable	Sealed	Comments
1	Sheet metal shop	CO_2 #15	yes	yes	yes	Hanging bracket should be replaced

Date ___ 3-2-04 ___ Signed ___ J. Pendleton ___

FIGURE 11-4 Fire extinguisher safety checklist.

defective equipment (see Figure 11-4). It is a good idea to check with your extinguisher service people to know when routine service requires the fire extinguishers to be discharged or emptied. This is a good time to do employee/fire brigade training because the company will be paying for refill service. It is a good cost-saving measure.

Fire Alarm Signaling Systems

Early notification is the key to fast, loss-reduction, and lifesaving extinguishment of fire. If a fire is discovered in the very first stages before it develops, it can be extinguished with a minimal amount of exposure to personnel and use of extinguishing agents. For example, should a fire develop in a waste can and activate a sensor quickly, a person could easily and safely approach it with a handheld extinguisher and extinguish it. This is the general concept of detection/signaling systems and fire extinguishers on the premises that could be used by trained personnel. The simple rule is small fire, small extinguishment, large fire, large extinguishment. Should that waste can fire go undetected and develop, lateral extension would not allow personnel to safely approach it and on-premises extinguishing units would not be adequate for extinguishment. Delays in fire detection and alarm notification have been one of the major causes of large loss of life and property fires.

Most fires are discovered by our senses. We see them, smell them, and sometimes hear them. Normally we are left with nothing more than the chance for human discovery. A fully functional high-tech alarm system (see Figure 11-5) gives early detection 24 hours a day, 365 days a year. Whether someone is on the premises or not, notification can occur.

A total fire alarm system, like intrusion alarms, can be viewed as consisting of the signaling system and the alarm and sensor system. The alarm system discovers the fire and activates a circuit, and then the signaling device notifies those concerned of the danger.

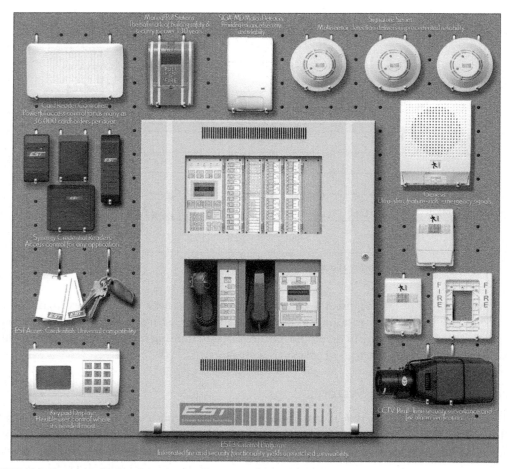

FIGURE 11-5 Elements of a fire alarm system. *(Courtesy of EST, www.est.net/std.)*

When determining the type of alarm system to utilize, one should check with code requirements, seek recommendations from insurance carriers, and evaluate unique features of the premises and operation. One will discover that some systems have better sensitivity, reliability, maintainability, and stability than others. Though all detectors may on the surface appear to do what you want, it must be realized that some do a better job in certain situations. For example, it may not be advantageous to have a highly sensitive sensor in a highly contaminated area. This could cause numerous false alarms. Also in areas such as computer rooms, one would want a sensor that would activate before heat or smoke develop.

Sensors

Sensors (detectors) can be categorized by what they are sensing. By looking at what fire produces, we can class them as flame detectors, smoke detectors, and heat detectors. Again, an evaluation should be made of the premises to determine what type of detector meets code and the specific needs of the area to be protected.

Today the terms sensor, detector, and initiating device are synonymous. This is a device that "initiates" a signal to activate the alarm system. This device should be viewed as nothing more than a switch, similar in principle to a light switch. When we turn on a light switch the light comes on. When initiating devices are activated they can cause alarms to sound, signals to be sent, and even safety features of the premises to be activated. These devices can be automatic: smoke detectors, heat detectors, or flame detectors. They could also be manual, such as a pull station. Alarm systems will be incorporated with other safety systems such as a sprinkler system. Flow switches are initiating devices that are placed in the piping of sprinkler systems to send a signal indicating water is flowing and the system has activated. Sprinkler systems can be incorporated into the alarm system as any initiating device.

Flame detectors are used in locations where signals must be sent before heat and smoke develop. They are often found in computer and high-tech areas. Flame detectors are "line of sight" detectors, meaning they must be located in varying spots in the area to be protected because they must "see" the flame. They respond to either ultraviolet or infrared light. Some detectors today are produced with both ultraviolet and infrared sensors in them to reduce false alarms.

Smoke detectors are broken into two groups, photoelectric and ionization. Photoelectric are either beam or refraction type. Beam smoke detectors operate on the principle of a light and a receiver. Once enough smoke or fire by-product breaks the beam of light, the device activates. The refraction type detector has a blocker between the light and the receiver and operates with the principle that once there is enough smoke in the detector the light signal refracts (reflects) around the blocker to the receiver. Ionization type detectors monitor the air around them constantly. Once enough fire by-products enter the detector, the contaminants will complete a circuit that sends the signal.

Heat detectors can be grouped as fixed temperature and rate of rise. Fixed temperature detectors activate at a predetermined temperature. Rate of rise detectors activate at changes in heat in the area they are located. Heat detectors are slower to react than other detectors, but they have less chance for false alarms and are durable in many applications.

Fixed temperature detectors can be the following types: fusible link, frangible bulb, continuous line, and bimetallic.

Rate of rise detectors are categorized as either pneumatic, rate compensation, or thermoelectric.

Signal Devices

The signal sent by initiating devices will send a signal to the system control unit that processes the alarm. It is often referred to as the main control panel. These main panels are electrically powered and have a secondary backup source of battery power or a generator.

The system control unit will send signals via circuitry to various components the system may have. It could retransmit the signal to a signaling system and activate on-premises safety features. For example, a system may be designed such that when a heat detector is activated, horns will go off in the building to notify the occupants, a signal will be sent via telephone wires to an alarm monitoring location, fire doors in the building will shut to compartmentalize the fire, and heating and cooling units will shut down to stop the spread of smoke and toxic gases.

Signaling systems are broken into five categories: local, auxiliary, remote station, proprietary, and central station.

Local systems do not retransmit the alarm anywhere. This is always a concern because there is a fear that people will assume that a response to the alarm is coming. Workers, residents, and employees must be made aware of the operation of this system. Most codes today will require systems to be retransmitted, but many are still in place since their installation.

Auxiliary systems are utilized in communities that have a municipal alarm system. Building alarms are connected to a system owned and operated by the municipality. The signal may be transmitted to a fire station or central receiving location.

Remote station systems are ones that lease phone line service to transmit the alarm signal to a location. This is nothing more than a phone line, the same as you have in your home.

Proprietary systems are those that are owned and operated by the building or complex owner. A series of buildings or warehouses on the same location would have their alarm systems send a signal to an owner-controlled location, such as a security office. This office would monitor the entire area. Once a signal is received, company personnel via phone would retransmit the alarm.

A central station system is one that is independently owned and operated by an alarm company. It may be located in the immediate town or city, or it may be located across the country. Signals are transmitted via phone lines to a continually manned location that then contacts the proper responding agency. This is one of the more popular systems due to the fact that many cities have decided not to take on the responsibility of receiving alarms and instead have privatized systems.

Whenever alarm and signaling systems are on the premises, representatives of the property should view an acceptance test after the installation of the system. Service tests should be conducted at required or recommended intervals and documentation should verify the testing.

Automatic Sprinkler Systems

Similar to alarm systems, sprinkler systems provide around-the-clock protection to a building. Ninety-six percent of fires in buildings equipped with sprinkler systems will either be extinguished or held in check until responding agencies arrive. The 4 percent of failure is due to human error, design, maintenance, or the result of an explosion that has rendered the system inoperative.

Concerns of sprinkler systems are the high cost of installation, especially retrofitting them to existing buildings, and water damage. As the popularity of plastic pipe increases and the fire codes allow more of their use, the cost of sprinkler system installation is coming down. If one were to calculate the cost of smoke and heat damage in a fire and the amount of water damage done by fire personnel with fire hoses, it would be realized that the fears of sprinkler system water damage are unfounded. Good sprinkler system design will eliminate those fears. Many codes will require sprinkler systems in certain occupancies and it can be found that over a period of time the savings in insurance coverage justifies their use. Interfacing the sprinkler system with a detection system can control water damage.

Sprinkler systems are classified into five groups: wet, dry, pre-action, deluge, and cyclic.

Wet systems will have water in them all the time, right up to the sprinkler head mounted in the ceiling or wall. The sprinkler heads are made to open individually at a given temperature, determined by code or requirements of the occupancy. Once they open, they do not close again. To discontinue water flow the source of water must be shut off.

A dry system has no water in it. The water is held back by a valve that is kept in the closed position by air pressure or an electric valve. This system is used mainly to protect unheated areas. It is more cost-effective to heat the valve room than an entire warehouse used for cold storage.

Pre-action systems incorporate a detection system. Concerns for water damage and false activation of the system can be eliminated with this system. A valve holds water in the system back. The valve is activated by initiating devices (detectors) in the area. Once the detector senses the fire, the valve is opened and the system is charged. The sprinkler head must also open for water to be discharged. In this case, two steps are required to discharge water. Multi-detector circuits could be added to require multiple steps for system activation.

A deluge system is similar to a pre-action system except all the sprinkler heads are in the opened position. It is obvious that water damage in this case is not a concern. Due to the nature of the situation or occupancy, extinguishment far outweighs other issues. Once the detector senses a fire, the valve opens and all sprinkler heads discharge.

A cyclic sprinkler head is more of a head type than a system. The opening and closing of a cyclic sprinkler head is controlled by a thermostat that is part of the sprinkler head. When the thermostat reaches a predetermined temperature, it opens; once it falls below that temperature, it closes. Cyclic sprinkler heads are one of the rare sprinkler heads that do not need to be replaced after a fire unless damaged. One drawback to the cyclic sprinkler head is that, although many thought it to be a help in reducing water damage, it was troublesome in that it often just discharged enough water to keep the fire in the smoldering state, causing smoke generation.

Education in Fire Prevention and Safety

Educating employees about fire prevention, fire protection, and evacuation procedures should be a continuous program. Ignorance and carelessness are the causes of most fires and of much loss of life. An ongoing fire-safety program will inform all employees and help keep them aware of the ever-present, very real danger of fire.

Such a program would ideally include fire and evacuation drills. Since such exercises require shutting down operations for a period of time and lead to the loss of expensive productive effort time, management is frequently cool toward them. Indoctrination sessions for new employees and regular review sessions for all personnel, however, are essential. Such sessions should be brief and involve only small groups.

In many businesses visitors and guests should also be made aware of fire and evacuation plans. Pamphlets describing plans should be available as well as placards that indicate the location of fire exits in relation to the placards. In the case of hotels and other facilities that have resident clientele, the law demands such actions. In the recent nightclub deaths in Chicago (2003), West Warwick, Rhode Island (2003), and Perm, Russia (2009), patrons were unable to escape the facilities due to panic and lack of orientation to their surroundings. Perhaps better marking of emergency exits with properly working emergency exit hardware may have reduced the number of fatalities.

Employees in Fire Fighting

Because the danger of fire with its concomitant risk to life and property affects every employee, many experts feel that the responsibility in case of fire is a shared one. The exception to this rule is when the firm has a well-trained fire unit. Few experts would disagree that everyone must be educated in the principles of fire prevention and protection, including indoctrination in evacuation procedures and on reporting a fire. But beyond this, there is little agreement on what employees should be asked to do.

Some business offices set up a system of floor wardens whose job it is to pass the word for evacuation and who then sweep their area of responsibility to see that it is clear of personnel, that papers are deposited in fireproof containers, and that high-value, portable assets are removed from the premises.

Others take the view that their employees were not hired to act as emergency supervisors. Many firms of this latter persuasion ask that certain minimal functions be performed by those persons who are on hand but do not assign roles to specific people. Examples might be a policy of returning tapes to a fireproof container in the computer area or securing fire-resistant safes in the accounting or cashier's office before evacuation. This responsibility would fall on the personnel in these areas at the time of an alarm and would or should take little time to accomplish. When the signal for evacuation is given, no time should be lost in vacating the building.

Many professionals feel that office employees should never be asked to do more than see to their own safety by making an orderly retreat along predetermined escape routes. Only in the most extreme emergency—and then only if they are otherwise trapped—should employees engage in fighting a fire of any magnitude. They can be expected as a normal reaction to make an effort to put out a wastebasket fire or a small blaze in a broom closet, but even in these cases the alarm must be given as first priority. Any fire that threatens to involve a major part of an office or other parts of the building should be left to professionals (company fire units, security personnel, or the fire department). Obviously all such situations are matters of on-the-spot judgment. Policies covering every situation are difficult, if not impossible, to predetermine.

The situation is quite different in industrial fire operations. In such facilities, the formation of a fire brigade composed of a few selected and trained employees is a fairly common practice. There is general agreement that the nature of their employment in industrial areas makes these employees more competent to handle fire-fighting assignments, which are in many cases not that far removed from their regular work.

The exact size of each fire-protection organization will vary according to the size and function of each facility. Very large facilities or those whose fire risks are high because of the nature of the operations may have a full-time fire department. In smaller or less hazardous facilities, regular employees are organized into fire brigades that are broken down, usually by departments or areas, into fire companies. These companies are assigned to given areas for purposes of fire protection, fire prevention, and fire fighting. They are also available, as part of the fire brigades, in any other area of the plant if a fire occurs that requires more personnel than the assigned company can provide.

The size of the brigade will depend on the size of the plant, the nature of the risk involved, and the willingness to take on that risk. It will also be affected by the general availability, size, competence, and response capability of the public fire-fighting facilities in the neighboring areas. Whatever the size, however, it must consist of people sufficiently well trained and familiar with the plant operation and layout to fight fires effectively in any part of the facility if such a need arises. They may be an incipient fire brigade or an interior fire brigade. Incipient fire brigades will be trained to handle fires in their earliest stage. The use of fire extinguishers by employees may be all the involvement this group will have. Interior fire brigades will be trained and equipped to the level of many fire departments. OSHA standards as well as other standards should be evaluated to determine the level of training required for an interior fire brigade.

The plant engineer and the maintenance crew should certainly be included in the brigade. Their knowledge in servicing valves, pumps, and other machinery is invaluable in emergency situations.

Evacuation
EVACUATING INDUSTRIAL FACILITIES
Evacuation plans for an industrial facility are relatively simple to design because most buildings within the perimeter are one-, two-, or in rare cases three-story buildings. Fire exits can be readily identified for such a plan of action. Because they are occupied by personnel of a single company or under that company's jurisdiction, a single plan involving all personnel can be drawn up. And because in most cases aesthetic considerations are not a prime concern in the design of the industrial building, fire escapes can be constructed in any way for the greatest safety.

Although many of these buildings have elevators, most of which serve the dual purpose of hauling freight and personnel, these elevators are not necessarily the prime means of moving to and from the upper level as they are in high-rise office buildings.

Generally, these buildings can be cleared in minutes. This is not to say that an evacuation plan is not needed. It is, and it must be widely distributed and clearly understood. Most industrial buildings are more open, the exits are more visible—more a part of the unconscious orientation of the employees and, therefore, more a part of the natural traffic flow—than these elements are in many other types of construction. Again, accountability of all personnel is necessary once evacuation is completed. Part of the evacuation plan should address and define the process to be used for accountability.

EVACUATING HIGH-RISE BUILDINGS
Evacuation from high-rise urban buildings is quite a different story. In this situation, employees come to work and leave regularly by elevator. Yet if there should be a fire, they are told they must not use the only means of entrance and exit they really know.

In most cases, they have never been on the fire stairs. They may have only a vague idea where those stairs are located, but in a time of emergency—a time of anxiety bordering on panic when instinctual behavior is the most natural—they are asked to vacate the premises in a way that to them is very unnatural indeed.

Even in wide-open industrial facilities where orientation is quick and easy, there will always be some people who will pass a clearly marked exit to get to the employee door they are used to using. This is much more likely to happen in buildings with windowless corridors and fire exits—however clearly lighted and marked they are—that are well off the normal traffic pattern used by employees. A good emergency lighting system can do much to reduce this potentially life-threatening problem.

To overcome this problem, which must be overcome until such time as elevators are made safe to use in the case of fire, it is advisable to walk all employees as a drill from their desks to the nearest fire exit and to the nearest secondary exit they would use in case their first escape route was cut off by fire or smoke. In addition, the use of "You Are Here" placards can assist those who become disoriented in a building to find the nearest safe exit. This orientation could be done over a period of time with small groups at each drill. It is important that the drill actually start at the desk or office of the employee so that the route as well as the location of the exit is made clear.

People need to be educated that in some occupancies elevators are programmed not to operate once the fire alarm system activates. And though dramatic on the news, roof evacuations are difficult, slow, and subject to many factors that make it the least desirable evacuation. People should be taught to move down, and if down is not possible, laterally to another exit and then down. Roof evacuation should be the last consideration.

PLANNING AND TRAINING

Evacuation plans must be based on a well-considered system and on thorough and continuing education. They should also be based on indoctrinating employees in the principles of fire safety, stressing that they are to make their own way to the proper exit and leave as quickly and as calmly as possible.

Adults do not respond to being lined up like children at a fire drill and marched down the fire stairs. Though they might be inclined to follow a leader under many circumstances, when it comes to a concept as simple as vacating the premises, a leader has no purpose or place. They will rebel or even panic if they feel restrained or regimented in their movements toward the exit.

In setting up plans for evacuation, it might be well to review and evaluate the circumstances of a given facility and then ask a few questions.

1. Are routes to exits well lighted, fairly direct, and free of obstacles?
2. Are elevators posted to warn against their use in case of fire? Do these signs point out the direction of fire exits?
3. Are disabled persons provided for?
4. Do corridors have emergency lighting in the event of power failure?
5. Who makes the decision to evacuate? How will personnel be notified?
6. Who will operate the communication system? What provisions have been made in case the primary communication system breaks down? Who is assigned to provide and receive

information on the state of the emergency and the progress of the evacuation? By what means?

7. How do we account for everyone? Do we know if everyone is out of the building? How do we verify that?

Safety and Loss Control

Safety consciousness in business and industry did not begin with the establishment of the OSHA in 1970, but it is largely a product of the 20th century. Prior to the Industrial Revolution early in the 19th century, workers were independent craftspeople. If they suffered economic loss because of accident or illness rising out of prolonged exposure to a particular work environment, the problem was the craftsperson's, not the employer's. This attitude generally prevailed during the rapid expansion of the factory system in America throughout the 19th century. Only toward the latter part of the 19th century did it begin to become obvious that factories were far superior in terms of production capability to the small handicraft shops, yet they were often inferior in terms of human values, health, and safety.

The atmosphere of reform that gained impetus after the turn of the century resulted in, among other new laws, the first effective Workmen's Compensation Act in Wisconsin in 1911. Compulsory laws on workmen's compensation followed in many states after the U.S. Supreme Court upheld their constitutionality in 1916. Even the most hardheaded employers found that their costs dictated compliance with the spirit of the law.

As a result of this growing concern for industrial safety, there followed a long downward curve in work-connected accidents and injuries that lasted through the period between the two world wars and continued into the 1950s. By 1958 this trend had leveled off, and by 1968, for the first time in more than 50 years, the curve began to rise again.

Fourteen thousand occupational fatalities and more than two million disabling, work-connected injuries each year seemed to be considerably more than the number that might one day be arrived at as the irreducible minimum. The result throughout the 1960s was increasing federal concern with establishing standards of occupational safety and health. Prior to that decade, only a few federal laws, such as the Walsh-Healey Public Contracts Act, had been enacted, with most legislation in this area being left to the states. During the 1960s a number of laws were passed—the McNamara-O'Hara Service Contracts Act, the Federal Construction Safety Act, and the Federal Coal Mine Health and Safety Act among others—all dealing with safety and health standards in specific fields and under specific circumstances. Public Law 91-596, known as OSHA, which was signed into law on December 29, 1970, was the first legislation that attempted to apply standards to virtually every employer and employee in the country.

OSHA Standards

Generally speaking, OSHA requires that an employer provide a safe and healthful place for employees to work. This is spelled out in great detail in the act to avoid leaving the thrust of the legislation in any doubt. Though much of the language in the act is technical in nature and largely couched in legalese, the thrust of the legislation is absolutely clear and unambiguous in what is known as the "General Duty" clause that states that each employer "shall furnish to each of his employees a place of employment free from recognized hazards that are causing or likely to cause death or serious physical harm to his employees" and that, further, the employer "shall comply with all occupational safety and health standards promulgated under this Act."[9] Much of the rest of the act deals with procedures and standards of safety and is in places difficult to follow.

It speaks of free and accessible means of egress, of aisles and working areas free of debris, of floors free from hazards. It gives specific requirements for machines and equipment, materials, and power sources. It specifies fire protection by fixed or portable systems, clean lunchrooms, environmental health controls, and adequate sanitation facilities. Whereas in past years employers might contend in all sincerity that their facilities met community standards for safety and cleanliness, with the enactment of OSHA these standards have been formalized to describe minimum levels of acceptability. Although employers might also contend that some specific demands of the act are unclear, there is no mistaking the purpose of the act: "The Congress declares to be its purpose and policy to assure so far as possible every working man and woman in the nation safe and healthful working conditions and to preserve our human resources."

Perhaps the strongest resistance to OSHA in its first years was the complaint that some of the basic standards went too far or were unnecessary. In May 1978 the U.S. Supreme Court ruled that the agency could not conduct surprise workplace inspections without a proper warrant.[10] And with growing criticism from that time period, the OSHA administration has continually sought the elimination of "Mickey Mouse" standards that have no direct bearing on improving safety in the workplace.

In 1988 OSHA issued the Hazard Communication Standard, which states that all employees have the right to know what hazards exist in their place of employment and what to do to protect themselves from the hazards. Simple labels and warnings on containers are not enough. Employers must have a program to communicate more detail on all hazards, including a Material Safety Data Sheet (MSDS) that must be available for each chemical at the work site. Each MSDS contains seven sections:

1. Product identification and emergency notification instructions
2. Hazardous ingredients list and exposure limits
3. Physical and chemical characteristics
4. Physical hazards and how to handle them (that is, fire, explosion)
5. Reactivity—what the product might react with and whether it is stable
6. Health hazards—how the product can enter the body, signs and symptoms of problems, and emergency first-aid steps
7. Safe handling procedures

Setting Up the Safety Program

H. W. Heinrich, an outstanding pioneer in safety studies, held that unsafe acts caused 85 percent of all accidents and that unsafe conditions caused the remaining 15 percent.[11] Therefore if these acts could be modified, the accidents would be sharply reduced. Today safety supervisors agree that unsafe acts are the principal villain and that the system's approach to safety is the only real way to control losses. It is necessary, however, for management to get the system together and implement a strong, active program so that it is effective. Safety problems are caused, they do not just happen, and each problem can be identified and controlled.[12]

The three Es need to be addressed when developing a safety program. They are engineering, education, and enforcement. First, one must build and develop a good program. Second, everyone needs to be educated to the program and the part he or she plays in it. Third, the program must be enforced to see that it is followed.

Accidents, by our definition, refer to property damage as well and in aggregate can amount to substantial cost for the company that fails to keep them under control. In fact, an effective loss-control program can be an organization's best moneymaker, when it can be shown that the actual cost of accidents may be anywhere from 6 to 50 times as much as the money recovered from insurance. Uninsured costs in building damage, production damage, wages to the injured for lost time, clerical costs, cost of training new workers and supervisors, and extra time all mount up. By controlling such incidents through careful study of hazards and the introduction of safety programs to deal with the hazards, the profit picture will be immeasurably improved. In a company operating at a 4 percent profit margin, the sales department would have to generate sales of $1,250,000 just to compensate for an annual loss of $50,000 in incidents.

Finding the Causes of Accidents

The causes of accidents should be determined before they occur. Because accidents are caused, the conditions that cause them can be known and controlled. It is therefore of the greatest importance that management deal vigorously with what can cause an accident. Unsafe acts and unsafe conditions will ultimately cause accidents if they are allowed to continue.

Unsafe acts will be discovered and corrected only when immediate supervisors are alert to the problems. They must set up systems for observing all workers closely while they are performing their jobs, especially those in hazardous jobs. To do this, they must have a job-safety analysis at their disposal. This analysis divides each job into component parts, and each part is studied for the hazards it may present.

Unsafe conditions are uncovered by constant inspection. Such conditions do not disappear entirely because they have been taken care of once. Unsafe conditions are continuously created by the operation of the facility. Normal wear and tear, careless housekeeping, initial bad design, or simply the deterioration that results from inadequate maintenance caused by a cost-cutting management—all create unsafe conditions that have high potential loss factors. Early discovery of unsafe conditions is essential to good loss control, and the procedure is simply inspection, inspection, inspection.

Identification and Control of Hazards

OSHA standards (or equivalent state standards) provide the baseline for a company's safety program. A bewildering catalog of standards has already developed, and new ones are constantly being added. Checklists (available from OSHA, the National Safety Council, insurance carriers, and other sources) can provide the starting point for detailed inspections to identify hazards. The confusion that might accompany a consideration of all the standards begins to sort itself out when inspections zero in only on those standards that apply to specific operations and conditions.

A safety program should include periodic inspections scheduled at regular intervals. Figure 11-6 is an example of a monthly checklist for inspection. In addition, looking for safety hazards and violations should be part of the day-to-day activity of both safety professionals and security personnel. Some hazards that might be present in any business facility are shown in Table 11-1.

A Hazardous Materials Program

In addition to the seven steps in safety planning, particular types of businesses dealing with hazardous substances should have a hazardous materials program. As a minimum, it is necessary to:

1. Identify what hazardous materials you have and where they are
2. Know how to respond to an accident involving hazardous materials
3. Know how to deal with spills
4. Set up appropriate safeguards
5. Train employees in dealing with hazardous materials

As has been discussed earlier, MSDSs are designed specifically to help identify the nature of potential hazards. These data sheets are obtainable from vendors of hazardous materials or equipment.

Management Leadership

Management's attitude toward safety filters down throughout the entire company. Top management's concern will be reflected in that of the supervisors; in turn, the supervisor's attention to safety will affect the individual employee's attitude. Management is responsible not only for a basic policy for providing a work environment free of hazards, which should be embodied in an executive policy statement, but also for active leadership. This can be expressed by holding subordinates responsible for accident prevention and in such visible ways as plant tours, letters to employees, safety meetings, posters, prompt accident investigations, and personal example. (In a hard hat area, for example, the president of the company should also put on a hard hat.)

General safety rules must be established and published in the employee handbook or manual. Safety rules should be continually reviewed and updated.

Assignment of Responsibility

Responsibility for the safety program should be clear and personal. In the small company, it may rest on the owner. It will generally be an added responsibility of the supervisors in companies with fewer than 100 employees.

Monthly Safety Check

Dept._____Date_____

Supervisor _____

Indicate discrepancy by ⊠

General area	
Floor condition	
Special purpose flooring	
Aisle, clearance/markings	
Floor openings, require safeguards	
Railings, stairs temp./perm.	
Dock board (bridge plates)	
Piping (water-steam-air)	
Wall damage	
Ventilation	
Other	
Illumination—wiring	
Unnecessary/improper use	
Lights on during shutdown	
Frayed/defective wiring	
Overloading circuits	
Machinery not grounded	
Hazardous location	
Other	
Housekeeping	
Floors	
Machines	
Break area/latrines	
Waste disposal	
Vending machines/food protection	
Rodent, insect, vermin control	
Vehicles	
Unauthorized use	
Operating defective vehicle	
Reckless/speeding operation	
Failure to obey traffic rules	
Other	
Tools	
Power tool wiring	
Condition of hand tools	
Safe storage	
Other	

First aid	
First aid kits	
Stretchers, fire blankets, oxygen	
Fire protection	
Fire hoses hung properly	
Extinguisher charged/proper location	
Access to fire equipment	
Exit lights/doors/signs	
Other	
Security	
Doors/windows, etc. secured when required	
Alarm operation	
Department shut down security	
Equipment secured	
Unauthorized personnel	
Other	
Machinery	
Unattended machines operating	
Emergency stops not operational	
Platforms/ladders/catwalks	
Instructions to operate/stop posted	
Maintenance being performed on machines in operation	
Guards in place	
Pinch points	
Material storage	
Hazardous & flammable material not stored properly	
Improper stacking/loading/securing	
Improper lighting, warning signs, ventilation	
Other	

FIGURE 11-6 Monthly safety checklist.

In larger companies, safety should be a responsibility assigned to a ranking member of management who may delegate the authority to oversee the program to a safety director (who may be called the safety professional, safety engineer, or safety supervisor, depending on his or

Table 11-1   Common Safety Hazards

1. Floors, aisles, stairs, and walkways

   Oil spills or other slippery substances that might result in an injury-producing fall.

   Litter-obscuring hazards such as electrical floor plugs, projecting material, or material that might contribute to the fueling of a fire.

   Electrical wire, cable, pipes, or other objects crossing aisles that are not clearly marked or properly covered.

   Stairways that are too steep, have no nonskid floor covering, inadequate or nonexistent railings, poor lighting, or those that are in a poor state of repair.

   Overhead walkways that have inadequate railings, are not covered with nonskid material, or that are in a poor state of repair.

   Walks and aisles that are exposed to the elements and have not been cleared of snow or ice, that are slippery when wet, or that are in a poor state of repair. Aisles may be blocked with stock or items that reduce the ability to exit safely or get emergency equipment where needed.

2. Doors and emergency exits

   Doors that are ill-fitting, stick, and that might cause a slowdown during emergency evacuation.

   Panic-type hardware that is inoperative or in a poor state of repair.

   Doors that have been designated for emergency exit but that are locked and not equipped with panic-type hardware.

   Doors that have been designated for emergency exit but that are blocked by equipment or by debris.

   Missing or burned-out emergency exit lights.

   Nonexistent or poorly marked routes leading to emergency exit doors.

3. Flammable and other dangerous materials

   Flammable gases and liquids that are uncontrolled, in areas in which they might constitute a serious threat.

   Radioactive material not properly stored or handled.

   Paint or painting areas that are not properly secured or that are in areas that are poorly ventilated.

   Gasoline pumping areas located dangerously close to operations that are spark-producing or in which open flame is being used.

4. Protective equipment or clothing

   Workmen in areas where toxic fumes are present who are not equipped with or who are not using respiratory protective apparatus.

   Workmen involved in welding, drilling, sawing, and other eye-endangering occupations who have not been provided or who are not wearing protective eye covering.

   Workmen in areas requiring the wearing of protective clothing, due to exposure to radiation or toxic chemicals, who are not using such protection.

   Workmen engaged in the movement of heavy equipment or materials who are not wearing protective footwear.

   Workmen who require prescription eyeglasses who are not provided or are not wearing safety lenses.

5. Vehicle operation and parking

   Forklifts that are not equipped with audible and visual warning devices when backing.

   Trucks that are not provided with a guide when backing into a dock or that are not properly chocked while parking.

   Speed violations by cars, trucks, lifts, and other vehicles being operated within the protected area.

   Vehicles that are operated with broken, insufficient, or nonexistent lights during the hours of darkness.

   Vehicles that constitute a hazard due to poor maintenance procedures on brakes and other safety-related equipment.

   Vehicles that are parked in fire lanes, block fire lanes and emergency exits, or fire protection equipment and system access.

(Continued)

Table 11-1   (Continued)

6. Machinery maintenance and operation
   Frayed electrical wiring that might result in a short circuit or malfunction of the equipment.
   Workers who operate, process, or work near or on belts, conveyors, and other moving equipment who are wearing loose-fitting clothing that might be caught and drag them into the equipment.
   Presses and other dangerous machinery that is not equipped with the required hand guards or with automatic shutoff devices or dead-man controls.
7. Welding and other flame- or spark-producing equipment
   Welding torches and spark-producing equipment being used near flammable liquid or gas storage areas or being used in the vicinity where such products are dispensed or are part of the productive process.
   The use of flame- or spark-producing equipment near wood shavings, oily machinery, or where they might damage electrical wiring.
   Follow-up inspection should be done periodically at intervals to make sure there are no smoldering heat sources.
8. Miscellaneous hazards
   Medical and first-aid supplies not properly stored, marked, or maintained.
   Color coding of hazardous areas or materials not being accomplished or that is not uniform.
   Broken or unsafe equipment and machinery not being properly tagged with a warning of its condition.
   Electrical boxes and wiring not properly inspected or maintained, permitting them to become a hazard.
   Emergency evacuation routes and staging areas not properly marked or identified.

Source: Eugene Finneran, *Security Supervision: A Handbook for Supervisors and Managers* (Boston: Butterworth-Heinemann, 1981).

her qualifications and the nature of the operation). In many companies, safety is a responsibility of the security director, who will often have a safety specialist as a subordinate. (In virtually all circumstances, there is a close relationship between safety and security.)

Training

All employees must be initially and periodically trained both in general safety principles and in safe work practices in their specific jobs. Safety rules such as the wearing of protective clothing (gloves, headgear, respirators, shoes, eye protection, and such) should be clearly explained and promptly enforced. The importance the company attaches to safety should particularly be emphasized in new employee training, but it is also important to pay attention to regular employees including the "old-timers" who did not grow up with safety awareness as part of their conditioning. Certificates for completion of classes should be given to employees at the workplace to show the importance of the program.

In addition to the above, there are specific training requirements in the OSHA standards (such as those involving the operation of certain types of equipment). Employers and employees should be aware of those standards that apply in the specific workplace.

Emergency Care

Under OSHA, all businesses are required, in the absence of an infirmary or hospital in the immediate vicinity, to have a person or persons trained in first aid available, along with first-aid supplies. Where employees are exposed to corrosive materials, procedures for drenching or flushing the eyes and body should be provided in the work area.

Procedures should be established for handling injury accidents without confusion or delay. The extent of these preparations will, of course, depend on the nature of the business and the types of hazards.

Employee Awareness and Participation

Developing safety and health awareness is one of the primary goals of OSHA. Active steps by management, such as those suggested previously, are essential to involve all employees in the need to create a safe work environment.

Safety awareness has an added benefit for both the employer and employees in that it tends to carry over into a concern for off-the-job safety. Accidents away from the work environment account for more than half of all injuries, and the ratio of deaths is three-to-one higher in off-the-job accidents. Carrying safety practices from the job to activities away from the job is an aspect of safety training that is receiving increasing emphasis from today's safety professionals. Table 11-2 lists potential problems associated with disasters and the agency charged with providing potential assistance.

Summary

With the advent of OSHA, the attention focused on safety in the workplace created many new attitudes about the place of loss control within the organization. Many companies that had at best paid lip service to concepts of safety that are commonplace today came to see that safety, like security, is good business and that a well-managed loss-control program would produce gratifying savings in a potentially costly area of company operations. But it must be noted that a well-managed safety program goes well beyond simply complying with OSHA standards.

Table 11-2   Typical Human Problems as a Result of Disaster and Potential Agencies for Assistance

Shelter	Civil Defense and Red Cross
Food	Red Cross and civic groups
Water	Civil government
Emergency care	Hospitals and clinics
Medical evaluations	Hospitals and health agencies
Personal protection	Police and National Guard
Illumination	Public utilities
Communications	Citizen band radio and National Guard
Transportation	National Guard, local trucking and bus companies
Property protection	Police, auxiliary police, and National Guard

Source: Dennis Sigwart, "Disaster Planning Considerations for the Security/Safety Professional: A Historical Interface," in John Chuvala III and Robert Fischer, eds., *Suggested Preparation for Careers in Security/Loss Prevention,* ed. 2 (Dubuque, IA: Kendall/Hunt 1999).

In addition to recognizing OSHA standards, many companies have also placed more emphasis on fire prevention and protection. Some companies have even established their own fire departments, which are often better equipped than some municipal departments. Although there is some recognition of the importance of contingency planning, far too few firms have anything beyond a contingency plan that sits on a shelf in the CEO's office. Even in those companies with crisis management teams, the members often do not meet to discuss how the team would function in an actual situation.

The most progressive firms offer the team members, fire brigades, and employees an opportunity to preplan (contingency planning) through mock exercises that replicate industrial disasters, explosions, fires, or tornado alerts. The end result is a better-prepared team of employees. Unfortunately many firms have not gone this far.

Contingency planning may not have been a traditional security process, but in today's global business environment the security organization is assuming a much greater role and responsibility for its implementation. Even prior to the events of September 11, 2001, many organizations were becoming more conscious of the need to have contingency plans. A complete contingency planning program has three major elements:

1. Emergency response
2. Crisis management
3. Business continuity: business recovery and business resumption

Emergency response activities involve responding to an incident, crisis or disaster and managing that incident at the scene. Should an incident escalate to the crisis or disaster stage, a CMT should take over managing the crisis to its conclusion. If the crisis or disaster does cause damage to a company building, facility or operation, the CMT should hand over to a business continuity team the responsibility of recovery and resumption. After a disaster, it is critical that the business recovers and resumes normal (pre-event) operations as soon as possible. Customers, shareholders and stakeholders expect nothing less. Executive management has the obligation to ensure contingency planning is properly considered and addressed within their company. The consequences of not planning for contingencies can be catastrophic, with numerous liability issues.

■ ■ CRITICAL THINKING ■

What is the relationship between safety issues, fire prevention and fire fighting, other emergencies and the process of contingency planning? Can a business be successful without having contingency plans?

Review Questions

1. What are the classes of fire, the fuels needed to ignite each, and the extinguishing agents that can be used in each class?
2. In what ways is an ionization detector different from a smoke, infrared, or thermal detector?

3. What are the key elements of any contingency plan?

4. What should be the role of security in developing a contingency plan?

5. When management is developing a plan for emergency evacuations, what things need to be considered?

6. What is OSHA, and what effect has it had on company safety operations?

References

[1] Block R. Pushing disaster preparedness the lieberman way. Wall St J Online 02/09/2007 and ANAB Accreditation for Private Sector Preparedness Voluntary Certification, download 6/17/12, www.anab.org/accreditation/preparedness.aspx.

[2] Sigwart DF. Disaster planning considerations for the security/safety professional: a historical interface. In: Chuvala III J, Fischer R, editors. Suggested preparation for careers in security/loss prevention (2nd ed.). Dubuque, IA: Kendall/Hunt; 1999.

[3] <www.acp-international.com/> For contact information mail to:chairman@acp-international.com.

[4] http://www.flu.gov/individualfamily/about/h5n1/#what

[5] http://en.wikipedia.org/wiki/Spanish_flu

[6] Pugh T. Rating system develop to gauge pandemics. Houst Chron February 2, 2007:A10.

[7] Chapter 13, Firefighter's handbook. 2nd ed. Clifton Park, NY: Thompson Delmar Learning; 2004.

[8] Bryan JL. In: Fire suppression and detection systems. New York: Macmillan; 1982. pp. 11–12.

[9] General Industry: Safety and Health Regulations, Part 1910, U.S. Department of Labor, OSHA; 1974.

[10] Marshall v. Barlow's, 98 Sct. Rptr. 1816 (1978).

[11] Heinrich HW. In: Industrial accident prevention. New York: McGraw-Hill; 1959.

[12] Ibid.

12

Internal Theft Controls/ Personnel Issues

OBJECTIVES

The study of this chapter will enable you to:

1. Understand the complexity of relative honesty.
2. Develop strategies to manage employee honesty.
3. Understand the relationship between security and human resources personnel.
4. Know the basics of background checks.
5. Understand the operation of various lie detection devices and integrity tests.
6. Know the rights of employees when confronted with wrongdoing in the workplace.
7. Identify problem areas where employee theft is most likely.

Introduction

It is sad but true that virtually every company will suffer losses from internal theft—and these losses can be enormous. Early in this new century even the large corporate giants such as Enron, WorldCom and Martha Stewart have been affected by internal corruption that reached the highest levels of the organization In addition, the name of Bernie Madoff will long be associated with perhaps the greatest customer and company theft of all times. In its 2010 report, "The Cost of Occupational Fraud," the Association of Certified Fraud Examiners estimate that fraud (employee theft) cost the world business community $2.9 trillion or 5% of the estimated Gross World Product in 2009.[1] While this figure is startling, it must be remembered that there is no accurate way to calculate the extent of fraud. In 2002, *Security* reported that in the retail business alone 1 in every 27 employees is apprehended for theft from their employer. Internal theft in the retail business outstrips the loss from shoplifting by approximately 7.9 times.[2]

The significance of employee theft is pointed out in a 2010 University of Florida and National Retail Federation Report. Dr. Richard Hollinger, lead author, reported that $14.4 billion was lost to retailers thanks to thieving employees, down slightly from earlier studies.[3]

What Is Honesty?

Before considering the issue of dishonest employees, it is helpful to understand the concept of honesty, which is difficult to define. Webster says that honesty is "fairness and straightforwardness of conduct, speech, etc.; integrity; truthfulness; freedom; freedom from fraud." In simple terms, honesty is respect for others and for their property. The concept, however, is relative. According to Charles Carson, "Security must be based on a controlled degree of relative honesty" because no one fulfills the ideal of total honesty. Carson explores relative honesty by asking the following questions:

1. If an error is made in your favor in computing the price of something you buy, do you report it?
2. If a cashier gives you too much change, do you return it?
3. If you found a purse containing money and the owner's identification, would you return the money to the owner if the amount was $1? $10? $100? $1,000?[4]

Honesty is a controllable variable, and how much control is necessary depends on the degree of honesty of each individual. The individual's honesty can be evaluated by assessing the degree of two types of honesty—moral and conditioned. Moral honesty is a feeling of responsibility and respect that develops during an individual's formative years; this type of honesty is subconscious. Conditioned honesty results from fearing the consequences of being caught; it is a product of reasoning. If an honest act is made without a conscious decision, it is because of moral honesty, but if the act is based on the conscious consideration of consequences, the act results from conditioned honesty.

It is vital to understand these principles because the role of security is to hire employees who have good moral honesty and to condition employees to greater honesty. The major concern is that the job should not tempt an employee into dishonesty.

Carson summarizes his views in the following principles:

No one is completely honest.
Honesty is a variable that can be influenced for better or worse.
Temptation is the father of dishonesty.
Greed, not need, triggers temptation.

Unfortunately, there is no sure way by which potentially dishonest employees can be recognized. Proper screening procedures can eliminate applicants with unsavory pasts or those who seem unstable and therefore possibly untrustworthy. There are even tests that purport to measure an applicant's honesty index. But tests and employee screening can only indicate potential difficulties. They can screen out the most obvious risks, but they can never truly vouch for the performance of any prospective employee under circumstances of new employment or under changes that may come about in life apart from the job.

The need to carefully screen employees has continued to increase. In today's market, there are many individuals who have been called the "I deserve it!" generation. According to a study by the Josephson Institute for the Advanced Study of Ethics, cheating, stealing and lying by high school students have continued an upward trend, with youth 18 and younger five times more likely than persons over 50 to hold the belief that lying and cheating are necessary to succeed. The 2008 report showed that 64% of American high school students cheated on an exam, 42% lied to save money and 30% stole something from a store. The Institute, which conducts nonpartisan ethics programs for the Internal Revenue Service, the Pentagon, and several major media organizations and educators, states that their findings show evidence that a willingness to cheat has become the norm. The 2008 study found that young people believe that ethics and character are important but are cynical about whether a person can be ethical and succeed.[5]

The Dishonest Employee

Because there is no fail-safe technique for recognizing the potentially dishonest employee on sight, it is important to try to gain some insight into the reasons why employees may steal. If some rule of thumb can be developed that will help identify the patterns of the potential thief, it would provide some warning for an alert manager.

There is no simple answer to the question of why heretofore honest people suddenly start to steal from their employers. The mental and emotional processes that lead to this are complex, and motivation may come from any number of sources.

Some employees steal because of resentment over real or imagined injustice that they blame on management indifference or malevolence. Some feel that they must maintain status and steal to augment their incomes because of financial problems. Some may steal simply to tide themselves over in a genuine emergency. They rationalize the theft by assuring themselves that they will return the money after the current problem is solved. Some simply want to indulge themselves, and many, strangely enough, steal to help others. Or employees may steal because no one cares, because no one is looking, or because absent or inadequate theft controls eliminate the fear of being caught. Still others may steal simply for excitement.

The Fraud Triangle

A simplified answer to the question of why employees steal is depicted in the fraud triangle. According to this concept, theft occurs when three elements are present: (1) incentive or motive , (2) attitude/rationalization or desire, and (3) opportunity.

In simple terms, incentive or motive is a reason to steal. Motives might be the resentment of an employee who feels underpaid or the vengefulness of an employee who has been passed over for promotion. Attitude or desire builds on motive by imagining the satisfaction or gratification that would come from a potential action. "Taking a stereo system would make me feel good, because I always wanted a good stereo system." Opportunity is the absence of barriers that prevent someone from taking an item. Desire and motive are beyond the scope of the loss-prevention manager; opportunity, however, is the responsibility of security.

A high percentage of employee thefts begin with opportunities that are regularly presented to them. If security systems are lax or supervision is indifferent, the temptation to steal items that are improperly secured or unaccountable may be too much to resist by any but the most resolute employee.

Many experts agree that the fear of discovery is the most important deterrent to internal theft. When the potential for discovery is eliminated, theft is bound to follow. Threats of dismissal or prosecution of any employee found stealing are never as effective as the belief that any theft will be discovered by management supervision.

Danger Signs

The root causes of theft are many and varied, but certain signs can indicate that a hazard exists. The conspicuous consumer presents perhaps the most easily identified risk. Employees who habitually or suddenly acquire expensive cars and/or clothes and who generally seem to live beyond their means should be watched. Such persons are visibly extravagant and appear indifferent to the value of money. Even though such employees may not be stealing to support expensive tastes, they are likely to run into financial difficulties through reckless spending. Employees may then be tempted to look beyond their salary checks for ways to support an extravagant lifestyle.

Employees who show a pattern of financial irresponsibility are also a potential risk. Many people are incapable of handling their money. They may do their job with great skill and efficiency, but they are in constant difficulty in their private lives. These people are not necessarily compulsive spenders, nor do they necessarily have expensive tastes. (They probably live quite modestly since they have never been able to manage their affairs effectively enough to live otherwise.) They are simply people unable to come to grips with their own economic realities. Garnishments or inquiries by creditors may identify such employees. If there seems a reason to make one, a credit check might reveal the tangled state of affairs.

Employees caught in a genuine financial squeeze are also possible problems. If they have been hit with financial demands from illnesses in the family or possibly heavy tax liens, they may find the pressures too great to bear. If such a situation comes to the attention of management, counseling is in order. Many companies maintain funds that are designated to make low-interest loans in such cases. Alternatively some arrangement might be worked out through a credit union. In any event, employees in such extremities need help fast. They should get that help both as a humane response to the needs and as a means of protecting company assets.

In addition to these general categories, some specific danger signals should be noted:

- Gambling on or off premises
- Excessive drinking or signs of other drug use
- Obvious extravagance
- Persistent borrowing
- Requests for advances
- Bouncing personal checks or problems with creditors

What Employees Steal

The employee thief will take anything that may be useful or that has resale value. The thief can get at the company funds in many ways—directly or indirectly—through collusion with vendors, collusion with outside thieves or hijackers, fake invoices, receipting for goods never received, falsifying inventories, payroll padding, false certification of overtime, padded expense accounts, computer records manipulation, overcharging, undercharging, or simply by gaining access to a cash box or company goods.

This is only a sample of the kinds of attacks that can be made on company assets using the systems set up for the operation of the business. It is in these areas that the greatest losses can occur because they are frequently based on a systematic looting of the goods and services in which the company deals and the attendant operational cash flow.

Significant losses do occur, however, in other, sometimes unexpected areas. Furnishings frequently disappear. In some firms with indifferent traffic control procedures, this kind of theft can be a very real problem. Desks, chairs, computers and other office equipment, paintings, rugs—all can be carried away by the enterprising employee thief.

Office supplies can be another problem if they are not properly supervised. Beyond the anticipated attrition in pencils, paper clips, note pads, and rubber bands, sometimes these materials are stolen in case lots. Many firms that buy their supplies at discount are in fact receiving stolen property. The market in stolen office supplies is a brisk one and is becoming more so as the prices for this merchandise soar.

The office equipment market is another active one, and the inside thief is quick to respond to its needs. Computers always bring a good price, as well as equipment used to support high-tech offices.

Personal property is also vulnerable. Office thieves do not make fine distinctions between company property and that of their fellow workers. The company has a very real stake in this kind of theft because personal tragedy and decline in morale follow in its wake.

Although security personnel cannot assume responsibility for losses of this nature because they are not in a position to know about the property involved or to control its handling (and they should so inform all employees), they should make every effort to apprise all employees of the threat. They should further note from time to time the degree of carelessness the staff displays in handling personal property and send out reminders of the potential dangers of loss.

Methods of Theft

A 2007 report by Gaston and Associates reported that the American Management Association believes that 20% of business failures were the result of employee dishonesty.[6] And a 2010 report by the Association of Certified Fraud Examiners estimates that 5% of total revenue losses for most companies are from employee fraud of some type.[7] Therefore, there is a very real need to examine the shapes the dishonesty frequently takes. There is no way to describe every kind of theft, but some examples may serve to give an idea of the dimensions of the problem:

1. Payroll and personnel employees collaborating to falsify records by the use of nonexistent employees or by retaining terminated employees on the payroll.
2. Padding overtime reports and kicking back part of the extra unearned pay to the authorizing supervisor.
3. Pocketing unclaimed wages.
4. Splitting increased payroll that has been raised on signed, blank checks for use in the authorized signer's absence.
5. Maintenance personnel and contract-service people in collusion to steal and sell office equipment.

6. Receiving clerks and truck drivers in collusion on falsification of merchandise count. (Extra unaccounted merchandise is fenced.)
7. Purchasing agents in collusion with vendors to falsify purchase and payment documents. (The purchasing agent issues authorization for payment on goods never shipped after forging receipts of shipment.)
8. Purchasing agent in collusion with vendor to pay inflated price.
9. Mailroom and supply personnel packing and mailing merchandise to themselves for resale.
10. Accounts payable personnel paying fictitious bills to an account set up for their own use.
11. Taking incoming cash without crediting the customer's account.
12. Paying creditors twice and pocketing the second check.
13. Appropriating checks made out to cash.
14. Raising the amount on checks after voucher approval or raising the amount on vouchers after their approval.
15. Pocketing small amounts from incoming payments and applying later payments on other accounts to cover shortages.
16. Removal of equipment or merchandise with the trash.
17. Invoicing goods below regular price and getting a kickback from the purchaser.
18. Manipulation of accounting software packages to credit personal accounts with electronic account overages.
19. Issuing (and cashing) checks on returned merchandise not actually returned.
20. Forging checks, destroying them when they are returned with the statement from the bank, and changing cash account records accordingly.
21. Appropriating credit card, electronic bank account, and other electronic data.

The Contagion of Theft

Theft of any kind is a contagious disorder. Petty, relatively innocent pilferage by a few spreads through the facility. As more people participate, others will follow until even the most rigid break down and join in. Pilferage becomes acceptable—even respectable. It gains general social acceptance that is reinforced by almost total peer participation. Few people make independent ethical judgments under such circumstances. In this microcosm, the act of petty pilferage is no longer viewed as unacceptable conduct. It has become not a permissible sin but instead a right.

In the last century, the docks of New York City were an example of this progression. Forgetting for the moment the depredations of organized crime and the climate of dishonesty that characterized that operation for so many years, even longshoremen not involved in organized theft had worked out a system all their own. For every so many cases of whisky unloaded, for example, one case went to the men. Little or no attempt was made to conceal this. It was a tradition, a right. When efforts were made to curtail the practice, labor difficulties arose. It soon became evident that certain pilferage would have to be accepted as an unwritten part of the union contract under the existing circumstances.

This is not a unique situation. The progression from limited pilferage through its acceptance as normal conduct to the status of an unwritten right has been repeated time and again.

The problem is, it does not stop there. Ultimately, pilferage becomes serious theft, and then the real trouble starts. Even before pilferage expands into larger operations, it presents a difficult problem to any business. Even where the amount of goods taken by any one individual is small, the aggregate can represent a significant expense. With the costs of materials, manufacture, administration, and distribution rising as they are, there is simply no room for added, avoidable expenses in today's competitive markets. The business that can operate the most efficiently, and offer quality goods at the lowest prices because of the efficiency of its operation, will have a huge advantage in the marketplace. When so many companies are fighting for their economic life, there is simply no room for waste—and pilferage is just that.

Moral Obligation to Control Theft

When we consider that internal theft accounts for at least twice the loss from external theft (that is, from burglars, armed robbers and shoplifters combined), we must be impressed with the scope of the problem facing today's businesspeople. Businesses have a financial obligation to stockholders to earn a profit on their investments. Fortunately there are steps that can be taken to control internal theft. Setting up a program of education and control that is vigorously administered and supervised can cut losses to relatively insignificant amounts.

It is also important to observe that management has a moral obligation to its employees to protect their integrity by taking every possible step to avoid presenting open opportunities for pilferage and theft that would tempt even the most honest people to take advantage of the opportunity for gain by theft.

This is not to suggest that each company should assume a paternal role toward its employees and undertake their responsibilities for them. It is to suggest strongly that the company should keep its house sufficiently in order to avoid enticing employees to acts that could result in great personal tragedy as well as in damage to the company.

Program for Internal Security

As for all security problems, the first requirement before setting up protective systems for internal security is to survey every area in the company to determine the extent and nature of the risks. If such a survey is conducted energetically and exhaustively, and if its recommendations for action are acted on intelligently, significant losses from internal theft will be a matter of history. Security surveys and their companion, the operational audit, have been discussed in detail in Chapter 7.

Need for Management Support

Once concerns have been identified, it is especially important that the strong support of top management be secured. In order to implement needed security controls, certain operational procedures may have to be changed. This will require cooperation at every level, and cooperation is sometimes hard to get in situations where department managers feel their authority has been diminished in areas within their sphere of responsibility.

The problem is compounded when those changes determined to be necessary cut across departmental lines, and even serve to some degree to alter intradepartmental relationships. Affecting systems under such circumstances will require the greatest tact, salesmanship, and executive ability. Failing that, it may be necessary to fall back on the ultimate authority vested in the security operation by top management. Any hesitation or equivocation on the part of either management or security at this point could damage the program before it has been initiated.

This does not, of course, mean that management must give security carte blanche. Reasonable and legitimate disagreements will inevitably arise. It does mean that proposed security programs based on broadly stated policy must be given the highest possible priority. In those cases where conflict of procedures exists, some compromise may be necessary, but the integrity of the security program as a whole must be preserved intact.

Communicating the Program

The next step is to communicate necessary details of the program to all employees. Many aspects of the system may be proprietary or on a need-to-know basis, but because part of it will involve procedures engaged in by most or all of company personnel, they will need to know those details in order to comply. This can be handled by an ongoing education program or by a series of meetings explaining the need for security and the damaging effects of internal theft to jobs, benefits, profit sharing, and the future of the company. Such meetings can additionally serve to notify all employees that management is taking action against criminal acts of all kinds at every level and that dishonesty will not be tolerated.

Such a forceful statement of position in this matter can be very beneficial. Most employees are honest people who disapprove of those who are criminally inclined. They are apprehensive and uncomfortable in a criminal environment, especially if it is widespread. The longer the company condones such conduct, the more they lose respect for it, and a vicious cycle begins. As they lose respect, they lose a sense of purpose. Their work suffers, morale declines, and at best effectiveness is seriously diminished. At worst, they reluctantly join the thieves. A clear, uncompromising policy of theft prevention is usually welcomed with visible relief.

Continuing Supervision

Once a system is installed, it must be constantly supervised if it is to become and remain effective. Left to their own devices, employees will soon find shortcuts, and security controls will be abandoned in the process. Old employees must be reminded regularly of what is expected of them, and new employees must be adequately indoctrinated into the system they will be expected to follow.

There must be a continuing program of education if expected results are to be achieved. With a high turnover within the white-collar workforce, it can be expected that the office force, which handles key electronic data and related paperwork, will be replaced at a fairly consistent rate. This means that the company will have a regular influx of new people who must be trained in the procedures to be followed and in the reasons for these procedures.

Program Changes

In some situations, reasonable controls will create duplication of effort, cross-checking, and additional work. Because each time such additional effort is required there is an added expense, procedural innovations requiring it must be avoided wherever possible, but most control systems aim for increased efficiency. Often this is the key to their effectiveness.

Many operational procedures, for a variety of reasons, fall into ponderous routines involving too many people and excessive record keeping. This may serve to increase the possibility of fraud, forgery, or falsification of documents. When the same operational result can be achieved by streamlining the system and incorporating adequate security control, it should be done immediately.

Virtually every system can be improved and should be evaluated constantly with an eye for such improvement, but these changes should never be undertaken arbitrarily. Procedures must be changed only after the changes have been considered in the light of their operational and security impact, and such considerations should further be undertaken in the light of their effect on the total system.

No changes should be permitted by unilateral employee action. Management should make random spot checks to determine if the system is being followed exactly. Internal auditors and/or security personnel should make regular checks on the control systems.

Violations

Violations should be dealt with immediately. Management indifference to security procedures is a signal that they are not important, and where work-saving methods can be found to circumvent such procedures, they will be. As soon as any procedural untidiness appears and is allowed to continue, the deterioration of the system begins.

It is well to note that, while efforts to circumvent the system are frequently the result of the ignorance or laziness of the offender, a significant number of such instances are the result of employees probing for ways to subvert the controls in order to divert company assets to their own use.

Personnel Policies for Internal Security

There is no greater need for cooperation between two departments than that needed between human resources and security/loss prevention. The job for loss prevention is much simpler when the right person is hired for the right job. While human resources reviews all applications and maintains all records on employees, the security department should be responsible for conducting thorough background checks on specific employee positions. While it would be desirable to background all employees, the amount of employee turnover or volume of new hires often makes such a comprehensive program an impossibility.

Human Resources and the Screening Process

Internal security's objective is preventing theft by employees. If all employees were of such sterling character that they could not bring themselves to steal, security personnel would have little to do. If, on the other hand, thieves predominate in the mix of employees, the system will be sorely tried, if indeed it can be effective at all. Basic to internal security effectiveness is the cooperation of that majority of honest personnel performing as assigned and in so doing refusing to initiate or collaborate in conspiracies to steal. Without this dominant group, the system is in trouble from the start.

The first line of internal security defense is the human resources department, where bad risks can be screened out by use of reasonable security procedures. Screening is the process of finding the person best qualified for the job in terms of both skills and personal integrity. This process may or may not involve a background check (which will be discussed later), but it must include at least a basic check of an applicant's references and job history.

In some industries, especially those with high technical requirements, screening can be a problem because qualified personnel may be difficult to find. Resistance can come from the human resources department when an otherwise qualified applicant is disqualified on the marginal grounds of a potential security problem. Security management must undertake the job of convincing objectors that a person who may later embezzle from the company is a poor risk from many points of view, no matter how highly qualified that person may be in the specific skills required.

Rejecting bad risks must be made on the basis of standards carefully established in cooperation with the human resources director. Once established, these standards must be met in every particular circumstance, just as proficiency standards must be met. Obviously such standards require reviewing from time to time to avoid dealing with applicants unjustly and placing the company in the position of demanding more than is either realistically possible or available. Even here, however, a bottom line must be drawn. At a certain point, compromises and concessions can no longer be made without inviting damage to the company.

Such a careful, selective program will add some expense to the employment procedure, but it can pay for itself in reduced losses, better employees, and lower turnover. And the savings in terms of potential crimes that were averted, though incalculable, can be thought of as enormous.

Employment History and Reference Checking

The key to reducing internal theft is the quality of employees employed by the facility. The problem, however, will not be eliminated during the hiring process, no matter how carefully and expertly selection is made. Systems of theft prevention and programs of employee motivation are ongoing efforts that must recognize that elements of availability, susceptibility, and opportunity are dynamic factors in a constant state of flux. The initial approach to the problem, however, starts at the beginning—in the very process of selecting personnel to work in the facility. During this process, a knowledgeable screener who is aware of what to look for in the employment application or résumé can develop an enormous amount of vital information about the prospective employee. Some answers are not as obvious as they once were, and the ability to perceive and evaluate what appears on the application or résumé is more important than ever as applications are more restrictive in what they can ask.

The increased focus on screening and background checks over the past decade is a direct result of the following:

- A rise in lawsuits from negligent hiring.
- An increase in child abuse reporting and abductions which have results in new laws that require criminal background checks for anyone who works with children, including volunteers.
- September 11, 2001 has resulted in heightened security and required identity verification.
- The Enron scandal has increased scrutiny of corporate executives, officers and directors.
- Increasing use of inflated and fraudulent résumés and applications.
- New federal and state laws requiring background checks for certain jobs, i.e., armored-car employees.
- The information age has added to the increase in checks as information is now available through many computer databases.

Privacy legislation coupled with fair employment laws drastically limits what can be asked on the employment application forms. The following federal legislation relates directly to hiring and dealing with employees:

- Title VII of the Civil Rights Act of 1964
- Pregnancy Discrimination Act of 1978
- Executive Order 11246 (affirmative action)
- Age Discrimination in Employment Act of 1967
- National Labor Relations Act
- Rehabilitation Act of 1973
- Vietnam Era Veterans' Readjustment Assistance Act of 1974
- Fair Labor Standards Act of 1938 (The Wage and Hour Law)
- The Federal Wage Garnishment Law
- Occupational Safety and Health Act of 1970
- Immigration Reform and Control Act of 1986
- Employee Polygraph Protection Act of 1988
- Consolidated Omnibus Reconciliation Act of 1985 (COBRA)
- Worker Adjustment and Retraining Notification Act (Plant Closing Law)
- EEOC Sexual Harassment Guidelines
- The Americans with Disabilities Act of 1990
- Family Educational Rights and Privacy Act (FERPA)
- Bankruptcy Act
- Fair Credit Reporting Act (FCRA)
- Equal Pay Act 1963
- Privacy Act of 1976

Table 12-1 summarizes the protected classes covered under these and other federal acts, and Table 12-2 provides examples of acceptable and unacceptable inquiries for preemployment screening based on these laws.

Table 12-1 Who Is Protected/Who Is Affected
Federally Covered Employers and Protected Classes

Legislation	Race/Color	National	Origin/Ancestry	Sex	Religion	Age	Disabled	Union	Covered Employers	Federal Agency
Title VII Civil Rights Act	X	X	X	X					Employers with 15+ EEs; unions, employment agencies	EEOC
Equal Pay Act (EPA) as amended			X						Minimum wage law coverage ("administrative employees" not exempted)	EEOC
†Age Discrimination in Employment Act (ADEA)						40+			20+ EEs (unions with 25+ members), employment agencies	EEOC
*Age Discrimination Act of 1975 (ADA)						X			Receives federal money	EEOC
*Executive Order 11246.11141	X	X		X	X	X			All federal contractors and subcontractors	OFCCP
*Title VI Civil Rights Act	X	X		X	X				Federally-assisted program or activity—public schools and colleges also covered by Title IX	Funding Agency and EEOC
*Rehabilitation Act of 1973							X		Receives federal money; federal contractor, $2,500+	OFCCP
National Labor Relation Act (NLRA)	X	X		X	X	X		X	ER in interstate commerce	NLRB
Civil Rights Act of 1866	X	X							All employers	Courts
Civil Rights Act of 1871	X	X		X					Private employers usually not covered	EEOC
Revenue Sharing Act of 1972	X	X		X	X	X	X		State and local governments that receive federal revenue sharing funds	OFCCP

Law			Coverage	Enforcement
Education Amendments of 1972 Title IX	X		Educational institutions receiving federal financial assistance	Dept. of Education
Vietnam Era Vets Readjustment Act—1974		X	Government contractors—$10,000+	OFCCP
Pregnancy Discrimination Act of 1978	X		All employers 15+ EEs	EEOC-OFCCP
Fair Labor Standards Act			Includes minimum wage law and equal pay act with DOL complex method of coverage	
*Rehabilitation Act of 1973		X	Receives federal money; federal contractor, $2,500+	OFCCP
Americans with Disabilities Act of 1990		X	Covers employers with 15 or more employees	EEOC
Federal Privacy Act of 1976			Federal agencies only	
Freedom of Information Act			Federal agencies only	
Family Educational Rights and Privacy Act			Schools, colleges, and universities, federally assisted	
Immigration Reform Act of 1986			All employers	INS

*Applies to federal agencies, contractors, or assisted programs only.
†Mandatory retirement eliminated except in special circumstances.
EE = Employee
ER = Employer
EEOC = Equal Employment Opportunity Commission
OFCCP = Office Federal Contract Compliance Programs
NLRB = National Labor Relations Board
DOL = Department of Labor
INS = Immigration and Naturalization Service

Table 12-2   Examples of Acceptable and Unacceptable Inquiries for Preemployment Screening

Subject	Unacceptable Inquiries	Acceptable Inquiries
Address or duration of residence	Do you own or rent your home?	What is your place of residence? How long have you resided in this state or city?
Age	How old are you? What is your birth date? What are your children's ages? Dates of attendance or completion of elementary or high school	Are you 18 years of age or older? If not, state your age If hired, can you show proof of age? If under 18, can you submit a work permit after employment?
Arrest and convictions records	Have you ever been arrested?	Have you been convicted of a crime? If so, give details.
Birthplace, citizenship	Of what country are you a citizen? Are you naturalized or a native-born citizen? What date did you acquire citizenship? Please produce your naturalization or first paper Are your parents or spouse naturalized or native-born United States citizens? What date did your parents or spouse acquire United States citizenship?	Are you authorized to work in the United States? Can you, after employment, submit verification of your legal Right to work in the United States? Statement that such proof may be required after employment
Disability	What is your corrected vision? Have you ever been unable to cope with job-related stress? Do you have a disability that would interfere with your ability to perform the job? When will your broken leg heal? Can you stand? Can you walk? How many days were you sick last year? Have you ever been treated for mental illness? Do you have asthma? Do you have any physical disabilities or handicaps? Questions regarding receipt of workers' compensation	Do you have 20/20 corrected vision? How well can you handle stress? How did you break your leg? Can you stand for 5 hours? Can you walk 20 miles in one day? Can you meet the attendance requirements of this job?

(Continued)

Table 12-2   (Continued)

Subject	Unacceptable Inquiries	Acceptable Inquiries
Discharge from military service	Did you serve in the armed forces of another country? Did you receive a discharge that was less than honorable?	Have you ever been a member of the United States armed services or in a state militia? If so, what branch? If so, explain your experience in relation to the position for which you are applying. Did you receive a dishonorable discharge? (A statement should accompany inquiries regarding military service that a dishonorable discharge is not an absolute bar to employment and that other factors will affect the final decision.)
Education		Describe your academic, vocational, or professional education as it relates to this position. What private or public schools did you attend?
English language skills	What is your native language? How did you acquire your foreign language skill?	What foreign language do you read, write, and/or speak fluently?
Experience		Inquiries include those regarding work experience
Marital status, number of children, child care, sex	Are you married? Single? Divorced? Separated? What is your spouse's name? Where is your spouse employed? What is your spouse's salary? What are your child care plans? Who can we contact in case of emergency? (This question can be asked after a person has been hired.) Do you wish to be addressed as Miss? Mrs.? Ms.? Questions regarding pregnancy, childbearing, or birth control.	Information such as this, which is required for tax, insurance, or Social Security purposes, may be obtained after hiring. Lawful inquiries include those regarding one's ability to travel if the job required it. However, all applicants must be asked the same question.
Notice in case of emergency	Name and address of person to be notified in case of accident or emergency. (Information obtained after the applicant has been hired.)	Name and address of person to be notified in case of accident or emergency. Name and address of a relative or spouse to be notified in case of accident or emergency.

(Continued)

Table 12.2   (Continued)

Subject	Unacceptable Inquiries	Acceptable Inquiries
Name	Please state your maiden name. If you have worked under another name, state that name and dates.	Have you ever worked for this company under a different name? Is additional information relative to change of name, use of an assumed name or nickname necessary to enable a check on your work record? If yes, explain.
Organization, activities	List all clubs, societies, and lodges to which you belong.	Please state your membership in any organization(s) that you feel is/are relevant to your ability to perform this job.
Physical description, photograph	Questions as to applicant's height and weight.	Any questions that have an impact on one's ability to perform the job requirements.
	Require applicant to affix a photograph to application. Request applicant at his or her option to submit a photograph. Require a photograph after interview but before employment. Inquiries include those that are not related to job requirements.	Statement that photograph may be required after employment.
Race, color, religion, or national origin	Questions as to applicant's race or color. Questions regarding applicant's complexion or color of skin, eyes, hair. Questions regarding applicant's religion. Questions regarding religious days observed. Does your religion prevent you from working weekends or holidays? Questions as to nationality, lineage, ancestry, national origin, descent, or parentage of applicant, applicant's parents, or spouse.	Statement by employer of regular days, hours, or shifts to be worked.
References	Questions of applicant's former employers or acquaintances that elicit information specifying the applicant's race, color, religious creed, national origin, ancestry, physical handicap, medical conditions, marital status, age, or sex.	Who referred you for a position here? Names of persons willing to provide professional and/or character references for applicant.

In some aspects, these regulations had a streamlining effect, eliminating irrelevant questions and confining questions exclusively to those matters relating to the job applied for. The subtler kinds of discrimination on the basis of age, sex, and national origin have been largely eliminated from the employment process. In making these changes to protect the applicant, state and federal laws have created new dilemmas for employers and their security staffs.

Various federal and state laws prohibit criminal justice agencies (police departments, courts, and correctional institutions) from providing information on certain criminal cases to non–criminal justice agencies (for example, private security firms or human resources departments). The Fair Credit Reporting Act requires that a job applicant must give written consent to any credit bureau inquiry.

All states have some type of privacy legislation meeting the guidelines set forth in the Federal Privacy Act of 1976:

The most controversial portion of the Act states "that information shall only be used for law enforcement and criminal justice and other lawful purposes." The crux of the issue is the way that "other lawful purposes" is defined. Does this meaning include human resources departments and private security operations? The verdict is mixed. Human resources, security, and loss-prevention operations must be aware of the interpretation of the privacy legislation in each state in which they operate. Recent legislation as discussed in Chapter 5 regarding the Department of Homeland Security has allowed for greater access of government agencies and private security firms to criminal histories, financial records and medical records.

Understandably there is some confusion regarding the rules governing employment screening. In spite of such confusion, the preemployment inquiry remains one of the most useful security tools employers can use to shortstop employee dishonesty and profit drains. Security management should consult with legal counsel to determine which laws relate to their locality and establish firm and precise policies regarding employment applications and hiring practices. An employer should be as familiar as possible with the federal Fair Credit Reporting Act (FCRA) which governs what employers must do when contracting record checks with third parties.

Generally speaking, look for, and be wary of, applicants who:

1. Show signs of instability in personal relations
2. Lack job stability: a job hopper does not make a good job candidate
3. Show a declining salary history or are taking a cut in pay from the previous job
4. Show unexplained gaps in employment history
5. Are clearly overqualified
6. Are unable to recall or are hazy about names of supervisors in the recent past or who forget their address in the recent past

In general all or some of the following information might be included in a background check.

- Driving records
- Vehicle registration

- Credit reports
- Criminal records
- Social security number
- Educational records
- Court records
- Workers' compensation
- Bankruptcy
- Character references
- Neighbor interviews
- Medical records
- Property ownership
- Military records
- State licensing
- Drug tests
- Past employment
- Personal references
- Arrest records
- Sex offense lists

If the job applied for is one involving the handling of funds, it is advisable to get the applicant's consent to make a financial inquiry through a credit bureau. Be wary if such an inquiry turns up a history of irresponsibility in financial affairs, such as living beyond one's means.

Application forms should ask for a chronological listing of all previous employers in order to provide a list of firms to be contacted for information on the applicant as well as to show continuity of career. Any gaps could indicate a jail term that was "overlooked" in filling out the application. When checking with previous employers, dates on which employment started and terminated should be verified.

References submitted by the applicant must be contacted, but they are apt to be biased. After all, since the person being investigated submitted their names, they are not likely to be negative or hostile. It is important to contact someone—preferably an immediate supervisor—at each previous job. Such contact should be made by phone or in person.

The usual and easiest system of contact is by letter or in recent years by email, but this leaves much to be desired. The relative impersonality of these forms of communication, especially one in which a form or evaluation is to be filled out, can lead to generic and essentially uncommunicative answers. Because many companies as a matter of policy, stated or implied, are reluctant to give someone a bad reference except in the most extreme circumstances, a written reply to a letter will sometimes be misleading.

As noted, over the past several years the letter has often been replaced with an e-mail. What has been noted as pitfalls of the letter apply to e-mail, although the ease of use and virtually no cost benefits often puts this form of reference check at the top of many budget-conscious managers' preferred means of contact.

On the other hand, phone or personal contacts may become considerably more discursive and provide shadings in the tone of voice that can be important. Even when no further information is forthcoming, this method may indicate when a more exhaustive investigation is required.

Backgrounding

It may be desirable to get a more complete history of a prospective employee—especially in cases where sensitive financial or supervisory positions are under consideration. In-depth investigations will involve extra expense, but may prove to be well worthwhile. Backgrounding, which involves a discreet investigation into the past and present activities of the applicant, can be most informative.

An estimated 90 percent of all persons known to have stolen from their employers were not prosecuted. While this is a disturbing estimate the U.S. Chamber of Commerce estimates that approximately 75% of all employees steal at least once. But 50% do it a second time. Experts believe that we generally catch only chronic offenders.[8]

A thorough investigation of potential employees is certainly justified, especially the backgrounds of persons being considered for jobs with responsibility for significant amounts of cash or goods, or those seeking management positions in shipping, receiving, purchasing, or paying.

Also significant to note is that there is agreement among personnel and security experts that 20 percent of any given workforce is responsible for 80 percent of personnel problems of every variety. If backgrounding can turn up this kind of recorded or known employee behavior, it is well worth the expense.

Backgrounding is also employed to investigate employees being considered for promotion to positions of considerable sensitivity and responsibility. Such persons may have been the very model of rectitude at the time of employment but may since have had financial reverses threatening their lifestyles. If a background investigation uncovers such information, a company is in a position to offer assistance to that employee if it so desires, relieving both strain and need, which can lead to embezzlement. Such action can boost company morale as well as reduce the potential for theft committed out of desperation.

With today's access to Internet resources, security staff can subscribe to any number of Internet investigative resources. The cost associated with these services ranges from as low as $25 per inquiry to several hundred dollars. The cost is often a reflection of the depth of the inquiry. Common resources include:

- www.instantpeoplecheck.com
- www.uscriminalcheck.com
- www.accuratecredit.com
- www.web.public.records-now.com

In addition to paying for Internet searches, employers are also making use of Internet networking sites, such as Facebook, Xing and Linkedin, as well as Google. While the information can be valuable, it may also lead the company into problem areas. These may include legal risks that can cost the company. It is therefore wise to follow established practices and remember the limitations of information posted on public sites. This practice and developing trends are worth monitoring. The *New York Times* reported in August of 2010 that Germany had drafted a law that would place restrictions on the use of Facebook profiles for hiring purposes.[9]

Integrity and Lie Detection Tests

Another option in lieu of a full background investigation is the use of various integrity and lie detection tests. In states where their use is legal, these tests, according to some experts, can be useful tools in determining the past and current records of candidates for employment or promotion to sensitive positions.

Using integrity and lie detection tests is controversial, as is the discussion among practitioners about the relative merits of different types of tests. Many security people look on integrity tests as invaluable tools, but generally agree that their use should be restricted to the hands of competent, trained professionals for productive results.

The three main categories of integrity and lie detection tests are (1) the polygraph, (2) the Psychological Stress Evaluator (PSE), and (3) the Personal Security Inventory (PSI). The first two are machines that operate on the same basic physiological principle. The third is a psychological pencil-and-paper test. Controversial recent entries into the area of truth/stress detection are infrared heat scanner and MRI brain scans. In addition, recent research has indicated that eye-tracking technology might be a cost-effective alternative method of lie detection. Current research at the University of Utah reports that accuracy in measuring deception ranges between 82 and 91 percent and requires only 20% of the amount of time needed for polygraph tests.[10] Whether these new technologies will survive is speculative.

The polygraph, PSE, MRI scan and infrared face scan operate on the premise that lying creates conflict, which in turn causes anxiety leading to stress reactions. Stress reactions typically include increased respiration, increased pulse rate, higher blood pressure, digestive disorders, perspiration, temperature change, muscle tension, and pupil dilation. The polygraph and the PSE measure some of these changes. Most polygraphs measure galvanic skin response, blood pressure, and respiration. The PSE measures changes in voice quality from tension in the vocal cords. The readouts of both devices may be recorded on paper tape and the responses to various questions are analyzed in terms of the charted reactions, which are compared with reactions to simple test questions establishing an individual's normal response to lying. Digital readouts are now more common, and although the PSE has not been as well publicized as the polygraph, portable PSEs are now available. Manufacturers claim that the digital device is the world's first portable lie detector.[11]

Do these machines really work? The controversy over this question continues to draw attention in professional publications. The Employee Polygraph Protection Act of 1988 strictly limits the use of the polygraph to certain industries and for specific purposes. It prohibits preemployment polygraphs by all private employers except those whose primary business is providing security personnel for the protection of currency, negotiable securities, precious commodities or instruments, or proprietary information. Companies who manufacture, distribute, or dispense controlled substances may use the polygraph for any employee having direct access to the manufacture, storage, distribution, or sale of these controlled substances.

To determine whether the polygraph or PSE is worth using in a security operation, various factors need to be considered, such as "What is a lie?" According to the *American Heritage Dictionary of the English Language,* a lie is making a statement or statements that one knows to be false, especially with the intent to deceive. Because a lie is really a subjective evaluation, what these machines record is what a person believes to be a lie. In addition, because the machines measure only stress, the subject must believe that the machine works so that enough stress is produced to record.

Because these machines simply measure the physiological symptoms of stress, other stressors may also be recorded. For example, a question concerning drug history may evoke a dramatic response from a subject who has lost a close relative to drug abuse. The subject's physical condition at the time of the test can also influence the tracings. Fatigue, drugs, and alcohol are the biggest culprits in this category, but even simple things like needing to use the bathroom or suppressing a cough can affect the tracings.

Can someone beat the machine? The answer is both simple and complex. A subject can affect the tracing of the machine by saying prayers or counting holes in acoustic tiles, but a good examiner can and will note these changes. Although the machine registers only physiological changes and thus cannot be beaten, the accuracy of the machine is determined solely by the quality of the examiner.

Problematic to the issue of accuracy, then, is the training of examiners. Training can range from as little as 6 weeks to as long as 6 months. Still a 2003 National Academy of Sciences report concluded that research on the polygraph's efficacy was inadequate.[12] Today the polygraph and related lie detection tests continue to generate discussion about their use and reliability.

What is the role of these devices if such doubts can be cast on their accuracy and on the skill of examiners? Jerome Skolnick, noted criminologist, suggests that their results should be used "to open up leads to further investigation of information rather than being itself prima facie evidence."[13]

Much case law regarding the use of the polygraph has been added since the 1988 Polygraph Act. Still, the polygraph's advocates are striving for national recognition for the professional use of the instrument through improving standards for operators. In Illinois, a licensed school must certify polygraph operators before they can operate a machine. Even so, Illinois does not allow polygraph evidence to be presented in its courts.

As for the PSE, the question remains whether polygraph case law will apply to this type of testing as well. Since both machines operate on the same principles, it is likely that much of the established case law relating to the polygraph might be applicable to the PSE. It is interesting to note, however, that during the past 10 years, little has changed in reference to the use of the PSE, but the polygraph continues to struggle to survive under legal fire.

As previously mentioned, the PSI is a psychological pencil-and-paper exam. Many variations of this test exist throughout the United States (for example, the California Personality Inventory [CPI], the Minnesota Multiphasic Personality Inventory [MMPI], Inwald Personality Inventory [IPI], Wonderlic), but they all have common traits. These tests are designed to evaluate prospective employees for honesty and integrity, drug or alcohol abuse, and violence or emotional instability. Most of these tests can be administered on company property and are relatively inexpensive. The test questions are designed to make consistency (veracity) checks by comparing responses and to provide clues for the psychologist evaluating the personality of the potential employee tested. For example, a typical question might be as follows:

You are riding on a bus. A sign indicates "No Smoking."

A person sits down next to you and lights a cigarette. What would you do?

1. Inform him that he is not allowed to smoke
2. Point out the "No Smoking" sign
3. Move to another seat
4. Get off the bus
5. Say or do nothing

Adherents of this system claim a high degree of accuracy for such tests when they are conducted by properly trained personnel. Table 12-3 provides a list of pointers that can be used in evaluating pencil-and-paper tests.

It is important for security managers to research the limitations on the use of such instruments in their state. Some states forbid their use as a requirement for employment but permit such testing to be used on a voluntary basis. Other states forbid their use under any circumstances.

In general, organized labor has lobbied diligently for legislation banning the use of so-called integrity testing and lie detectors for all industrial or commercial applications, their efforts proving successful with the passage of the 1988 Polygraph Act. The American Polygraph Association, which opposes the use of the PSE on grounds that it has not yet proved itself, has endorsed legislation setting stricter standards for polygraph operators, but naturally fights labor's stand on the issue.

Table 12-3 Scrutinizing Vendor Pencil-and-Paper Tests
Twelve Pointers in Evaluating a Test's Relevance for Meeting Your Hiring Goals

1. Beware of tests for which little or no validation research exists.
2. Beware of tests that claim they "only replace the polygraph."
3. Beware of studies that are not based on the predication model of validation.
4. Beware of studies that do not tell you how many people were incorrectly predicted to have job problems.
5. Beware of tests that claim to predict dangerous or violent behavior or tendencies.
6. Beware of studies that report "significant" correlations as evidence of their validity.
7. Beware of studies that use small numbers of people to predict important job-performance outcomes.
8. Beware of studies that have not been cross-validated.
9. Beware of claims that tests are valid for use with occupational groups for which validation studies have not yet been conducted.
10. Beware of studies based on questionnaires or tests filled out anonymously.
11. Beware of studies that have not used real job candidates as subjects in their validation efforts.
12. Beware of tests whose validation studies have been designed, conducted, and published only by the test developer or publishing company without replication by other, totally independent, psychologists or agencies.

Source: From materials developed by Dr. Robin Inwald, Hilson Research Inc., February 1988.

Despite the 1988 Polygraph Act limitations on preemployment usage, many firms continue to use the polygraph in various types of investigations. Firms using these machines in this manner have generally found them to be useful. The federal government still performs tens of thousands of polygraph tests each year.[14]

According to the 1988 Polygraph Act, the polygraph may be used on existing employees under the following circumstances:

1. In connection with an ongoing investigation involving economic loss or injury to the employer's business (that is, theft, embezzlement, or misappropriation)
2. If the employee has access to the property that is the subject of an investigation
3. If the employer has reasonable suspicion that the employee was involved in the accident or activity under investigation
4. If the employer provides to the employee *before* the test the specifics of the inquiry—that is, what incident is being investigated and the reason the employee is being tested

Even if the employer meets the above criteria, employees cannot be disciplined or discharged solely on the basis of the results of the polygraph examination or for refusing to take the polygraph test.

Such examinations can cost from $50 to $150, depending on the length of the test, geographic location and firm, so not every firm will find using them practical. Firms finding the expense acceptable usually limit the polygraph's application to particularly sensitive investigations and then only after deciding whether the risk is such that the expense is warranted. Even so, these companies first explore whether they can be satisfied by an evaluation based on other, less expensive methods.

Of recent interest is the use of MRI technology to scan brain activity. Functional MRI technology expands the traditional static picture MRI to allow for a series of scans that show changes in the flow of oxygenated blood preceding neural events. The theory is that lying requires more cognitive effort than telling the truth. The test is too expensive for general commercial use at this time, with an estimated cost of $10,000 for an examination. For more information see www.noliemri.com.

Whatever decision the security manager makes regarding the use of integrity tests, he or she would be well advised to consult a reputable firm to learn about what such examinations involve.

Concerns following the September 11, 2001 bombing of the World Trade Center have researchers looking for ways to detect deception at airports. The problems associated with traditional polygraphs and the PSE have been discussed. Recent work by scientists has taken the same theory of lying and stress and applied it to new technology. In this instance a heat sensitive camera spots possible deceit by looking for telltale increases in body temperature around the eyes. Such a device could be of value to airport security personnel. Although the research team was led by experts from the Mayo Clinic and Honeywell Laboratories, polygraph authorities question the utility of the device until further testing is completed.[15] As of 2007 it appears that the same stress pitfalls are present in this new technology. TSA personnel might be targeting nervous, jangly, harmless travelers as well as those who might be terrorists.[16] As of 2010 there is little new published information on the developments on this front.

Americans with Disabilities Act

The ADA is a federal statute requiring employers to focus on the *abilities* of applicants rather than on their disabilities. Because of this, inquiries about medical conditions or medical tests before a job offer is made are prohibited. Offers can, however, be made contingent on successful completion of medical examinations. Employers must make certain that medical, psychological, and physical agility examinations come at the end of the hiring process rather than at the beginning. Currently the ADA does not protect the following categories: drug abusers; homosexuals, bisexuals; persons who engage in aberrant sexual behavior; compulsive gambling, theft, or pyromania; persons whose disorders have been caused by drug abuse; or persons whose disabilities are only temporary. It should be noted that tests for illegal drugs are not subject to the ADA's restrictions on medical examinations.[17]

Job announcements, descriptions, and applications should be carefully reviewed to ensure that they conform to the ADA requirements concerning the description of each position's essential functions. All staff involved in recruiting, hiring, and personnel processing and decision making generally should be retrained to ensure that they understand and conform to all ADA requirements. In the end, employers should hold persons with disabilities to the same performance standards as other employees.[18]

Drug Screening

There are few areas of preemployment screening that provoke the strong reactions that drug screening does. Two major issues usually raised about this type of testing concern invasion-of-privacy arguments and the problem of the risk of false positives. The issues related to drug screening will be dealt with in more detail in Chapter 14.

Other Screening Options

In addition to the screening tools already discussed, other sources of information can provide details of the applicant's background that might prove valuable in making a hiring decision.

Credit reports not only reflect an applicant's financial situation and stability but also provide other useful information such as past addresses and previous employers. The legal restrictions of the Federal Fair Reporting Act must be complied with if a person is denied employment as the result of information discovered through this source.

Motor vehicle records are easily obtained and can aid in identifying high-risk employees by noting the number of driving violations and types.

Civil litigation records provide detailed, documented records of an applicant's personal history, background, and financial relationships. These records also document previous injury complaints. They may also provide clues to other employment problems not filed as criminal charges, such as theft, fraud, or serious misconduct.

There are many other possible sources of information, but time and space considerations do not allow for total coverage in an introductory text. As noted earlier there are numerous web based sites willing to provide information for a price.

Hiring Ex-Convicts and Parolees

It should be strongly noted here that rigid exclusionary standards should never be applied to ex-convicts or parolees who openly acknowledge their past records. Such a policy would be at best unjust and at worst irresponsible. These people have served their sentences. Their records are available in situations in which employment is being sought. To turn them away solely on the basis of their past mistakes would be to force them back into a criminal pattern of life in order to survive.

While people with criminal backgrounds might well be unacceptable to certain companies, a rigid policy refusing all such people employment would be denying companies many potentially good employees deserving a chance to show that they have rehabilitated themselves. Experience has shown that these employees, knowingly hired in the right positions and properly supervised, are not only acceptable but are frequently highly responsible and trustworthy. They should be given an opportunity to reestablish themselves in society.

Employee Assistance Programs

One innovation in coping with employee problems relating to, among other things, honesty, alcohol/drug problems, and depression has been the advent of the employee assistance program. "At a time when substance abuse, mental health problems, and other stresses beset the American work force, an effective employee assistance program (EAP) can be a wise investment."[19] Today more than 90 percent of the *Fortune* 500 companies have introduced EAPs.[20] Programs vary with the type and size of company. Some companies provide full-service, in-house operations, whereas smaller firms may restrict the type of service and contract with professional EAP firms.

The fact is that EAPs are proving valuable for some firms. McDonnell-Douglas reports a 34 percent reduction in absences in comparison to its corporate counterparts. In the area of attrition of drug users, McDonnell-Douglas displays a job drop-out rate from 40 percent in their control group to 7.5 percent in the EAP participants. Other firms also report success: Chicago Bell, General Motors, HARTline, and so forth. There is a belief among many of these employers that improving family life will also reflect positively in the workplace.[21] For a discussion on EAPs see Chapter 14, Violence and Drug Use in the Workplace.

Continuity of the Screening Program

When a company makes a systematic and conscientious effort to screen out dishonest, troublesome, incompetent, and unstable employees, it has taken a first and significant step toward reducing internal theft. It is important that the program continues on a permanent basis.

Care must be taken to avoid relaxing standards or becoming less diligent in checking the backgrounds and employment histories of applicants. There is a tendency to lose sight of all the dimensions of the problem if the security program makes substantial inroads into the loss factor. Past problems are too soon forgotten, and carelessness follows closely behind. Active supervision is always necessary to maintain the integrity of this essential aspect of every security program.

CASE STUDY

A security survey conducted for the Assets Corporation recommended a complete revamping of the activity and aims of the Human Services Department. At the present time the personnel department deals only with plant personnel hiring. The office manager finds and hires all office personnel, sidestepping Human Services. The Human Services manager merely records these employees after they are hired.

The Assets Corporation uses one application form for all employees, regardless of job classification. The legal-size application form was copied 10 years ago from one used by the human services manager in a prior job. It contains one line to record the applicant's race and another to record religion. It also asks for three non-relative references and a listing of all prior employment with only the years of such employment required. No skill-related tests are given. The Human Services Department bases the employment decision on a review of the application and an interview with Human Services staff.

In a pending court case, the Assets Corporation has been charged with discharging a secretary for reasons of race. This employee stated on her application form that she had a high school diploma and could type 60 words a minute. After employment, her typed letters showed many uncorrected errors, including typing errors, misspelled words, and poor knowledge of business terms. This poor performance continued despite the availability of a comprehensive word processing program. She responded to all criticism by claiming racial discrimination on the part of her supervisor. She was verbally discharged, and no written record was kept of her poor performance.

- How could Assets protect itself from allegations such as those made by the discharged stenographer?
- What changes in the practices of the personnel department should the survey have recommended regarding:
  - Make-up of the application form?
  - Decision to hire employees?
  - Clerical applicant tests?

While employee screening and background checks can reduce later problems with employee performance and dishonesty, time and money often restrict the amount of effort in these important areas. Even the tools, such as polygraph and paper-and-pencil tests, that are used in place of personal checks of records are not universally accepted.

One thing, however, is clear. Employees who have options and feel treated like a part of a team are usually not going to become security problems.

Procedural Controls

Auditing Assets

Periodic personal audits by outside auditors are essential to any well-run security program. Such an examination will discover theft only after the fact, but it will presumably discover any regular scheme of embezzlement in time to prevent serious damage. If these audits, which are normally conducted once a year, were augmented by one or more surprise audits, even the most reckless criminal would hesitate to try to set up even a short-term scheme of theft.

These audits will normally cover an examination of inventory schedules, prices, footings, and extensions. They should also verify current company assets by sampling physical inventory, accounts receivable, accounts payable (including payroll), deposits, plant assets, and outstanding liabilities through an ongoing financial audit. In all these cases, a spot check beyond the books themselves can help to establish the existence of legitimate assets and liabilities, not empty entries created by a clever embezzler.

Separation of Responsibility

The principle of separation of responsibility and authority in matters concerning the company's finances is of prime importance in management. This situation must always be sought out in the survey of every department. It is not always easy to locate. Sometimes even the employee who has such power is unaware of the dual role. But the security specialist must be knowledgeable about its existence and suggest an immediate change or correction in such operational procedures whenever they appear.

An employee who is in the position of both ordering and receiving merchandise or a cashier who authorizes and disburses expenditures are examples of this double-ended function in operation. All situations of this nature are potentially damaging and should be eliminated. Such procedures are manifestly unfair to company and employee alike. They are unfair to the company because of the loss that they might incur; they are unfair to the employee because of the temptation and ready opportunity they present. Good business practice demands that such invitations to embezzlement be studiously avoided.

It is equally important that cash handling be separated from the record-keeping function. Cashiers who become their own auditors and bookkeepers have a free rein with that part of company funds. The chances are that cashiers will not steal, but they could and might. They might also make mathematical mistakes unless someone else double-checks the arithmetic.

In some smaller companies, this division of function is not always practical. In such concerns, it is common for the bookkeeper to act also as cashier. If this is the case, a system of countersignatures, approvals, and management audits should be set up to help divide the responsibility of handling company funds from that of accounting for them.

Promotion and Rotation

Most embezzlement is the product of a scheme operating over an extended period of time. Many embezzlers prefer to divert small sums on a systematic basis, feeling that the individual thefts will not be noticed and that therefore the total loss is unlikely to come to management's attention.

These schemes are sometimes frustrated either by some accident that uncovers the system or by the greed of the embezzlers, who are so carried away with success that they step up the ante. But while the theft is working, it is usually difficult to detect. Frequently the thief is in a position to alter or manipulate records in such a way that the theft escapes the attention of both internal and external auditors. This can sometimes be countered by upward or lateral movement of employees.

Promotion from within, wherever possible, is always good business practice, and lateral transfers can be effective in countering possible boredom or the danger of reducing a function to rote and thus diminishing its effectiveness. Such movement also frustrates embezzlers. When they lose control of the books governing some aspect of the operation, they lose the opportunity to cover their thefts. Discovery inevitably follows careful audits of books they can no longer manipulate. If regular transfers were a matter of company policy, no rational embezzler would set up a long-term plan of embezzlement unless a scheme was found that was audit-proof—and such an eventuality is highly unlikely.

To be effective as a security measure, such transfers need not involve all personnel since every change in operating personnel brings with it changes in operation. In some cases, even subtle changes may be enough to alter the situation sufficiently to reduce the totality of control an embezzler has over the books. If such is the case, the swindle is over. The embezzler may avoid discovery of the previous looting, but cannot continue without danger of being unmasked.

In the same sense, embezzlers dislike vacations. They are aware of the danger if someone else should handle their accounts, if only for the 2 or 3 weeks of vacation, so they make every effort to pass up holidays. Any manager who has a reluctant vacationer should recognize that this is a potential problem. Vacations are designed to refresh the outlook of everyone. No matter how tired they may be when they return to work, vacationers have been refreshed emotionally and intellectually. Their effectiveness in their job has probably improved, and they are, generally speaking, better employees for the time off. The company benefits from vacations as much as the employees do. No one should be permitted to pass up authorized vacations, especially one whose position involves control over company assets.

Computer Records/Electronic Mail and Funds Transfer/Fax

The computer has become the most powerful tool for record-keeping, research and development, funds transfer, electronic mail, and management within most companies today. It is essential that the computer and its support equipment and records be adequately protected from the internal thief.

Besides the computer, the transfer of information using Wi-Fi, wireless phones and other electronic devices is an everyday occurrence. This is such an important area of security that additional coverage is provided in Chapter 17.

Physical Security

It is important to remember that personnel charged with the responsibility of goods, materials, and merchandise must be provided the means of properly discharging that responsibility. Warehouses and other storage space must be equipped with adequate physical protection to secure the goods stored within. Authorizations to enter such storage areas must be strictly limited, and the responsible employees must have means to further restrict access in situations where they feel that the security of goods is endangered (see Chapter 10 for a discussion of pass systems).

Receiving clerks must have adequate facilities for storage or supervision of goods until they can be passed on for storage or other use. Shipping clerks must also have the ability to secure goods in dock areas until they are received and loaded by truckers. Without the proper means of securing merchandise during every phase of its handling, assigned personnel cannot be held responsible for merchandise intended for their control, and the entire system will break down. Unreasonable demands, such as requiring shipping clerks to handle the movement of merchandise in such a way that they are required to leave unprotected goods on the dock while filling out the rest of the order, lead to the very reasonable refusal of personnel to assume responsibility for such merchandise. And when responsibility cannot be fixed, theft can result.

The Mailroom

The mailroom can be a rich field for a company thief. Not only can it be used to mail out company property to an ally or to a prearranged address, but it also deals in stamps—and stamps are money. Any office with a heavy mailing operation must conduct regular audits of the mailroom.

Some firms have taken the view that the mailroom represents such a small exposure that close supervision is unnecessary. Yet the head of the mailroom in a fair-sized eastern firm got away with more than $100,000 in less than 3 years through manipulation of the postal meter. Only a firm that can afford to lose $100,000 in less than 3 years should think of its mailroom as inconsequential in its security plan. In addition, recent events related to bioterrorism make mailroom security an even greater responsibility.

Trash Removal

Trash removal presents many problems. Employees have hidden office equipment or merchandise in trash cans and have then picked up the loot far from the premises in cooperation with the driver of the trash-collecting vehicle. Some firms have had a problem when they put trash on the loading dock to facilitate pickup. Trash collectors made their calls during the day and often picked up unattended merchandise along with the trash. On-premises trash compaction is one way to end the use of trash containers as a safe and convenient vehicle for removing loot from the premises.

Every firm has areas that are vulnerable to attack. What and where they are can only be determined by thorough surveys and regular reevaluation of the entire operation. There are no shortcuts. The important thing is to locate the areas of risk and set up procedures to reduce or eliminate them.

When Controls Fail

There are occasions when a company is so beset by internal theft that problems seem to have gotten totally out of hand. In such cases, it is often difficult to localize the problem sufficiently to set up specific countermeasures in those areas affected. The company seems simply to "come up short."

Management is at a loss to identify the weak link in its security, much less to identify how theft is accomplished after security has been compromised.

Undercover Investigation

In such cases, many firms similarly at a loss in every sense of the word have found it advisable to engage the services of a security firm that can provide undercover agents to infiltrate the organization and observe the operation from within.

Such agents may be asked to get into the organization on their own initiative. The fewer people who know of the agents' presence, the greater the protection, and the more likely they are to succeed in investigations. It is also true that when large-scale thefts take place over a period of time, almost anyone in the company could be involved. Even one or more top executives could be involved in serious operations of this kind. Therefore secrecy is of great importance. Because several agents may be used in a single investigation and because they may be required to find employment in the company at various levels, they must have, or very convincingly seem to have, proper qualifications for the level of employment they are seeking. Over- or under-qualification in pursuit of a specific area of employment can be a problem, so they must plan their entry carefully. Several agents may have to apply for the same job before one is accepted.

Having gotten into the firm's employ, agents must work alone. They must conduct the investigation and make reports with the greatest discretion to avoid discovery. But they are in the best possible position to get to the center of the problem, and such agents have been successful in a number of cases of internal theft in the past.

These investigators are not inexpensive, but they earn their fee many times over in breaking up a clever ring of thieves. It is important to remember, however, that such agents are trained professionals. Most of them have had years of experience in undercover work of this type. Under no circumstances should a manager think of saving money by using employees or well-meaning amateurs for this work. Such a practice could be dangerous to the inexperienced investigator and would almost certainly warn the thieves, who would simply withdraw from their illegal operation temporarily until things had cooled down, after which they could return to the business of theft.

Prosecution

Every firm has been faced with the problem of establishing policy regarding the disposal of a case involving proven or admitted employee theft. They are faced with three alternatives: to prosecute, to discharge, or to retain the thief as an employee. The policy that is established is always difficult to reach because there is no ready answer. There are many proponents of each alternative as the solution to problems of internal theft.

However difficult it may be, every firm must establish a policy governing matters of this kind. And the decision about that policy must be reached with a view to the greatest benefits to the employees, the company, and society as a whole. An enlightened management would also consider the position of the as-yet-to-be-discovered thief in establishing such policy.

Discharging the Thief

Most firms have found that discharge of the offender is the simplest solution. Experts estimate that most of those employees discovered stealing are simply dismissed. Most of those are carried in the company records as having been discharged for "inefficiency" or "failure to perform duties adequately."

This policy is defended on many grounds, but the most common are as follows:

1. Discharge is a severe punishment, and the offender will learn from the punishment.
2. Prosecution is expensive.
3. Prosecution would create an unfavorable public relations atmosphere for the company.
4. Reinstating the offender in the company—no matter what conditions are placed on the reinstatement—will appear to be condoning theft.
5. If the offender is prosecuted and found not guilty, the company will be open to civil action for false arrest, slander, libel, defamation of character, and other damages.

There is some validity in all of these views, but each bears some scrutiny.

As to learning (and presumably reforming) as a result of discharge, experience does not bear out this contention. A security organization found that 80 percent of the known employee thieves they questioned with polygraph substantiation admitted to thefts from previous employers. Now it might well be argued that, because they had not been caught and discharged as a result of these prior thefts, the proposition that discharge can be therapeutic still holds or at least has not been refuted. That may be true, and it should be considered.

Prosecution is unquestionably expensive. Personnel called as witnesses may spend days appearing in court. Additional funds may be expended investigating and establishing a case against the accused. Legal fees may be involved. But can a company afford to appear so indifferent to significant theft that it refuses to take strong action when it occurs?

As to public relations, many experienced managers have found that they have not suffered any decline in esteem. On the contrary, in cases where they have taken strong, positive action, they have been applauded by employees and public alike. This is not always the case, but apparently a positive reaction is usually the result of vigorous prosecution in the wake of substantial theft.

Reinstatement is sometimes justified by the circumstances. There is always, of course, a real danger of adverse reaction by the employees, but if reinstatement is to a position not vulnerable to theft, the message may get across. This is a most delicate matter that can be determined only on the scene.

As far as civil action is concerned, that possibility must be discussed with counsel. In any event, it is to be hoped that no responsible businessperson would decide to prosecute unless the case was a very strong one.

Borderline Cases

Even beyond the difficulty of arriving at a satisfactory policy governing the disposition of cases involving employee theft, there are the cases that are particularly hard to adjudicate. Most of these involve the pilferer, the long-time employee, or the obviously upright employee in financial difficulty who steals out of desperation. In each case, the offender freely admits guilt and pleads being overcome by temptation. What should be done in such cases? Many companies continue to employ such employees, provided they make restitution. They are often grateful, and they continue to be effective in their jobs.

In the last analysis, individual managers must make the determination of policy in these matters. Only they can determine the mix of toughness and compassion that will guide the application of policy throughout.

It is hoped that every manager will decide to avoid the decision by making employee theft so difficult and so unthinkable that it will never occur. That goal may never be reached, but it is a goal to strive for.

Review Questions

1. What are some of the common danger signals of employee dishonesty?
2. Discuss procedural controls for decreasing the incidence of employee theft in specific departments.
3. What should management's role be in effecting internal security?
4. Should employees be prosecuted for stealing? Why?
5. Discuss the differences between personnel screening and backgrounding.
6. What areas of federal legislation must be considered when conducting reference checks and employment histories?
7. Discuss the role of lie detection in backgrounding of employees.
8. What impact has the Americans with Disabilities Act had on employee selection?

References

[1] Association of Certified Fraud Examiners. The cost of occupational fraud. 2010.

[2] Zalud B. 2002 Industry Forecast Study Security Yin-Yang: Terror Push, Recession Drag. Security, January 2002.

[3] Grannis K. Retail Losses Hit $41.6 Billion Last Year, According to National Retail Security Survey. National Retail Federation, <www.nrf.com>.

[4] Carson CP. In: Managing employee honesty. Boston: Butterworth-Heinemann; 1977.

[5] Josephson M. No baloney: stronger ethics an anti-fraud antidote. J Insur Fraud Summer 2010;1(1):2.

[6] <www.insurecast.com/html/crime_insurance.asp>, downloaded 7/9/2007.

[7] Association of Certified Fraud Examiners. 2010 Report to the Nation on Occupational Fraud and Abuse. p. 4.

[8] Merchants Information Solutions, Inc. Employment Screening, September 2009, <http://employment-screening-services.blogspot.com>, downloaded 9/13/10.

[9] Jolly D. Germany Plans Limits on Facebook Use in Hiring. The New York Times, August 25, 2010.

[10] Mahoney S. Seeing through Lies. Secur Prod September 2010:110.

[11] <www.pimall.com>, downloaded 7/9/2007.

[12] www.newyorker.com/2007/007/02.

[13] The use of polygraphs and similar devices by federal agencies: hearing before a subcommittee on government operations, house of representatives. Washington, DC: Government Printing Office; 1974. p. 29.

[14] <www.newyorker.com/reporting> 2007/07/02.

[15] Shachtman N. Liar, Liar, Eyes on Fire? Wired January 2002 [downloaded 3/6/2003, <www.wired.com/news/technology/0,1282,49458,00.html>]

[16] www.newyorker.com

[17] U.S. Equal Employment Opportunity Commission. Facts about the Americans with Disabilities Act; U.S. Equal Employment Opportunity Commission, The ADA: Your Responsibilities as an Employer; U.S. Equal Employment Opportunity Commission, The ADA: Questions and Answers.

[18] Fyfe JJ, Greene JR, Walsh WF, Wilson OW, McLaren RC. In: Police Administration, 5th ed. New York: McGraw-Hill Companies; 1997. Pre-employment Screening Considerations and the ADA (1997), <http://janweb.icdi.wvu.edu/kinder/pages/pre_employment_screening.html>U.S. Equal Employment Opportunity Commission, Facts about the Americans with Disabilities Act; Gilbert Casella (1995, October 10), ADA Enforcement Guidance: Preemployment Disability-Related Questions and Medical Examinations (EEOC Notice Number 915.002), <www.eeoc.gov/docs/preemp.txt>.

[19] Pope T. An Eye on EAPs. Secur Manage October 1990:81.

[20] Contact, "Is an Employee Assistance Program (EAP) A Good Idea for My Client Companies?" downloaded 1/14/2003, <www.contactbhs.com/brokers/brokers/brokers10.html>.

[21] Pope.

Specific Threats and Solutions

In this final section, attention is directed at specific security threats ending with a look toward the future. Many of the threats are common to all types of organizations, whereas some, such as retail, are limited to a subcategory. While there are many threats, we have dedicated their own chapters to the areas of terrorism, retail theft, transportation and cargo security, workplace violence and drugs, and information security. In addition, we briefly cover traditional problems such as burglary, robbery, labor disputes, espionage and piracy.

The final chapter, titled "Security: The Future," is, as you might expect, speculative in nature. However, the speculation is based on a sound foundation of past events, industry knowledge of evolving and current trends and expert "guesswork"!

13

Transportation Security Issues and Regulation

OBJECTIVES

The study of this chapter will enable you to:

1. Understand the growing importance of security in the transportation industry.

2. Discuss the changes that have occurred in the airline industry since September 11, 2001.

3. Identify some of the technology that is being used to improve security in the transportation industry.

4. Discuss the role of the federal government in establishing security standards in the area of transportation security.

Introduction

The first thought that most Americans have regarding transportation security has to do with airlines and the hijackings that resulted in the destruction of the World Trade Center in 2001. Still, transportation security has been a growing security problem since 1970, when the U.S. Senate Select Committee on Small Business estimated that almost $1.5 billion in direct loss was attributable to cargo theft.[1] By 1975 the U.S. figure had reached $2.3 billion; by 1990 it had climbed to $13.3 billion, according to Federal Bureau of Investigation (FBI) figures. These figures do not reflect pilferage and unreported crimes, which represent a minimum of 5 to 10 times the amount of the reported crimes. Worldwide, cargo losses have been estimated at $30 billion.[2] In 2006 the FBI reported that an estimated $15–$30 billion was lost to cargo theft in the United States. In March 2006, the category "cargo theft" was added to the FBI's Uniform Crime Reporting (UCR) System through a provision of the USA Patriot Improvement and Reauthorization Act.[3]

The indirect costs of claims processing, capital tied up in claims and litigation, and market losses from both nondelivery and underground competition from stolen goods can cost between two to five times the direct losses.[4] This equates to an estimated $2 to $5 for every $1 of direct loss—a $30 billion to $150 billion annual loss in the U.S. national economy. Worldwide, this may amount to another $60 billion to $150 billion.

The problem continues to increase in severity. According to the FBI, cargo theft has become a significant concern as noted by the government's focus on including it as its own category in the UCR. The shipment of goods is vital to the economy and ultimately to the survival of the country. Since the 1970 figures were established, the growth of international transportation systems using containers that can be transported by truck, ship, and rail has developed to the extent that land bridges, particularly in the United States, have been thoroughly established. These land routes carry millions of dollars of goods from other countries over our rail system on stack trains through the United States. The liability for the contents of these containers, moving via land bridge, is shared among a multitude of ocean carriers, rail companies, and truckers. The security concerns for the safe handling of this movement of goods are many and include many different groups: rail police, state and local police, and customs officials as well as other federal authorities.

Still, cargo security is only one part of the concern in the 21st century. The Department of Homeland Security has made the transportation business one of many priorities in efforts to protect citizens from the threat of terrorism. As noted in Chapter 1, transit agencies have taken serious steps to improve security for passengers. The problems are enormous in a system that relies on public accessibility and rapid movement.

The Role of Private Security

Because it appears that cargo theft is more a problem of "inside jobs" than "highway robbery," it would appear that by far the greatest burden falls on the security apparatuses of the private concerns involved.[5] It is true that public law enforcement agencies must make a greater effort to break up organized fencing and hijacking operations, and they must find a way to cut through their jurisdictional confusion and establish more effective means of exchanging information. But the bulk of the problem lies in the systems now employed to secure goods in transit.

On the other hand, public transportation problems require carefully planned cooperative ventures between public law enforcement and private security. The very nature of public transportation requires that private security protect the infrastructure of transit companies, while public law enforcement provides protection of the public's highways. Heightened security of the early 21st century due to increased threats of terrorism resulted in public law enforcement personnel providing security for railroad bridges and at public transit terminals.

There is no universally applicable solution to this problem. Every warehouse, terminal, and means of shipment has its own particular peculiarities. Each has weaknesses somewhere, but certain principles of cargo security, when they are thoughtfully applied and vigorously administered, can substantially reduce the enormous losses so prevalent in today's beleaguered transport industry.

A good loss-prevention manager must recognize that the key to good cargo security is a well-organized cargo-handling system. As Louis Tyska and Lawrence Fennelly note, cargo loss exists whenever the "three C's of cargo theft" are present.[6] The three C's are confusion, conspiracy, and the common denominator (the dishonest employee). Confusion is a primary ingredient and represents the loss-prevention specialist's opportunity to reduce theft. Confusion arises when an adequate policy does not exist or, if it does exist, when it is not followed. Tyska and Fennelly identify the following activities as great contributors to the confusion variable:

1. Personnel entering and exiting the specific facility. These people include everyone from repair people to regular employees.
2. Movement of various types of equipment, for example, trucks, rail cars, and lift trucks.
3. The proliferation of various forms of paper—freight bills, bills of lading, manifests, and so on.

Conspiracy builds on confusion. Two or more people take advantage of their positions and the confusion to steal. Many major cargo security losses would not occur, however, without the common denominator, the dishonest employee. When the security manager is dealing with more than cargo pilferage, monitoring the employee is essential. The manager must be aware of the preceding variables. By eliminating any one variable, opportunities for theft are reduced.

Accountability Procedures

The paramount principle is accountability. Every shipment, whatever its nature, must be identified, accounted for, and accounted to some responsible person at every step in its movement. This is difficult in that the goods are in motion and there are frequent changes in accountability, but this is the essence of the problem. Techniques must be developed to refine the process of accountability for all merchandise in transit. Technology has now advanced to the point where much merchandise can be tracked using geographic positioning systems (GPS).

Invoice System

This accountability must start from the moment an order is received by the shipper. As an example of a typical controlled situation, we might refer to a firm that supplies its salespeople with sales slips or invoices that are numbered in order. In some cases paper invoices are still used, but in today's world of technology, the computer form has replaced much of the old paper world. Whether invoiced using paper or the computer, this step in establishing a record is very important if any control is to be maintained over product order and delivery. Without proper records, whether numbering of hard copies or computer entries, sales slips could be destroyed and the cash, if any, pocketed; or they could be lost so that the customer might never be billed. When these invoice forms are tracked, every invoice can and should be accounted for. Even those forms that have been spoiled by erasures or physical damage should be voided and returned to the billing department.

Merchandise should only be authorized for shipment to a customer on the basis of the regular invoice. This form is filled out by the salesperson receiving the order and sent to the warehouse or shipping department. The shipping clerk signs one copy, signifying that the order has been filled, and sends it to accounts receivable for billing purposes. The customer signs a copy of the invoice indicating receipt of the merchandise, and this copy is returned to accounts receivable. It is further advisable to have the driver sign the shipping clerk's copy as a receipt for the load. In some systems, a copy of the invoice is also sent directly to an inventory file for use in inventory and audit procedures. Returned merchandise is handled in the same way but in reverse.

In this simplified system, there is a continuing accountability for the merchandise. If anything is missing or unaccounted for at any point, the means exist whereby the responsibility can be located. Such a system can only be effective if all numbered invoices are strictly accounted for and merchandise is assembled and shipped only on the basis of such invoicing. The temptation to circumvent the paperwork for rush orders or emergencies frequently arises, but if a company succumbs, losses in embezzlement, theft, and/or lost billing can be substantial.

Similarly, transport companies and freight terminal operators must insist on full and uninterrupted accounting for the goods in their care at every phase of the operation—from shipper to customer.

The introduction of the computer, bar coding, scanners, radio frequency identification (RFID), and electronic signature technology has made the process of accounting for merchandise from the point of purchase by the wholesaler to delivery at the retail outlet a reality. Bar coding provides not only the information discussed in the previous paragraphs but also additional information that allows security to trace losses to the specific persons with responsibility for the merchandise at the time of the loss.

Separation of Functions

Such a system can still be compromised by collusion between or among people who constitute the links in the chain leading from order to delivery unless efforts are made to establish a routine of regular, unscheduled inspection of the operations down the line and, depending on the nature of the operation, regular inventories and audits. It is advisable, for example, for the shipper to separate the functions of selecting the merchandise from stock from those of packing and loading. This will not in itself eliminate the possibility of collusion, but it will provide an extra check on the accuracy of the shipment. And as a general rule, the more people (up to a point) charged with the responsibility and held accountable for merchandise, the more difficult and complex a collusive effort becomes.

In order to fix accountability clearly, it is advisable to require that each person who is at any time responsible for the selecting, handling, loading, or checking of goods sign or initial the shipping ticket that is passed along with the consignment. In this way, errors, which are unfortunately inevitable, can be assigned to the person responsible. Obviously any disproportionate number of errors traced to any one person or to any one aspect of the shipping operation should be investigated and dealt with promptly.

Similarly, at the receiving end, the ticket should be delivered to a receiving clerk. Appropriate personnel who verify the count of merchandise received without having seen the ticket in advance should then unload the truck. In this way, each shipment can be verified without the carelessness that so frequently accompanies a perfunctory count that comes with the expectation of receiving a certain amount as specified by the ticket. It will also tend to eliminate theft in the case of an accidental, or even intentional, overage. Since the checkers do not know what the shipment is supposed to contain, they cannot rig the count.

Driver Loading

Many companies with otherwise adequate accountability procedures permit drivers to load their own trucks when the driver is taking a shipment. This practice can defeat any system of theft prevention because drivers are accountable only to themselves when such a method of loading is in effect. The practice should never be condoned. It is worth the investment of time to have others check the cartons tendered to the driver. The potential losses in theft of merchandise by overloading or future claims of short deliveries would almost certainly exceed the costs involved in the time or personnel involved in instituting sensible supervisory procedures.

Theft and Pilferage

Targets of Theft

An analysis of claims data from the transportation industry shows that a few very specific commodities attract the attention of pilferers and thieves. These items are often referred to as "hot products." These are items that are in demand and easily disposed of such as computers, entertainment equipment, name-brand clothing and footwear, perfume, jewelry, cigarettes, and prescription drugs.[7]

But the kinds of things stolen vary considerably, depending on the location and the nature of the merchandise. Thieves might prefer to steal part of a shipment of clothing, but if none is available, they might just as vigorously pursue a truckload of dog food as long as there is a means of disposing of it. Anything can be stolen if the thief is given an opportunity. Anything that has a market—and nothing would be shipped unless a market existed—can be considered attractive to a thief.

It is important for transportation industry managers to recognize that all merchandise is susceptible to theft but at the same time to know the high-loss items at their locations so that they can exert extra efforts to secure such goods. It is also useful to know that motor carriers are the victims of more than 74 percent of the theft-related losses, rail and maritime carriers account for about 24 percent, and air cargo losses make up the difference.[8]

Pilferage

According to the U.S. Department of Justice, as much as 80 percent of cargo theft losses are the result of a series of minor thefts.[9] The exact amount, however, cannot be established because of extensive nonreporting of these crimes. Thefts of this nature are generally held to be impulsive acts, committed by persons operating alone who pick up an item or two of merchandise that is readily available when there is small risk of detection. Typically the pilferer takes such items for personal use rather than for resale because most of those who are termed pilferers are unsystematic, uncommitted criminals and are unfamiliar with the highly organized fencing operations that could readily dispose of such loot.

Such pilferage is always difficult to detect. Because it is a crime of opportunity, it is rarely committed under controlled circumstances that can either pinpoint the culprit or gather evidence that would later lead to discovery. Generally the items taken are small and readily concealed on the person or easy to transport and conceal in a car.

Pilferage is usually aimed at items in a freight or cargo terminal awaiting transshipment. In such instances, merchandise may be left unprotected on pallets, handcarts, or dollies awaiting the arrival of the next transport. In this mode, it is highly susceptible to pilferage as well as to a more organized plan of theft.

Broken or damaged cases offer an open invitation to pilferage if supervisory or security personnel fail to take immediate action. Accidental dropping of cases to break them open is a common device used to get at the merchandise they contain. If each such case is carefully logged—listing the name of the person responsible for the damage as well as the names of those who instantly gather at the scene—a pattern may emerge that will enable management to take appropriate action.

Whatever form pilferage takes, it can be extremely costly. Although each individual instance of such theft may be relatively unimportant, the cumulative effect can be enormous. Eighty percent of an estimated total direct loss of $30 billion worldwide adds up to a staggering bill for petty theft.

Deterring Pilferage

Here again, accountability controls can provide an important deterrent to pilferage loss. If some one person is responsible for merchandise at every stage of movement or storage, the feasibility of this kind of theft can be substantially reduced.

In cases where it does occur, a rapid and accurate account of the nature and extent of such loss can be an invaluable tool in indicating the corrective action to be taken. Properly supervised accountability controls can locate the point along the handling process where losses occurred and, even in those cases where they will not identify the culprit, these controls will underscore any weaknesses in the system and indicate trouble areas that may need more or different security application.

Movement of personnel in cargo areas must be strictly controlled. All parcels must be subject to inspection at a gate or control point at the entrance to the facility. Private automobiles must be parked outside the area immediately encompassing the facility, and beyond the checkpoint. All automobiles should be subject to inspection on departure if a parking area inside the boundaries of the facility is provided.

Every effort should be made to keep employee morale high in the face of such security efforts. Though some managements have expressed an uneasiness about inspections and strict accountability procedures, fearing that they might damage company morale, it should be pointed out that educational programs aimed at acquainting employees with the problems of theft and stressing everyone's role in successful security have resulted in boosting morale and in enlisting the aid of all employees in the security effort.

Here, as in other areas of security, employees should be encouraged to report losses immediately. They must never be encouraged to act as informers or asked to report on their coworkers. If they simply report the circumstances of loss, it is the job of security to carry forward further investigation or to take such action as may seem indicated.

Large-Quantity Theft

While the point-to-point trafficking of merchandise utilizing storage containers has reduced the amount of petty theft, the number of load thefts has increased. According to a report prepared by Todd Roehrig and Treven Nelson, at least two cargo containers disappear from the Port of Los Angeles every day.[10] The FBI reports that freight trailer losses range between $12,000 and $3 million depending on the type of load.[11] Thefts in such amounts are no longer in the category of pilferage, but rather theft engaged in by one or more persons who are in it for profit—for resale through traffickers in stolen goods.

Thieves may or may not be employees, but because in either case they need information about the nature of the merchandise on hand or expected, they will usually find accomplices inside who are in a position to have that information. They are interested in knowing what kinds of cargo are available in order to make a decision about what merchandise to hit, depending on its value and on the demands of the fencing organization with which they deal.

Dealers in stolen goods are subject to the vagaries of the marketplace in the same way legitimate businesspeople are. Whereas a certain kind of merchandise may find a ready market today, it may move slowly tomorrow. Such dealers are anxious to move their goods rapidly not so much because of fear of detection (even if they are found, mass-produced goods of any nature are difficult if not impossible to identify as stolen once they find their way into other hands) but because they generally want to avoid the overhead and the attention created by a large warehousing operation.

Removal of Goods

Once the thieves have the information, they need to arrange to take over the merchandise and remove it from the premises. To do this, they will usually try to work with some employee of the warehouse or freight terminal.

In cases where accountability procedures are weak or inadequately supervised, thieves have few problems. In tighter operations, they may try to bribe a guard, or they might forge the papers of an employee of a customer firm. They might even create confusion—such as a fire in a waste bin or a broken water pipe—in order to divert attention for those few moments needed to accomplish the actual theft. Generally the stolen goods are taken from the facility in an authorized vehicle driven either by the thieves, who have false identification papers and forged shipping documents, or by an authorized driver who is more often than not working with the thieves.

Disposal of Stolen Goods

Disposal of the goods usually presents no problem because it is customary for thieves to steal from an order placed for any one of certain kinds of merchandise. The loot is normally presold. In its July 2006 press release on cargo theft FBI Unit Chief Eric Ives describes a cargo theft enterprise that is well organized at regional and even national levels. The organization includes the thieves, brokers (fences), drivers (Lumpers) and specialists at foiling anti-theft locks.[12]

It might be noted at this point that it is for this reason principally that cooperation between private security and public law enforcement is so vital in the war against cargo theft. No thief will continue to steal unless there is a ready market for the stolen goods. In the hearings of the Select Committee on Small Business, one of the most inescapable conclusions was that the fences were the kingpins behind the majority of the thefts taking place. They dictated the nature of the merchandise to be taken, the price to be paid, and the amount wanted. They directed the thefts without ever becoming involved—in many cases never even seeing the merchandise they bought, warehoused, and sold. Without the fences—many of whom are otherwise legitimate businesspeople—the markets would shrink, the distribution networks would disappear, and losses would be dramatically reduced. Another possibility discovered by a major retailer has been the theft of merchandise by employees who either sell merchandise at local flea markets or advertise in local papers. One individual reportedly stole electronic items and developed a business through advertising in local papers for some time before getting caught.

Unfortunately, only sporadic efforts have been made to break up the big fencing operations, and the business of thievery thrives. Any assistance that private security can give to public law enforcement by way of instant and full reports on thefts can help combat these shady operations, and all private industry will benefit immeasurably in the long run.

Terminal Operations

Terminal operations are probably more vulnerable to theft than other elements of the shipping system. Truck drivers mingle freely with personnel of the facility, and associations can readily develop that lead to collusion. Receiving clerks can receipt goods that never arrive; shipping clerks can falsify invoices; and checkers can overload trucks, leaving a substantial percentage of the load unaccounted for and therefore disposable at the driver's discretion.

Here again, these thefts can be controlled with a tight accountability system, but too often such facilities fail to install such a procedure or to follow up on it after it is in effect. This is poor economy, even under the most difficult situations when seasonal pressures are at their highest.

Railway employees on switching duty at a freight terminal can also divert huge amounts of goods. They can easily divert a car to a siding accessible to thieves who will unload it at an opportune time. This same device can be employed to loot trucks that have been loaded for departure the next morning. Unless these trucks are securely locked and parked where they are under surveillance by security personnel, they can be looted with ease. Drivers can park their trucks unlocked near a perimeter fence for later unloading unless the positioning and securing of the vehicle is properly supervised.

In all cases, a professional thief is someone with a mission. Such people cannot be deterred by the threat of possible detection. They recognize the possibility, accept the risk, and make it their business to circumvent detection. Only alert and active countermeasures will serve to reduce losses from their efforts.

Surveillance

There must be strict surveillance (human and/or CCTV) at entrances and exits, and there must be patrol activity at perimeters and through yards, docks, and buildings. Key control must be tight and painstakingly supervised. Cargo should be stored in controlled security areas that are enclosed, alarmed, and burglar-resistant. High-value cargo should be stored in high-security areas within the cargo area. Special locks, alarms, and procedures governing access should be employed to provide the highest possible security for these sensitive goods. The use of CCTV (closed-circuit TV) and other surveillance technology is covered in Chapter 8.

Shipments of unusual value should be confidential, and only those employees who are directly concerned with loading or unloading or transporting such shipments should be aware of their schedules. Email and fax information about such movements should be restricted, and trailer numbers should be covered while the vehicle is in the terminal. Employees involved in any way with such shipments should be specially selected and further indoctrinated in the need for discretion and confidentiality.

A Total Program

Adequate physical security installations supported by guard and alarm surveillance will go a long way toward protecting the facility from the thief or terrorist, but these measures must be backed by proper personnel and cargo movement systems, strict accountability procedures, and continuing management supervision and presence to ensure that all systems are carried out to the letter. Regular inspections of all facets of the operation, followed by prompt remedial action if necessary, are essential to the success of the security effort.

Planning for Security

It is important to the security of any transportation company, shipper, or freight terminal operation to draw up an effective plan of action to provide for overall protection of assets, whether general cargo or people. The plan must be an integrated whole wherein all the various aspects are mutually supportive.

In the same sense, in large terminal facilities occupied by a number of different companies, all individual security plans must be integrated to provide for overall security as well as for the protection of individual enterprises. Without full cooperation and coordination among participating companies, much effort will be expended uselessly and the security of the entire operation could be threatened.

Such a plan should establish area security classifications. Designated parts of the building or the total yard area of the facility should be broken down into controlled, limited, and exclusion areas. These designations are useful in defining the use of specific areas and the mounting security classifications of each.

Controlled, Limited, and Exclusion Areas

Controlled areas are those restricted as to entrance or movement by all but authorized personnel and vehicles. Only part of a facility will be designated a controlled area because general offices, freight receiving, personnel, restrooms, cafeterias, and locker rooms may be used by all personnel, some of whom would be excluded if these facilities were located within an area where traffic was limited. Within this area itself, all movement should be controlled and under surveillance at all times. A fence or other barrier should additionally mark it, and access to it should be limited through as few gates as possible.

Limited areas are those within the controlled area where an even greater degree of security is required. Sorting, handling of broken lots, storage, and reconstituting of cases might be vulnerable functions handled in these areas.

Exclusion areas are used only for the handling and storage of high-value cargo. They normally consist of a crib, vault, cage, or room within the limited area. The number of people authorized to enter this area should be strictly limited, and the area should be under surveillance at all times. Since such areas should be locked whenever they are not actually in use, careful key control is of extreme importance.

Pass Systems

All employees entering or leaving the controlled area should be identified, and their authorization to be there should be checked. Each employee should be identified by a badge or pass, using one of the several systems discussed in Chapter 10.

Vehicle Control

The control of the movement of all vehicles entering or leaving a controlled area is essential to the security plan, as is checking the contents. All facility vehicles should be logged in and out on those relatively rare occasions when it is necessary for them to leave the controlled area. They should be inspected for load and authorization.

All vehicles entering the controlled area should be logged and checked for proper documents. The fastest and most efficient means of recording necessary data is by use of optical readers that record information digitally in the company's computer system or the camera system, such as a Regiscope, which will record all of the required information digitally on a single photograph/record. This should include the driver's license, truck registration, trailer or container number, company name, waybill number, delivery notice, a document used to authorize pickup or delivery, time of check, and the driver's picture if it is not on the driver's license.

The seal on inbound loaded trailers should be checked, and the driver should be issued a pass that is time-stamped on entering and leaving the area and that designates the place for pickup or delivery.

All vehicles leaving the area should surrender their passes at the gate where their seals will be checked against the shipping documents. Unsealed vehicles will be inspected as will the cabs of all carriers leaving the facility. Partial load vehicles should be returned to the dock from time to time on a random basis for unloading and checking cargo under security supervision. They should also be sealed while they are in the staging area.

Loading and unloading must be carefully and constantly supervised since it is generally agreed that the greater part of cargo loss occurs during this operation and during the daylight hours.

Other Security Planning

The security plan must also specify those persons having access to security areas, and it must specify the various components necessary for physical security such as barriers, lighting, alarm systems, fire protection systems, locks, and communications. It must detail full instructions for the guard force. These instructions must contain both general orders applicable to all guards and special orders pertaining to specific posts, patrols, and areas.

There must be provision for emergency situations. Specific plans for fire, flood, storm, or power failure should be part of the overall plan of action. Who to call in an emergency should also be specified.

After the security plan has been formulated and implemented, it must be reexamined periodically for flaws and for ways to improve it and keep it current with existing needs. Circulation of the plan should be limited and controlled. It must be remembered that such a plan, however well conceived, is doomed at the outset unless it is constantly and carefully supervised.

Security Surveys

Security managers of freight terminals or companies engaged in shipping will be continually occupied with surveys of the facility under their security supervision. Security surveys were discussed in detail in Chapter 7.

An initial survey must be made to formulate the security plan governing the premises. It should be thorough enough to detect the smallest weaknesses in the operation and to provide the information needed to prepare adequate defenses. Further surveys will be necessary to evaluate the effectiveness of the program established, and follow-up surveys should determine whether all regulations and procedures are being followed. Additional surveys may be necessary to reevaluate the security picture following changes in operational procedures in the facility or to make special studies of particular features of the security plan.

Inspections

In addition to these surveys, essentially designed to evaluate the security operation as a whole or to reevaluate it in the light of changing conditions, the security manager should make regular inspections of the facility to check on the performance of security personnel and to check the operating condition of the facility (see Figure 13-1). Such inspections should include potential trouble areas and should not overlook a check of fire equipment and alarm systems.

Education

If the security plan is to succeed, it must have the full cooperation and support of all employees in the facility. A continuing program of education in the meaning and importance of effective security can only achieve this in every phase of the business.

All personnel should be indoctrinated at the time of their employment, and a continuing program should be instituted to update the staff on current and anticipated problems. More advanced courses on procedures might be instituted for management personnel. Part of this

FIGURE 13-1 A truck cargo inspection. *(Photo courtesy of Zistos Corporation, Hauppauge, NY, www.zistos.com.)*

program should be devoted to educating employees about the importance of security to each individual and job.

Security reminders are also important to keep the subject of security constantly alive in the minds of everyone. Posters, placards, and notices prominently posted are all effective devices for getting the message across. Leaflets or pamphlets covering more details can be distributed to employees in their pay envelopes.

Cargo in Transit

The Threat of Hijacking/Sabotage

Although the crime itself is dramatic and receives much publicity, armed hijacking of an entire tractor-trailer with a full load of merchandise represents only 1 percent of the losses suffered by the shipping industry as a whole. This is clearly not to say that it is a minor matter. On the contrary, such a crime is of extreme importance to the carrier taking the loss because the enormity of the theft represents a huge financial blow in one stroke, whatever its significance in the overall percentages.

There is, however, little that can be done by a driver who is forced over by a car carrying armed and threatening hijackers. When it comes to that point, the driver has little choice but to comply. The load is lost, but hopefully the driver is unhurt. There is little that private security can do in such cases, and it is indeed fortunate that the incidence of such crimes is as low as it is. The matter is in the hands of public law enforcement agencies and must be handled by them.

If there is cause to believe that hijacking of a load is an imminent possibility, trucks should be scheduled for nonstop hauls and rerouted around high-risk areas. Schedules should be adjusted so that carriers do not pass through high-risk areas at night. In extreme cases, trucks might be assigned to travel in pairs or in larger convoys. Company cars can follow very high-value loads that are deemed especially vulnerable. Two look-alike trucks might make it more difficult to pinpoint the specific desired shipment. Such procedures constitute selective protection at best since they can be used only infrequently and are impractical for general application. The second driver, however, would be assigned the task of identifying the persons, the vehicle used, and so on and of reporting the hijacking if it occurs.

Other aspects of theft or sabotage on a line haul can be dealt with, however, and it is important that procedures be established that will serve to protect the cargo.

Global Positioning System (GPS)

With the development of accurate, commercially available tracking systems, it is now possible for trucking firms and individuals to track vehicles in a cost effective, real-time, mapping and reporting environment. Such systems make it possible to know when a truck is not following planned routes or spending too much time stopped.

Personnel Qualifications

The cardinal rule in the management of a transportation concern is that those assigned to line hauling duties must be of the highest integrity. Drivers and helpers must be carefully screened for job suitability and trustworthiness before they are hired, and personnel and security managers must carefully evaluate them before they are given this critical responsibility. An irresponsible driver can cost the company all or part of a load, and whether the loss is unintentional or the result of the driver's carelessness, the cost to the company will be the same.

Many cases have been reported where drivers set themselves up as victims of a hijack. This is difficult if not impossible to prevent unless the honesty of the driver is unquestionable—an attitude that can only be determined by a careful screening process and regular analysis of the person's behavior and performance. Any changes in a driver's demeanor or lifestyle should be noted, as deterioration of morale or a basic change in attitude toward the job might lead to future problems.

Procedures on the Road

All employees should receive specific instructions about procedures to be followed in every predictable situation on the road. The vehicle should be parked in well-lighted areas where it can be observed. It should be locked at all times, even if the driver sleeps in the cab. Trailers should be padlocked as well as sealed. A high-security seal, requiring a sizable tool to remove it, should always be used on valuable shipments that are parked overnight.

Drivers must be instructed never to discuss the nature of their cargos with anyone. Thieves frequently hang out at truck stops hoping to pick up information about the nature of loads passing through. All too often, the most innocent conversations among truckers can lead to the identification of a trailer containing high-value merchandise that, when it is spotted, becomes a target.

The driver should never deviate from the preplanned route. In the case of a forced detour or a rig breakdown or if in any way the schedule cannot be met, the driver must notify the nearest terminal immediately.

Trucks should be painted on the top as well as on the sides to facilitate identification by helicopter in the event of theft.

Seals

Among the many seals available today, the one in most common use is the metal railroad boxcar type. This is a thin band of metal that is placed and secured on the trailer in such a manner that the door or doors cannot be opened without breaking it, thus revealing that the doors have been opened and a theft has, or may have, taken place. They are easy to break and must be broken when the destination has been reached and the merchandise is unloaded. They in no way secure the doors but are placed there simply as a device to indicate whether the doors have been opened at any point between terminal stops. Each seal is numbered and should identify the organization that placed the seal (see Figure 13-2).

All doors on a trailer must be sealed. Those trucks or trailers with multiple doors may habitually load by only one, in which case those doors not in regular use may carry the same seal for months at a time with only the rear being regularly sealed and unsealed.

Seal numbers must be recorded in a permanent log as well as in the shipping papers. The dock superintendent or security personnel should be responsible for recording seal numbers and affixing seals on all trucks. The seals should be positioned in such a way that locking handles securing the door cannot be operated without actually breaking the seal. Some truck or railway car locking devices are so large that several seals may have to be used in a chain to properly seal the carrier. In this case all seal numbers must be recorded.

Resealing

Trailers loaded to make several deliveries along the route must be resealed after each stop. To accomplish this, enough seals must be issued to be placed on the truck after each delivery is made. In this case, the truck is sealed at the point of origination and the seal number logged

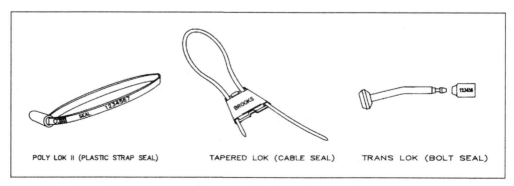

POLY LOK II (PLASTIC STRAP SEAL)   TAPERED LOK (CABLE SEAL)   TRANS LOK (BOLT SEAL)

FIGURE 13-2 Security seals in common use. *(Courtesy of E.J. Brooks Company.)*

and entered in shipping documents. The additional, as yet unsealed, seals are also logged, and the numbers to be used at designated points of delivery are entered in the shipping documents.

When the first checkpoint is reached, the receiving clerk verifies the seal number of the truck. After the merchandise is unloaded, the consignee, not the driver, as directed by the shipping documents, affixes the next seal. This procedure is followed at all stops, including the last one. There, a seal is affixed to the now-empty vehicle, which may return to the point of origin, where the seal will again be checked against the shipping documents.

Empty trailers should also be sealed immediately after being unloaded. This practice will discourage the use of empties to remove unauthorized material from dock areas, and it does not preclude the necessity of physical inspection of all vehicles leaving the controlled areas.

Seal Security

All seals must be held under the tightest possible security at all times. Previously unissued seals should be logged and secured on receipt. They should then be issued in numerical order, and the assignment of each should be duly noted in the log. The seal supply should be audited daily; a careful check to account for each seal, issued or not, should be made at the same time. Without such regular inventories, the entire system can be seriously compromised. As a further part of such audit, all seals that are damaged and cannot be used, as well as seals taken from incoming vehicles, should be logged and secured until they can be destroyed.

Modern Technology and Seals

As with most areas of security, technology has entered the picture. Whereas metal and plastic seals can be tampered with, the new electronic seals require advanced technological information before criminals can compromise their integrity. In addition, the new system avoids the problems associated with the distribution of stocks of seals to depots and ensuring their safe storage. The electronic seals work much like electronic locks and can be monitored by the push of a button.[13]

Special Issues in Airline Security

In the aftermath of September 11, 2001, the airline industry fell under close scrutiny by the U.S. government. What had gone wrong with passenger screening? How did hijackers board the plane? These questions prompted other questions relating to the security provided for airlines in general (see Figure 13-3).

The problems associated with security for airlines and overlapping responsibilities among airlines, customs, police, and airport security have confounded airports. Still, most airlines have a well-developed security program to combat cargo thefts.

Increased Security Measures

Even before September 11, 2001, airport security had been increased throughout the world. The United States, while increasing security, often lagged behind security operations in other countries. However, the U.S. Federal Aviation Administration has for many years provided training for airport police officers. In recent years the Airport Law Enforcement Agency

FIGURE 13-3 Airport security. *(Courtesy of BEI Security, Inc.)*

Network (ALEAN) has taken a lead in organizing programs and communication among airport police departments. The FBI has also increased its interest in aviation security issues. However, even with the increased emphasis on security issues, the perceived quality of security screening, of passengers in particular, has been found inadequate. As a result the U.S. government passed the Airport Security Federalization Act of 2001.

Specific Solutions for a Specific Problem

On November 6, 2001, the U.S. House of Representatives passed Senate Bill 1447, "An Act to Improve Aviation Security." The act, known as the Airport Security Federalization Act of 2001, soon set into motion the federalization of airport security. The act placed the responsibility for civil aviation security functions under the Under Secretary of Transportation. The Secretary's specific functions include the following:

- Receive, assess, and distribute intelligence information related to transportation security
- Assess threats to transportation
- Develop policies, strategies, and plans for dealing with threats to transportation security
- Make other plans related to transportation security, including coordination of countermeasures with appropriate departments, agencies, and instrumentalities of the U.S. government
- Serve as the primary liaison for transportation security to the intelligence and law enforcement communities
- Supervise all airport security and screening services using Federal uniformed personnel[14]

There are other functions listed in the act, but it is clear that airport security is now firmly under the control of the U.S. government. This approach follows that developed in other countries, which have also been combating air crimes for several decades.

In March 2003 the FBI announced that selected airports would receive antimissile technology to protect against shoulder-launched missiles fired at planes taking off or landing. The program was prompted by the attack on an Israeli plane over Kenya in November 2002.[15]

In November/December 2006, the TSA began a program of additional screening of employees at Chicago's O'Hare International and Midway airports. Workers, including delivery truck drivers, city workers, and food vendor employees, had been required to show identification or swipe security badges through electronic readers to gain access to security areas; they are now required to face extra measures to include inspection of vehicles and pat-downs. The TSA is using this increased security at more than 100 airports.

The TSA is also developing additional personnel to serve as "bomb-appraisal officers." These individuals are part of the TSA "bomb-appraisal officer program, which is based on a quick psychological screening of passengers through the use of observation techniques.[16]

In 2007 the U.S. government began asking foreign counties to allow pilots to carry guns in the cockpit when they fly overseas. While the program, now over 4 years old, allows thousands of U.S. pilots to carry guns on domestic flights, some countries are wary of allowing armed pilots on international flights. Congress cut $11.5 million from the last two budget requests for armed pilots because the program was not spending all of its allocation.[17] While thousands of pilots have opted for the training, many are not interested in being armed.

Even with the emphasis on airline security measures, problems continue to be found. While passenger suitcases loaded onto flights are regularly screened for explosives, other cargo that fill the bays generally are not screened. The disparity has been noted in Congress, where some representatives are calling for all air cargo to be inspected at the same level as baggage. The estimated cost of the program would be $6 million. The TSA argues that current standards are sufficient to prevent a bomb from getting on board. Security officials argue that a security zone surrounding the cargo packages during their journey from source to destination is sufficient.[18]

In November, 2001, the U.S. government launched an effort to secure the global supply chain. Customs-Trade Partnership Against Terrorism (C-TPAT) is a voluntary government–business initiative intended to strengthen and improve overall international supply chain and U.S. border security. This initiative is intended to bring together, in a cooperative relationship, the owners of the international supply chain, including importers, carriers, consolidators, brokers and manufacturers.[19] Essentially, businesses are asked to ensure the integrity of their security program and practices in coordination with their supply chain partners. Through participation in this program trade-related businesses receive priority processing for inspections, reducing border delay times. Businesses interested in participating can contact the U.S. Department of Homeland Security.

Other Transportation Industry Responses to Terrorism

Bus Transportation

The American Bus Association (ABA) established its Anti-Terrorism Action Plan in 2002. The association sought to increase awareness of potential hijackers and improve security plans for transit companies, including protection of the bus industry infrastructure. As noted in Chapter 1, cooperative ventures among the federal government, local law enforcement, and local bus companies have placed cameras throughout the bus system, including on-board buses and in terminals. At the 2005 ABA Board meeting, Board members were informed that the DHS had provided $50 million in security grants to the bus industry.[20] In 2007 the TSA provided support for intercity bus operations through the Intercity Bus Security Grant Program as part of the DHS Infrastructure Protection Program. The Intercity Bus Security Grant Program is designed to assist fixed route intercity and charter bus services.[21]

Maritime Operations

The U.S. Coast Guard has provided major leadership in this arena. It has called for the inspection of high-risk cargos at the point of shipment. Agreement among more than 100 governments has lead to the International Ship and Port Facility Security Code.

The Coast Guard has also stepped up its inspections efforts. Sea marshals are assigned to "high interest" vessels arriving and departing from U.S. ports. Additional Department of Homeland Security measures are discussed in Chapter 1.

Amtrak/Local Commuter Trains

The public rail system is struggling with controls. Terminals have been secured to some extent by limiting access to underground areas. Increased presence of security personnel is apparent. Tickets are now checked before passengers are allowed into final waiting areas and, in some operations, positive identification is being requested.

The Department of Homeland Security has on occasion requested local, state, or National Guard units to assist in monitoring key bridges and terminal facilities.

In response to questions regarding the application of airline security measures to rail passengers, the Department of Homeland Security has spent $7 million to test screening equipment that would reliably stop terrorist attacks similar to those on the Madrid and London rail systems. The 2006 tests showed that the technologies tested each had significant problems. Many devices triggered excessive false alarms, and some took too long to screen passengers. Robert Jamison, Deputy Director of the Transportation Security Administration, said there are no plans to put the tested technology to work. However, the assessment of the testing program concluded that the effort to screen rail passengers needed significant investments. With over 12 million passengers traveling on the subways and rail lines each day, the need is great.[22]

In Los Angeles the Metropolitan Transportation Authority, which manages the city's subway system, is considering adding security. To date the system, which has been known for a lack of traditional subway components such as turnstiles, gates, attendants and police officers, has relied on surveillance cameras. However, a late 2006 incident where mercury was spilled on a platform and remained undetected for eight hours has convinced MTA officials that additional security may be needed.[23]

Oil and Gas

Transportation of oil and gas, as well as other liquids and gases, is completed through millions of miles of pipelines. The Department of Transportation's Office of Pipeline Safety (OPS) began inspecting the system in 2002. The OPS is also requiring pipeline operators to develop plans to reduce risks and respond to disasters in areas where pipeline failures would have the greatest impact on populations or ecological systems.

In 2006 a liquid natural gas facility in Massachusetts had its security breached by intruders who slipped through the fence and took pictures of themselves standing on the storage tank stairwell. While the incident was recorded by CCTV the company waited for 5 days to file a report with Massachusetts authorities. Massachusetts Representative Edward Markey noted, "This incident raises serious questions about the adequacy of the perimeter security and surveillance monitoring in place at this facility." Markey is a senior member of the House Homeland Security Committee and the House Energy and Commerce Committee.[24]

Summary

One of the fastest-growing crimes appears to be theft of merchandise in transit. Even with gains in reducing petty pilferage, new containerized transportation has brought on new problems, including the theft of entire containers of merchandise. Control of merchandise during transit is often difficult but, through proper employee screening and security applications, the problem may be controlled. However, it is unlikely, given the financial incentives for thieves, that the problem will ever be eliminated.

While airport security has undergone federalization in the United States due primarily to the increased threat of terrorist attacks, similar to operations in other countries, it would be unwise to consider applying all of these security concepts to all transportation/cargo security problems. Still, given the events of the past few years where passenger trains and buses have been attacked in European countries, additional scrutiny is certainly called for.

■ ■ CRITICAL THINKING ■

What are the real issues that need to be considered in implementing additional security measures on public transportation in the United States?

Review Questions

1. Why is it said that the greater burden in preventing cargo thefts falls on private security rather than on public law enforcement?
2. Describe the operation of an invoice system that would establish accountability for all merchandise in shipments.
3. What are some policies and procedures that would be effective in deterring pilferage?
4. Define controlled areas, limited areas, and exclusion areas.
5. Offer procedures for the control of the movement of and contents in all vehicles entering or leaving a controlled area.
6. Describe the measures taken by governments and companies in response to terrorism.

References

[1] Crime against small business: A report of the small business administration transmitted to the select committee on small business, United States Senate. Washington, DC: Government Printing Office; 1969.

[2] Salkin S. Safe and Secure? Warehousing Manage, December 1999.

[3] Cargo Theft's High Cost. FBI press release 07/21/2006, downloaded 7/11/2007.

[4] U.S. General Accounting Office. In: Report by the comptroller general of the United States: promotion of cargo security receives limited support. Washington, DC: U.S. General Accounting Office; 1980.

[5] It's a Crime. *Viewpoint* 1999;**24**(2), downloaded 12/12/2002, <www.aais.org/communications/viewpoint/vp24_2.htm>.

[6] Tyska L, Fennelly L. In: Controlling cargo theft. Boston: Butterworth-Heinemann; 1983. pp. xxvii–xxix.

[7] Atkinson W. How to protect your goods from theft. Logistics Manage Distrib Rep, March 2001.

[8] It's a Crime.

[9] United States Department of Justice. 1332 Charging Theft from Interstate Shipment—Dollar Thresholds, Local Efforts. *Criminal Resource Manual* 1332 (October 1997), downloaded 12/12/2002, <www.usdoj.gov/usaoo/eousa/foia_reading-room/usam/title9/crim01332.htm>.

[10] Roehrig T, Nelson T. Cargo Theft, downloaded 12/12/2002, <www.ucalgary.ca/MG/inrm/industry/theft.htm>.

[11] Cargo Theft's High Cost, FBI.

[12] Ibid.

[13] Smart tags, seals computerized advantages. *Security* March 1997:65.

[14] Mayhew C. The detection and prevention of cargo theft. *Trends and issues in crime and criminal justice.* Australian Institute of Criminology; September 2001.

[15] FBI: Airports to Take Anti-missile Measures, CNN.com March 31, 2003, downloaded 4/4/2003, <http://cnn.travel.printthis.clickability.com/pt/cpt?actions=cpt&expire=-1&urlIC=5857474&f>.

[16] Security Management Daily, February 28, 2007 excerpted from *Chicago Tribune*, (02/27/2007).

[17] Frank T. U.S. Asks to Arm Pilots Abroad. *USA Today*, <www.usatoday.com/travel/news/2007-02-07-us-pilots-guns_x.htm>.

[18] Lipton E. Security debate centers on tougher standards for inspections of air cargo. The New York Times February 8, 2007.

[19] U.S. Department of Homeland Security, CBP.gov, C-TPAT Overview, dtd. 12/13/2007. <http://www.cbp.gov/xp/cgov/trade/cargo_security/ctpat/what_ctpat/ctpat_overview.xml>.

[20] Chairman Young, Pundit Tony Blankley, DHS Officical & Fuel Expert Address ABA Board Members at Fall 2005 meeting, <www.buses.org/Press_Room_the_report_newsletter/2126.cfm>.

[21] Intercity Bus Security Grant Program, <www.tsa.gov/join/grants/ibsgp.shtm>.

[22] Frank T. USA Today, Security devices falter in rail tests, usatoady.com/news/nation/2006-02-13-rail-security_x.htm.

[23] Guccione J, Blankstein A. Call for more security inside Los Angeles's subway system. Christian Sci Monit, 02/12/2007.

[24] Crocker M. Platforms, pipelines, and pirates. Secur Manage June 2007:77–86.

14

Violence and Drug Use in the Workplace

OBJECTIVES

The study of this chapter will enable you to:

1. Understand the complexity of violent and/or aberrant behavior in the work environment.
2. Gain a working knowledge of how to identify potential workplace violence problems.
3. Discuss the prevention of workplace violence.
4. Know the role of violence intervention education and incident management teams.
5. Identify the problems associated with drug use in the workplace.
6. Know how to spot potential drug users.
7. Discuss the components of a comprehensive substance abuse program.

Introduction

Workplace violence has become a growing concern for human resources and security managers over the past decades. There have always been some problems associated with violence in the workplace, but recent decades have focused additional attention on this serious problem. Whether the violence is the result of a personnel problem such as disciplinary action, salary dispute, workforce reduction or termination or if it is of a domestic nature, such as a "love triangle," employees do bring their problems to work. Far too often the pressures result in some form of violence or aberrant behavior against fellow employees, employers, or the work facility. To a degree, any work environment can be like a small city and in that city you will have aberrant behavior issues arise that require a response from a security team, not unlike a city's police force.

In addition to traditional concepts of workplace violence, the late 1990s saw serious problems with violence in our schools. Columbine High School and Little Rock will be remembered for many years, as will recent shootings at an Amish school in 2006 and Virginia Tech University in 2007—the worst mass shooting in school history. No one had thought about the possibility of violence becoming murderous on our school properties. The murders at these and other schools brought the public's focus onto schools. Schools and government bodies responded with safe school studies and grants. Many schools have dramatically improved their access systems, locking doors, assigning identification systems, and in some cases installing metal detectors.

Though the response to school violence was swift, the reality is that schools remain the safest environment for children. According to the findings of *The Final Report and Findings of the Safe School Initiative* (May 2002), the Department of Education (DoE) reported that there are almost 60 million children in American schools. The Safe School Initiative study was able to identify only 37 incidents of targeted school-based attacks committed by 41 individuals over a 25-year period. The increased security can do nothing but improve safety, but caution must be taken not to create an environment that children and parents view as draconian.

This chapter also discusses drug use in the workplace. Unfortunately, drug abuse and violence in the workplace are often found together, which can produce a volatile mixture of emotions. For smaller companies, workplace substance abuse costs Americans over $100 billion annually and causes companies to incur a 300 percent increase in medical benefits.[1] Substance abuse is also an ongoing problem in the school setting. In 2004 the U.S. public school systems reported 32,641 incidents of distribution of illegal drugs and 131,267 incidents of possession or use of alcohol or illegal drugs. The forces that cause frustration that leads to violence also contribute to substance abuse. In turn, substance abuse often leads to a lack of control and violent behavior.

Webster defines *violence* as an "[unwanted] exertion of physical force so as to injure or abuse" or a "vehement feeling or expression." Encompassed within this broad definition are subtle forms of harassment, threats, intimidation, and sabotage, as well as overt acts of violence and temper tantrums. According to Joseph Kinney, workplace violence includes four broad categories:

- Threat
- Harassment
- Attack
- Sabotage

Before defining each category of violence, it is very important to remember from the onset of this chapter that an over-response can unduly escalate a situation. It is equally critical to understand that a range of emotions can be expected when individuals are affected by a workforce reduction, for example. Venting, crying and expressions of frustration, surprise, anger, and sadness, among other human characteristics, are well within the boundaries of normal responses and can be expected by people who are truly nonviolent. In short, an emotional response doesn't lead to an extreme. Finally, organizations are fast becoming a global ecosystem of different nationalities, religions and cultural backgrounds, all of which should be taken into consideration when attempting to diagnosis a workplace civility issue.

Threat

Threat involves an expression of an intention to inflict injury. A threat can be an intimidating stare, posture, or verbal exchange. An intimidating stare or posture is less obvious and, therefore, can be more subtle; a verbal exchange is more direct and obvious. The key is to determine whether the threat was made in jest or with malice aforethought. In all cases, threats should not be tolerated in the workplace.

Harassment

Harassment in general involves a behavior designed to trouble or worry someone. For example, sexual harassment often causes people to fear the loss of their jobs if they resist or report it. Harassing behavior can be something like putting grease on a coworker's chair or phone, feces in or on their desk, or graffiti on bathroom walls or making phone calls with immediate hang-ups.

Attack

Attacks involve the use of unwanted force against someone in order to cause harm. To attack is to make contact in an unwanted manner such as spitting, choking, punching, slapping, and grabbing. The key word is *unwanted*. Like threats, attacks, even in harmless fun, should not be tolerated in the workplace.

Sabotage

Sabotage involves the destruction of an employer's property, tools, equipment, and products to hinder the manufacturing process, which can ultimately affect a company's profits. For example, take the case of a factory worker who attacked a conveyor and shut down production for a half day. Although this action occurred due to drug ingestion, the initial factor leading to this incident was employee game playing. Another example is the General Motors employee nicknamed "Edward Scissorhands" by other factory workers, who would often cut power to the plant to halt production. The worker was motivated by frustration and anger over the GM workforce reductions that caused this employee and others to work longer hours and weekends to meet production schedules.

Employees do bring their problems to work!

Violence and the Workplace

The Phenomenon of Workplace Violence

Workplace violence is not new, and in fact reached a high point during the late 1890s and early 1900s with the growing union movements; however, the focus of the violence has changed greatly over the years. Today's workplace is too often the focus of random acts of violence that on the surface appear to have no logical cause. A study of the phenomenon and underlying factors makes it possible for security managers to prevent violence or proactively intervene in a potential problem area before violence occurs.

Each day a newspaper, magazine, radio, or television reports another act or occurrence of workplace violence in America. According to the March 2011 National Crime Victimization Survey which comes from the US Department of Justice, the rate of violent crime against employed persons has declined since 1993. In fact, according to the same survey, from 2002 to 2009 the rate of nonfatal workplace violence has declined by 35%, following a 52% decline in the rate from 1993 to 2002.

It is perhaps coincidental that the decline parallels major companies' efforts to combat the increase in workplace violence that occurred in the 1980s and early 1990s.[2] In any case, the victims are new and the lives shattered are real. Who are the victims? Victims range from those directly involved to those indirectly involved, such as first responders, family, friends, and colleagues. Between 2005 and 2009, law enforcement officers, security guards and bartenders had the highest rates of nonfatal workplace violence. According to the U.S. Department of Justice, more than 572,000 nonfatal violent crimes occurred at work or on duty in 2009. This level of nonfatal crime is about a quarter of the 2.1 million nonfatal violent crimes that occurred at the workplace in 1993. Along with the decline in nonfatal workplace violence, the number of homicides decreased by 51% from a high of 1,068 homicides in 1993 to 521 in 2009. A majority of workplace violence (both fatal and nonfatal) occurs in private companies. Of the victims, middle aged men are more likely than women to experience violence. The overall cost of workplace violence is hard to determine. However, the Department of Justice estimates that it can cost employers well over 6 billion in lost wages, medical and support costs.[3] Naturally, the loss of life and overall suffering cannot be measured in dollars but make no mistake, there is a cost to the health and well-being of an organization.

As it relates to the most serious violent incidents in the workplace, specifically homicides, Tables 14-1, 14-2 and 14-3 below show incidents by occupation of the victims, offender and incident type between 2005 to 2009.

Regardless of the decline in workplace violence since 1993, there are three primary reasons it exists in the first place. First, there is society's general acceptance of using violent means to deal with emotions and negative feelings. In other words, those who use violence as a form of personal communication believe their behavior is an acceptable way to deal with conflict, emotions and problems. Today's children, who are now becoming the employees of America, are generally viewed as an aggressive and potentially violent bunch. A study of teachers in 1949 revealed their primary concerns to be student tardiness, smoking, and ditching class on occasion. The same study conducted in 1995 reveals a much different and more alarming picture. The primary concerns of today's teachers include the availability of weapons and their use on campus, violence in general, drugs and alcohol use and abuse among students, and finally, the breakdown of the family structure. This still holds true today.

A second factor is the general availability of guns and the mass media's glamorous and accepted portrayal of their use to remedy a wrong done or to seek revenge. According to Joseph Kinney, "The availability of guns, the experience that people have in using such weapons, and the perception that such use is legitimate have created circumstances encouraging weapons use."[4] Eighty percent of all workplace homicides were committed with firearms.[5]

Between 1993 to 2009, workplace homicides decreased by 51% from 1,068 in 1993 to 526 in 2009. The three most risky environments to work based on chance of death through homicide are retail operations, service industries, government and transportation operations. The most dangerous occupation was that of a taxicab driver. In 1998 a taxicab driver risked dying on the job at a rate of 36 times that of the national average.[6] This still holds true today.

Finally, economic factors also contribute to workplace violence. According to Michael Mantell, "the rising tide of workplace violence incidents points to two carefully linked factors:

Table 14-1 Workplace Homicides, by Occupation of Victim, 2005–2009

Occupation	Percent of Workplace Homicide Victims Age 16 or Older
Total	100.0%
Management, business and financial	9.6%
Management	9.2
Business and financial operations	0.4
Professional and related	5.1%
Computer and mathematical	--*
Architecture and engineering	0.3*
Life, physical, and social science	0.1*
Community and social services	1.2
Legal	0.6
Education, training, and library	0.6
Arts, design, entertainment, sports, and media	1.1
Healthcare practitioners and technical	1.2
Service	30.4%
Healthcare support	0.7
Protective service	17.2
Food preparation and service-related	7.2
Building and grounds cleaning and maintenance	2.4
Personal care and service	2.9
Sales and office	32.8%
Sales and related	27.9
Office and administrative support	4.9
Natural resources, construction, and maintenance	6.5%
Farming, fishing, and forestry	0.8
Construction and extraction	3.0
Installation, maintenance and repair	2.7
Production, transportation, and material moving	15.7%
Production	2.4
Transportation and material moving	13.2

Note: The National Crime Victimization Survey and Census of Fatal Occupational Injuries use different categories of occupations. See *Methodology*. Includes 2009 data which are preliminary. Excludes homicides where the victim occupation was unknown. Details do not sum to total due to rounding.
*Based on 10 or fewer cases.
-- Rounds to less than 0.05%.
Source: Census of Fatal Occupational Injuries, U.S. Department of Labor, Bureau of Labor Statistics.

people and money."[7] Today, perhaps more than ever, workers feel vulnerable, especially in corporate America. Even a "secure" government job is becoming a less secure place to work as pension funding becomes more problematic and the privatization and reengineering movements take hold. Teams of employees often find themselves processing each other out of a job as unnecessary steps are eliminated. Workers are more apt to turn to violence to deal with emotions and negative feelings toward co-workers and the employing organization.

Table 14-2   Workplace Homicides, by Offender Type, 2005–2009

Offender Type	Percent of Workplace Homicide Victims Age 16 or Older
Total	100.0%
Robbers and other assailants	70.3%
Robbers	38.3
Other assailants	32.0
Work associates	21.4%
Co-worker, former co-worker	11.4
Customer, client	10.0
Relatives	4.0%
Spouse	2.9
Other relatives	0.8
Other personal acquaintances	4.3%
Current or former boyfriend or girlfriend	2.0
Other acquaintances	2.3

Note: Excludes strangers or assailants who were unknown. Includes 2009 preliminary data. Details do not sum to total due to rounding. See *Methodology*.
Source: Census of Fatal Occupational Injuries, U.S. Department of Labor, Bureau of Labor Statistics.

Table 14-3   Workplace Homicides, by Incident Type, 2005–2009

Incident Type	Percent of Workplace Homicide Victims Age 16 or Older
Total	100.0%
Hitting, kicking, beating	6.1
Shooting	80.0
Stabbing	8.1
Unknown	5.8

Note: Includes 2009 preliminary data. Unknown category includes homicides not elsewhere classified and homicides with unspecified types. See *Methodology*.
Source: Census of Fatal Occupational Injuries, U.S. Department of Labor, Bureau of Labor Statistics.

It is unfair, however, to suggest that all employees will act out violently against coworkers or the organization. To the contrary, some aggressors just commit suicide. According to the Bureau of Labor Statistics, 447 employees committed suicide on the job between 2007–2008.[8] Even when grievance or employee appeal processes are used to redress the sources of problems, acts of violence continue to rise. For example, a former postal employee in the Royal Oaks, Michigan, post office killed four postal workers and himself although an arbitration process, albeit lengthy, had been invoked to help him get his job back. No matter the type of workplace violence, it is destructive and often senseless. It not only attacks the very fabric of the organization, it also serves to polarize and frighten the workforce, which can negatively affect productivity and employee satisfaction.

According to Dr. Charles Labig, there are six common sources of violence on the job:

- Strangers, who are typically involved in the commission of a crime or who have a grudge against the business or an employee
- Current or past customers
- Current or former co-workers who commit murder
- Current or former co-workers who threaten an assault
- Spouses or lovers involved in domestic disputes
- Those infatuated with or who stalk employees[9]

The Work Environment and Violence

One key finding in a Northwestern National Life Insurance survey on workplace violence was a strong correlation between job stress and workplace violence. Many factors must be considered in this formula—for example, employee/employer relations, leadership styles, communication patterns, and job security. These factors need to be explored and understood in the context of potential workplace violence. Demeanor and tone can contribute to an employee's feelings and job satisfaction. The traditional McGregor Theory X leader often contributes to work-related problems such as stress attacks, headaches, insomnia, ulcers, nightmares, and anxiety bouts.

In the 1990s, William Lunding stated: "To survive and thrive . . . [leaders] need to shift their thinking from 'kick butt' to compassion, from suspicion to trust, from a 'no-brainer' to a learning environment."[10] Generally speaking, employees like to work in an environment where leaders view their employees as an integral and important part of the organization in furthering its mission. Working for a management structure that trusts and respects its employees' opinions will naturally make the work environment less stressful. Still, workers cannot be left to their own imaginations and direction. Mantell's analogy says it well:

> *Employees in [an organization] are much like blades of grass that together make up a vast green lawn. Given the proper amount of attention, "care and feeding" if you will, nurturing, and exposure to warmth, this "lawn" will flourish. Left to grow unchecked, without careful supervision, control and planning for the future, many parts of the lawn will wither and die or grow completely out of control.[11]*

As a society, we have become increasingly attached to our work. Often, the nature of our work defines who we are, what we are, and what social status we enjoy in the community. In most social conversations, soon after "How are you?" comes "What do you do for a living?" More than ever, people are judged not so much by who they are but by what they do for a living. People who are unemployed often avoid social functions to avoid answering the inevitable questions. Simply stated, for many employees, success at work means success at life.

An organization provides many human needs, from pay to provide for food and shelter to benefits to provide for protection of not only the employee but his or her family as well. In essence, the organization provides security, stability, and structure, which, in turn, provide friendships and sometimes love in the workplace, self-respect and a sense of competency, and ultimately belonging. After a while, and particularly if one is a long-term employee, one begins to count on the organization to provide a standard of living. Consider what happens when an employer takes a person's job away for cause or due to downsizing. This type of rejection, particularly for an emotionally unstable person, dependent in some cases on drugs or alcohol and who identified his or her self-worth and self-esteem with the job, can become potentially explosive.

"We watch in amazement as people make requests of their employer they probably wouldn't make of their own mother, including … education, recreation, [specialized] medical care, psychological care, and plenty of tangible and intangible 'warm fuzzies' that help people pull themselves out of bed and head out to work."[12] We need to recognize that the loss of stature (whether real or perceived), income, or opportunity, such as with a job change, job loss, or demotion, can be devastating to a person's sense of well-being. A job loss can severely attack an individual's self-esteem and sense of identity, causing the person to lash out either overtly or covertly at the organization or individual who caused the pain.

Profiling Violent Behavior

The well-established profile for violent behavior in the workplace, and perhaps the prevailing view, according to Tom Harpley of National Trauma Services, is as follows: "The workplace murderer is likely to be a Caucasian male [between 25 and 40 years of age], using an exotic weapon, such as an Uzi, an AK-47…legally acquired."[13] Although this may be the prevailing view, there is little supporting evidence that it is an accurate profile or that certain kinds of persons can be identified with violent behavior. Typecasting is unrealistic in the work world and can lead to a grossly false diagnosis of potentially violent behavior. Violence is not the result of a particular type of person but rather a mixture of experiences and emotions reinforced over time, sparked by some event that causes violence. Still, there are certain common behavioral characteristics or predictors that can be used to recognize a person's potential for violent behavior:

- *Disgruntled over perceived injustices at work.* This type of employee will be angry, upset, and annoyed about such things as pay, benefits, working conditions, discipline, and the way executive management especially operates. It is not uncommon for such an employee to feel paranoid, persecuted, or conspired against. This type of employee is readily recognizable as one who takes up causes almost to the extreme either on his or her own behalf or for a co-worker who is reluctant to come forward.
- *A loner who is socially isolated.* This type of employee does not appear to have any outside interests; he or she identifies their self-worth and self-esteem with the job and avoids socializing during lunchtime, breaks, and other social functions. When someone attempts to seek them out to invite them, they seem more than just shy.

- *Poor self-esteem.* This type of employee lacks the self-esteem necessary to move ahead and will often become easily frustrated and has difficulty accepting constructive criticism. It is not uncommon for this type of employee to be extremely pessimistic, carrying around a personal collection of stories of hurt, rejection, and powerlessness.
- *Angry.* This type of employee is easily angered and often blows his or her cool for even the smallest of reasons. It is not uncommon for this type of employee to escalate into a full-blown rage from a seemingly normal conversation. It would not be uncommon for this type of employee to have a criminal history.
- *Threatening.* This type of employee takes pleasure in directly threatening, harassing (including sexually), or intimidating co-workers that he or she does not like and the organization as a whole. Statements such as "you will be sorry for what you said" or "revenge is sweet" are not uncommon among the many statements this type of employee will make.
- *Interested in media coverage of violence.* This type of employee has an excessive interest in the mass media's coverage of violence and can often be heard quoting articles about workplace violence episodes. It is not uncommon for this type of employee to suggest that if the same act occurred where he or she worked, management would finally take notice. An employee of this nature might even attempt to copycat other acts of workplace violence.
- *Has asked for help before.* This type of employee has indirectly or directly asked for help from the organization's employee assistance program, a co-worker, or a supervisor.
- *Collects weapons.* This type of employee collects weapons, particularly guns, and may often brag about his or her collection. It would not be uncommon for this employee to have subscriptions to such magazines as *Soldier of Fortune* or *Survivalist.* This employee might also have a fascination with the military.
- *Unstable family life.* This type of employee has either grown up in a dysfunctional family, had a chaotic childhood, or has no support system on which to fall back. This type of employee may disrespect animals and may have abused them as a child.
- *Chronic labor/management disputes.* This type of employee has a long history of ongoing labor/management disputes or has numerous unresolved physical or emotional injury claims. It is not uncommon for this employee to take management's instructions as suspect. This employee will routinely violate organizational policies and procedures.
- *Stress.* This type of employee shows constant signs of stress or is a chronic complainer who always seems to feel overburdened by the pace, the workload, or the physical or psychological demands of the job. It is not uncommon for this employee's true personality to come out under stress; this may be the exact person one sees each day: aggressive, uncompromising, and belligerent.
- *Migratory job history.* This type of employee has bounced from job to job in a relatively short time. In fact, a history of migratory jobs should be caught at the pre-employment interview and rigorously questioned.

- *Drug and alcohol abuse.* This type of employee will show signs of alcohol and other drug abuse, which is traditionally characterized by bloodshot, drooping, or watery eyes; impairment in speech or motor skills; and an unusually disheveled appearance.
- *Vindictive.* This type of employee will be vindictive in his or her actions or words. This type of employee will not leave well enough alone and will often attack the character of a person or organization even though the problem has been resolved. This employee is a typical "organizational sniper" who takes pleasure in watching others dodge the virtual bullets. It is not uncommon for this employee to feel little or no remorse after hurting someone and in some cases actually taking a perverted pleasure in it.

Violence is not the result of a particular type of person but rather a mixture of experiences and emotions reinforced over time.

These above characteristics are not all-inclusive and will require updating as new clues are developed. Unfortunately, the behavioral patterns of a typical perpetrator of workplace violence are frequently apparent, though often noted only in retrospect. These behavioral characteristics alone or in combination with one another do not necessarily guarantee that an individual will become violent. In other words, they should not be considered a guarantee of violent behavior. However, they can, and often do, act as an early warning system, so that preventive intervention techniques can be used before it is too late. All supervisors, human resource professionals, and staffing specialists should be trained to identify and properly handle these behavioral characteristics when they manifest themselves either independently or jointly in an employee.

Basic Levels of Violence

Once a person decides to act out, violence can take many forms. Experts generally agree that it manifests itself in three levels of intensity:

- *Level 1.* Subject actively or passively refuses to cooperate with superiors; spreads rumors and gossip to harm others; frequently argues with co-workers; is belligerent toward customers and clients; constantly swears; and, finally, makes unwanted sexual comments.
- *Level 2.* Subject argues increasingly with customers, co-workers, and supervisors; refuses to comply with the organization's policies and procedures; sabotages equipment and steals the organization's property for revenge; verbalizes the wish to hurt co-workers and supervisors; sends sexual or violent messages to co-workers and supervisors; and, finally, regards self as victimized by management—"me against them."
- *Level 3.* Subject frequently displays intense anger; recurrent suicidal threats; recurrent physical fights; destroys or sabotages company property; uses weapons to harm others; and, finally, commits murder, rape, or arson.

> *"Violence does not occur in a vacuum. It is the result of an escalating process, rather than of one sudden event."*[14]
>
> —*Charles Labig*

Preventing Workplace Violence

An ounce of prevention is worth a pound of cure. The need for prevention is so apparent that the Centers for Disease Control (CDC) issued an alert in 1993 requesting organizations to prevent workplace violence, particularly workplace homicide. The purpose of the alert was to (1) identify high-risk occupations and workplaces, (2) inform organizations and employees about the risk, and (3) encourage organizations to gather statistics and to take active intervention education measures.

> *"The single biggest deterrent to violence in the workplace is careful hiring [and screening]."*
>
> —*Joseph Kinney*

Both the CDC, through its division the National Institute for Occupational Safety and Health (NIOSH), and OSHA are deeply involved in research and training initiatives in the area of violence in the workplace. Both groups have produced major works reporting statistics as well as suggesting methods to reduce the potential for work-related violence. (See www.osha.gov/SLTC/workplaceviolence/index.html or www.cdc.gov/niosh/homepage.html for additional resources.)

Investing in the prevention of violence in the workplace is as vital to a business as investing in research and development. According to Joseph Kinney, the key step in preventing workplace violence is to look for a history of violence in a person's background.[15] Mantell supports Kinney's suggestion: "The single biggest deterrent to violence in the workplace is careful hiring [and screening]."[16] In short, one of the best ways to prevent workplace violence is not to hire a potentially violent person into the fabric of the organization.

Even with proper hiring processes, employees can still become disillusioned and violence may still occur. An organization that fails to prepare for the likelihood of violence can be faced with regulatory sanctions, costly litigation, and the loss of faith among once trusting employees, partners and customers. Under no circumstances should an organization believe that it is immune from workplace violence. An organization that is proactive on the issue of workplace violence should consider the development of a Violence Intervention and Contingency Team (VIACT). Such recognized employers as the United States Postal Service, General Dynamics, International Business Machines (IBM) Corporation, Honeywell, Minnesota Mining and Manufacturing (3M), Kraft General Foods, General Motors, and the Elgar Corporation have formed VIACTs to address violent acts because of past violent incidents. At the least, the teams will communicate to all employees that the organization is genuinely concerned about their welfare and is doing everything possible to prevent and defuse potentially violent situations before they might occur.

The Violence Intervention and Contingency Team (VIACT)

Since all organizations are different, forming a VIACT to prevent and respond to violent situations will require some degree of tailoring to fit the organizational structure. Ideally, members of the team should include, but not be limited to, the following: management, human resources, health and safety, medical, legal, public relations, and security. The team's primary goal is to ensure that all available resources are used at the *earliest opportunity* to prevent and respond to potentially violent situations. Independent of an actual incident, the VIACT must meet regularly and should become subject-matter experts.

Finally, one of VIACT's primary goals is to ensure that all available resources are used at the earliest opportunity to prevent and respond to potentially violent situations.

Pre-Violence Prevention Mission

The team should lead the way in developing policies and procedures that make it absolutely clear to all employees that the organization will not tolerate threats, acts of sabotage, intimidation, harassment, stalking, or violent acts in the workplace. A key component is to communicate to employees that they will be held accountable for their actions and that the organization will cooperate fully with local law enforcement and public prosecutors in dealing with any person involved in workplace violence.

The team should also work to develop a cooperative liaison and open communication with local law enforcement, fire support, and emergency medical services. The team should examine the capabilities and responsiveness of these agencies and detect shortfalls, and if any exist, arrange for an alternative or coordinated response.

Finally, the team should identify intervention education processes to prevent workplace violence, formulate education and awareness programs, and ensure that employees have access to knowledgeable resources that can help prevent or defuse violence in the workplace.

Post-Violence Mission

Once an incident has occurred, the team should convene to review incidents involving the potential for violence or to recommend corrective actions or intervention strategies. If a violent situation arises or the potential for violence is imminent, the team should convene to review the possibility and seriousness of a violent episode; examine administrative, disciplinary and medical options; and examine legal alternatives such as seeking arrest, committing to a medical or mental facility, issuing no trespass warnings, or obtaining an injunction against harassment (known as a restraining order). In all situations involving workplace violence, the team should recommend a timely and decisive response to any violent behavior or act of sabotage.

The VIACT should know when to meet. Prematurely convening a team may frustrate team members and discredit the entire mission. The team should convene when the nature of the threat, harassment, attack, or act of sabotage is unique and falls outside the scope of the organization's normal progressive disciplinary policy. In sum, the VIACT should be the organization's single point of contact with local law enforcement, regulatory bodies, and the media. They should be the information-gathering center for the organization, in turn separating and disseminating factual information.

Strategies for Dealing with Potential Violence

An important part of the VIACT mission is to protect employees from harm and limit liability and regulatory sanctions against an organization. Toward that end, the team should recommend a swift protection strategy after a violent incident occurs. An organization must consider serious all acts of violence, veiled threats and sabotage until proven otherwise and appropriate action must be taken to protect employees and property from further harm or damage. In other words, an organization should take whatever action is reasonable and necessary to contain the violent act and minimize the risk of harm to employees and, secondarily, to property. The violent perpetrator should be managed or removed from the organization's premises as quickly as possible.

Particular attention should be given to employee(s) who is (are) directly threatened. Appropriate measures will vary according to the circumstances of each threat and may include, but are not limited to, the following:

- Involving local law enforcement
- Protecting the threatened individual's work environment (e.g., increasing security)
- Staggering or changing the individual's work shift
- Allowing the individual to park inside the facility or plant
- Transferring the individual to another work area, building, or site
- Having the individual escorted to and from his or her vehicle or home
- Relocating the individual to another facility out of the region, temporarily or permanently
- Advising the individual to alter his or her daily routine and remove its predictability

The team should consider the following options concerning a violent perpetrator:

- Changing the shift or transferring to another location
- Suspension with pay pending further investigation
- Immediate referral to a medical department or the organization's employee assistance program (EAP)
- Retirement
- Voluntary mutual separation
- Progressive discipline
- Involuntary termination of employment (for cause)

Perpetrators' Rights

Alleged perpetrators have rights. The case law in this area suggests that an employer can be found liable for defamation of character if it mistakenly reports the perpetrator as violent when the evidence suggests otherwise. To avoid a claim of defamation of character, an organization should use terms such as aberrant behavior over violent behavior and begin its investigation by discussing the allegations with the individuals who have personal knowledge of the violent incident; it should not rely on hearsay information. Moreover, if an employer discharges an employee without validating the fact that the employee is violent, the employer can be found liable for wrongful discharge. After a threat or actual act of violence occurs, employers should suspend with pay the alleged perpetrator pending further investigation.

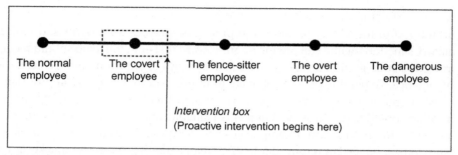

FIGURE 14-1 Workplace violence spectrum. *(From Michael Mantell, Ticking Bombs: Defusing Violence in the Workplace [Burr Ridge, IL: Irwin Publishing, 1994]. Intervention box and strategy developed by the researcher.)*

Intervention Strategy

Mantell offers what he calls the *workplace violence spectrum* (see Figure 14-1), a proactive strategy for recognizing, dealing with, and defusing a potentially violent act before it is too late. The key to using the spectrum is for an organization to look at all employees to determine as accurately as possible where on the spectrum each employee falls:

- *The normal employee.* The normal employee is a person who gets along with others and resolves conflict in a constructive manner and, therefore, does not pose a threat of workplace violence.
- *The covert employee.* The covert or closet employee engages in silent, hidden, or behind-the-scene activities designed to be disruptive to the workplace. This employee might sabotage equipment, leave notes, vandalize property, or leave disturbing voicemail messages or threatening faxes.
- *The fence-sitter employee.* The fence-sitter is on the border between covert tactics and actual violence. This employee's actions will be more direct and confrontational. In some cases, he or she might not take steps to hide his or her identity.
- *The overt employee.* The overt employee will act out directly and openly against the organization or person perceived to have caused him or her harm. This employee can be highly volatile and ready to strike at any moment.
- *The dangerous employee.* The dangerous employee is the potentially homicidal employee, bent on causing destruction and threatening the lives of not only him- or herself but of others as well. In short, this employee is a ticking bomb waiting to explode.

Proactive intervention begins with the covert employee. The strategy is to identify the covert employee before he or she reaches the fence-sitter stage of the continuum and to direct the employee back to the normal range of behavior. An organization should apply as many resources as necessary to identify the covert employee. Anonymous hotlines, investigative techniques such as covert surveillance cameras, and education and awareness programs for employees are all methods to identify the covert employee. Once the covert employee is positively identified, the organization can prescribe many forms of intervention to push the employee back to a normal employee state. Forms of intervention include psychological counseling, employee assistance programs, grievance hearings, time away from work to relieve stress and pressure (e.g., vacation, personal or medical leave), intervention sessions, or progressive discipline designed to coach.[17]

Intervention Education

Intervention education is defined as reaching out to employees before it is too late to do so no matter where they fall on the continuum. More specifically, intervention education means immediately recognizing and correcting unsatisfactory behavior and performance patterns before they get out of hand. Once they're out of control, the result can be termination of employment or worse—violence. Intervention education is the key to eliminating the sparks of future workplace violence. Violent behavior clues are usually evident, as noted earlier. Prevention begins with careful observations and proactive communication so the aberrant behavior problem does not fester.

Intervention begins with establishing open communications with all employees and providing an anonymous reporting channel for complaints about aberrant behavior. If employees believe they can speak openly and honestly with the organization, it can serve as a pressure release valve, blowing off steam when the pressure and stress get to be too much to handle. In other words, allowing employees to vent their feelings is an effective way to reduce job pressure and stress that can lead to violence.

To reduce stress in the organization, employees should believe they have five inalienable rights:

1. Freedom of speech
2. Credit for work performed
3. Strong support
4. Reliable guidance
5. Solid leadership

Humor that is not harassing or combative should also be encouraged in the workplace. Research shows that the positive effects of laughter can be enormous. Employees should also be encouraged to relax.

The Intervention

Any form of intervention must be tempered by the realization that violence could occur. You should quickly learn to expect the unexpected when it comes to an intervention session. Don't be fooled into thinking that violence always involves physical contact. Verbal threats, abusive language, and acts of intimidation can sometimes be more threatening than physical contact.

Always consider personal safety before an intervention session. If you believe that the intervention session could be confrontational, contact the organization's security department so the necessary security countermeasures can be put in place, or consult the guidance manual for addressing violent acts in the workplace.

During the intervention session, always face the employee and sit as close to the exit as possible; never allow the employee to sit between you and the door in case an escape is necessary. Remove all potentially dangerous items from the intervention area. If appropriate, have a second person in the hearing room during the intervention session. Not only can this provide a calming effect, but it can also act as a force presence. Women provide a calming effect when it comes to male-to-male hearings. Finally, have available tissues and business cards with the organization's EAP phone number. Intervention must be tempered by the realization that violence could occur.

Handling an intervention session can be difficult, given the fact that people in general do not like or handle confrontation well. It is important to recognize that the intervention session can be laden with emotion on both sides. The session should be made as much a positive experience for the employee as possible. It is also helpful to offer a beverage as an icebreaker.

There is no one best way to conduct an intervention session. However, how you begin the session will often determine its outcome. As a result, begin with something like the following statement: "Listen, the organization and I do not want to fire you. However, you are here because of your unsatisfactory behavior. Do you understand how you arrived at this stage?" It is appropriate to wait for a response at this point, and then say something like "The organization's goal is to help you become the best employee you can be. However, the organization needs your help to accomplish this goal. Do you understand what we are saying?"

Intervention can be a positive experience if you quickly remove the immediate threat of being fired from the employee's mind. Once the threat has been removed, move to get the employee to accept responsibility for improving his or her unsatisfactory behavior. If the employee responds, "I understand what you mean and I am sorry," you have arrived at the acceptance stage. Now move quickly to solidify ideas for improvement. Employees who believe their suggestions are included in the improvement solution are more apt to take ownership of their behavior and voluntarily seek to improve it. It is also important to recognize when an expert is needed. Some employees bring deep-seated emotional problems to the intervention session. If this is the case, refer the employee to the organization's EAP.

If the intervention session involves progressive discipline that requires some form of documentation, such as a written reminder, the session be rehearsed prior to the employee's arrival, to ensure a smooth delivery of the letter's contents. Always remain genuine and authentic during the session. Do not speed through the written reminder. Take the time necessary to explore each paragraph with the employee. Wait for questions, and answer them as you can. The session should not be dictatorial, in which the organization imposes sentence on the employee. The session should be designed to get the employee to accept and voluntarily correct his or her behavior. Be firm but polite and respectful. Do not be surprised if the employee gets emotional.

As a rule, do not give the written reminder to the employee when he or she walks into the intervention session. In the employee's mind, this could be an ominous start to an otherwise productive session. Instead, talk each paragraph out, as an actor would a script. In this way, the employee will be focusing not on the letter but the session's intent.

If the employee is represented under a collective bargaining agreement and requests union representation, wait until representation arrives before proceeding. At the conclusion of the intervention session, give the employee a copy of the written reminder as a capstone of the conversation. For written material, it is wise to have the employee co-sign the letter. If a union representative is present, have the representative sign the letter as well.

Research and the authors suggest that an effective system of disciplinary due process will reduce workplace violence. Many instances of workplace violence have been the result of some perceived injustice at work. Due process gives the employee a chance to work out his or her differences. Using the intervention strategy and recommendations, the organization can effectively intervene in a positive and productive manner. Finally, leaders within an organization must learn to understand human behavior. It is not what you say that makes employees respond, but rather *how* you say it.

Finally, do not expect all employees to behave alike. Learn to recognize their strengths and weaknesses and respond accordingly. To be a successful leader, you must adopt a separate and sometimes distinct approach for each employee.[18]

Drugs in the Workplace

While many U.S. citizens view drugs as a recent problem, the use of drugs in the workplace has been a problem for many years. In fact, historically, private enterprise has pioneered most of the programs in drug detection, prevention, and rehabilitation. The federal government has adopted a policy that makes some drugs illegal and drug usage unacceptable. Support for the federal policy has been forthcoming from all segments of society, including federal government, state government, and business.

American business leaders share a consensus that illegal drugs have become pervasive. The exact costs associated with the drug problem in the workplace are not known, but the estimates are staggering. Using data from 2007, the National Drug Intelligence Center[19] issued a report that estimated illicit drug use has created a $193-billion economic impact on American Society. Approximately $120.3 billion of the total burden relates to losses in productivity.[20] When examining alcohol, the National Institute on Alcohol Abuse and Alcoholism[21] estimates the cost of alcohol abuse at approximately $184.6 billion yearly.

Of the 19.3 million current illicit drug users (18 or older), approximately 66.6 percent are employed full or part time. When rate of use is examined by employment status, approximately 8.0 percent of full-time and 11.5 percent of part-time workers used an illicit drug last month. In addition, 42.7 million employed workers reported alcohol binge drinking (drinking five or more drinks on the same occasion on one day in the last month) and 12.4 million full- and part-time employees were categorized as heavy drinkers (drinking five or more drinks on the same occasion on five or more days in the last month).[22]

Drug use and abuse is broadly distributed across occupations and industries; however, prevalence varies. Table 14-4 shows the prevalence of illicit drug use and levels of heavy drinking among 12 industries.[23] Combined data from 2002–2004 show that illegal drug use was greatest among workers in the occupations of accommodations and food services; construction; and arts, entertainment, and recreation. Heavy alcohol use was also relatively high among these occupations, as well as among mining and wholesale trade.

While alcohol is the most commonly abused drug, 7.3 million (6.4%) full-time employees reported using marijuana.[24] The use of psychotherapeutics (e.g., prescription painkillers, stimulants, tranquilizers, or sedatives) also has become a concern in the workplace. A study published

Table 14-4 Percentage of Full–Time Workers, Ages 18–64, Reporting Illicit Drug Use and Heavy Alcohol Use, by Industry Category, 2002–2004 Combined

Industry Category	Percent Illicit Drug Use	Percent Alcohol Use
Accommodations and Food Service	16.9	12.0
Construction	13.7	15.9
Arts, Entertainment, and Recreation	11.6	13.6
Information	11.3	10.4
Management of Companies and Enterprises, Administrative Support, Waste Management, and Remedial Services	10.9	10.4
Retail Trade	9.4	8.8
Other Services (Except Public Administration)	8.8	9.9
Wholesale Trade	8.5	11.5
Professional, Scientific, and Technical Services	8.0	7.1
Real Estate, Rental, Leasing	7.5	9.8
Mining	7.3	13.3
Utilities	3.8	10.1

Source: Office of Applied Studies, SAMHSA (2007). National Survey on Drug Use and Health Report: Worker Substance Use, by Industry Category.

in the *Journal of Applied Psychology* reported marijuana and psychotherapeutics have been found to be the most commonly used drugs prior to and during work hours.[25] Additionally, of the 2.2 million drug tests performed by Quest Diagnostics for the general United States workforce (January through June, 2010), the greatest number of positive results were found for marijuana, followed by amphetamines, prescription painkillers, and cocaine, respectively.[26]

Impact of Drug Use

Substance abusers, as compared to nonabusers, are more likely to have a negative impact on the company or business in a variety of ways, including increased theft, decreased productivity and quality control, increased incidence of accidents and injuries, increased absenteeism, high turnover rate, and increased costs due to personal problems.

Theft

Many security administrators say that drug abuse and theft go hand in hand. Drug users often steal from employers and fellow employees to support their habits. Criminal activities, including vandalism, often result in increased legal costs for the firm. Lawsuits, legal fees, and court costs are added expenses when drug usage leads to theft, vandalism, and other criminal behavior.

Productivity, Quality Control, Accidents, and Injuries

Several years ago during a break period, one line employee for a major corporation was slipped drugs in his drink for laughs. The joke got out of control when the employee threw a bucket into a conveyor. Fifty percent of the operation was shut down. The company lost several hours of work while still paying employees who were idled by the incident. In general, substance abusers, as compared to nonabusers, are involved in many more job mistakes and are more likely to have lower output, and work shrinkage.[27]

Substance abusers also experience three to four times more on-the-job accidents and are responsible for 40 percent of all industrial fatalities.[28] A study reported in the *Journal of American Medical Association* found those employees who tested positive for cocaine or marijuana had a significantly higher rate of on-the-job accidents and injuries.[29]

Instances of workplace substance use have set the stage for potential disaster, and in some cases, death. In 1991 a New York subway operator crashed his train near a station in Manhattan; 5 people were killed and 215 others were injured. The operator admitted that he had been drinking prior to the crash. His blood alcohol content was found to be 0.21—more than twice the legal limit.[30] In July, 2002, two America West pilots were fired after being stopped at the gate by security personnel. The security personnel smelled alcohol on their breath and stopped the two pilots from boarding the 124-passenger airliner. Both pilots possessed blood alcohol levels over the legal limit for pilots (0.04).[31] In 2010, thirteen Detroit auto workers were dismissed after being caught on video smoking and drinking during their lunch hour.[32] Most recently, an air traffic controller was removed from duty after failing a random drug test for alcohol.[33]

Absenteeism and Turnover Rates

Overall, substance abusers are late to work three to four times more often, absent five to seven times more often, and are three to four times more likely to be absent for longer than eight consecutive days. Also, they use three times more sick leave.[34] When specific drug use is examined, a study found that employees who were marijuana or cocaine users had a higher rate of absenteeism (78 percent and 145 percent more absenteeism, respectively) as compared with other employees.[35]

The substance that causes the greatest problem with tardiness, absenteeism and lost work days is alcohol. In an attempt to personalize the amount of loss for an industry, The George Washington University Medical Center[36] has developed an online "Alcohol Cost Calculator," in which a company can enter specific information about their industry and geographic location to calculate the number of potential alcohol abusers, health care costs and work day losses. For example, according to the calculator, a West Virginia mining company with 500 employees would have an estimated 52 problem drinkers, who would create 11 days work loss per month at a cost of $24,077. Alcohol related health-care costs are estimated at $302,864. The Alcohol Cost Calculator can be accessed at: www.alcoholcostcalculator.org/business.

Substance abusers also have higher turnover rates. According to a nationwide study of full-time workers ages 18 to 64, of those who reported past month illicit drug use, 12.3 percent reported working for three or more employers in the past year, whereas only 5.1 percent of workers with no past month drug use reported working at three or more jobs in the previous year. Variation occurred depending on the industry and occupation.[37]

Personal Problems

Approximately half the employees who have personal problems are substance abusers of one type or another. Substance abusers use three times more medical benefits, file five times as many workers' compensation claims, and increase premiums for the entire company for medical and psychological insurance. Also, substance abusers are seven times more likely to experience wage garnishments and are more often involved in grievance procedures, taking up both union and organization time and resources.[38]

Using a Comprehensive Substance Abuse Program

In order to address substance use problems in the workplace, the federal government passed the Drug Free Workplace Act in 1988. Companies administering federal grants and contracts are obligated to follow the Act's guidelines, listing countermeasures to reduce the costs associated with employee drug uses and abuse. The Drug Free Workplace Act, which was subsequently updated in 1994, 1998, and 2004, sets forth the following requirements:

1. Development of a clear drug-free workplace policy. This policy must clearly state the reason it is needed (for example, safety, product quality), the company's expectations regarding employee behavior, the rights and responsibilities of employees, and the type of action that will be taken if drug use or possession is suspected or discovered.
2. Establishment of continuing drug education and awareness programs. Firms are expected to provide educational programs aimed at ensuring that employees understand the drug policies and their consequences. In addition, supervisors must be trained to identify and deal with employees suspected of drug use or possession.
3. Implementation of an employee assistance program (EAP) or other appropriate mechanism. Through the employee assistance program, the employer is able to help employees who have a substance abuse problem rather than resort to immediate termination. This is accomplished by referral for evaluation and treatment/rehabilitation. There are many ways to set up an EAP, including establishing a program at or near the work site or buying EAP services from an outside provider.
4. Reporting to the federal government any convictions related to drug crimes.

While not all private industries are required to follow these guidelines, companies have found it cost effective to implement a comprehensive drug program and provide employee drug education and treatment. In their book, *Preventing Workplace Substance Abuse: Beyond Drug Testing to Wellness,* Bennett and Lehman[39] outline various components of drug programs that demonstrate promise in dealing with workplace substance abuse. The authors believe when substance abuse is addressed from a holistic and humanistic perspective, the cost-benefit ratio of substance abuse education and treatment is greatly improved.

By providing treatment and retaining an employee who abuses drugs, a company not only reduces employee replacement costs, but provides major savings with reduced interpersonal conflicts, enhanced workplace productivity, and lowered drug-related accidents. Dealing with a substance abuse problem can yield up to a 12 to 1 cost-savings ratio when expenses related to drug-related crime, criminal justice costs, theft, and health care are considered.[40] Information on development of policies and procedures for a comprehensive substance abuse program can be obtained from the Center for Substance Abuse Prevention Workplace Helpline (1-800-WORKPLACE).

Drug Testing

To strengthen national safety related to workplace drug use, the Omnibus Transportation Testing Act was passed in 1991. As a result of the Act, public safety and national security personnel involved in aviation, trucking, railroads, mass transit, and other transportation industries are subject to drug testing requirements. Additionally, the Department of Defense (DOD), United States Department of Energy, and the Nuclear Regulatory Commission have developed drug testing policies due to the safety-sensitive nature for a number of their contracted industries.

Many other private-sector businesses have initiated some type of drug testing (pre/posthire, random, or following an incident) in order to reduce problems associated with employee drug abuse. The words "Must have a clean drug history" may discourage those who have drug problems from ever applying. In SAMHSA's findings on worker substance use,[41] 1 out of 5 workers who reported illicit drugs last month would be less likely to work for an employer who conducted pre-hire testing than workers who did not report illicit drug use (18.2% vs. 3.7 %, respectively).

According to a 2004 survey of 506 businesses conducted by the American Management Association,[42] nearly 63 percent of those surveyed said they had a workplace drug testing program. Additionally, a national survey of workers [43] reported that of the workers coming from an industry with over 500 employees, 70 percent said drug testing programs were in place; however, percentages did vary by occupational category (see Table 14-5).

For companies who do drug testing, the rate of positive results has declined over the years. In 1988, 13.6% of drug tests were positive, but in 2010, there was a positive return rate of only 3.6%.[44]

Preemployment Drug Testing

When companies conduct preemployment screening they should adopt the following guidelines:

Table 14-5   Percentage of Full-Time Workers, Ages 18–64, Reporting Workplace Drug Testing Programs, by Major Occupation Categories, Selected: 2002–2004, Combined

Occupation Category	During Hiring	Random
Protective Services	76.2	61.8
Transportation and Material Moving	73.3	62.9
Production Occupations	63.1	40.9
Installation, Maintenance, and Repair	57.4	42.0
Engineering, Architecture, and Repair	52.7	35.1
Healthcare Practitioners and Technical Occupations	52.5	30.9
Office and Administrative Support	46.6	29.5
Construction and Extraction	34.7	31.9
Food Preparation and Serving Related	26.3	20.6
Arts, Design, Entertainment, Sports, and Media	17.3	9.9

Source: Larson, Eyerman, & Gfroerer, Office of Applied Statistics, SAMHSA, (2007).

1. Notify the applicant of the company's policy of drug screening.
2. Make sure the test results are valid by using a reputable laboratory. The Substance Abuse and Mental Health Administration maintains a list of certified laboratories.[45]
3. Ensure confidentiality.

Postemployment Drug Testing

As with all policies, the key is that the policy is well written and communicated to the employees in a clear manner. Expectations regarding the use of drugs and the penalties associated with that use must be clearly stated. The policy should specify under what conditions an employee will be expected to submit to drug testing (for example, after an on-the-job accident).

All policies and procedures for drug testing should: (1) be consistently administered; (2) explain prescription drug use, including what types of drugs need to be declared to company supervisors; (3) require substantive proof of drug use; and (4) be consistent with statutory or regulatory requirements, collective bargaining agreements, and disability discrimination provisions.

Types of Drug Tests

Through drug testing it is possible to identify the use of a variety of drugs, both illicit and licit. Substances that may be tested for are alcohol, amphetamines, barbiturates, benzodiazepines, marijuana, cocaine, methadone, methaqualone, natural and synthetic opiates, phencyclidine (PCP), and testosterone-like anabolic steroids.

Once introduced into the body, drugs are biotransformed and eventually eliminated from the body through excretion. The presence of drugs or their metabolites can be detected in urine, blood, sweat, saliva, hair, and nails through various types of testing technology.

The most widely used method is urine testing. Urine is easy to collect and concentrations of drugs and metabolites tend to be higher, allowing longer detection times. Table 14-6 provides a listing of common drugs and the number of days that they are typically detectable in urine.[46]

Unfortunately, the high-profile publicity surrounding drug testing has allowed violators to develop a counter-educational movement in an attempt to beat the screening systems. Smuggling urine into the bathroom is widely practiced. With careful monitoring, this should be almost impossible; however, some employees have devised very clever schemes that may escape notice, such as carrying a plastic bag of "clean" urine under the arm. In addition, users of cocaine know that it stays in the system for a maximum of 72 hours; thus it is possible to abstain prior to urine testing so that the test results are negative for cocaine use.

Any one of three basic formats for drug testing is common. The first and most popular is enzyme immunoassay (EIA). The EIA uses urine samples and can be used to screen for up to 10 drugs. A second format is the radioimmunoassay (RIA). While similar to EIA, it uses radioactivity as an indicator for a positive test rather than a color change. While the costs of drug tests can vary greatly, a conservative estimate of costs for these two tests is between $25–$75 per test.[47]

The third format is a combination of gas chromatography and mass spectrometry (GC-MS). This test is 99.9 percent accurate and is the only test accepted by most courts as prima facie evidence of drug use. Although it is an extremely accurate test, the GC-MS is not widely used

Table 14-6   Detection Period Range Chart (Urine Only)

Drug	Time
Alcohol	7–12 hours
Amphetamine	48 hours
Methamphetamine	48 hours
Barbiturate	
Short-acting (e.g., phentobarbitol)	24 hours
Long-acting (e.g., phenobarbitol)	3 weeks
Benzodiazepines	
Short-acting (e.g., lorazepam)	3 days
Long-acting (e.g., diazepam)	30 days
Cocaine metabolites	2–4 days
Marijuana	
Single use	3 days
Moderate use (4 times/wk)	5–7 days
Daily use	10–15 days
Long term heavy smoker	>30 days
Opioids	
Codeine	48 hours
Heroin (morphine)	48 hours
Hydromorphone	2–4 days
Methadone	3 days
Morphine	48–72 hours
Oxycodone	2–4 days
Propoxyphene	6–48 hours
Phencyclidine	8 days

Source: Moeller, et al. (2008).

for screening purposes due to cost. However, it is often used as a confirming test when positive results are obtained using EIA or RIA.

Other types of screening tests are:

- Thin layer chromatography
- Agglutination
- Fluorescence polarization immunoassay

Another test is RIAH (radioimmunoassay of hair). This test requires that the individual provide several strands of hair, which are then tested for drugs. The procedure initially involves testing using the radioimmunoassay technique. If the results are positive, a confirming test is performed using the extremely accurate gas chromatography/mass spectrometry technique.

Major advantages of this test are listed below.

- The collection, transportation, preservation, and storage of hair samples are simple and relatively nonintrusive processes as compared to other types of drug testing.
- It is less prone to tampering.
- Human hair maintains a 90-day record of drugs ingested by the body, and thus drug use can be detected as far back as 3 months.

- A laboratory can determine the date when a drug was last used within 7 days.
- Hair can also be matched to the owner with the exactness of fingerprints.
- Contrary to popular belief, the use of special shampoos cannot "beat the test," because the analysis is performed on the cortex of the hair.

Although RIAH test results have been challenged in court, it has predominantly been judged to be a valid and reliable test.[48] Prevalence of RIAH testing in the workplace has been increasing in the past few years, in part because of the advantages mentioned above, as well as the fact that it has become less expensive. The average cost is approximately $45. Companies such as General Motors and Anheuser-Busch routinely use RIAH for drug testing.

Another relatively noninvasive test is oral swabbing. Swabbing the mouth for oral fluids/saliva offers the advantage of an observed collection and is easily administered. There are no known adulterants that can be used to tamper with oral fluid testing.[49] Because swabbing can detect drug use one hour after ingestion, it may be a preferred method following a workplace incident.

Drug Testing Process

Businesses have a number of ways they accomplish the task of drug testing. One option is to contract with a certified laboratory to collect all specimens and oversee the testing. Others have employees report to health facilities where specimens are collected and forwarded to certified laboratories. A final option for initial drug screening is to utilize company medical personnel to collect specimens from the employee.

The advent of disposable urine test kits that provide quick results has made internal drug screening an economical choice for many companies. The Georgia Department of Corrections uses a urine test cup that provides immediate results and reduces their cost from $22.50 for a laboratory test to $7 per test cup.[50] Non-confirmed negative results are forwarded to a laboratory for further testing.

Drug testing is an effective method to reduce drug problems in the work environment. It not only screens out drug abusers at the time of hiring, but allows employers to detect drug abuse among current employees. The rights of individual privacy are often of less significance than the importance of public health and safety. With appropriate policies and procedures, the accuracy of drug testing is greatly enhanced. This includes having a chain of custody for handling the specimen, a policy for a second confirming test for all positive results from the initial screen, and use of a medical officer to review the results.

Additional Methods

In addition to drug testing, some companies have also adopted the use of undercover operatives to discover possible drug trafficking and use. Some firms use camera systems, ranging from simple hidden cameras to more elaborate hidden or open observations.

Spotting Drug Use

Watching for drug use or intoxication on the job is not easy because there are many different types of drugs and drug-dependent people. Reactions differ with the type of drug and often

Table 14-7   Indications of Possible Drug Abuse

Arriving late, leaving early and/or often absent
Unreliable and often away from assigned job
Careless and repeatedly making mistakes/reduced productivity
Argumentative and uncooperative
Unwilling or unable to follow directions
Avoiding responsibility
Indifference to personal hygiene
Making excuses that are unbelievable or placing blame elsewhere
Taking unnecessary risks by ignoring safety and health procedures
Frequently involved in mishaps and accidents or responsible for damage to equipment or property
Overt physical signs such as unexplained exhaustion, hyperactivity, dilated pupils, bloodshot or glassy eyes, slurred speech, or an unsteady walk

---

Source: America Council for Drug Education and US Depart of Labor.

with the personality and problems of the individual user. In addition, there are medical conditions that present symptoms similar to those displayed by drug abuse. Table 14-7 provides a list of signs and behaviors related to possible drug abuse.[51,52] It is important to remember that, if an employee displays these signs, it may or may not mean there is a drug problem, but the possibility should be considered.

Following is a sample of paraphernalia used by drug abusers:

Marijuana
Pipes in cylinders made to look like a permanent marker.

Crack Cocaine
A simple aluminum drink can is crushed in the middle, and holes are punched in the can with a nail. Crack is placed over the holes and lighted. The user smokes the can by inhaling the smoke through the open tab end of the can. When the user is finished, the can is discarded.

Cocaine
Small vials are often placed in the employee's wallet, along with supporting paraphernalia such as a straw, a pocket mirror, and a razor blade. In addition, small safes made to look like soft drink cans or fire extinguishers are becoming popular hiding places.

Other Methods
Drug users have long used tin foil, prescription bottles, and zip-top bags.

Summary

As the statistics clearly indicate, workplace violence has decreased 62% from 1993 to 2002. In fact, from 2002 to 2009, workplace violence has further decreased by an additional 35%. Unfortunately, shootings account for 80% of workplace homicides with strangers committing the greatest number of workplace violent acts. In 2009, there were 521 workplace homicides mostly against middle-aged males. Regardless of the number, these figures are still too high. Organizations' efforts through VIACT programs, manager training and other prevention plans have had the desired positive impact. However, organizations large and small continue to look for better ways to diagnose early signs of aberrant behavior the fostering the goal of reducing major workplace violence incidents.

Drugs in the workplace continue to be a concern as is alcohol abuse, many times contributing to the violence issue as well as to absenteeism and substandard performance. Testing of applicants and tenured employees is becoming more widespread. The loss of time on the job as well as job-related injuries caused by drug and alcohol impairment has brought these issues to the forefront. Security must continue to monitor the workplace for illegal drug use and make sure that employees with problems are either placed into employee assistance programs or, when illegal activity is discovered, discharged and criminally prosecuted.

■ ■ CRITICAL THINKING ■

Although drug use is illegal, some people believe it is a personal decision and should not be penalized. Shouldn't workers have the option of using drugs whenever or wherever they choose? If they are impaired at work or become violent, the employer should terminate their employment, not send them to assistance programs! Consider the above statement and comment.

Review Questions

1. What is a VIACT?
2. Can the potential for violence be identified?
3. What are the steps that should be taken before an intervention is initiated?
4. What is an EAP?
5. Name the various typologies of possible problem employees.
6. Name at least three things that an employer should look for that could indicate possible drug use on the job.
7. Is there any difference in the incidence of drug abuse among various types of jobs? If so, which occupations have the greatest problems?
8. What are the major problems associated with drug abuse and the workplace?

References

[1] www.sba.gov

[2] Violence in the workplace. National Crime Victimization Survey March 2011.

[3] Ibid.

[4] Kinney J. Violence at work. Englewood Cliffs, NJ: Prentice-Hall; 1995. p. 1.

[5] Violence in the workplace. December 2001.

[6] Workplace violence. Occupational Safety and Health Administration, U.S. Department of Labor, downloaded January 28, 2003, <www.osha.gov/oshinfo/priorities/violence.html>. BLS August 2009.

[7] Mantell M. Ticking bombs: defusing violence in the workplace. Burr Ridge, IL: Irwin Publishing; 1994. p. 53.

[8] Fatalities to young workers and all workers by event and exposure. U.S. Bureau of Labor Statistics, downloaded January 28, 2003, <www.stats.bls.gov>.

[9] Labig C. Preventing violence in the workplace. NewYork: AMACOM; 1995.

[10] Yates RE. The healing manager. San Diego Union 1993. p. Cl.

[11] Mantell, Ticking bombs. p. 136.

[12] Ibid., p. 33.

[13] Baron A. Violence in the workplace: a prevention and management guide for business. Ventura, CA: Pathfinder Publishing of California; 1993. p. 88.

[14] Labig, Preventing violence in the workplace. p. 16.

[15] Kinney, Violence at work. p. 125.

[16] Mantell, Ticking bombs. p. 1.

[17] Baron, Violence in the workplace. p. 139.

[18] Walters DC. San Diego state masters project. Violence in the workplace–a prevention and intervention education guide for management; 1996.

[19] National drug intelligence center (2011). The economic impact of illicit drug use on American society. Washington D.C.: United states department of justice. Retrieved July 7, 2011 from <http://www.justice.gov/ndic/pubs44/44731/44731p.pdf>.

[20] Ibid.

[21] Harwood H. Updating estimates of the economic costs of alcohol abuse in the united states: estimates, update methods and data. 2000. Report prepared by the Lewin group for the National Institute on Alcohol Abuse and Alcoholism. Retrieved July1, 2011 from <http://pubs.niaaa.nih.gov/publications/economic-2000/alcoholcost.pdf>.

[22] Substance Abuse and Mental Health Services Administration. Results of the 2009 national survey on drug use and health: Vol I. Summary of national findings (Office of applied studies, NSDUH series H-38A, HHS publication No. SMA 1004586 findings). Rockville, MD; 2010.

[23] Substance Abuse and Mental Health Services Administration, Office of applied studies (2007). The NSDUH report: worker substance use by industry category. Rockville, MD. Retrieved June 30, 2011 from <http://oas.samhsa.gov/2k7/industry/worker.htm>.

[24] Larson SL, Eyerman J, Foster MS, Gfroere JC. Worker substance use and workplace policies and programs. (DHHS Publication No. SMA 07-4273, Analytic Series A-29). Rockville, MD: Substance Abuse and mental Health Services Administration, Office of Applied Studies; 2007. Retrieved July 9, 2011 from <http://www.oas.samhsa.gov>.

[25] Frone MR. Prevalence and distribution of illicit drug use in the workforce and in the workplace: findings and implications from a U.S. national survey. J Appl Psychol 2006;91(4):856–69.

[26] Quest diagnostics (2011). Drug testing index. Retrieved July 14, 2011 from <http://www.Questdiagnostics.com/employersolutions/dti/2011_01/dti_index.html>.

[27] Inaba DS, Cohen WE. Uppers, downers, all arounders. 6th ed. Medford, Or: CNS Productions; 2007.

[28] Inaba DS, Cohen WE. 2007.

[29] Zwerling C, Ryan J, Orva E. The efficacy of preemployment drug screening for marijuana and cocaine in predicting employment outcomes. J Am Med Assoc 1990;264(20):2639–43.

[30] Cunradi CB, Ragland DR, Greiner B, Klein M, Fisher JM. Attributable risk of alcohol and other drugs for crashes in the transit industry. Inj Prev 2005;11:378–82.

[31] Polk County, Florida. Airline tells pilots in alcohol arrests they will be fired. (2002), <www.polkonline.com/stories/070402/sta_pilots.shtml>.

[32] MyFOXDetroit.com (2010, September). Chrysler autoworkers busted. Retrieved July 29, 2011 from <http://www.myfoxdetroit.com/dpp/news/chrysler-auto-workers-busted_20100923_dk>.

[33] Air Safety Week (2011, July). Air traffic controller found drunk on site. Retrieved from <http://www.aviation-today.com/asw/topstories/Air-Traffic-Controller-Found-Drunk-On-Site_73863.html>.

[34] Inaba DS, Cohen WE. 2007.

[35] Zwerling C, Ryan J, Orva E. 1990.

[36] George Washington University Medical Center (n.d.). Ensuring solutions to alcohol problems: what can your company do about costly alcohol problems? Retrieved July 28, 2011 from <http://www.alcoholcost-calculator.org/business/>.

[37] Larson SL, Eyerman J, Foster MS, Gfroere JC. 2007.

[38] Inaba DS, Cohen WE. 2007.

[39] Bennet JB, Lehman WEK, editors. Preventing workplace substance abuse: beyond drug testing to wellness. Washington, D.C: American Psychological Association; 2003.

[40] National Institute on Drug Abuse (n.d.). Principles of drug addiction treatment: a research based guide. Retrieved on July 26, 2011 from <www.drugabuse.gov/PODAT/Faqs.html#faq4>.

[41] Larson SL, Eyerman J, Foster MS, Gfroere JC. 2007.

[42] American Management Association (2004). AMA workplace testing survey: medical testing. Retrieved July 12, 2011 from <http://www.Amanet.org/training/whitepapers/2004-Medical-Testing-Survey-17.aspx>.

[43] Larson SL, Eyerman J, Foster MS, Gfroere JC. 2007.

[44] Quest diagnostics. Drug Testing Index; 2011.

[45] Substance Abuse and Mental Health Services Administration (2011, July 12). Current list of laboratories and instrumented initial testing facilities which meet minimum standards to engage in urine drug testing for federal agencies. Fed Regist 76(133). Retrieved from <http://workplace.samhsa.gov/DrugTesting/pdf/FR071211.pdf>.

[46] Moeller KE, Lee KC, Kissack JC. Urine drug screening: practical guide for clinicians. Mayo Clin Proc 2008;83(1):66–76.

[47] Substance Abuse and Mental Health Services Administration (n.d.). Drug testing facts and statistics. Retrieved June 15, 2011 from <http://workplace.samhsa.gov/WPWorkit/pdf/drug_testing_facts_and_stats_FS.pdf>.

[48] McBay A. (1996). Legal challenges to testing hair for drugs: a review. Retrieved from <http://www.big.stpt.usf.edu/~journal/mcbay2.html>.

[49] Quest Diagnostics. Drug Testing Index; 2011.

[50] Gater L. Managing the workplace drug testing process. Corrections Forum 2007;16(6):20–3.

[51] American Council for Drug Education (n.d.). Drugs and Alcohol in the workplace: facts for employers. Retrieved June 6, 2011 from <http://www.acde.org/employer/DAwork.htm>.

[52] United States Department of Labor (n.d.). First line fact sheet: keeping your worksite drug and alcohol free. Retrieved July 5, 2011 from <http://www.dol.gov/asp/programs/drugs/workingpartners/First-Line_Fact_Sheet.pdf>.

15 ⸬⸬⸬

Retail Security

OBJECTIVES

The study of this chapter will enable you to:

1. Define shrinkage, its causes and its impact on the retailer.

2. Define the retail supply chain and discuss a bigger picture approach to protecting it.

3. Be aware of the various risks that present themselves in a retail environment and that must be tracked and eliminated, mitigated or controlled.

4. Demonstrate how the loss prevention function contributes to the financial success of a retail store.

Introduction

At the retail sales end of the distribution chain, merchants are beset on all sides by assaults on their profit and loss positions. The very nature of retailing demands that quantities of merchandise be attractively displayed in easily accessible areas. The public can roam at will and handle much of the merchandise. Every effort is made to create a desire to possess the displayed merchandise and to make possession as effortless as possible.

Of course, the merchant expects payment for the goods. Others, customers and associates alike, sometimes overlook that aspect of the transaction, and the merchant takes a loss. Generally, each loss is relatively small, but the aggregate damage to the business from such erosion of inventory can be enormous. At retail, 2010 losses were estimated to be $37.1 billion, up from $33.5 billion in 2009.[1]

Add to these problems the issues of targeting by terrorists, and retailers have their hands full. The 1999 bombings of three Lowe's Home Improvement stores in North Carolina and the 2002 sniper attacks in the Washington, D.C. area required retailers to take added precautions to protect their customers.[2] Additionally, in 2007, Robert A. Hawkins shot 13 people with a rifle at a Von Maur store in a mall located in Omaha, Nebraska, killing eight. Workplace violence risks, coupled with the risk of robbery, for many segments of the retail industry have challenged retail security executives in the way they go about protecting their customers.[3] Chains vary in their response to threats, ranging from dispatching loss prevention personnel to rooftops in affected areas to monitoring parking lot activity and deploying rapid response to potential problems. Although these actions might seem extreme, consumers have come to expect a safe shopping experience. Those retailers that fail to provide a safe shopping experience may find that they are front page news and tomorrow's bankruptcy statistic.

As in the past, the associate contributes more to losses than the shoplifter. According to the University of Florida's 2010 National Retail Security Survey's preliminary results, 43.7% of retail loss is due to theft by associates, whereas 32.6% is from shoplifting.[4] In recent years, drugs have become associated with both shoplifters and associates, with 46% of shoplifters and 55% of associates showing evidence of drug use when apprehended.[5]

Most businesses are subject to inventory shortage, but few feel the problem as acutely as retailers. They necessarily deal in merchandise that must be received, stored and possibly even moved from company warehouse to the store receiving area and then onto the sales floor. All of these operations pose a risk of loss from theft or damage. Inadequate or careless record keeping will also contribute to inventory shrinkage.

The three principal sources of loss to the retailer are external losses from theft, internal losses from associate dishonesty, and losses from carelessness or mismanagement. In the aggregate, these losses are referred to as inventory shrinkage. As previously noted, loss prevention executives report that over 75% of losses are due to theft by associates and customers. The other, approximately 25%, of the inventory shrinkage problem, is due to theft by administrative error (carelessness), theft by vendors and, lastly, from unknown causes.

Rolled up into shoplifting is the ever-increasing problem of organized retail crime or ORC—highly organized groups sometimes associated with international crime groups. These groups know exactly what they want to steal and have established theft and fencing networks.

A commonality of associate and customer theft is return fraud. Return fraud is estimated to cost retailers billions of dollars each year. The retailer is caught between two opposing desires—making the return process as hassle-free as possible, and controlling that process to limit the opportunity for fraud by customers and associates alike.

All of these losses contribute to shrinkage, whether internally caused or as a result of external factors. The term "shrinkage" in a retail environment refers to inventory loss. Inventory refers to the merchandise a store sells. Retailers keep inventory records so they can effectively manage all of the items within the store. Shrinkage is the difference between the value of the merchandise that a store's records depicts, commonly referred to as "book" inventory, versus what is actually present in the store. If a store's inventory records state that the store should have ten of a particular item, but there are only eight of those items, the store has experienced a loss, or shrinkage, of two of those items.

Shrinkage is calculated as both a dollar figure and as a percentage of sales. Most retailers calculate shrinkage at retail, not cost. This would be the actual selling price of the item to the consumer, not the landed cost of the item prior to establishing the retail price. Landed costs refer to the retailer's cost to get the item on the store shelf for sale. The difference between the cost and the retail sale price of the item is referred to as the margin. For example, an item with a landed cost of $100 that sells to the consumer for $130 would have a margin of 30%. In this example, if we are missing two of these items, our loss at retail would be $260.

When using the retail accounting method, retailers compare their shrinkage calculation to total retail sales based on a set time period. Industry averages for shrinkage, based on surveys of major US retail organizations typically range between 1.5 to 2% of total annual sales. For example, if a store is realizing annual sales of $1,000,000 and the store is experiencing a shrinkage rate of 2%, the store is losing $20,000 annually.

Formulating a solid shrinkage reduction plan starts with calculating shrinkage for each store. It is imperative that each store conduct an annual inventory of their products in order to establish the true level of shrinkage. You cannot rely on benchmarking data alone. Some stores calculate their shrinkage on an ongoing basis (cycle counts) and others do so based on a one-time annual inventory. A store may count certain high shrinkage stock keeping units (SKUs) on a daily basis. Whatever method is employed, there must be a firm understanding of the opportunities for loss in order to establish meaningful, effective controls to reduce those opportunities, and thus reduce losses.

According to the 2010 National Retail Security Survey, the estimated shrinkage of all participating retailers averaged 1.56% of total annual sales (at retail).[6] It should be noted that most retailers that report their shrinkage statistics have established retail loss prevention programs, which are designed to reduce the impact of shrinkage on that retailer's bottom line. Those retailers that do not have established programs to address the causes of shrinkage should expect their shrinkage averages to be significantly higher than those statistics reported in the National Retail Security Survey and similar surveys.

Why is it so important to manage retail inventory shrinkage? The net profit of most major retailers is fairly thin. For example, the North American Retail Hardware Association tracks bottom-line profits for the hardware/home center industry. On average, bottom-line profits fall in the range of roughly $0.03 for every $1.00 sold of merchandise within the hardware/home center industry. This has been such a generally accepted standard in this particular industry that the NRHA created training awareness literature coined "The Three Pennies of Profit" to engage store associates in the impact of shrinkage on store profits.[7] As stated above, in 2010, the average retail shrinkage for those retailers involved in the NRSS was 1.56% of sales. As one could surmise, controlling all losses to preserve bottom line profit is imperative. A sustained shrinkage of 3% could conceivably put a retailer out of business.

Supply Chain

When we think of retail, most of us think about our personal shopping experiences. This may include shopping for everyday items or perhaps shopping for electronics, appliances or more expensive items. Very few of us think of all of the effort and cost involved in getting the item we just purchased to the retailer. The process involved in getting the product to the store shelf is referred to as the supply chain. The supply chain represents raw product sourcing, where the product was manufactured, packaged, stored, shipped, delivered and then ultimately placed on a store shelf for sale to the consumer. Most major retail chains that effectively control their shrinkage look at the entire supply chain when considering an effective asset protection program.

It is myopic thinking, unless you have multiple third-party sourcing options, to view retail security as focused on physical retail locations only. When we talk about "retail security," our bigger picture focus will ultimately be on the entire supply chain. There are risks that can only be addressed by taking a much larger view of the retail supply chain. For example, if you are a retailer that sources product internationally, depending on your supply contracts, you may be responsible for a loss that occurs at the time of shipment, versus receipt at a physical company-controlled location. Shifting that loss responsibility back to your supplier or your transportation company through negotiations would be one way of transferring your loss risk. Additionally, if you are a product manufacturer or you engage in private label arrangements (third-party manufacturer of product with your brand name), issues of brand protection may become a risk for your organization.

It is recommended that retail security professionals whose responsibility is the protection of assets for their retail organizations "map out" their supply chain. A supply chain map is just what it sounds like: a world map that depicts the point of origin of raw products, end products, packaging, manufacturing, transportation methods, storage, etc., and that then finishes with the product being received at the retail location. Once mapped out, the chart should be methodically reviewed and the risks identified, and not only the risks to inventory, but all potential losses as identified by all those responsible in each stage, as the inventory moves through the supply chain. These risks should include risks to personnel and assets alike. Next, the risks should be ranked based on severity or impact to the organization, and likelihood of the occurrence. A well thought-out asset protection program will identify the risks to the retail organization so that a strategy can be crafted to address all risks to the retail organization through mitigation or appropriate response techniques.

The focus of this chapter will be to discuss the causes of shrinkage and the impact it has at retail.

Shoplifting

Retailing today demands that merchandise be prominently displayed and exposed so that customers can see it, touch it, pick it up, and examine it. Because the open display of merchandise is such a successful sales technique, there is little likelihood that merchandise will ever go back behind the counter. To combat the risk of loss associated with making the product more accessible, some stores have found ways to make it more difficult to steal openly displayed items. These techniques and procedures are discussed later in this chapter in the section titled "Preventive Measures." Displaying merchandise in this way is hardly a theft-proof practice, but theft must be controlled if profit margins are to be maintained. Shrinkage from shoplifting is not likely to be eliminated, but it can be reduced through a well thought-out and implemented program.

Extent of Shoplifting

Shoplifting is defined as "the taking or carrying away of the property of another, without the consent of the owner and with the intent to permanently deprive the owner of the property in whole or in part," and generally classified as a "larceny" in most states, punished as a misdemeanor or felony, depending on the dollar value of the item(s) taken.

It is difficult to accurately assess the full dimension of the shoplifting problem. Few shoplifters are apprehended, and of those who are, even fewer are referred to police. In fact, the most optimistic studies of shoplifting prevention effectiveness indicate that stores that do a good job apprehend no more than 1 out of every 35 shoplifters.

Information provided by one major retailer from its Midwestern region provided an excellent profile of shoplifting activity. The number of shoplifting apprehensions was consistent between April and September. Beginning in October, this number climbed until it reached a peak level in December. In contrast, January provided the lowest number of apprehensions, which increased gradually until April. This same Midwestern retailer reported shoplifting apprehensions by day of the week are consistent, with the exception of Saturdays, when apprehensions are approximately 50% higher than they are during the week. Sundays are the slow days. As far as time of day, most shoplifting apprehensions are made between 1:00 p.m. and 7:00 p.m., with the period between 4:00 p.m. and 7:00 p.m. contributing the greatest amount of activity.

According to Peter Berlin, founder of the National Association of Shoplifting Prevention (NASP), most people steal for one reason—to get something for nothing. Additionally, one out of every 11 Americans is a shoplifter, according to NASP, a fact that the National Retail Security Survey claims cost U.S businesses $12.1 billion in lost retail sales in 2010.[8]

Methods of Shoplifting

Shoplifting is conducted in every imaginable way; the ingenuity of the shoplifter is legendary. By and large, the great majority of thefts are simple and direct, involving nothing more sophisticated than putting the stolen merchandise into a handbag or a pocket. However, there are certain methods beyond the simple taking of items that are in general use and should be anticipated, including:

1. The "bloomer" technique—using large, baggy clothes such as bloomers or pantyhose that can be filled like shopping bags.
2. The clothing technique—slitting pockets in coats or jackets, hiding merchandise inside a coat or up a sleeve, sewing large cloth storage areas inside a coat or jacket, or sewing hooks to the lining of coats or jackets.
3. The fitting room technique—wearing stolen clothes under the thief's own clothing.
4. Hiding items in purses or umbrellas.
5. Palming—placing small items in the palm of the hand.
6. The bag technique—using shopping bags, sometimes even using the store's own bags.
7. The packaging technique—hiding items with other prepackaged items being purchased.
8. Wearing the item in plain view or walking out with large items.
9. Grabbing an item and running.
10. Booster boxes and cages—boxes designed with special spring lids in which merchandise can be concealed, or cages worn by a woman to make her appear pregnant.
11. Hiding items in books, newspapers, or magazines.
12. "Crotching"—a technique used whereby an item is held in place by the thighs under a full skirt, dress, robe, or long garment.
13. Price changing, switching the price tags or bar codes on items. One person may switch the tag or alter the bar code on an item, leave the store and the other person buys the item; or it is done by the same person. (Note: This activity can be quite sophisticated, utilizing printers, scanners and other computer equipment to create bar codes and other pricing media to accommodate their thefts.)

Common Red Flags Displayed by Shoplifters

1. Continuously looking around to see where associates may be, commonly referred to as displaying "furtive movements."
2. Acting nervous for no apparent reason.
3. Lingering in an area, but never really appearing to be looking at merchandise.
4. Group enters the store together and then splits up, with some members engaging sales associates, while the other members steal.
5. Decline assistance from sales associates.
6. Clothing appears inappropriate for the weather.
7. Carrying large boxes, bags, purses, backpacks, etc.
8. Picks up two items and returns only one to the shelf.
9. Repeatedly handles the same item over and over, putting it back and picking it up again.
10. Enters the store right when the store opens or just before the store closes to take advantage of unprepared or reduced associate staffing.

According to research by Commercial Services Systems, Inc., the techniques used vary with individual preference and type of establishment. For example, purses are most commonly used for concealment of merchandise in supermarkets, followed by the use of pockets and carrying items under clothing. It is estimated that the purse is used 26–33% of the time, pockets 17–30%, and clothing 12–24% of the time.[9]

Who Shoplifts

Types of shoplifters can be broken down into two distinct groups: professionals and amateurs. The amateur group can be further broken down into opportunists, the impoverished, substance abusers, thrill seekers and kleptomaniacs. Each is described below.

Professionals

These are individuals or groups of individuals who are very organized in their approach. They make their living from the sale of shoplifted merchandise. Professionals will use devices to maximize the amount of merchandise that they can steal at one time. For example, some will go as far as to push a baby stroller (either containing a real baby, a doll, or a padded area to give the appearance of a baby) and use portions of the baby stroller to conceal merchandise. Still others have clothing modified that will hold large amounts of concealed merchandise. Others will use special bags that are foil lined, place the merchandise into the foil-lined bag and then exit the store. These specially constructed bags are used by the shoplifter to defeat electronic security devices that are placed on the merchandise they are stealing. When attached devices are not removed or deactivated, they will cause an alarm when they leave a designated area. The foil-lined bags prevent the security devices at store exits from working properly, thus allowing the shoplifter to remove the item(s) from the store without audible detection.

Organized Retail Crime

Falling into the category of professionals are Organized Retail Crime (ORC) groups (see Figure 15-1). ORC refers to professional shoplifting, cargo theft, retail crime rings and other organized crime occurring in retail environments. Two defining characteristics of ORC involve premeditation of the crime by two or more assailants. Additionally, the items stolen are intended for resale, trade, or other value, and not for use by the individual who steals them. Federal and state legislation involving steeper penalties and lower thresholds for felony classification has been passed and is continually reviewed for needed changes. On November 2010, the House of Representatives passed H.R. 5932, the Organized Retail Theft Investigation and Prosecution Act of 2010. Again, varying by state, lobbyists for stricter regulation against ORC pushed hard to increase the awareness and legal support to track and punish offenders.

These thieves typically work in teams and travel from state to state stealing select product. These teams will typically target specific retailers and literally clean them out of predetermined products. They tend to stay in motel rooms and will fill the motel rooms with the product they have stolen. They will move from city to city, not staying more than a week or two before moving along. In many cases, they have an elaborate distribution network and have been known to either ship the stolen merchandise overseas for resale or have established connections to re-introduce the stolen product back into the supply chain through either legitimate or illegitimate sources. Product stolen from one retailer may end up on the shelf of another retailer the next day.

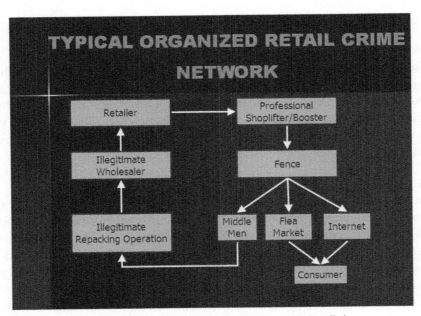

FIGURE 15-1 Typical Organized Retail Crime network. *(Courtesy of Theresa L. Tapella.)*

Opportunist

The opportunist is an individual who takes advantage of shoplifting situations as they become available, thereby seizing the opportunity. This individual will steal if the opportunity presents itself and the risk of getting caught is perceived to be low. Most retailers will come to find that this type of shoplifter may also be one of their best customers. Opportunists come from all walks of life. Opportunists usually have the means to pay for the item(s) they steal. Income levels for this type of shoplifter range from low to extremely wealthy. When this type of shoplifter is caught they show guilt, express remorse, are embarrassed, will offer to pay for the item and will often become very emotional.

Impoverished

The impoverished are individuals who steal because they need the particular item they are stealing. These individuals are usually very poor and may be homeless. They are easy to spot because of their unkempt physical appearance and lack of hygiene. They are not very cautious in their approach and generally act very suspicious. However, this type of shoplifter can be dangerous when caught. They tend to be boisterous and will attempt to flee.

Substance Abuser

Individuals who steal to support their addiction to alcohol and/or drugs are substance abusers. The substance abuser will steal items that can be easily sold or traded to support his or her addiction. This type of shoplifter can be extremely dangerous depending on the type of dependency and is considered one of the most dangerous shoplifters when caught, as he or she may also be in possession of drugs or drug paraphernalia.

Thrill Seeker

The thrill seeker is an individual who steals for the emotional "high" associated with the crime. Many juvenile shoplifters fall into the thrill seeker category. They steal to show off to their peer group. However, as time progresses, the thrill seeker will take bigger risks and begin to perfect his or her shoplifting techniques as he or she continues to steal. The thrill seeker will tend to steal until caught. Once caught, much like the opportunist, thrill seekers will show remorse, shame and regret.

Kleptomaniac

A kleptomaniac is an individual who will steal just about anything from anywhere. The sole purpose of a kleptomaniac is to fulfill the desire to get something for nothing. This individual is slightly associated with the thrill seeker; however the kleptomaniac doesn't usually need the item he or she is stealing. This type of shoplifter will steal items that don't seem to be typical stolen items. When caught, they will attempt to talk their way out of being turned over to law enforcement. They will claim that they don't remember taking the item, don't need the item, unintentionally concealed the item, and in extreme cases will produce a doctor's note explaining that they have a medical problem, and that is why they took the item.

Prevention and Response

The best deterrent to shoplifting is good customer service. Individuals attempting to shoplift do not want to be engaged with a retailer's sales staff. A sales staff that offers a greeting and assistance will deter many amateur shoplifters.

Training sales associates on how to respond when they believe they are encountering a shoplifter or have witnessed a shoplifting situation is critical. The principles that are taught to sales associates are fairly consistent. You must observe the shoplifter approach the product, take the product, conceal or carry away the product; you must maintain continuous observation of the shoplifter, observe the shoplifter fail to pay for the item and depart the store. If a sales associate cannot attest to all of the above in a positive manner, then it is recommended that associates not approach, attempt to detain or accuse the individual of shoplifting.

While good customer service is the number one deterrent to shoplifting, the second is how the store is designed, how the aisles are configured, and where and how your product is displayed.

It is essential that retail stores limit the number of uncontrolled exit points. A controlled exit point is one that is staffed or at least monitored by store personnel. There should be a point of sale terminal positioned in close proximity to all customer exits. Aisles should be arranged to allow the sales staff to easily move about the store and observe customer activity.

Placing high pilferage prone merchandise in plain sight and in full view of check-out areas is a very effective approach to reducing the loss of such items. Additionally, keeping merchandise "fronted" on the shelves and peg hooks filled will allow your sales associates to quickly observe when merchandise is missing. "Fronting" is defined as keeping merchandise close to the front portion of the display shelf, which gives the appearance of a well-stocked store.

Securing a product in a secure area with limited access, or locking the product up and retrieving the item for the customer are two effective ways to limit shoplifting losses of theft-prone items. However, locking up merchandise on display may result in reduced sales of that item too. This is a balancing act that each retailer will need to experiment with and is a matter of individual risk tolerance. If it is decided to place theft-prone merchandise in a locked display case, there should be a "lock it up – walk it up" policy; that is, a sales associate must retrieve the secured item and carry it to the point of sale. The premise is that if an item is theft prone to the extent that it must be displayed in a locked display case, it should receive that same degree of protection to the point of sale.

Technology can greatly assist the retailer in shoplifting prevention. One of the biggest deterrents to shoplifting is the placement of an anti-shoplifting tag on the theft-prone items sold in the store. Tags are either placed inside the product packaging (most desired) or affixed to the product's surface or package. The tags will interact with a device, usually a pedestal, at the exit point if they are not properly deactivated. Deactivation will typically take place at the point of sale area at the time of payment.

The use of a closed circuit television system (cameras) can also assist retailers in shoplifting prevention. Placement of cameras should be well planned. Entry and exit points should be covered. Areas that cannot be easily monitored by sales associates, commonly referred to as "blind spots," should also be monitored. Shoplifters will leave clues to where your blind spots are, in the form of empty product packaging. The packaging is removed to make concealment easier. If the store has a security staff that monitors the camera system on a real-time basis, it will be the primary detection tool when used in conjunction with store detectives on the floor. In smaller businesses, the camera system serves primarily as a deterrent, not a detection tool. It does provide, in both scenarios, the capability to review video surveillance to either confirm or refute a suspicion. It can also be used as a training tool for sales associates, as well as a means to identify and document individuals who stole, but were not detained, so that the next time the individual enters the store and is seen, management and/or the security staff may be alerted. Lastly, it also serves as irrefutable evidence in a criminal trial.

Detaining Shoplifters

Approaching and detaining shoplifters takes a considerable amount of training. It is up to each individual store to determine protocol related to how they intend to handle confronting shoplifters. The number one concern to all retailers should be the safety of their associates. While laws vary from state to state, it is generally agreed that shoplifters can be detained at the point where they have physically gone beyond the last point of payment. Sometimes states clearly define this location (which may, in fact, be at any point outside the department where the item is displayed) and other times it is left to interpretation. It is up to each retailer to know what the rules are for each state in which they are doing business.

When detaining a shoplifter, the sales associate should identify him or herself as an associate of the store. They should immediately identify where the stolen item is located and ask the shoplifter to produce the item and then take possession of it. The sales associate should then escort the shoplifter to an area away from other customers and contact local law enforcement for assistance. While there are many more detailed steps in this process, the above represents a general overview of how most retailers handle shoplifting detentions. Some retailers have set a dollar threshold at which they will contact law enforcement. Still other retailers hire and train retail security personnel, commonly referred to as "store detectives" to handle shoplifting detentions exclusively. The mishandling of a shoplifting detention can have legal repercussions, and it is therefore extremely important to have a policy in place that is in line with local shoplifting laws.

Internal Theft

Of all the different causes of shrinkage, theft by a trusted associate is the most insidious. It is the definitive act that reveals an associate's acute displeasure with the company, supervisor or manager, or even fellow associates. If an associate respects his/her employer, supervisors and managers, and enjoys being part of a cohesive team that receives proper recognition for work accomplished, that associate is far less apt to steal from his or her employer.

Common Associate Theft Justifications

- The associate is living beyond his/her means.
- The associate is not able to manage his or her finances and maintain a sensible budget.
- The associate desires more lavish clothing, more food, higher living conditions, etc., for dependents than he or she can afford, and believes that theft is justified to obtain this improved standard of living.
- The associate has no self-control and steals on impulse.
- The associate was raised by parents who did not endorse honesty and may have, themselves, committed a variety of dishonest acts, thus sending the message that stealing is acceptable.
- A family illness has created a need for money that the associate feels he or she cannot satisfy honestly.
- The associate believes that he or she has not been treated fairly or given proper credit for work accomplished.
- The associate feels that his or her wage is too low.
- The associate believes that the company expects losses and doesn't really care as long as it doesn't get out of hand.
- The associate is frustrated or dissatisfied about something in his or her personal life and theft from the store is a way of diverting attention away from addressing the personal problem.
- A supervisor has criticized the associate and the theft is justified as a way to "get even."
- The associate believes that one or more other associates get better treatment and/or recognition.
- The associate knows of co-workers who are stealing and getting away with it. They say to themselves, "everyone else is stealing, why not me?"
- The associate knows of co-workers who have stolen, been caught and while being fired, were not prosecuted.
- The store's internal controls are insufficient and or not enforced; therefore "management must not care."

The above justifications/excuses are but a few of the many that associate-thieves use to rationalize why it is justifiable in their mind to steal from their employer. And while the employer has little or no control over what goes through their associates' minds, the one thing an employer does have control over is reducing the opportunities for losses to occur.

If internal controls are insufficient on their face, or are adequate but not enforced, it is not a matter of whether retail associates will steal, but rather how many will steal and how much they will steal. Theft by an associate typically starts with small items or small amounts of cash, and will grow steadily as the associate becomes emboldened or increasingly dependent and stops only when caught or the employer goes out of business.

Associate theft is everything from a single dishonest act by one associate to an organized scheme that involves a single or multiple associates, and multiple thefts of cash and/or merchandise. There are numerous annual and ad hoc studies conducted and reported each year that speak to the magnitude of associate theft; and while specific statistics within those studies vary, it is a generally accepted premise that associate theft accounts for almost half of a retailer's overall annual shrinkage. Other studies indicate that as many as one out of three retail associates have stolen, are stealing, or will steal from their employer.

Prevention and Response

While there are widely accepted human, mechanical, electronic and procedural controls that should be implemented to effectively deter (and detect) associate theft, it is similarly acknowledged that those retailers that are successful in developing a "culture of honesty" among their associates have a leg up on this annual multi-billion dollar problem. That culture manifests itself in a workforce that truly cares about the success of the business, with all associates committed to managing the various aspects of the operation to ensure its success. Shrinkage control then becomes one more aspect of the retail operation that gets equal attention. So just what is their secret?

Developing and maintaining a culture of honesty permeates the attitudes and daily work habits of associates. It pays benefits that can be measured by the degree of associate job satisfaction and demonstrated work ethic, both of which have a direct impact on customer service and sales. The added benefit is reduced shrinkage and as a result, enhanced bottom-line profits.

Creating this culture of honesty requires a commitment on the part of the corporate staff and managers at all levels and, ultimately, all associates. For additional information on internal security and honesty see Chapter 12.

Methods of Internal Theft

The following are the leading types of thefts by retail associates, along with summaries of the internal control processes that must be in place to either prevent or deter them. However, it must be noted that this information is not all-inclusive; unique operations and or duties, as well as the continual evolution of retail environments and systems, require that the store manager formulate internal controls to be both job and store specific.

Cash

No matter what type of desirable merchandise may be available in a retail environment, there is nothing as tempting for an associate to steal as cash (see Figure 15-2). It is, therefore, necessary to hold associates who have access to cash to a higher standard than, say, a sales floor or stockroom associate, and to have enhanced controls in place to safeguard against the theft of cash. Cash losses at the point of sale (POS) can quickly grow to large amounts when appropriate controls are not in place.

A basic POS control that should be in place is the "one cashier per drawer" policy. When more than one cashier operates out of the same till, it creates a situation in which all of those cashiers fall under suspicion when an unexplained shortage occurs and you lose individual accountability. Even if each cashier has a unique system identification and password to sign

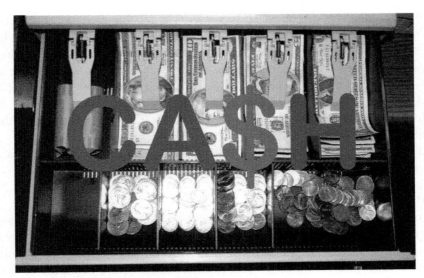

FIGURE 15-2 Cash drawer. *(Courtesy of William H. Cafferty.)*

on to the system before ringing a sale, there is still the dilemma of who among the multitude of assigned cashiers made the honest mistake or stole the missing money. Let's say, for example, a retail location has three registers, of which two are assigned to designated cashiers. No one other than those associates should be allowed to ring on their respective terminals. The third terminal can be designated the "overflow" or "manager" terminal, on which more than one cashier, associate or manager rings, but only when both of the assigned terminals are busy or unavailable. When an associate on an assigned terminal goes on break, no one else should be allowed to ring on that terminal. In this process, the vast majority of sales will occur on the two assigned terminals and if an error or unexplained loss occurs at either of those terminals, the manager will have increased accountability for follow up. This policy will provide an excellent deterrent to theft, as a dishonest cashier will know that suspicion cannot be spread out among the entire store.

It is also important that POS terminals be counted down and the cash and sales reconciled at the end of the assigned cashier's shift. If there is a delay and the sales are not reconciled until the following morning and there is missing cash, it could be days before there is a chance to talk face-to-face with the cashier responsible. Counting down refers to tallying up of all of the media in the drawer. Media is defined as the various methods customers use to pay for merchandise (cash, checks, credit cards, debit cards, in-store credit, house accounts, etc.). All of these media should be totaled and placed in a sealable cash bag. Normally the cash drawer will be counted down to the starting change fund amount set by the store, usually $200 to $500, depending on the type of retail business.

Control of cashier (and manager) passwords is also essential. If a dishonest cashier knows another cashier's password, it is easy to log onto a POS terminal using that cashier's password and hide the fact that the dishonest cashier was ever on that terminal. Cashiers must be advised to protect their password and to change it when it is suspected that it has been compromised. This is even more important for managers. Managers should always take precautions against being "shoulder surfed" (cashier observes which keys the manager presses to enter his/her password) by putting his or her body between the cashier and the POS keyboard. Creation of scan cards for use by managers is preferable to entering passwords; this is also advisable for cashiers, if possible. If scan cards are used, it is, of course, important that they be protected at all times.

Following are the major POS associate embezzlement schemes, along with the minimum controls that must be in place to deter them.

Fraudulent Refunds

This is the most often used method of theft of cash by dishonest associates. A cashier will simply scan or ring in a nonexistent item of merchandise for a refund, which creates a corresponding cash overage in the till. The cashier will then remove that overage when the opportunity presents itself. As long as the dishonest associate does not remove any more or less than the overage in the till, that register will not be "over" or "short" when counted down at the end of the cashier's shift. Store managers who use the daily over and short total as their only indicator of POS issues will not suspect that thefts are occurring. There are a number of controls and procedures that will provide the retailer with both an effective deterrent to this type of embezzlement, as well as an enhanced potential for detection when a dishonest cashier attempts to circumvent the controls.

First, a decision should be made as to the level of independent authority a cashier has to process a refund. The retailer may want to set a maximum dollar amount that a cashier can ring, above which a manager override must occur in order for the transaction to proceed. Most modern day POS systems provide this capability via a security bit or option that can be set to halt the transaction when the dollar value of the refund exceeds the set amount, and allow it to proceed only after a manager override is performed. In such cases, it is further recommended that the approving manager be required to actually initial the hard copy refund document. Should a cashier learn a manager's code, gain access to a manager's swipe card or key, he or she must now also forge the manager's initials, which adds another layer of protection to deter this type of theft.

Second, if cash refunds are given when the customer does not have proof of purchase (a receipt), a decision should be made as to whether cashiers have the independent authority to process such refunds. It is important that the back office associate whose job it is to reconcile the previous day's POS activity has a prescribed standard to confirm that all procedures were followed. In cases where the customer does not have proof of purchase, the chance of the refund being fraudulent (perpetrated either by the customer, the associate, or the customer and associate in collusion) is much greater. If a manager must approve all such refunds, that greatly reduces the opportunity for wrongdoing. Again, the manager should initial the refund document. If a store credit or gift card is given in lieu of cash when the customer does not have proof of purchase, it is important that a policy be in effect that the subsequent purchase with that store credit or gift card be recorded as such so that the customer cannot return with that particular receipt later and obtain a cash refund.

The disposition of the refunded item—returning it to stock, returning damaged product to the vendor for credit, or other disposition—must involve an associate other than the cashier who processed the actual refund. Further, there should be some sort of documentation that provides an audit trail of this separation of duties. The goal is to ensure that there actually is a refunded item and that someone other than the cashier who rang the refund transaction verifies, through initials on the refund document, that the item was, in fact, present and retrieved from the cashier's location. These product pickups should occur no less frequently than daily, ideally before the end of the cashier's shift.

Fraudulent Voids

If not properly controlled, fraudulent voids can also be used by a dishonest cashier to embezzle cash. Keep in mind that there are two types of voids: those that occur before the transaction is completed and those that occur after the transaction is completed. In both cases, if the cashier is allowed to independently process the void, it creates an overage in the drawer and, like the bogus refund, if only the exact amount of the void is removed by the dishonest cashier, the final cash count will equal what the terminal has tabulated; thus, there will be no "over" or "short" and a manager who does not look further into why voids occurred will not be the wiser to the losses taking place.

In a pre-transaction void, a dishonest cashier will not complete the transaction and therefore the customer will not be provided a receipt. Typically these types of transactions are done when a cashier knows that a customer is going to pay cash and would not likely ask for a receipt. The POS system calculates the total of the sale on the register, but then the cashier will void out the transaction, advise the customer of the total and open the drawer by ringing a "No Sale" or through another means, take the customer's money and make change. Once the overage is created in the drawer, the cashier will remove the cash when the opportunity presents itself. If the customer comments that he or she would like a receipt, the cashier will state that the register has malfunctioned and the transaction will have to be re-rung. Signs to look for at the point of sale are evidence of any "system" used to keep track of the dollar value of voids initiated. This might be scrap pieces of paper you find in the garbage with dollar amounts written on them, stacked coins on the outside of the register that represent certain amounts, a personal calculator that doesn't belong to the store. Additionally some cashiers will set the money aside in the register or face the bills in a different direction and at the end of their shift take all paper money faced in their direction.

On some of the older registers, the cash drawer can be left open and the register will still function properly, thus making the retrieval of the cash much easier. Newer registers will not work when the cash drawer is open, but only when configured that way. In order to retrieve the cash, the cashier can take the cash in front of other customers, take the cash during another transaction or ring a "no sale" transaction to open the drawer. The "No Sale" key is typically used to make change for customers or to remove excessive cash in the drawer. Managers should track both the cashier voids and the number of no sale transactions by cashiers.

If a cashier voids a transaction after it is completed, depending on the features of the cash register or POS terminal, it may be classified as a transaction void, a reverse transaction, or even as a refund. In any case, this type of transaction should not be allowed as an independent action by a cashier. There should always be a manager approval process, whether it is via an electronic manager override or a manual approval.

Most POS systems have electronic security features that allow the retailer to enable a manager override requirement that will not halt the post transaction void until a manager inputs his or her code. It is strongly recommended that retailers enable this feature on their systems. If the registers in use do not have this feature, it should, nevertheless, be a requirement that a manager be called to approve post-transaction voids. In both pre- and post-transaction voids, the cashier and approving manager should initial the void receipt, and the cashier should indicate the reason for the void on the receipt, which is then turned in for reconciliation by the back office.

Fraudulent Suspensions

If the POS system includes a transaction suspension feature, which means placing a transaction in a "hold" status, there is a potential for a dishonest cashier to use that to embezzle cash as well. The scenario would be similar to the fraudulent use of the "void" function, as previously described. It is used to give the appearance to the customer of a legitimate transaction. Cash is removed in the same fashion as previously described.

The defense against this type of fraud is to not allow cashiers to independently suspend a transaction. Further, there should be a regimen in which suspended transactions are reviewed and purged from the system no less frequently than weekly by management, with follow-up on all questionable suspensions. Lastly, there should be a periodic review of sales transactions to look for "no sale" rings or small purchase rings immediately following a suspended transaction. This will be done so the dishonest cashier can access the drawer to remove the cash.

"Sweethearting"

"Sweethearting" is the dishonest act of a cashier that allows a friend or relative to pass through the checkout and pay less than the actual value (or nothing at all) of the merchandise they may appear to be purchasing. The cashier can simply bag items without scanning them; provide an unauthorized discount; pull them around the counter scanner instead of over it; or even use a hidden barcode for an inexpensive item that has been taped to the bottom of his or her hand. The item is typically turned upside down so that the item's barcode is not scanned, but the one on the palm of the hand is. Unfortunately, there is no POS security bit or other single feature that will deter or control this type of crime; however, the typical POS system does have controls that address discounting. Cashiers can be restricted to applying just the minimum discount without a manager override (approval). Additionally, the typical POS system will provide a report of all items sold below margin, which identifies all such sales, to include the cashiers who processed them.

One of the best technological controls to combat sweethearting schemes is the closed circuit television-point of sale (CCTV-POS) interface. Retailers are now mounting cameras over the check-out areas that focus in on the scanning area and the customer. The camera is interfaced to the register and the data that is entered into the register is overlaid or displayed to the side of the video image in the form of the store's receipt tape. In our example above, if a bar code doesn't belong to the product scanned, it will show on the register overlay. Or, if the total amount on the overlay looks to be far less than the number and value of products being scanned, this is an obvious indicator of wrongdoing.

It is important that retail stores have a strong policy against cashiers ringing sales for relatives and close personal friends, as it is these two categories of customers for whom the vast majority of sweethearted sales are rung. It is thus recommended that cashiers be prohibited from ringing sales to relatives and close personal friends. It is understood that management does not know who the relatives and close personal friends are of each cashier, but usually other retail associates do know who they are. If it is discovered that a cashier has violated that rule, administrative action should be taken against that cashier.

Theft from the Safe

This type of cash theft is typically the act of a very desperate person, usually a manager or supervisor. The perpetrator is usually in dire need of cash and doesn't have the knowledge or opportunity to steal the cash through fraudulent POS transactions or other more sophisticated schemes. There are a number of basic controls that should always be in place to reduce the risk of theft from the safe.

Controls to Reduce the Risk of Safe Theft

1. Limit the number of persons who are given the combination to the safe to the absolute minimum, consistent with operational requirements.
2. Advise combination holders that they are not to record the combination on calendar pads, within desk drawers, or other "secret" locations within the office.
3. Require the safe door to be closed and locked when not in actual use.
4. Prohibit leaving the safe on "day-lock," which is turning the combination dial back to the last number in combination, thus locking the door, but also allowing the door to be re-opened by simply turning the dial back to "0." This cannot be done with the newer model electronic accessed safes.

Safes equipped solely with the traditional dial-type combination lock have many inherent weaknesses. Safes equipped with electronic locks, on the other hand, provide a greater degree of security and, depending on the level of sophistication, provide various features such as (1) the capability of assigning a different combination to each authorized associate (then you can simply delete an associate's combination when he/she terminates or is no longer authorized); (2) creating an audit trail that shows when each authorized associate entered the safe; and (3) setting "time windows" that restrict when an associate's assigned combination will work. Advanced electronic safe locks use both combination entry keypads and electronic "keys" that can store and transmit electronic data. These are the ultimate in safe entry devices for the typical retail store safe.

Another aspect that must be considered is the type of safe that is best suited for the business. A single door safe that allows access to everything stored within it provides the least effective control, as you must store everything within, but persons with access may not have the need to access certain items or cash stored therein. Multi-door safes, on the other hand, such as the "depository" safe, have two doors (or a separate lockable compartment inside the safe), and provide a means to "drop" deposit bags into the safe without opening it. The other compartment can be used to store tills. Thus the opening and closing managers can be given access to the tills in the lower compartment, but not to the money bags containing sales in the upper compartment. A small change fund can also be maintained in the lower compartment with the tills for use when office personnel are not present.

The sophistication does not end with electronic locks and double door safes. Convenience stores and other retail establishments in need of the ultimate in physical security, robbery deterrence and associate access control are typically equipped with what is known as a "cash dispensing safe." These safes have sophisticated electronic access controls, and can be configured to dispense small amounts of change in currency and coins, provide a time delay between transactions, record deposits and keep an accurate count on the day's transactions. Additionally, this type of safe provides a variety of reports for use in cash accountability.

Last, but not least, it is advisable to have one or more CCTV cameras covering the safe. One should be positioned to provide a frontal view of the associate accessing the safe or, at the very minimum, looks directly down on the associate and the safe. If the camera is positioned so that all you see is the back of the associate, you will not see his or her hands and therefore will not have the capability to see exactly what was removed from the safe or where it was placed on the associate's person.

Merchandise

As with the deterrence of cash thefts, it is equally necessary to have effective internal controls in place to deter the theft of merchandise by associates during their performance of duties. While the modus operandi varies, depending on exactly what the associate is doing (e.g., receiving, stocking, operating a POS terminal, trash removal, home delivery, etc.), there is normally an effective means available to control the opportunities to steal; it is just a matter of implementing and enforcing the controls.

A common denominator in a vast majority of associate merchandise theft scenarios is an uncontrolled back or side door. As with other aspects of the retail security business, there is an axiom about back doors: "If you don't control your back door, it's not a matter of if you will lose merchandise through it, but rather how much merchandise you will lose through it."

Shoplifters steal merchandise through front doors, but theft by associates is, for the most part, through back doors. Shoplifters must contend with the chance of being observed by sales associates, cashiers and honest customers, but dishonest associates who remove merchandise through a back door can take their time and wait for the opportunity to exit when they know there is no one watching. And while we use the term "back door," it applies to any door at the back or side of a store that is an associate-only door.

The stockroom normally contains an overhead door, often with a personnel door adjacent to it. It is not uncommon for a retailer to believe that the receiving door must remain unsecured because it is used often throughout the day. If so, it is imperative that someone in a supervisory or managerial role be in the area at all times the door is unsecured. Ideally, of course, the receiving door should be closed and locked when not in actual use, with the keys to the lock controlled by management. If the receiving door must be left open for ventilation reasons, it is recommended a security screen be installed.

The personnel door is, in many cases, a designated fire exit. As such, NFPA, Life Safety and local codes will usually demand that it remain unlocked while the building is occupied. This door should thus be equipped with a stand-alone exit alarm or an alarm-equipped panic bar. An exit alarm is a device that will sound an alarm when the door is opened, unless it has been deactivated prior to opening the door. There are also various brands of emergency exit door panic bars that come equipped with a built-in key-controlled exit alarm. If not deactivated by a key holder, it will alarm when the door is opened and will continue to sound until someone with a key resets it. Stand-alone exit alarms come in a variety of configurations, including those that are battery operated. The key to the exit alarm should, of course, be held only by management personnel. It is understood that keeping back doors secured will not be as convenient as leaving them unlocked. Unfortunately, convenience and good controls do not normally share the same space.

Associate parking should never be allowed near the receiving door or any back (or side) door that is not closely controlled. Associates should be required to park their cars as far away from those doors as possible. The reason for this is obvious. Uncontrolled back doors and close-by parking of associates' cars is a recipe for losses.

Removal of empty boxes through back doors is the number one method of product theft by associates if not properly controlled. Again, another simple rule applies. Management should require and enforce the flattening of all empty boxes before they are removed from the retail store. The ideal solution is to install a compactor in the stockroom and require its use for all empty boxes.

Trash removal can be a dirty business. It is a chore that no one likes, and is usually assigned to a junior associate, often a part-time person. While it is, indeed, a dirty business, it is also a prime candidate for use by a dishonest associate bent on stealing merchandise. It is very important that control of this daily chore not be neglected.

First, while the procedure may be to collect trash just prior to closing, trash bags should not be removed from the store until the following morning. There are two reasons for this. First, allowing trash to be carried out during hours of darkness increases the opportunity for associate theft; second, it also increases the opportunity for robbery. It is not uncommon for businesses to be robbed by someone who waits out back at closing time, accosts an associate who is taking trash to the dumpster and then forces that associate, at gunpoint, back into the store to steal the day's cash.

Second, clear trash bags should be used, both in trash receptacles and for the collection of trash. This increases the risk of detection for an associate who is thinking of placing merchandise in the trash bag for removal to the dumpster.

Third, the removal of trash should always be witnessed by a member of management. This would be a natural occurrence if the back doors are secured and the keys to door locks or alarms are kept by management personnel only.

And fourth, if possible, the Dumpster should be locked, again with the key retained by management personnel only. Most Dumpsters come configured so that they can be locked, and the refuse company may even supply a padlock to which their drivers have the key. Or they may just require that it be open at the times they come to empty it. Locking the Dumpster removes the incentive for an associate to put stolen merchandise in it for later pick-up, and it also prevents others from using the store's Dumpster.

If a back or side door is the designated associate entrance or exit, management should evaluate the feasibility of using the front door instead—that is, unless there is a security officer or loss prevention associate continually monitoring the associate entrance or exit. With a back or side door as the designated associate entrance or exit, it must necessarily remain unlocked, and associate vehicle parking is typically close by—again, a recipe for losses. Another associated risk is that this door is frequently left unsecured early in the morning for the convenience of arriving associates. Needless to say, no door should be left unlocked and uncontrolled prior to opening, as this presents an excellent opportunity for robbery. One solution is to install a cipher lock (or card reader) on the outside of the door for associate use. The cipher lock would be an additional lock for use when the deadbolt is unlocked, but would provide a means to prevent unauthorized persons from entering. The combination to that lock would have to be changed, of course, whenever someone with the combination terminates employment.

Closed circuit television plays an important role in reducing the threat of theft of merchandise through the back or side door. Strategically placed cameras enhance a store's theft deterrence efforts. Cameras should be placed on the wall immediately above the interior door frame or in the ceiling near the door at a location that covers the path to the door. It is important that the camera is able to record the face of the associate and anything that may be in the associate's hands; not the backside of the person as he or she is walking out. Also, by positioning the camera so that it points away from the door, there is less chance that high light levels from outside sunlight will impact the image. It is also recommended that at least one exterior, environmentally housed camera be mounted on the perimeter wall to observe the overhead and exit doors, the Dumpster and any associate parking areas at the back/side of the store.

Investigations

Most major retailers assign investigative responsibility for losses involving professional shoplifting rings, suspected ORC situations and internal and external loss scenarios to specialized personnel in their corporate office or field offices who are hired and trained to handle these types of issues. Investigators will typically be assigned auditing and training responsibility as well for a set number of stores. Retail loss-prevention investigators will typically work closely with law enforcement on the majority of these types of investigations. Smaller retailers that do not have their own investigators will hire third-party investigators familiar with retail operations and theft scenarios.

The investigator will first determine if a policy violation has occurred. Additionally, if theft is suspected, the investigator may gather more information to determine the extent of the theft or thefts. This could include interviewing other store associates, securing the appropriate documentation or perhaps relocating an existing camera or installing a hidden camera in the store after hours to document the extent of the loss. It is up to the investigator to prove or disprove the suspicion or allegation, determine the extent of the problem and work with local store management to resolve the issue.

The final phase of most retail investigations is interviewing all associates who may have involvement in the loss. This will usually be coordinated with local management, unless local management is involved. Retail loss-prevention investigators have very specialized training in the area of interviewing.

When good evidence exists that an associate has been involved in a theft, the investigator will interview the associate in a private setting, which may or may not include a silent witness. The ultimate goal of the interview is to determine the total extent of an associate's involvement, whether it is cash or merchandise, and whether there were any other associates involved. The first time an investigator becomes aware of a loss is generally not the first time the suspected associate has stolen something.

A secondary goal will be to determine how the associate inflicted the loss on the store. The investigator will want to determine how the associate was able to circumvent controls in order to inflict the loss. This will help the investigator in tightening up the operation to prevent losses from occurring in the future.

Third, is the associate aware of other associates who may be stealing? The investigator will want to determine if there are others and determine if collusion may be involved. Gathering this information may prove to uncover additional losses that the store must address.

Fourth, the investigator will ask the associate to voluntarily write an admission of the theft. The admission is typically written in the associate's own handwriting (if feasible). The statement will sum up the losses, describe how the associate was able to inflict the losses and should tie into specific evidence the investigator has obtained. For example, a cashier who may have been stealing by creating fraudulent refunds should initial and date all the fraudulent refunds and make reference to these same refunds in the written statement. The investigator should encourage the associate to include in the statement an acknowledgment that the associate knew what he or she was doing violated company policy. If a witness is in the room, the witness will then sign the associate's written admission as a witness. A copy should be provided to the associate and the original kept by the investigator.

Lastly, the investigator will want to recover any cash, merchandise or company property the associate indicates he or she has in his or her possession or at another location.

At this point the matter is usually turned over to the retailer's human resource group or the local store manager to determine disposition regarding the future employment of the associate. Many retailers decide to buy time to make a decision by placing the associate on "administrative leave pending further investigation." The benefit of such may be to finish the investigation, consult with legal counsel and or local law enforcement. With a consistent approach to investigations, the retailer now has recourse in civil court and also has documentation to support criminal action, if deemed appropriate, by the applicable criminal jurisdiction.

External Threats

Although the majority of losses at retail stem from internal vulnerabilities, external threats, or those things that originate or are caused by forces outside of the store, can impact the location's income, integrity, and safety. While external threats can never be fully mitigated, understanding the threats and preparing to respond will decrease the overall impact.

Burglary

Burglary constitutes any kind of illegal entry to a structure or property, forcibly or not, with the intent to commit a felony or theft; the exception to this is entry into a vehicle, which is larceny. Varying degrees of burglary have been defined based on the types of entry, but differ by state. Statistics available from the FBI Uniform Crime Reports indicate that "property crimes," of which burglary is included, dropped 2.8% on a year-by-year comparison (2010 versus 2009).[10] Retailers can reduce the likelihood of being victimized by this type of crime through physical security measures and electronic means, which would include the installation of a burglar alarm system. These systems can be basic or very sophisticated and are dependent upon local law enforcement to respond in a timely manner for appropriate apprehension. The level of physical security and systems will be driven by the geographical area in which the retail location is domiciled and the type of merchandise being protected. Data is readily available, which will assist the retail security executive in making these types of protection decisions.

Robbery

The defining characteristic that differentiates robbery from burglary is that robbery involves an individual or individuals attempting to steal another's property by seizing the property by force or violence, directly from the victim. Burglary may result in the deprivation of the same property, but not taken directly from a person, or by using force or violence toward a person. The degrees of robbery differ between states, as does the specific definition.

While this action can never be fully prevented, a properly evaluated and protected retail establishment can reduce their risk of falling victim to a robbery as well as decrease the risk of physical harm during a robbery. Not all precautions available are right for every kind of location. Therefore it is important to consider multiple factors before implementing any deterrent strategies. Factors such as actual vs. perceived threat, region, and product vulnerability, past incidents, opportunity, and federal regulations should all be weighed when formulating a robbery prevention and response plan.

Regardless of the associated risk, the basics of a solid prevention program include the integration of physical security (CCTV, alarm system, proper lighting, and proper locking mechanism). Properly placed and utilized physical security devices can provide an effective deterrent for a potential robber and can aid in evidence collection after a robbery. Cash protection at the point of sale, while in the safe, and during deposit must be implemented, and a monitoring system should be in place. A monitoring system could be as elementary as enforcing the "two-person rule" when money is unprotected for counting, transfer, or deposit. It can also be as complex as investing in armored car services and armed protection. Finally, access control procedures including stringent opening and closing regulations, monitoring of all ancillary doors, and inclusion of proper limitations to and monitoring of personnel entry and exit are essential.

Checks and Credit/Debit Cards

Credit and debit cards are the number one method of payment in brick-and-mortar retail stores and at "retailer" websites on the Internet. This fact has not gone unnoticed by the criminal element. Attacks against both retailers and their customers continue to be a major threat and present an ever-increasing technological challenge to banks, credit/debit card issuers, retailers and businesses in general.

Compliance with Payment Card Industry (PCI) Data Security Standards is now required for all brick-and-mortar retailers. Violation or noncompliance with those standards can result in large fines against a retailer.

The Nature of the Check

The use of checks for payment in retail stores is continuing to diminish and may soon be obsolete. A check is nothing more than an authorization to the holder of the funds or the bank to debit the account of the authorizer and to credit the named person or the bearer in the amount specified. Provided that sufficient funds are available in the debited account, the exchange is made either in cash or by crediting the account of the payee.

Checks and the Retailer

Bad checks of various kinds add to the merchant's cost of doing business. Forgeries or fraudulent checks are a direct loss of both merchandise and cash. Checks that are ultimately collectable, but that are returned by the bank for any number of reasons present a huge administrative headache in making the collection. Even though a majority of such checks are soon made good, often after a single letter or phone call, the costs of such follow up and double handlings, as well as the additional cost of collection agencies, if such are required, may bring the cost of each returned check to between $25 and $30. The charge for any returned check is usually added to the eventual collection, but most retailers feel that they do not collect the full amount. Such returns actually cost them in direct and indirect losses.

Check and Credit/Debit Card Approval

The two most important and relatively simple steps that merchants can and must take in approving checks are a thorough examination of the check itself and a positive identification of the check casher. Neither of these steps will screen out the accomplished forger or the skilled bank check artist, nor will they eliminate the problem of honest, but careless, people issuing checks that are returned because of insufficient funds. But they will reduce, or possibly eliminate, a great deal of the most persistent losses.

Retailers are not obliged to cash checks—and, in fact, the number of retailers who have decided not to accept checks is on the rise. They do so in their own interests, as a service. They must assure themselves that the customer is properly identified. Generally any official document that describes the holder and bears a picture can be accepted as adequate identification. An artful forger can fabricate such documents, but because forging of this kind requires some equipment and skill, it is not a widespread general practice.

Driver's licenses, passports, national (major) credit cards, birth certificates, and motor vehicle registrations are all acceptable forms of identification.

Credit and Debit Cards

As with check-cashing policies, stores must properly identify card users and, if necessary, ask for identification and compare card signatures to signatures on other forms of identification such as a driver's license or state-issued ID. As the business environment turns more "cashless" and checkless, credit and debit cards will totally dominate the retail transactions. The good news is that with electronic data transfer, retailers can be assured that customer accounts are either adequate to cover the sales cost or guaranteed by the credit card company.

ID Equipment and Systems

Electronic check verification systems, using computer interlink capabilities, give some merchants the ability to check individual checks against lists of problem check writers; in some cases, this technology even allows for the verification of the solvency of the account on which the check is drawn.

Store Policy

Every store must set and maintain its own policy for handling checks. Certainly the policy should be reviewed and adjusted as necessary, but it must be strictly adhered to as long as it is in force. Associate indoctrination and continuing education are essential to the success of any check control program.

As a guide, though certainly not a rigid rule, merchants should consider certain limitations. The retailer should not accept the following:

1. Third-party checks. Payment can be stopped on such checks and the merchant's recourse is only to the customer, not to the payer.
2. Checks drawn on out-of-town banks. Such checks are difficult to verify, and the time involved in clearance is such that, if the check is fraudulent, the customer has long disappeared before the check is returned.
3. Checks over a certain amount above purchase.
4. Checks for cashing, other than government or payroll checks.
5. Checks for cashing, without adequate predetermined documents of identification.
6. Checks for cashing drawn on other than personal checks imprinted with the name and address of the customer.

If such rules are followed, the loss from bad checks should be small, provided all checks are themselves carefully examined by the cashiers.

Policies, Procedures and Practices

Many people who are new to the retail business feel that controls and procedures send the message to all associates that they don't trust their work force. At the same time, when a large loss occurs, associates are flabbergasted that such an enormous loss could have happened without management knowing. Additionally, after a retail establishment becomes the victim of loss over and over again, they quickly understand that policies and procedures need to become part of the daily operation. They also understand that policies and procedures are in place to prevent losses, losses that turn into profits, profits that ultimately may turn into pay raises and benefits. The reason for policies and procedures needs to be communicated to all the associates of the store. There needs to be a TRUST BUT VERIFY mentality, which will, in turn, send the message of individual accountability versus lingering suspicion when policies and procedures don't exist.

Establishing written and enforced policies and procedures is very important. Violating a policy should be considered a very serious misstep. A procedure generally describes how your policies are to be implemented. A procedure may be a separate document from your store's written policies. A practice is generally less stringent than a procedure and it is generally informal and unwritten. It may not match what your policies and procedures say, but it may be a general practice of the store. This is an important distinction, because you cannot have a solid system of accountability if a policy says one thing, the actual procedure says another, and the general practice of the store associates doesn't come close to either. Additionally, when there are violations, managers will not enforce the rules on a consistent basis if there is inconsistency in adherence.

Internal policies (controls) are important for any business, but are critical in the retail business. The retail business gives team members ready access to both desirable products and cash and, as such, has risks of loss not shared by other types of businesses. An ideal policy (control) consists of three components. Think of it as a three-legged stool (see Figure 15-3). If any one of the three legs breaks or doesn't support its fair share of the load, the person sitting on the stool is at risk of falling, as is your policy or control at risk of failing.

Policy Components—The Three-legged Stool

- The first leg is physical, mechanical or electronic. It can be something as simple as a padlock or as complex as a security bit (setting) in the point of sale (POS) system.
- The second leg is the human component, and involves that aspect of a policy (control) in which a person, normally a supervisor or manager, plays a key role. An example would be a policy that states that a supervisor must approve refunds over $50. If a supervisor becomes lax and knowingly allows refunds over $50 to occur, without approval, that is now a store practice and the policy is useless.

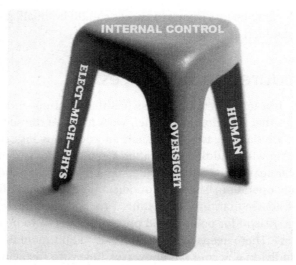

FIGURE 15-3 Three-legged stool. *(Courtesy of William H. Cafferty.)*

- The third leg of an ideal control is the oversight component. That can be anything from the back office daily reconciliation of POS summary reports against supporting documentation, to a quick review of digital CCTV records to verify that a refund with no supporting documentation was legitimate. Back office personnel are your final line of defense to ensure policy compliance.

Policies and procedures must be enforced. As can be seen, instituting and maintaining a culture of honesty is a daunting task that is a constant management battle and difficult in the fast moving pace of retail. Management is frequently challenged by personnel shortages and unexpected events, which make it more difficult to ensure that policies don't get pushed aside and sloppy practices begin to take their place.

Training

A well-rounded retail security program involves a top-down approach, including proper program support, organization, implementation, maintenance, and continual improvement. It will also take a commitment on the part of management to train associates on the proper way to execute their job duties, while adhering to applicable policies and procedures. A complete and ongoing training program should be developed and implemented. Topic-specific training and ongoing refresher training will play an active role in ensuring ongoing compliance.

Topic-specific training may also be provided to aid in enhancing an associate's understanding and ability to respond to various events, security issues included. A classic example would be how associates should react and respond to a robbery situation. This is topic specific and will require refresher training periodically to ensure that the principles are not forgotten. It is up to each retail location to determine the training topics and frequency based on the risks to the operation. Training records should be maintained and updated.

Auditing for Compliance

One of the most important tenets of a proactive retail security program includes having proper policies and procedure in place, and conducting a periodic audit of those policies and procedures to ensure they are being maintained. Depending on the size of the store, this may be the store manager or, if a multi-store operation, a dedicated security manager may be responsible for reviewing policy and procedure compliance.

The store manager and/or security representative should, at a minimum, conduct an annual audit of loss-specific policies and procedures. Policies and procedures that are considered critical will require more frequent review. When a policy or procedure is not being followed, corrective action must be taken to ensure ongoing compliance.

Technology

Today's retail security executive must embrace technology and then use it to benefit the asset protection program. Internally, this means understanding the intricacies of the point of sale system, inventory adjustments that may impact shrinkage and various other technological advances that track money and product. These technological advances have allowed retailers to become more efficient and eliminate staff and overhead that have kept them competitive. Additionally, these same technologies can also be used by potential thieves and it is up to those in the retail security world to understand the various risks that each of these technologies have on money and product. Externally, technology has advanced in light years compared to the technology available to retail security executives just 10 years ago. The following is a sampling of technology and the uses in today's retail environment.

Point of Sale

The Point of Sale (POS) system is the heart and engine of the retail merchandise inventory control and sale processes. It is the system that maintains an ongoing accounting of the "book" inventory; that can automatically place replenishment orders for every SKU in the store based on preset reorder points; that typically has other features that can accommodate many other related tasks such as accounts receivable and accounts payable; and it is the system used to record all sales, returns and other actions (e.g., layaway, special orders, suspended sales, etc.) related to merchandise movement.

Stores that do not have a point of sale system, but rather rely on simple electronic cash registers to record sales, returns, etc., and on separate computer systems (not interfaced with the cash registers) or even manual methods of tracking store inventory levels, payables and receivables, are at a distinct disadvantage. Of course, like any automated system, the POS system is vulnerable to actions by users who would use the system to commit dishonest acts. It is the responsibility of the business owner to be knowledgeable of and apply the electronic safeguards offered by the POS system, along with implementing associated operational controls, to prevent or deter dishonest users from the opportunity to commit a dishonest act, and provide an alert and/or audit trail when a suspicious act is committed.

Exception-based Reporting (Software)

Exception-based reporting is software that ties into the transaction data of the point of sale system. Retailers program the software to retrieve information based on pre-established exceptions. The requirements may be unique to each retailer, but the end result is the same: to ferret out areas where internal theft may be occurring. This is a major tool used by retailers to combat theft losses at their point of sale. Once programmed, reports are generated listing out all the exception detail. For example, a location may average 3 percent of their daily sales transactions as refunds to customers. The set point for the software would be to tell you when any cashier is processing a return percentage greater than 3 percent. The individual assigned to look into these issues would then pull all of the refund documentation to ensure proper protocol was followed. If not, investigative action would then be initiated. The software is used to identify internal theft issues early, before they grow into larger problems.

Electronic Article Surveillance

Electronic Article Surveillance (EAS) is a technology used in the retail sector to deter and detect shoplifting. It consists of disposable soft tags and reusable hard tags embedded with electronic technology that are attached to items of merchandise or inserted inside packaging. If the tag is not properly deactivated or removed at the point of sale, an alarm will be triggered as the tagged item passes through a store's exit door.

Disposable tags come in a variety of forms, everything from price labels embedded with the technology, to tags hidden within sealed packaging, to tags with strong adhesive that are attached to the outside of the item or packaging. These tags are deactivated by proprietary devices located at the POS, typically integrated with the scanning process. Hard tags must be removed at the POS through use of a proprietary tool. There are hard tags that if not removed with the proper tool will rupture and spill ink onto the item. This type of tag is referred to as a "product denial tag" and is normally used on clothing items. It should be noted that if a soft tag is applied to the outside of an item or packaging, shoplifters may pry or cut it off. If the item is protected by a hard tag, professional shoplifters may have tools to forcibly remove it from the item. Also, professional shoplifters will use bags or other containers lined with materials to thwart the effectiveness of the sensing devices (pedestals) at store exits. The primary EAS technologies are radio frequency (RF), acousto-magnetic (AM) and electromagnetic (EM).[11]

Alarm Systems

There are a number of alarms in use in retail establishments that protect the physical structure, the merchandise displayed and stored within, and the exits from the building. All are essential components of a good loss-prevention program.

Intrusion Detection Alarm (IDA)

Also known as a "burglar alarm," an intrusion detection alarm is a system that notifies the alarm company when there is a forced entry to a building through a door, window, roof or other access point. It is critical that sufficient and proper alarm devices (e.g., door and window contacts, motion, sound or glass breakage sensors, infrared devices, etc.) be placed in all vulnerable areas in order to thoroughly protect a building. For example, if motion detection devices are not placed in proximity to the store safe and in the ceiling or areas outside the room where the safe is located, burglars could forcibly enter through the roof, drop down into the office where the safe is located and never trigger any of the door or window alarms. Notification to a UL approved central station and/or a local law enforcement agency is typically through an automatic landline telephone dialer. There are more sophisticated communication methods available to the retailers to ensure intrusion signals are received. It is important that the system provide an alternative means of notification should the primary means be disrupted.

Duress Alarm

This is the feature of an IDA, primarily used in robbery-in-progress situations. A key holder would activate this device, which can be fixed or mobile. These are typically used by those opening or closing a store or installed in a fixed position at a cashier station or in a back office where cash counting occurs or where the safe is located. This also may be in the form of a handheld device that a key holder would carry with his or her everyday car and house keys. Additionally, a duress code can be assigned to all authorized store personnel. The alarm system is programmed to know that the person entering that particular code is in imminent danger. This unique code will deactivate or activate the alarm system, but will also send the message to your central station; "I'm under duress," which would cause the alarm-monitoring station to notify local law enforcement to respond to the retail location.

Merchandise Alarm

A merchandise alarm is designed to protect expensive and/or small items of merchandise that have an increased risk of being shoplifted. The alarm can be a simple contact alarm system consisting of wired leads from an alarm control box to the items, with adhesive or lasso-type contacts attached to the merchandise, which, when cut or pulled off the item, will cause an alarm to sound, to more sophisticated wireless motion detection systems that alert sales associates when a customer opens a display. Use of merchandise alarms normally provides an effective deterrent to shoplifting; however, it is important that the keys and/or combinations to the alarms be protected and that sales associates react immediately when the alarm sounds.

Exit Alarm

Exit alarms can be mounted on the inside of perimeter doors that must remain unlocked because they are designated fire exits. These doors, which are located both on sales floors and in stockrooms and other associate-only areas, provide a convenient avenue of exit for customer and associate thieves if not controlled. Exit alarms are available as integral elements of emergency exit panic bar door hardware and as stand-alone devices that are easily installed. They are normally battery operated, so it is important that they be tested regularly, the batteries replaced when necessary, and the keys to the alarm modules be strictly controlled.

Closed Circuit Television (CCTV) Systems

There was a time when closed circuit television CCTV coverage in a retail store was considered a luxury. That day has long passed and it is now recognized that CCTV coverage of POS terminals, high value/theft prone/sensitive merchandise displays, critical areas containing safes, communications and IT equipment, back offices where money is counted, perimeter doors, and exterior areas containing theft prone product and/or equipment is essential. Only with CCTV coverage in these areas will a retailer have an effective deterrent to theft and the capability to review and determine if a theft occurred or an action was taken by a customer or associate that requires detailed investigation. Video coverage of a theft by an associate or customer is rarely challenged in court—it is, indeed, the best evidence a retailer can have in the prosecution of a customer or associate.

Today's CCTV systems provide a wealth of features and capabilities that can be customized to fit the precise needs of the retailer. Remote viewing capabilities through the Internet also provide business owners a capability to monitor their systems from around the corner to around the world. Additionally, as mentioned previously in this chapter (see sweethearting) is the capability to interface your CCTV system to your point-of-sale system.

Access Control

Access control devices do just what the term implies—provide a means to control a customer's or an associate's access to certain areas within the store. They run the gamut from simple deadbolt locking mechanisms to devices that use keypads, biometric scanners and other identification methods to allow only those with proper authorization to enter. The more sophisticated devices can also transmit alerts via e-mail or telephone when certain actions occur and keep a detailed audit trail of those individuals who accessed the controlled area. Like alarm and CCTV systems, access control systems are essential elements of an effective loss-prevention program.

Radio Frequency Identification (RFID)

Radio frequency identification tagging is a technology that uses radio waves to transfer data from an electronic tag that is attached to an object through a reader for the purpose of identifying and tracking the object by means of a unique serial number. It has become a very important component of large retail organizations' inventory control and loss-prevention programs. It allows retailers to track individual items (referred to as "item level") through the supply chain: from the warehouse, to the delivery truck, to the store's receiving dock, then to the shelves, all the way to the point of sale. This technology can also serve as reliable confirmation that an item presented by a customer for return was actually sold.

Perhaps the greatest benefit of RFID technology is its assistance in improving inventory accuracy information and gaining a more correct image of the sources of inventory shrinkage. To efficiently and strategically manage inventory, a store must have an accurate picture of the actual inventory on hand. For example, knowing when a product is out of stock is important for knowing when to order more. RFID tagging allows the retailer to gain a real-time perspective on inventory because each item is tracked electronically versus relying on the computer inventory numbers alone.

In the future, a move toward greater use of RFID technology will make retailers more efficient and proactive in loss assessments as well as inventory management. Since RFID technology is not suitable for all kinds of products (for example, it is difficult to use on liquids and some metals), a balanced approach to use of RFID with Electronic Article Surveillance (EAS) technology may prove to be the ultimate efficiency tool at retail.[12]

Summary

Retail security, facing some of the same problems in the 21st century that it always has, is also faced with the new challenges in the areas of e-commerce, social media and organized retail crime (ORC), to name a few. E-commerce has allowed thieves to quickly move stolen merchandise with some degree of anonymity, reaching international buyers via the Internet. Inroads have been made by retail organization associations in building relationships with major e-commerce trade companies that have historically been known to host the sale of stolen product. These e-commerce companies are now working in conjunction with retailers and local law enforcement to shut down auction or sale postings, which specifically can be traced to stolen or counterfeited product.

Add to that the recent shoplifting multiple offender crimes, sometimes associated with flash mob activity, which have used social media to coordinate group crime attacks on retailers. Additionally, more sophisticated and aggressive ORC groups are pushing the resources of the retailer and law enforcement alike to take aggressive action as well, which pulls the investigator role away from the handling of traditional in-store issues.

All of these issues continue to challenge retailers to provide resources, people and technology, to combat the ever-changing landscape of retail security. However, some of the traditional retail security issues remain the same, as discussed in this chapter; not much has changed related to shoplifting and internal theft. The tools for discovering, preventing and apprehending internal thieves and shoplifters continue to improve each year. Still, improved technology has not drastically reduced retail shrinkage; thieves still find methods to defeat even the best electronic devices.

The same challenge will exist for retailers in the future; ensure risks are property identified and then ranked. Match up the sometimes limited resources to determine how retailers can cost effectively address these risks, all the while remaining competitive and profitable.

Review Questions

1. What are two broad categories of shoplifters?
2. What is the most effective tool for use in deterring internal theft known as "sweethearting"?
3. Describe the difference between a policy, procedure and a practice?
4. What are the three components of a balanced policy control system?
5. Identify the most common method of associate theft of cash at the point of sale (register) and what is the best way to reduce the opportunity for this loss to occur?

References

[1] University of Florida, National Retail Security Survey, 2009 and 2010.

[2] Lowe's Bombing Suspect Arrested by Monica Toriello, Home Channel News, December 13, 1999.

[3] Associated Press, Raw Data: Past Deadly U.S. Mass Shootings (Robert. A. Hawkins), April 3, 2009 (<www.foxnews.com>).

[4] University of Florida, 2009 and 2010.

[5] Substance abuse linked to increased problems on the job. Security April 1991:10.

[6] University of Florida, 2009 and 2010.

[7] Our Three Pennies of Profit, Retail Profitability Training for Employees—North American Retail Hardware Association & Russell R. Mueller Retail Hardware (<http://www.nrha.org/>).

[8] Berlin P. (n.d.). Root causes of shoplifting: why do shoplifters steal? Theft class by NASP. Retrieved September 6, 2011 from <http://www.shopliftingprevention.org/WhatNASPOffers/NRC/Understanding TheRootCauses.htm>.

[9] Commercial Services Systems, Inc., <http://www.ncjrs.gov/App/Publications/abstract.aspx?ID=83837>.

[10] FBI—Uniform Crime Report (2010) <http://www.fbi.gov/about-us/cjis/ucr/crime-in-the-u.s/2010/preliminary-annual-ucr-jan-dec-2010>.

[11] Electronic article surveillance definition in the Free Online Encyclopedia. (n.d.). Encyclopedia. Retrieved September 6, 2011 from <http://www.thefreedictionary.com/Electronic+article+surveillance>.

[12] <www.RFIDJournal.com>, downloaded 09/09/2011.

16 Terrorism
A Global Perspective

This chapter is primarily the work of Dr. Vladimir Sergevnin, Editor of the Law Enforcement Executive Forum *journal and Research Manager of the Illinois Law Enforcement Training and Standards Board Executive Institute, along with the previous work on terrorist tactics by Dr. Robert Fischer, with coauthors Colonel Fred Berger, U.S. Army ret., and the late Dr. Bruce Heininger. Portions of this chapter were published as "Terrorism: Nothing New—A Predictive Model for Handling Terrorist Incidents,"* Illinois Law Enforcement Executive Forum Special Edition, *2(1): 2002.*

OBJECTIVES

The study of this chapter will enable you to:

1. Understand the problems associated with defining terrorism.
2. Briefly discuss the evolution of terrorist tactics.
3. Identify major terrorist organizations.
4. Know the various motivations for adopting terrorist tactics.
5. Identify some of the tools being used to combat terrorism.

Introduction

With the events of September 11, 2001, the U.S. and world community became explosively aware of the power of terrorism as a tool of war. Although terrorism is not a new tool of war, the sheer magnitude of the attack on the World Trade Center has assured terrorist groups recognition beyond that horrendous event. This attack, the first since the 1812 attack on the continental U.S., made a deep impact on the American government, the people's collective psychology and the national security industry. Today, there is virtually no organization, agency or business in the U.S. that does not have, in one way or another, a security plan for terrorist attack. Nationally there are thousands of agencies and institutions, training entities, universities and colleges, and websites providing information, research, education, and recommendations related to terrorist incidents and appropriate security. It can be explained only because the threat of terrorist acts against the U.S. has significantly elevated since the end of the 1980s along with an increased level of public awareness. All this has created a tremendous increase in counter-terrorism investments on individual, local and governmental levels. According to recent data, in 2010, the U.S. total military expenditure is 41 percent of the world military spending at $ 1.7 trillion, with China at 9 percent, Russia at 4 percent and Iran at 0.5 percent.[1]

The Western world has heavily invested in a so-called "war on terror" with mostly vague and predominately controversial results. Ideologically, the last century generally saw decentralized terrorist groups converge into anti-Western and anti-Christian movements that do not require constant operational communications, but produce an unlimited stream of individual terrorist recruits and cells by the means of self-indoctrination and radicalization via advanced channels of communications. Geopolitically, the USA transformed from a firm ally of the Muslim world fighting against the Soviet Union during the war in Afghanistan to a definite enemy for many Muslim countries. Iraq has since become a new epicenter for *jihad* against the Western powers. Two consecutive invasions by Christian superpowers, with different political systems, insured Muslim's public opinion that these aggressions are religiously motivated. Tactically, the terrorist movement shifted from targeting governmental officials and military to killing civilians in an unprecedented scale, using effective, precise and flexible potential of the suicide missions. Al-Qaeda and its leaders not only survived the widest and best-equipped hunt, but became ideological and tactical leaders and icons for thousands and thousands of followers.

Terrorism will be a major world concern for the visible future, although the hype factor may be more significant than the actual probability of becoming a victim. In 1986 only 12 U.S. citizens were killed and 100 wounded in worldwide international terrorism. In 1987 the number killed dropped to 7, with only 40 wounded. By 1989, 16 U.S. citizens were killed and only 19 wounded.[2] In 1995, 70 attacks were directed toward U.S. businesses abroad[3] and 10 were killed. The following numbers of the U.S. citizen deaths caused by international terrorism from 1995 to 2000 provides some explanation as to why terrorism has not been a priority for the U.S. government prior to 9/11: 1996—25, 1997—6, 1998—12, 1999—5, 2000—19.[4] Between January 2005 and May 2006 there were only 9 American civilians wounded as a result of terrorist attacks and in 2006, 28 American citizens were killed while overseas. In 2009, there were 25 U.S. noncombatant fatalities from terrorism worldwide. This is less than the number of dogs fatally killing humans (32 deaths) in the U.S during the same period.[5]

In 2006, in the European Union countries there were less than 500 terrorist incidents, while Islamist-authored attacks accounted for less than 1 percent of the recorded incidents over the same period.[6] In 2009, 60 percent of all terror attacks occurred in Iraq, Afghanistan and Pakistan.[7]

Still, the loss of even one life and the fear associated with becoming a victim cannot be overestimated. With the events of the mid-1990s, including the bombing of the Federal Building in Oklahoma City, the bombing of the World Trade Center in New York, the capture of the Unibomber, and the bombing at the Olympics in Atlanta, terrorism reached the forefront of the American public's interest. Experts agreed that terrorism in the United States was just beginning. Their predictions were too accurate as evidenced by perhaps the most spectacular terrorist attack to date, the destruction of the twin towers and World Trade Center complex using hijacked commercial airplanes.

The impact on corporate United States is very significant. Terrorists have targeted 20 of the top 25 U.S. firms. And terrorism has had an impact on almost all international firms. The largest firms have taken security precautions. An analysis of American business victimization by terrorist attack demonstrated that almost every type of U.S. business overseas has been targeted.[8] Terrorism threats made against the Fortune 1000 companies increased from 17 in 1998 to 40 in 2003.[9]

One of the newer (but certainly not new) trends in attacking companies has been assault on executives—primarily kidnapping for ransom. Risk insurance, and the provision of related risk and crisis management services, has become an established industry and it is estimated that as much as $100 million in premiums is paid globally each year.[10] (Chapter 8 provides additional information on terrorism insurance.)

This practice has been common in many foreign countries such as sub-Saharan Africa, Eastern Europe, Central Asia, the Balkans and the Middle East, and is also increasing in the United States; the problem seems to be related to the increase in terrorism throughout the world. To be able to develop adequate protection for executives, security managers must understand the terrorist problem and the strategies of various terrorist leaders.

Although the number of terrorists may be small, their actions have a great deal of impact. Throughout the world there are more than 42 major terrorist organizations, which vary in size from a few members to several hundred.[11] In all, there are more than 3,000 members. According to the Office of International Criminal Justice at the University of Illinois–Chicago, there were 300 extremist groups operating around the globe less than 10 years ago.[12] However, only four or five such groups have enough members and support to be transnational threats. The real problem arises from terrorist sympathizers. Most of these groups have support from various governments and thus have access to automatic weapons, surface-to-air missiles, anti-tank weapons, and sophisticated bombs. In 1988 Pan Am Flight 103 was blown from the air over Scotland by a bomb. In 2002 SAM handheld missiles targeted a commercial jetliner flying over Africa. Luckily the aircraft took evasive action and survived.

Why does terrorism exist? Although political terrorism often appears irrational and unpredictable to victims and observers, it is very rational from the terrorist point of view. Basic to most terrorist theory is that violence will bring the uncommitted masses into the conflict on the side of right—of course, the terrorist point of view. Terrorist actions are violent because:

1. They show the strength of the group.
2. They are provocative, causing the general population to pay attention to the group's activities.

Today terrorism is a very viable method of attracting not only local, but also national or world attention through the mass media.

In the last few years, terrorist organizations have shifted their focus from governments to business and other soft targets. Since 1975, targeting of businesses has become more widespread even in the United States. According to Global Terrorism Database, between 1998 and 2003, terrorists targeted 17 businesses in the USA.[13] Attacks on private businesses in New York, Chicago, and on the West Coast are indicative of this trend. Terrorists use the rich nation/poor nation issue whereby multinational and other large companies are portrayed as exploiters of the poor. Many of these large companies are rich enough to pay enormous ransoms, and in the past, have paid the ransoms. Coupled with government target-hardening (increased security measures such as state-of-the-art hardware) stances, this makes companies more attractive targets. Because of these trends, businesses are more vulnerable to terrorism than at any time in the past.

Despite this vulnerability, the cost of the additional security measures must always be justified because businesses are profit-oriented. In many cases, chief executive officers (CEOs) view the cure as worse than the disease, and many CEOs are unwilling to give up their personal liberty and freedom for protection. These same CEOs generally are unwilling to concede that their firms could be targets for terrorists. In addition, many corporations are willing to pay ransoms rather than invest in protective measures because they believe that a possible ransom will work out cheaper than constantly spending money on security.

The United States and many other countries, however, have adopted a posture of not making concessions to terrorists. Following the destruction of the World Trade Center, President George W. Bush declared a worldwide war on terrorism. The vast majority of the world's governments joined him.

Current Issues

The last decade has seen tremendous changes in the international terrorism activities all over the world. The recent history of international terrorism attacks, from the World Trade Center to the hostage situation in Beslan, Russia (2004), and Glasgow airport bombing attempt (2007) has activated mass media and expert attention, reminding citizens over the globe that international terrorism is not distant from each and every one, and is capable of affecting national and global security. International cooperation in countering terrorism is becoming very strong. When the United States launched Operation Enduring Freedom in October 2001 a total of 136 countries offered a range of military assistance.

There is definite abuse in the use of the term *terrorism* by mass media and various political groups and movements. Despite the tremendous amount of research and publications, international terrorism remains a phenomenon that is not clearly understood, adequately analyzed, and effectively controlled. The limited scope of international terrorism analysis can be explained by political, ideological, and behavior approaches, which easily can overshadow the real substance of the phenomena.

Present-day international terrorism is quite different from old-fashioned terrorist acts, such as assassinations (Archduke Franz Ferdinand in 1914) or bombing (Russian Emperor Alexander II in 1881). Modern forms of international terrorism are more lethal and suicidal, more ideologically religious, and technologically advanced. The incidents of international terrorism caused around 1,500 deaths worldwide in the period 1991–1996.[14] Generally speaking, the human life value vacuum of international terrorists has struck humankind at the beginning of the 21st century.

According to the Department of State there are three trends in terrorism that were identified in 2006:

1. emergence of more and more decentralized operatives or groups in terrorism activities;
2. "sophistication" of terrorist activities through informational technologies and Internet finance;
3. convergence of terrorist activity and international crime.[15]

The lack of coordinated efforts at the international level means that different countries' security agencies must assume some of the initiative in establishing partnerships on the basis of similar approaches and standards.

This chapter is an attempt to make a contribution to America's ongoing international preparedness effort. It is designed for academic and pedagogical purposes to provoke interest and controversy. This chapter will demonstrate to students how international terrorism is defined and will shape the conclusions reached about terrorist characteristics and counterterrorism implications.

Historical Background

International terrorism has existed throughout the history of man and society. There are several epicenters from which modern terrorism has evolved.

Europe

Europe is a motherland of modern terrorism. Members of a radical society or Jacobin Club of revolutionaries promoted the Reign of Terror and other extreme measures were active mainly from 1789 to 1794. French revolutionaries used terror as a remedy for political transformation:

> *One of the original justifications for terror was that man would be totally reconstructed; one didn't have to worry about the kinds of means one was using because the reconstruction itself would be total and there would be no lingering after-effects ...*[16]

The "Terror the Order of the Day" or *Que la Terreur soit a L'ordre du jour* (Carlyle, 2002) was designed as a temporary domestic policy oriented on suppression of the enemies of the French Revolution, but its legacy provided international implications for more than two centuries. The original purpose of terror was to eliminate any opposition to the revolutionary Jacobins' regime and to consolidate power. The latest applications of governmental or state terrorism can be found in Soviet Russia (Civil War, 1918–1921), Communist China (Great Proletarian Cultural Revolution, 1966–1969), and Cambodia (Khmer Rouge Regime, 1975–1979).

These early experiments with state or governmental terrorism outlined several important objectives of this method of governing.

From the French revolutionaries who employed the strategies of international terrorism against European countries to Russian terrorists that carried us into the 19th and 20th centuries, we can observe a steady trend to gain political and ideological objectives. Marxism reinforced this orientation.

Considerable numbers of leftist and right-wing terrorist organizations were formed in the late 1960s in Europe, including Germany's Red Army Faction (RAF), France's Action Directe, Italy's Red Brigades, and German neo-Nazism.

Marxism

Karl Marx along with Friedrich Engels developed the communist doctrine of the class struggle, which has been a major agency of historical change. They theorized that the capitalist system would inevitably, after the period of the dictatorship of proletariat, be superseded by a socialist state and classless communist society. A dictatorship of the proletariat is necessary to ensure the removal of the capitalist society. The dictatorship is above the law because it is the law and therefore can be unlimited. By introducing the First International Working Men's Association of communist organizations in 1864 Marx and Engels launched the idea of international or global socialist revolution employing any means of class struggle, including terrorist tactics against dominant classes. The international character of the proletarian revolution was derived from the international development of the capitalist society.[17]

Russia

Russia has a long history of coexisting with political terrorism and lives under fear of terror. Russian anarchist Peter Kropotkin promoted the fundamental philosophical basis for utilization of terrorism as the tool for revolution proclaiming the concept "propaganda of the deed." Sergei Nechaev might be called the extremist forerunner of modern Russian terrorism; Dostoevsky used him as a model for the revolutionary protagonist of *The Devils*. He was the father of political terror, which he developed as a revolutionary tool as early as 1869, when he published the *Revolutionary Catechism,* in which he defined a revolutionary as: A man who is already lost … he has broken all links with society and the world of civilization, with its laws and conventions, with its social etiquette and its moral code. The revolutionary is an implacable enemy, and he carries on living only so that he can ensure the destruction of society.[18]

For Nechaev, when it came to revolution, the end always justified the means. He believed that a terrorist must be:

> … *hard towards himself, he must be hard towards others also. All the tender and effeminate emotions of kinship, friendship, love, gratitude, and even honor must be stifled in him by a cold and single-minded passion for the revolutionary cause. There exists for him only one delight, one consolation, one reward and one gratification—the success of the revolution. Night and day he must have but one thought, one aim—merciless destruction. In cold-blooded and tireless pursuit of this aim, he must be prepared both to die himself and to destroy with his own hands everything that stands in the way of its achievement.*[19]

This trend became even stronger 10 years later when the rebel group named itself the People's Will *(Narodnaya Volya),* the name under which the radicals were responsible for the assassination of Alexander II in 1881. The objective of the group was to cause a coup or overthrow the Russian government. They believed that the assassinations would be the trigger for revolution and would finally change the order of the regime.

The Bolsheviks and Lenin inherited terrorist approaches and converted them into the state policy. The communist state developed two main types of terrorism. First, there was the internal policy of using terror for the benefit of establishing a so-called "dictatorship of proletariat." The goal was to suppress and physically eliminate opposing forces in the country and convince the population to be loyal to the new regime. In September 1918 the Russian communist government officially announced the policy of "Red Terror." Hundreds of thousands died; millions were scared.

Second, there was support for international terrorism with the goal of causing destruction and chaos, resulting in a world communist revolution. There were many cases of state-supported terrorist actions. Soviet secret police (NKVD, OGPU, KGB) even established a special department, which was in charge of the elimination of popular political figures worldwide (e.g., the assassination of Leon Trotsky in August 1940).

Stalin developed terrorism as one of the most powerful tools of state policy, but individual and group terrorism were almost unknown under Stalin, Khrushchev, and Brezhnev. Isolated acts of terrorism by individuals (e.g., the explosion in Moscow's subway in January 1977) got the state security agencies' (KGB and MVD) attention, and terrorists were arrested and executed almost immediately.

Ireland

Ireland has been one of the longstanding centers of modern terrorism. At the end of the 19th century the Irish Republican Brotherhood (originally formed by Irish immigrants in New York City) had launched a campaign of assassinations and bombing against the British. The Easter Rising of 1916 helped to establish the Irish Republican Army, which was the main political and terrorist organization pushing for the formation of the Irish Free State.

Terrorism related to Northern Ireland has significantly diminished in recent years, with the Provisional IRA, Dissident Irish Republican terrorist groups and the main Loyalist groups ceasing their terrorist campaigns and engaging in the peace process. On 8 May 2007, the Northern Ireland Executive was successfully restored with the creation of a new government under a power-sharing agreement between nationalist and Unionist political parties.

Palestine

The new chapter of international terrorism was opened in the 1960s with the establishment of the Palestine Liberation Organization (PLO), led by Yassir Arafat. This Palestinian group vowed to fight the war of attrition against the occupying Israeli forces. In 1982 the Soviet Union initiated and sponsored the International Conference of the World Center for Resistance to Imperialism, Zionism, Racism, Reaction, and Fascism, which was held in Tripoli. This meeting resulted in the formation of a committee consisting of Libya, Cuba, Iran, Syria, and North Korea. The goal of this committee was the establishment of international terrorist training programs to prepare fighters to battle against all types of oppressors, primarily the United States.[20]

In 2007, an important transformation of power happened in Palestine. Hamas, a radical Islamist political group, forcibly seized control of Gaza from rival Fatah, an essentially secular Palestinian group. Therefore, for the first time, the significant part of Palestinian territories—the Gaza Strip—is under the authority of a radical group that regularly uses terrorist tactics.[21]

Chechnya

In 1999 Islamic justice was established in Chechnya. Terrorism, including a series of bombings in Moscow (several hundred people were killed there), erupted. After that, several thousand Islamic militants, armed members of a Chechen Muslim fundamentalist group whose aim was to merge Dagestan with neighboring Chechnya in a single Islamic state, invaded Dagestan. Russia responded with police and military attacks by federal forces, and the militants retreated; the incident contributed to Russia's decision to invade Chechnya later in 1999. International extremist organizations, including Osama bin Laden (in May, 2011, Osama bin Laden was killed by a U.S. Special Operations force) and other criminal associations, backed the Chechen terrorists. The territory of Chechnya is used to host and train terrorists from Arab countries and some Western European countries. The numerous terrorist groups are free and go unpunished and make raids in the territory of Russia. In 1999 there were 20 terrorist acts in the Russian Federation registered by the MVD security agency. In 2001 representatives of terrorist organizations were registered by the Federation of Small Businesses (FSB) in 49 of 89 states in Russia, and in 2001, the terrorist groups twice attempted to gain access to Russian nuclear munitions dumps. Security agencies do not exclude the possibility that terrorist groups may directly attack nuclear installations.

For the past several years terrorist attacks have mutated from Chechnya to other republics in the region. During this period there have been several extremely sophisticated and well-coordinated attacks. Internationally funded Chechen-led groups returned to hostage-taking as a terrorist tactic in the raid on Beslan's School in September 2004 in which 334 people were killed, including 186 children. Chechen terrorist groups shifted from mostly military operations to the following tactics: random attacks on military installations; random attacks on tourists, and the deliberate killing of foreign-aid workers; incidents of kidnapping, hostage-taking, and bombing of apartment buildings (in 1999–2005 more than 5,000 people were kidnapped by Chechen terrorists); and terrorist attacks on economic infrastructures, including energy distribution, transportation, and banking.

Terrorism in the 21st Century

State Sponsors of Terrorism

Since the beginning of the communist rule in Russia, state sponsors of terrorism provide political, financial, operational and military support to terrorist groups around the world. Without such states, terrorist groups would not have the same level of stability and ideological support. The limited success of international efforts materialized in Libya's renouncing, in 2003, terrorism and abandoning its WMD programs. North Korea shut down its nuclear program in 2007.

Diminishing the volume of state sponsorship for terrorist groups forces them to explore links with transnational and national organized crime entities. Several countries are taking steps in preventing this symbiosis. In June 2000, the Anti-Terrorist Center of the Commonwealth of Independent States (Armenia, Azerbaijan, Belarus, Georgia, Kazakhstan, Kyrgyzstan, Moldova, Russia, Tajikistan, Turkmenistan, Ukraine, Uzbekistan) was established with the purpose of coordination of counterterrorism measures in the territory of the former Soviet Union. In November 2001, the main organized crime administration of the criminal police service at the MVD established a special section for fighting terrorism and extremism. National police offices in the seven federal districts have already set up terrorism sections. The officers intend to cooperate with foreign law enforcement bodies in carrying out antiterrorist activities.

In 2007, the U.S. State Department included the following countries as sponsors of terrorism: Sudan, Cuba, Iran, and Syria. As of August, 2011, this list has not changed. Venezuela was certified by the Secretary of State as "not fully cooperating" with U.S. counterterrorism efforts.[22]

There is very little evidence that Cuba is currently active in international terrorism. The State Department described Iran as the most active state supporter of terrorism. Iran and Syria continue to support groups such as Hamas and Hizballah and, according to former President George W. Bush, Iran supports al-Qaeda.

Implications of a Changing World

Law enforcement officials around the world have reported a significant increase in the range and scope of international terrorist activity since the early 2000s. It is in contrast with the 1990s when the total number of terrorist incidents worldwide declined, but the percentage of terrorist acts resulting in fatalities had grown.[23]

Recently the number of incidents has grown from about 11,000 in 2005 to around 14,000 in 2006, and fatalities are up from about 14.5 thousand to about 20.5 thousand.[24] According to the Department of State globally, in 2006, the number of terrorist incidents increased 25 percent in comparison with 2005.[25] However, 2009 experienced a decline as the number of worldwide terrorists incidents dropped to 10,999.[26]

The level and severity of this activity and the accompanying growth in the power and influence of international terrorist organizations have raised concerns among governments all over the world—particularly in Western democracies—about the threat terrorists pose to democracy and stability in many countries and to the global economy. In 2006, mainly in Iraq and Afghanistan, terrorist incidents claimed more than 20,000 lives, which is 5,800 more deaths than in 2005. International terrorist networks have been quick to take advantage of the opportunities resulting from the revolutionary changes in world politics, business, technology, and communications that have strengthened democracy and free markets and brought the world's nations closer together.

The end of the Cold War resulted in the shift of political and economic relations not only in Europe but also around the world. This change opened the way for substantially increased trade, movement of people, and capital flows between democracies and free market countries and the formerly closed societies and markets that had been controlled by the Soviet Union. These developments have allowed international terrorists to expand their networks and increase their cooperation in illicit activities and financial transactions. Terrorists have taken advantage of transitioning and more open economies to establish financial ventures that are helpful in budgeting international terrorist activities: training camps, "sleeper cells," purchase of weaponry and explosives.

International terrorists have extended their reach by building globe-circling infrastructures. Lebanese Hizballah, whose presence now reaches most of the continents, has led the way. But other terrorist organizations, with agendas as diverse as the Palestinian group Hamas or the Sri Lankan Liberation Tigers of Tamil Eelam, have their active presence far from the lands where their objectives are focused.[27] And of course, following the bombing of the World Trade Center in September 2001, al-Qaeda has taken its place as a major world terrorist "player."

Present day terrorist networks are loosely structured, acting without an operational structure and with regular connections across the globe. They are bound by shared extremist ideology or experiences. Some of these networks are directly connected to al-Qaeda, but most are autonomous, and both work to carry out terrorist attacks, and are influenced by radical beliefs shared over the Internet. The terrorists draw their motivation from sporadically delivered messages articulated by leadership icons such as Osama bin Laden (now deceased).

Revolutionary advances in information and communications technologies have brought most of the world population closer together. Terrorist networks just as easily use modern telecommunications and information systems. Sophisticated communications equipment greatly facilitates international terrorist activities including coordination of terrorist acts and affords terrorists sufficient security from law enforcement counterterrorist operations.

Through the use of digital technologies, international terrorists have an unprecedented capability to obtain, process, and protect information from law enforcement investigations. They can use the interactive capabilities of advanced computers and telecommunications systems to plot terrorist strategies against U.S. representatives and institutions all over the globe, to find the most efficient routes and methods for financial transactions, and to create international virtual networks. Some terrorist networks are using advanced technologies for counterintelligence purposes and for tracking law enforcement operations.

The modern mega trend of globalization and the reduction of barriers to movement of people, commodities, and financial transactions across borders have enabled international terrorist networks to expand their global reach. International terrorist groups are able to operate increasingly outside traditional models, take quick advantage of new opportunities, and move more readily into the most vulnerable areas of the Western world. The major international terrorist groups have become more global in their operations and have developed more threatening goals. Since the end of the Cold War, terrorist groups from Middle Eastern countries have increased their international presence and worldwide networks or have become involved in more lethal terrorist acts. Most of the world's major international terrorist groups are aiming toward the United States.

In Search of a Definition for International Terrorism

Numerous attempts to produce the general definition of terrorism were more focused on terrorism as a concept (converging it with intentions and justifications) than a term and consequently include the damaging impact from ideologies and political partisanships.

The first recorded definition of *terrorism* was given in the 1795 supplement of the *Dictionaire of the Academe Francais* as *system regieme de la terror*. The Jacobins used the term when speaking and writing about themselves.

At present there is no precise or widely accepted definition of international terrorism, or terrorism in general. The difficulty in defining derives from: a) highly politicized usage of terrorism and terrorism-related terminology for national or ideological purposes; b) vagueness of boundaries for terrorist acts in comparison with crime, military actions, use of force, threat; c) attempts to infuse a concept into a term rather than defining a phenomenon. Alex Schmid surveyed more than a hundred definitions of terrorism and found two characteristics of the definition:

- An individual is being terrorized.
- The meaning of the terrorist act is derived from its target and victims.[28]

It shows that we are experiencing only the initial stage in scientific analysis of the phenomena. There are many approaches in analyzing and defining international terrorism and the most influential should be mentioned.

Political

Political analysis of international terrorism views it as one of the instruments in a political process. The Marxist–Leninist ideology and many other revolutionary factions accept terrorism, including international terrorism, as a legitimate instrument in the class struggle. For Marxist–Leninists, the political goal justifies the means. The political goal is above any law or moral code in the current society. These ideologies believe that it is not immoral for social revolutionaries to use terrorism because it leads to fulfillment of the political goal.

Governmental institutions of the United States are implementing the political approach in analyzing and defining terrorism. Information pertaining to the political definition of terrorism is contained in Title 22 of the United States Code, Section 2656f(d). That statute contains the following definitions:

- The term "terrorism" means premeditated, *politically* motivated violence perpetrated against noncombatant targets by subnational groups or clandestine agents, usually intended to influence an audience.
- The term "international terrorism" means terrorism involving citizens or the territory of more than one country.
- The term "terrorist group" means any group practicing, or that has significant subgroups that practice, international terrorism.

The U.S. government has employed this definition of terrorism for statistical and analytical purposes since 1983. At the same time not all terrorist acts have political motivation. This approach will exclude those whose motivation is religious or personal, and they are not trying to change any political institution in the foreign society.

Criminal

Most of the terrorists will claim to be not criminals, but freedom (religion, ideology, special interest) fighters. Almost all of the terrorist acts are characterized as criminal violence, which are violating criminal codes and punishable by the criminal justice system of the state. In this case terrorism is defined as "a *crime*, consisting of an intentional act of *political violence* to create an atmosphere of *fear*."[29] Some definitions of international terrorism will include just violence, but a few years ago, for example, the Islamic Army in Yemen warned foreigners to leave the country if they valued their lives but did not actually carry out its threat.[30] Using a criminal justice approach we will find no terrorism, because any of the acts will be just a crime.

Militaristic

Several international lawyers see a possible solution to the dilemma of a definition in the laws of war. If the laws of war are applied to terrorists they will be treated as soldiers who commit atrocities in international armed conflicts.[31] If this approach will be utilized, then we have to treat our own citizens as people in the war zone to prevent possible terrorist acts. Also, not all terrorist acts have international character.

The Physiological Approach

The physiological approach suggests a tendency to imitate behavior that is successful in delivering a message to the public. This approach analyzes the power of inspiration drawn from the media. David G. Hubbard concludes that much of terrorist violence is rooted not in the psychology but in the physiology of the terrorist, partly the result of "stereotyped, agitated tissue response" to stress.[32] From this approach, international terrorism is targeted more at an audience or the public in general rather than the immediate victims. But this definition will exclude those terrorist acts that targeted one person, or a very limited circle of the public, or the military. In this regard the violence against the American battleship *Cole* is not considered terrorism.

Psychological Approach

The psychological approach is not interested in the political or social contexts of international terrorists, but in the terrorists' personalities, their recruitment, and induction into terrorist organizations, beliefs, attitudes, motivations, and careers.

Some researchers view terrorism as a "syndrome." Recent results, thus far, found little support for the syndrome view. The psychological nature of terrorism's users is consistent with the tool view, affording an analysis of terrorism in terms of *goal and means* psychology.[33]

The weak points of this approach are that the analysis isolates the phenomena from the social and political context of international terrorism, and a slight presumption is made that international terrorists are born, not made.

Group Behavior

Another productive approach describes terrorism as rational strategy decided by group.[34] This approach is rather close to the political one because it requires collective decision making in the utilization of terrorist tactics to reach group goals. Group behavior is wider than the political approach because it can include, for example, criminal acts of international terrorism, in which the goals will not be political.

Very often the definitions of terrorism mention only "groups" or clandestine agents as an active core of this type of activity. But in this case we're excluding state-sponsored international terrorism.

In order to understand international terrorism, one must always assess what exactly constitutes terrorism and the definition in use. International terrorism or terrorism has to be analyzed as an instrument, not a concept.

Multifactor Approach

International terrorism preferably can be viewed as a multidimensional phenomenon. It would be misleading to analyze it by a single-cause approach. International terrorism can be attributed to the same type of events as social revolution, revolt, uprisings, and any kind of political, social, ideological, or religious unrest. Terrorism in general, and international terrorism in particular, is an instrumental phenomenon, and as a remedy could be utilized for multiple causes, forces, organizations, individuals, etc.

Terrorism as a Tactic, Instrument, or Method

Terrorism is a term describing a method or instrument of utilizing violence, intimidation, threat, and fear against an individual, a populace, or a government. Terrorists use a wide range of force, violence, and brutality with the purpose to manipulate human behavior and illegally reach goals.

Goals are diverse but can be grouped into the following categories:

- Political—change of regime, overthrow the government, *coup d'etat*, damage relations between the countries, disgrace to political system, and so on
- Social—upset social order
- Economic—damage to economic order; upset the budget; interrupt vital supplies like oil, gas, electricity
- Ethnic and religious—fundamentalist sects, racism, genocide, spread of new beliefs
- Ideological
- Personal

It depends on the target (domestic or international) or forces (individual, group, criminal, military, state sponsored) in categorizing and separating domestic from international terrorism. Any kind of individual, global, regional, or local player can use it. We have to separate terror as application of fear for criminal or insane purposes from political terrorism, which has a definite objective: political power. Application of fear is a universal method in military operations, criminal justice systems (punishment), and so on.

Terrorism in general is a wide range of criminal acts with a focus on use of violence and destruction or threat of violence and destruction to inculcate fear. The acts are undertaken by an individual, an organized group, or a state driven by generally ethnic, religious, nationalist, separatist, political (including governing), ideological, mentally deviant, and socioeconomical motivations. It is a vast area of crime with a flavor of nontraditional motivation.

Numerous publications describe international terrorism as terrorist acts with international goals, targets, and consequences,[35] or as acts committed by a group or individual that are foreign-based and/or directed by countries or groups outside the United States or whose activities transcend national boundaries.

These definitions are not accurate because, for example, stealing of diplomatic mail should be included in them. International terrorism is a term describing the utilization of fear and intimidation, including violence, brutality, and invasion of privacy, across national boundaries with the purpose of political, social, economic, ideological, religious, ethnic, or cultural change.

The concept of traditional "war" requires the mobilization of political, financial, and industrial resources for the development and production of modern weapons.

The concept of traditional "international terrorism" requires the mobilization of political, financial, and industrial resources for the development and production of modern homeland security and defense. It is much more expensive because instead of a front we have an unlimited number of potential targets.

Terrorism requires more efforts in the political sphere, which can eliminate potential grounds for terrorism activity.

Characterizing Modern Terrorism

Ideological Shift

At the end of the 19th century and beginning of the 20th century the dominant form of international terrorism was an ideological one. The phenomenon of ideological terrorism brought it to the global stage via bombings and extermination of "enemies of communism" beginning from around 1917, perpetrated by such states as the USSR, and lately by groups such as Red Army Faction, Red Brigades, Japanese Red Army, etc. The end of the Cold War has resulted in the shift from anti-democratic or anti-capitalist, Marxist-based ideologically motivated international political terrorists to ethnic and religious terrorism. One of the reasons why the ethno-religious type of international terrorism became dominant recently is the globalization of the Western type of economy and culture in traditionally culturally and economically endemic countries, such as in the Middle East region and Asia. International terrorism has an anti-American, anti-Western orientation because these countries view the spread of "global Western economy and culture," an increasing U.S. presence in the Middle East (Israel, Iraq, Afghanistan) and Pacific Rim, Western development of the Caspian oil reserves (Uzbekistan, Kazakhstan, Azerbaijan), and flourishing Western technological development in the Middle East and Pacific Rim as a threat to their powers and traditional ways of governing. It's not surprising then, that many of the international terrorist organizations are state sponsored.

Marxism was a dominant basis for terrorist ideology (1848 to the end of the 1980s). Communist ideology sought a global communist revolution through initiating riots, uprisings, and coups against imperialism and "weak bourgeois national governments." The period from 1848 to 1917 saw political parties and trade-union organizations sponsored by individuals and opposition groups. From 1917 to the end of the 1980s was a period characterized by state-sponsored ideological "warfare," the states being mainly Soviet Union and Warsaw Pact states.

International terrorism networking was established through the First, Second, Third, and Fourth International and national communist or totalitarian organizations.

The First International, as it was to be called, was formed in 1864 in London and composed of various elements: French, Italian, and British. Marx composed the Inaugural Address of the International, which is now considered a historic document. The organization stood for efforts of workers against anarchists and middle-class reformers of the time.

The Second International was founded in 1889 in Belgium. The primary members were German and Russian social democrats. The group worked to advance labor legislation and to strengthen the democratic socialist movement. It failed in its primary concern, the prevention of war. With the outbreak of World War I, the International collapsed.

The Third or Communist International was created in 1919 under the leadership of Lenin. The goal of this group was to bring about world revolution. The group was not supported by many mainstream socialists and was dissolved in 1943.

The role of ideology is still significant and old-fashioned "revolutionary" organizations continue to exist, such as the Turkish Revolutionary People's Liberation Party-Front and the Peruvian Sendero Luminoso (Shining Path). But the dominant terrorist ideology has ethno-religious character. The spread of ethno-religious ideology is a basis for international terrorism based on ethno-religious conflicts in the Caucasus, Balkans, Middle East, South Asia, and central Africa.

The comparison of communist and radical Islam ideology demonstrates some close similarities and explains the attractiveness in many former socialism-sympathetic countries.

Ethno-religious ideological activities are very often state-sponsored, which gives them more informational stability and coordination. Ethnic ideology is dealing with ethnic identity, solidarity, self-determination, and domination. Religious ideological activities usually are oriented toward establishing "pure and the only true religion" and aimed at the spreading of certain beliefs and defeating modern, Western ideology.

The 21st century is promising to be a century of an ideological war for radical religious indoctrination opposing Free World ideology, beliefs and values. It will be fought by all means of propaganda in the format of modern channels of communications.

Modern terrorist organizations realized much earlier than Western governments the power of propaganda. Unlike with communist propaganda, democratic governments are mostly hesitant in aggressive ideological response because of the religious underlining of the radical Islam propaganda. It should be realized that counter-ideological measures can be more effective than military and should focus on: promoting democratic values; providing a Western view on events inside Islamic countries: supporting education, social activity, especially for women; controlling terrorist virtual communications; depicting terrorists as criminals, not freedom fighters.

Terrorist organizations utilized websites to conduct psychological warfare, generating publicity and disseminating propaganda; data mining; fund-raising; recruitment and mobilization; networking; planning and coordination; and sharing information on their activities, organization, plans, and political, religious, and social goals to glorify their actions. If in 2005 there were more than 4,300 websites with terrorist affiliation and content, in 2006 the FBI counted as many as 6,000 extremist websites. Since January 2006, the primary clearinghouse for jihadist information releases has been an Internet outpost called the Al-Fajr Center. Al-Fajr Center has many divisions that are devoted to hacking, intelligence, propaganda, publications, multimedia and cybersecurity. Other influential online groups are the Ansar e-Group, the Jihadi Brigades and the Global Islamic Media Front, also known as GIMF.[36] Some recent reports conclude that Internet chat rooms are now major venues for recruitment and radicalization by terrorist groups like al-Qaeda, and video-hosting websites like YouTube broaden the outreach.[37] According to the FBI, in 2006 Al-Sahab, al-Qaeda's official media component, released 48 propagandist videos with various focus topics, the most al-Qaeda ever released in one year.[38] For the majority of terrorist organizations, the destruction of American and Western values and establishing "true religious order" worldwide is the main goal of these ideological activities. Law enforcement and security forces can concentrate their efforts on: isolating radicals in the community using a non-biased approach within and outside the department, outreach to ethnic groups (communication with ethnic representatives), and recruitment from ethnic groups.

Organizational Shift

There are three basic organizational levels of international terrorism:

1. Individual international terrorism often has criminal motivation (e.g., revenge, intimidation, and any other personal motives). It is close to organized crime activities. It is difficult to detect this form of terrorist. Mentally retarded individuals carried out some individual international acts.
2. Group terrorism requires organization and some type of leadership, recruitment, training, and retention of members.
3. State terrorism is one of the political tools utilized by a government, which establishes a specific agency or uses a legitimate state institution for gaining domestic or international benefits for the regime.

The current shift in the organizational sphere is increasingly away from state-sponsored international terrorist activities and to groups of terrorists. The process of decentralization of international terrorism was initiated by several factors. The modern world relies on very close economic and political interrelations between the countries. It is a tremendous political and economic disadvantage for any state to associate itself with international terrorist activities. Libya and Iraq are strong examples of governments that made all possible efforts to disassociate themselves with terrorist incidents. In May 2002 Libya agreed to pay $2.7 billion compensation for the Lockerbie bombing and has tied the money to the lifting of U.S. and United Nations sanctions. Libya continues to deny involvement in the explosion, which downed Pan Am Flight 103 in 1988 and killed 259 passengers and crew along with 11 Lockerbie residents. The methods by which states sponsor international terrorism and support these "isolated" groups are almost untraceable. In November 2002 the Federal Bureau of Investigation (FBI) was investigating whether a charitable contribution by Saudi Princess Haifa al-Faisal, wife of Bandar bin

Sultan, the Saudi ambassador to the United States, may have indirectly benefited two of the hijackers of the September 11, 2001, terrorist attacks. Because these groups cannot rely on open sponsorship from the state agencies, they turn more to involvement with international and domestic organized crime syndicates, and self-financing. International terrorist groups are more isolated and loosely organized than in the past, when under the influence of Soviet bloc sponsorship they were more or less interconnected and had centralized structure.

Geographical Shift

International terrorism once threatened Americans only when they were outside the country. The primary source of it was known as the Soviet bloc. Today international terrorists attack Americans on their home soil. There is more than just one major source of international terrorist activity. International terrorism gained a global character. Oceans or borders cannot stop it.

The Near East and South Asia are the regions responsible for 90 percent of the nearly 300 high-casualty attacks in 2006 that killed 10 or more people. Of the 14,000 reported attacks, 45 percent—about 6,600—of them occurred in Iraq where approximately 13,000 fatalities—65 percent of the worldwide total—were reported for 2006. Violence against civilians in eastern and sub-Saharan Africa, especially in Sudan and Nigeria, rose 64 percent in 2006, rising to 422 from the approximately 256 attacks reported for 2005. The number of reported incidents in 2006 fell in Europe and Eurasia by 15 percent from the previous year, for South Asia by 10 percent, and for the Western Hemisphere by 5 percent. The nearly 750 attacks in Afghanistan during 2006 are 50 percent more than the nearly 500 attacks reported for 2005 as fighting intensified during that year. The overall number of people injured in terrorists' incidents rose substantially in 2006—by 54 percent—with most of the rise stemming from a doubling of the reported number of injuries in Iraq since 2005.[39] Between 2005 and 2010, 88.4 percent of terrorist attacks took place in the Middle East, South Asia and South East Asia. Only 0.2 percent occurred in North America.[40]

Tactical Shift

In the past, international terrorism used more single assassinations and hostage situations. Airline hijackings have become unpopular among international terrorists because few countries will let them land, and chances are very high that they will be deported back to the country where the international terrorist incident was originated. Only 19 countries have extended their support to include asylum to aviation hijackers.[41]

According to the U.S. State Department the number of international terrorist attacks in 2001 declined to 346, down from 426 the previous year, 2000. A total of 3,547 persons were killed in international terrorist attacks in 2001. In 2000, 409 persons died in terrorist attacks.[42] International terrorism became more lethal. Most of the international groups are turning to indiscriminate killings of civilians. In the 1990s a terrorist incident was almost 20 percent more likely to result in death or injury than an incident 2 decades ago.[43] In the 21st century, terrorism has become more destructive and focuses on mass causalities, tremendous loss of property, and financial and economic downfall. Terrorists also shifted their focus on to more soft targets, such as schools, hospitals, etc. In 2006, attacks on children were up more than 80 percent, with more than 1,800 children killed or injured.[44]

International terrorist missions became more suicidal. In the past, terrorist groups did not exclude the possibility of becoming victims of counterterrorist security measures. Terrorist groups today are recruiting young volunteers to carry out violent acts. It is much harder to deal with this kind of terrorist because they don't utilize the natural concept of human life. Some of the law enforcement tactics in these cases will not work.

Most international terrorist groups shifted to the following tactics:

- Random attacks on military and diplomatic installations.
- Random attacks on tourists and the deliberate killing of foreign-aid workers.
- Incidents of kidnapping, hostage-taking, and bombing of apartment buildings have become frequent in the republics of the former Soviet Union.
- Terrorist attacks on economic infrastructures, including energy distribution, transportation, banking, and tourism have become routine in Colombia.
- Bomb threats.

Another significant tactical focus shift is suicide bombing. A suicide attack is an operational method in which the very act of the attack is dependent upon the death of the perpetrator. The typical type of bombing attacks carried out in the 1970s and 1980s were time bombs. This modus operandi often failed due to early or late detonation without many people concentrated in the immediate vicinity of the explosion, or as an outcome of early detection by law enforcement or civilians prior to detonation. Later, during the 1990s, terrorist organizations started to use remote controls to detonate their explosives. The remote control provided the bomber with the ability to detonate the bomb with optimal timing in order to cause the maximum number of casualties and damage. The suicide attack, like the smart bomb, is the most sophisticated tactic used by terrorist organizations since it provides the perpetrator with control of both the timing and location of the attack and, therefore, produces the maximum number of casualties and damage. Globally in the 1980s there were 4.7 suicide attacks per year while in: 2001—81; 2002—91; 2003—99; 2004—163; 2005—460. Eighty percent of suicide attacks since 1968 occurred after 9/11, with jihadis representing 31 of the 35 responsible groups.

Technological Shift

From relatively primitive means of technology (guns, explosives, and conventional weaponry), international terrorism shifted to highly sophisticated technologies such as weapons of mass destruction and chemical and biological weapons. According to the U.S. State Department, for instance, Iran, seen as the most active state sponsor of terrorism, has been aggressively seeking a nuclear arms capability.[45] North Korea decided in December 2002 to restart nuclear installations at Yongbyon that were shut down under the U.S.-North Korea Agreed Framework of 1994. Three atomic reactors will be able to produce 207 kilograms of plutonium annually, which is enough for the manufacture of nearly 30 atomic bombs per year.[46]

Modern Islamic terrorists have expressed interest in developing the capability to exploit cyber vulnerabilities to disrupt the banking communications, important infrastructures like transportation, medical services and others which will have economic costs, and undermine public confidence. Some terrorist websites provide instructions on how to create and spread viruses, develop code, and promote hacking plans. One of such online communities calling itself "Electronic Jihad" in October 2006 claimed it had designed and used a hacking program to synchronize a distributed denial-of-service attack against an Israeli website.[47] Businesses and corporations are addicted to easily compromised protocols and commercial Internet products to manage networks that increase the probability and vulnerability to cyber attacks by terrorists.

Organized Crime Shift

One of the more significant shifts since the early 1970s has been the growing involvement of organized crime groups with terrorist organizations. For example:

- In Peru from the late 1980s until the early 1990s the extremist Sendero Luminoso insurgents profited from protecting coca fields and extorting drug traffickers operating in the Andean region they controlled.
- In Western Europe, members of the terrorist Kurdistan Workers' Party (PKK) in Turkey have engaged in drug trafficking and other crimes to help finance local operations.
- In Colombia, since the late 1980s, Marxist insurgents have not been able to rely on financial support from Cuba and Russia. Some insurgent fronts of the Revolutionary Armed Forces of Colombia (FARC) and the National Liberation Army (ELN) generate substantial revenue by taxing and protecting coca cultivation, cocaine processing, and drug shipments in the areas they control. The U.S. government estimates that the FARC may earn as much as half of its revenue from involvement in the Colombian drug trade.[48]

With the substantial decline in state-sponsored international terrorism support, many terrorist networks reach out to criminal networks to acquire arms and supplies that cannot be obtained through more traditional or legitimate channels. International organized criminal groups are well-connected to outside gray arms merchants, transportation coordinators, money launderers, and other specialists who can provide the weapons and other logistics support once given by state sponsors. International organized crime groups cannot exist without corrupt contacts in law enforcement agencies, which are crucial in smuggling operations of weapons and other contraband for terrorist groups.

Characterizing International Terrorist Groups

This form of terrorist activity began to spread in the second half of the 20th century with the increase in the number of terrorist organizations and the expansion of social support for terrorism. In the recent Abu Mus'ab al-Suri and Umar Abd al-Hakim's 1,600 page publication, *The Call to Global Islamic Resistance (2005),* which is among the most frequently mentioned jihadi strategy books, they make a strong statement that the future of jihad is individual and small terrorist groups, with no chains connecting them to al-Qaeda leadership.[49]

This form of terrorist activity involves hiring participants to different terrorist organizations, personnel training on methods and ways of conducting terrorist acts, and preparation for the forthcoming terrorist acts—rehearsals of terrorist acts, separate stages of terrorist operations, and so on. The common practices in recruiting a jihadist are: 1) identifying individuals with appropriate ideology; 2) indoctrination; 3) cultivating the desire for jihad; 4) training and preparation for terrorist act.

This form also involves establishing contacts with other terrorist groups and creating and maintaining connections with organized crime institutions, representatives of illegal firearms businesses, and drug dealers. A high level of secrecy and specialization of terrorist group participants according to different functions (e.g., hiring personnel, recognizers, warriors, production specialists of combat substances and cover documents, secret apartment maintenance staff) are common for organized terrorist activity. There are several certain characteristics of international terrorist groups.

Seeking Political Gain

Hatred and the quest for political changes or status quo outside their own country dictate more international terrorist network decisions than any other single motive. It is this consuming desire for political power that typically drives and sustains international terrorism.

Motivation

In 1974 Dr. Frederick Hacker reported to the U.S. Congress, House Committee on Internal Security, Hearing on Terrorism, that all terrorists fall into three basic motivational groups:

1. Criminal—motivated mainly by personal gain
2. Mentally deranged—personal motivation, often accompanied by delusions or hallucinations
3. Political—motivation toward a real or imagined strategic goal

Islamic fundamentalists (i.e., Osama bin Laden) have political and often mentally deranged characteristics.[50] The current shift in motivation is from Marxist-based ideology to ethnic nationalism and religious extremism.

According to Jerrold M. Post (1984), religious international terrorists are more dangerous than the average political or social terrorists, who have a mission that is somewhat measurable in terms of public attention or government reaction; the religious terrorist can justify the most horrific acts "in the name of Allah," for example.[51] Terrorism that is religiously motivated is growing quickly, increasing the number of killings and reducing the restraints on mass indiscriminate murder. For religious terrorists, as for social revolutionary terrorists, violence is morally justified and legitimate.

Seeking Publicity

Almost all international terrorist groups are seeking publicity to promote themselves and their agenda, and to discredit those in opposition to them. International terrorist groups are highly motivated to publicize every act of terrorism in order to show the state's inability to control the terrorist activities. The propaganda of terrorism is connected with attempts to gain public approval of a terrorist activity as a form of political fight, with substantiation of its legal use and also with direct initiative calls to terrorist activities, which may lead to real commitment of criminal actions and involve separate individuals or groups committing severe violent crimes. These appeals are realized verbally or by distributing written or visually demonstrative materials.

Requiring Member Loyalty through Ideological, Religious, Ethnic, and Other Considerations

The purpose of this preference is that the terrorists generally believe that they have to accomplish some global mission, such as communist revolution, liberating Palestine, continuing jihad, etc.

Structure

Generally, international groups maintain a structure with defined leadership–subordinate roles, through which the group's objectives are achieved. Recently, because more groups are based on religious motives and may lack political or nationalistic agenda, they have less need for hierarchical structure. International terrorist groups have a tendency to rely on loose affiliations with like-minded groups in different countries.[52]

Tactical Diversity

Typically, international terrorist groups are utilizing more than one method of violent acts—suicide bombings, bombings, assassinations, sabotage communications, armed assaults, kidnappings, barricade and hostage situations, and hijackings.

Assistance

With the creation of large terrorist institutions, terrorist assistance is becoming more important in the overall system of terrorism. Main variations of this assistance are extremist groups financing and providing the means for terrorism, providing facilities for training their members, and harboring and hiding them after committing terrorist acts. This assistance can be used by states (so-called terrorism sponsors) as well as by representatives of business circles and ethnic and other social groups that express sympathy to terrorist organizations or support them because of their common political interests. There may also be direct involvement with extremist organizations in conducting tasks of legal political institutions to influence their enemies.

Organizational Maturity

In most cases international terrorist groups have a high level of organizational stability and do not depend on the continuing participation of one or a few individuals for their existence. For example, al Qadea was designed during the 1980s as a worldwide recruitment and support network for the purpose of resisting the former Soviet Union's occupation of Afghanistan that was backed with financial aid from the United States, Saudi Arabia, and other states. It is important to appreciate that al Qadea's organizational structure is complex by design. During the 1980s resistance fighters in Afghanistan developed a worldwide recruitment and support network with the aid of the United States, Saudi Arabia, and other states. In the early 1990s this network, which equipped, trained, and funded thousands of Muslim fighters, came under the control of Osama bin Laden. Al-Qaeda ("The Base") is a network of groups spread throughout the world, with a presence in Algeria, Egypt, Morocco, Turkey, Jordan, Tajikistan, Uzbekistan, Syria, Xinjiang in China, Pakistan, Bangladesh, Malaysia, Myanmar, Indonesia, Mindanao in the Philippines, Lebanon, Iraq, Saudi Arabia, Kuwait, Bahrain, Yemen, Libya, Tunisia, Bosnia, Kosovo, Chechnya, Dagestan, Kashmir, Sudan, Somalia, Kenya, Tanzania, Azerbaijan, Eritrea, Uganda, Ethiopia, and in the West Bank and Gaza.

As hierarchy, al-Qaeda is organized with the emir-general, at the top, followed by other al-Qaeda leaders and leaders of the different groups. Horizontally, it is integrated with 24 constituent groups. The vertical integration is formal; the horizontal integration, informal. Immediately below the emir-general is the Shura Majlis, a consultative council. Four committees report to the Shura Majlis: (1) military, (2) religious-legal, (3) finance, and (4) media. Members of these committees conduct special assignments for Saif al Adel and his operational commanders. Operational effectiveness at all levels is reached by compartmentalization and secrecy. While the organization has evolved considerably since the U.S. embassy bombings in Africa in 1999, the basic structure of the consultative council and the four committees remains intact.[53]

In addition, almost all of the international terrorist organizations have developed training functions.

Violent Nature

Terrorist groups routinely utilize violence to advance and protect their interests. These groups are ruthless and suicidal in protecting their interests from rivals and law enforcement alike. Unprecedented violence—bombings, hostage taking, contract killings, kidnappings, and large-scale massacres—has increased with competition for political, ideological, and religious purposes.

Weaponry

For many years international terrorism analysts did not believe that the terrorists were willing to use weapons of mass destruction (WMD). But present-day reality shows that religious extremists and sects with messianic or apocalyptic mind-sets have a tendency to use WMD or any kind of equivalent approaches. Such religious groups as al-Qaeda and Aum Shinrikyo are inclined to use equivalents of WMD. The September 11, 2001 attack by al-Qaeda and the sarin attack on the Tokyo (Japan) subway system on March 20, 1995 by Aum Shinrikyo demonstrated this new shift in the utilization of new terrorist technologies.

To complicate this already critical situation, with the collapse of the Soviet Union, the possibility of nuclear terrorism has increased. As reported by the Center for Strategic and International Studies (CSIS), "The current trafficking situation shows disturbing upward trends, substantial quantities of materials are likely to remain at large, and the potential for an accident or use of smuggled nuclear materials probably is increasing..." The seriousness of potential nuclear losses is evident from the number of thefts reported by the Russian Interior Ministry officials. In 1993 they reported 27 thefts of nuclear materials and 27 incidents in 1994. Globally from 1993 to 2003 there were registered 540 attempts of illegal nuclear materials trade including 182 incidents with materials that can be used in assembling nuclear weaponry.[54]

Dirty bombs pose an ongoing terrorist threat that the United States must be prepared to counter. The representative of the 9/11 Commission stated in 2004 that al-Qaeda is still interested in using a radiological weapon or "dirty bomb," which was confirmed in September 2006 by a statement by then leader of al-Qaeda in Iraq, Abu Hamza al-Muhajir.[55]

This tendency of utilizing unconventional weapons shows the international terrorism asymmetry—the usage of unconventional weapons against expected conventional weapons. This shift requires in many instances the development of connections with the arms dealers, or those who can manufacture arms (the Chechen terrorists have connections with a machine-gun manufacturer in Kovrov, Russia, for example). International terrorism groups are involved in such organized crime activity as weapons smuggling.

Finance

Financial support to international terrorist groups comes from many sources, including state sponsorship, organized crime, and drug and human trafficking. Most of Marxist and leftist terrorist organizations are suffering now from a lack of funding because of the disintegration of the USSR and Warsaw Pact countries. As was mentioned above, international terrorists' funding and logistical networks cross borders, are less dependent on state sponsors, and are harder to disrupt with economic sanctions.

Funds can be moved to terrorists in many ways. It can be done through financial institutions such as banks that have secret accounts. For example, half of 15,000 accounts of Clearstream Clearinghouse in Luxembourg are unpublished. This institution was suspected of moving Osama bin Laden's money. Among the international banks with the most secret accounts are: Citibank (271), Barclays (200), Crédit Lyonnais (23), and the Japanese company Nomura (12). Also, there are 2,000 investment companies, banks, and subsidiaries of banks—mainly British, German, American, Italian, French, and Swiss—with unpublished accounts.[56] Western Union and similar businesses are able to send money worldwide in 15 minutes and no bank account, background check, or ID is required to send less than $1,000. According to the U.S. Treasury, al-Qaeda, Hamas, and other terrorist groups use Muslim charities for financial transactions. The Holyland Foundation charity of Richardson, Texas, has been used to support the families of Arab suicide bombers on the West Bank affiliated with Hamas.[57]

Reducing the Risk of Terrorism

Since 9/11 the U.S. has made visible progress in terrorism prevention planning and education. However, the modern history of international terrorism shows that it is not realistic to eliminate it or to control it, but it is possible to reduce it. For this purpose, it is not enough to just improve antiterrorism legislation to solve the problems created by terrorism globally. This is because the laws on terrorism are aimed to suppress terrorist activity and to punish those who are responsible, but it is much more important to prevent such activity from occurring in the first place. Even a successfully conducted antiterrorist operation with the terrorists' capture and apprehension cannot compensate its damages and cannot be evaluated completely positively, because it shows missed opportunities to prevent such an act.

A lot of negative consequences occur during the preparation stage of a terrorist act. Usually other crimes are committed before a terrorist attack (e.g., burglary; illegal weapon possession; acquisition of explosives, toxins, and radioactive substances). A substantial number of groups and even layers of society are getting involved in different negative criminal consequences; social tension is increasing; international, interethnic, and religious controversies are growing; legal nihilism is spreading; and opponent aggression is increasing. The most effective prevention and of course the most expensive and difficult is early prevention. Today, the U.S. government focuses on several main activities that can reduce the risk of terrorism:

1. Analyze, localize, and minimize those social, political, financial, and other factors that create fertile ground for international terrorism. It is necessary for these purposes to study thoroughly this phenomenon and its roots and reasons. Studying and explaining motives of the widespread cases of international terrorism and subsequent data gathering and assessment can play an important role in determining sets of problems that need to be solved in the near future—economic, social, and political.

2. One of the effective remedies in terrorism prevention activities by security agencies is an implementation of programs that reward individuals for information that leads to the prevention of terrorist acts or that leads to apprehension of people committing such acts.

3. Launch an information campaign designed to disclose the criminal and violent nature of terrorist groups and organizations. Build public awareness about the legal consequences of participation in any activities related to terrorism.

4. Develop public safety programs to protect vulnerable objects and locations.

5. Enhance community participation in "terrorist watch" programs.

6. Improve intelligence by increasing the cooperation between law enforcement agencies worldwide.

7. Enhance information sharing by centralizing and increasing the protection of the integrated information system of security forces globally.

8. Improve training for security forces by developing realistic antiterrorist action scenarios and organizing regular exercises for security forces and citizenry.

9. Foster coordination between security forces and communities on the basis of model local, state, and federal plans of responding to terrorist attacks.

10. Develop a general policy of covering terrorism through the mass media. Legal issues of mass media participation in antiterrorism activities have not been thoroughly illustrated. Law enforcement agencies should try not to broaden the scale of psychological war. If pro-terrorism statements appear, and unfortunately they do, they strengthen terrorists' courage. Agencies shouldn't mix terrorism with politics and ideology and should not depict it as an international plot against the United States as a whole. Wrongful and contradictory understandings of causes and roots of international terrorism became a result of insufficient antiterrorism actions in Russia.

Political leaders of the United States consider counterterrorism as one of the most important state tasks. Some of the main trends in this activity are legislative improvement, strengthening cooperation between communities and law enforcement agencies, creation of special task forces, increasing federal agencies' personnel, dealing with terrorism problems, and providing better technical equipment.

American security agencies trying to put pressure on the forces supporting terrorism will use all available resources, including military ones, to the full extent to punish terrorism, to assist and collaborate with other countries, and to not allow any weaknesses in dealing with terrorists.

International intelligence gathering is coordinated through Interpol today. Interpol's involvement in the fight against international terrorism materialized during the 54th General Assembly in Washington in 1985 when Resolution AGN/54/RES/1[58] was passed calling for the creation of a specialized group within the then Police Division to "… coordinate and enhance cooperation in combating international terrorism …" Interpol's multinational police cooperation process has a three-step formula for dealing with terrorism, a formula all nations must follow: (1) pass laws specifying that the offense is a crime; (2) prosecute offenders, and cooperate in other countries' prosecutions; and (3) furnish Interpol with and exchange information concerning the crime and its perpetrators.[59] The challenges for the intelligence community are enormous—uncovering "sleepers" currently residing in both urban and rural settings is crucial to disrupt and prevent the next attack, as there seems little doubt that we must expect another attack. The truly global nature of the terror threat cannot and must not be confined to profiling of an ethnic group. The global nature of the organization means that greater efforts and communication must come from intelligence communities around the globe.

Specific Threats and Responses

Executive Protection

Countermeasures against terrorism for businesses are not as costly as executives may believe. Of course, no program can entirely guarantee protection against attack. But a business can take action to lessen its attractiveness as a target. In general, executive protection programs include target hardening, bodyguard operations, and training sessions to teach executives how to avoid being identified as targets, and what to do if they become targets. The key to success in executive protection is preplanning for a possible attack. One successful approach has been the crisis management team (CMT).

A company's CMT is made up of a carefully selected group of experts in a variety of fields who meet several times a year to discuss how to prevent attacks and what to do if an executive is kidnapped or another type of threat is received by the firm (for example, a bomb threat). In general these teams are composed of a senior executive, a team leader, a security executive, a police liaison, a medical consultant, a lawyer, a financial adviser, a communications expert, and a terrorist liaison. Each team member brings specific knowledge that will be crucial should a threat be received. The senior executive is responsible for making the final decisions. The security executive must coordinate security operations for the facilities involved to protect other employees or company property. The police liaison is responsible for seeing that the authorities are fully apprised of the situation and that the company cooperates with the civil authorities to its fullest ability. The medical consultant must have access to medical files on each executive so that a medical profile can be developed to help the kidnappers keep the victim in good health. The lawyer interprets what actions the company can take without violating company policy or various laws. This is particularly vital when multinational corporations are involved in negotiations with foreign countries. In the past, several firms that cooperated with terrorists in an attempt to recover kidnapped executives found that the governments involved prohibited such negotiations and subsequently confiscated the firms' assets. The financial adviser is necessary to determine how and if funds can be pulled together to meet demands. Ideally most firms do not have large sums of cash simply lying around. Rather, firms invest their capital in equipment, stocks, and bonds. The communications expert handles any direct communications desired by the terrorists. In many cases the authorities may handle this operation. The terrorist liaison may be the most difficult member to recruit; this person must understand terrorist organizations and their intentions. The terrorist liaison must analyze the terrorists' demands and predict their responses to actions by the company. In general, a psychologist, psychiatrist, or sociologist fills this role.

Regardless of what actions a firm takes to prevent or respond to terrorist attacks, it is most important that some action be taken. The problem must be recognized, and some type of planning must follow. According to Joseph Marog, manager of the Counterterrorism Training Program for the Defense Intelligence Agency's Joint Military Intelligence Training Center, there are six trends that security managers should consider when developing security programs:

1. Terrorism will remain a persistent international problem. "Not only will it not go away, but it is continually evolving."
2. There will be an emergence of transient groupings of terrorists such as those involved in the World Trade Center bombing.
3. Terrorists will increasingly use "soft targets" such as businesses.
4. Attacks will become more lethal.
5. More attacks will go unclaimed.
6. The lines will become less clear between domestic and foreign terrorism.[60]

Bombs and Bomb Threats

Any business, industry, or institution can become the victim of a bombing or a bomb threat. Most telephone bomb threats—approximately 98 percent—turn out to be hoaxes. The target of the threat, however, has no way of knowing whether a real bomb has been planted. Contingency planning is necessary for an organization to be able to protect its personnel and property from the hazards of an explosion. In the absence of a specific response plan, the bomb threat will often cause panic. This may be the precise result the caller seeks.

Controlling access to the facility; having adequate perimeter barriers and lighting; checking all parcels and packages; locking areas such as storerooms, equipment rooms, and utility closets; and taking note of any suspicious persons or of anyone not authorized to be in an area are all measures that can thwart a bomb planter as well as a thief or other kind of intruder.

Contingency plans should specify who will be responsible for handling the crisis and delegating authority in the event a bomb threat is received. The officer in charge of responding to a bomb threat must be someone who will be available 24 hours a day. A control center or command post provisioned for communication with all parts of the facility and with law enforcement agencies should also be designated. All personnel who will be involved in the bomb threat response must receive training in their assignments and duties. Plans should be in writing.

Telephone Operator's Response

The telephone operator's role is critical in handling the bomb threat call. The operator should receive training in the proper response so as to elicit from the caller as much information as possible. The two most important items of information to be learned are the expected time of the explosion and the location of the bomb. The operator should remain calm and attempt to keep the caller talking as long as possible in hopes of gaining information or clues that will aid investigators. The caller's accent, tone of voice, and any background noises should be noted. Many organizations provide telephone operators with a bomb threat report form for recording all information (see Figure 16-1).

After receiving a bomb threat call, the operator must inform the designated authority within the organization (the chief of security, for example). Law enforcement authorities, and others in the organization, will be notified in turn according to the written contingency plans.

Search Teams

A decision must be made whether to conduct a search of the premises and how extensive the search should be. If possible, employee teams rather than police or fire department officers should conduct the search. Employees are familiar with their work area and can recognize any out-of-place object. An explosive device may be virtually any size or shape. Any foreign object, therefore, is suspect.

Basic techniques for a two-person search team include the following:

1. Move slowly and listen for the ticking of a clockwork device. (It is a good idea to pause and listen before beginning to search an area to become familiar with the ordinary background noise that is always present.)
2. Divide the room to be searched into two halves. Search each half separately in three layers: floor to waist level, waist to eye level, and eye to ceiling level.
3. Starting back-to-back and working toward each other, search around the walls at each of the three height levels; then move toward the center of the room.

If a suspicious object is found, it must not be touched. Its location and description should be reported immediately to designated authorities. A clear zone with a radius of at least 300 feet should be established around the device (including the floors above and below). Removal and disarming of explosive devices should be left to professionals. Those assigned to search a particular area should report to the control center after completing the search.

**CHECKLIST WHEN YOU RECEIVE
A BOMB THREAT**

Time and date reported: _____

How reported: _____

Exact words of caller: _____

_____

_____

Questions to ask: _____

1. When is bomb going to explode? _____

2. Where is bomb right now? _____

3. What kind of bomb is it? _____

4. What does it look like? _____

5. Why did you place the bomb? _____

6. Where are you calling from? _____

Description of caller's voice: _____

Male _____ Female _____ Young _____ Middle age _____ Old _____ Accent _____

Tone of voice _____ Background noise _____ Is voice familiar? _____

If so, who did it sound like? _____

_____

Other voice characteristics: _____

Time caller hung up: _____ Remarks: _____

_____

_____

_____

Name, address, telephone of recipient: _____

_____

_____

FIGURE 16-1 Telephone bomb threat information form.

Evacuation

Evacuating the facility for any reason, particularly in response to a bomb threat, is a drastic reaction to the potential danger. There clearly are situations when such an extreme course is indicated, but such a decision should never be undertaken lightly. It is essential that a thorough and exhaustive dialogue relative to this complex problem be undertaken at the earliest opportunity so that plans and policies may be formulated prior to any actual pressure caused by such an emergency. Many experts in the field argue that a total evacuation is rarely if ever indicated. The argument concerns the risk of exposing a great number of people to the blast when the location of the bomb is unknown. Whenever personnel are moved about in large groups, the possibility of exposure to injury is increased. Moreover, the movement of large numbers under the threat of bombing can create panic—a very dangerous situation.

A bomber who has shown familiarity with the facility by placing a bomb on the premises can be assumed to be familiar with normal and perhaps even abnormal or emergency traffic patterns of that facility. The bomber has probably placed the bomb in such a way as to create the greatest possible injury and havoc. In such a situation, total evacuation might serve only to expose the greatest number of employees to injury and death.

It has also been found that hoaxers or mischief-makers may be encouraged by a mass evacuation to repeat such calls and to subject the facility to a string of bomb threats.

The decision whether to evacuate will be made by management of the threatened facility, often in conjunction with law enforcement officials. It may be decided to evacuate the entire facility, only the areas in the vicinity of the suspected bomb, or (unless a device has actually been discovered) not to evacuate at all. Detailed plans are necessary to ensure safe and orderly evacuation. Personnel should leave through designated exits and assemble in a predetermined safe area. Elevators should not be used during evacuation. Doors and windows should be left open for increased venting of the explosive force.

When authorities have determined that the bomb threat emergency has ended, all personnel who have been notified of the threat should be informed that normal operations are to resume.

Other Specific Response Issues in the United States

For additional information on specific responses to terrorist threats in the United States, see Chapter 5 on Homeland Security. Chapter 5 provides information on responses by the federal, state and local government, private sector companies, and joint responses.

Al-Qaeda

Al-Qaeda, the self-proclaimed terrorist group that destroyed the World Trade Center, continues to be a major concern for the Department of Homeland Security. In 2007, the U.S. intelligence community identified al-Qaeda as the most serious terrorist threat to the homeland, with plans of high-impact plots and attempts to acquire and employ chemical, biological, radiological, or nuclear material in attacks where they would not hesitate to use them. The Department of Homeland Security continued to encourage homeland businesses, security, and law enforcement agencies to be vigilant. Because the group has been careful and meticulous in its planning of attacks in an effort to inflict the greatest number of casualties, the Department believes that there are indicators that an organization or facility has been targeted by the group. In its Information Bulletin 03-004, "Possible Indicators of Al-Qaeda Surveillance," the department lists the following indicators:

- Unusual or prolonged interest in security measures or personnel, entry points and access controls, or perimeter barriers such as fences or walls
- Unusual behavior such as staring at or quickly looking away from personnel or vehicles entering or leaving designated facilities or parking areas

FIGURE 16-2 Nuclear power plant.

- Observation of security reaction drills or procedures
- Increase in anonymous telephone or email threats to facilities in conjunction with suspected surveillance incidents indicating possible surveillance of threat reaction procedures
- Foot surveillance involving two or three individuals working together
- Mobile surveillance using bicycles, scooters, motorcycles, cars, trucks, sport utility vehicles, boats, or small aircraft
- Prolonged static surveillance using operatives disguised as panhandlers, demonstrators, shoe shiners, food or other vendors, news agents, or street sweepers not previously seen in the area
- Discreet use of still cameras, video recorders, or note taking at non–tourist-type locations
- Use of multiple sets of clothing, identifications, or the use of sketching materials
- Questioning of security or facility personnel

Nuclear and Radiological Threats

While it is impossible to accurately predict the extent of threat from nuclear or radiological devices, the fear is real. An attack on a nuclear power facility (see Figure 16-2), while unlikely, could be catastrophic. Given the comments earlier in this chapter referring to the loss of radioactive materials from the now defunct Soviet Union, the fear is substantiated. To support this fear, on May 2, 2002, Jose Padilla was arrested at Chicago's O'Hare International Airport. Padilla, returning from Pakistan, was suspected of plotting to use a "dirty bomb" in the United States.[61] Still, the probability of any individual or company being the victim of such a fearsome device is minimal.

Bioterrorism

The threat of bioterrorism in the United States is real. The anthrax scares in Washington, D.C., on November 14, 2001, and Florida, New Jersey, and New York in 2001 and 2002 brought the reality home. Bioterrorism is the use of biological agents to produce illness or death in

FIGURE 16-3 Gas masks for civilian use. (*Courtesy of IBN Protection Property, www.domesticfront.com.*)

people, animals, or plants. These agents can be distributed by air or as contaminants in food or water. Gas masks (see Figure 16-3) can be used for protection against airborne agents by first responders, military personnel, and civilians in the event of a wide-scale event.

The Centers for Disease Control and Prevention (CDC) has identified biologics that can be used as weapons, including: anthrax, smallpox, botulism, Ebola, and several other lesser known diseases. Only with rapid identification and containment can these biologics be controlled. There are many excellent Internet sites that provide detailed information for anyone interested in learning more about bioterrorism (see Appendix B).

Chemical Agents

Although not biologics, chemical agents can perform the same actions as viruses, toxins, and fungi/bacteria. Chemical agents include poisonous gas, liquids, and solids. One of the best known chemical agents, used during World War I and most recently in Iraq on Kurdish resistance groups, is mustard gas. This gas attacks the skin, eyes, lungs, and even the gastrointestinal tract.

Cyber Terrorism

While almost everyone who uses the Internet can relay stories of virus problems, identity theft, and other problems, most are not aware of the possibility that terrorists might use viruses or physical attacks to shut down the Internet. The media has publicized the use of the Internet by terrorists for communication and electronic transfer of funds. But will terrorists ever consider attacking the Internet? The answer is a resounding yes. According to a CIA report to the U.S. Senate Select Committee, al-Qaeda has expressed the intention of developing skills necessary for an effective cyber attack.

Despite the CIA report, only three federal agency computer systems were able to pass General Accounting Office security evaluations. While some of the problems may be associated with agency problems, flawed software products have left holes in otherwise well-planned systems.

The solution may come from the recently established Common Criteria (CC), an international program attempting to establish standards against which security products may be evaluated to provide assurance of the products' security features.[62]

The Future/Summary

During 2002, despite the capture or death of many al-Qaeda and other extremist leaders, there were 12 major terrorist attacks worldwide, killing at least 300. According to Control Risks Group, a private consulting firm that evaluates international business risks, the global community should expect "more of the same." Car and truck bombings will be the most common method of attack. The firm predicts that the major targets will be the United States and British and Israeli military and diplomatic facilities. Other sites include businesses frequented by foreign visitors such as shopping malls, restaurants, bars, and supermarkets. Infrastructure facilities such as energy companies and transportation (including oil refineries) may continue to be targeted. Still other targets identified by Control Risks' report are commercial shipping and aviation, schools, and churches, as well as individual Western leaders.[63]

■ ■ CRITICAL THINKING ■

What is your definition of terrorism? Defend your position. Is it possible to find arguments to defend the use of terrorism as a tool in waging war? Consider the American Revolution and its use of tactics in combating the British.

Review Questions

1. Where did the first "terrorist" groups first appear? What was the reason for such activities?
2. Discuss trends in terrorism over the past two decades.
3. Outline a proper response to a bomb threat.

References

[1] World Military Spending, Global Issues, downloaded 6/19/12, http://www.globalissues.org.

[2] Cunningham WC, Strauchs JJ, Van Meter CW. In: The hallcrest report II: private security trends 1970–2000. Boston: Butterworth-Heinemann; 1990. p. 17.3.

[3] Reports shed light on terrorist threat. Secur Manage November 1996:12.

[4] Berry N. Targets of Terrorists. Online: <http://www.cdi.org/terrorism/moretargets.html)>; (2007).

[5] Jilani Z. Just 25 Americans Died as a Result of Terrorism Last Year. Online: THINKPROGRESS SECURITY, <http://thinkprogress.org/security/2010/08/10/112860/25-americans-terrorism-traffic/>; August 10, 2010.

[6] Richards J. Europe and the Nature of the Terrorist Threat in 2007. Online: <http://intellibriefs.blogspot.com/2007/07/europe-and-nature-of-terrorist-threat.html>; July 16, 2007.

[7] Christian Science Monitor. Last Year 10,999 terrorist attacks worldwide – A decline from 2008. Online: <http://www.csmonitor.com/USA/Foreign-Policy/2010/0805/Last-year-10-999-terrorist-attacks-worldwide-a-decline-from-2008>; August 5, 2010.

[8] Alexander D. In: Business confronts terrorism: risks and responses. The University of Wisconsin Press Terrace Books; 2004. p. 23.

[9] Top Security Threats and Management: Issues Facing Corporate America, 2003 Survey of Fortune 1000 Companies. Online: <http://www.asisonline.org/newsroom/surveys/pinkerton.pdf>. Retrieved July 16, 2007.

[10] Pharoah R. An unknown quantity: kidnapping for Ransom in South Africa. Institute for Security Studies. Online: <http://www.iss.co.za/index.php?link_id=24&slink_id=1029&link_type=12&slink_type=12&tmpl_id=3>; (2005).

[11] Country Reports on Terrorism: Released by the Office of the Coordinator for Counterterrorism. Online: <http://www.state.gov/s/ct/rls/crt/2006/82738.htm>; April 30, 2007.

[12] Terrorist Threat. Secur Manage November 1996:12.

[13] Global Terrorism Database. Online: <http://209.232.239.37/gtd2/browse.aspx?what=target>.

[14] Wilkinson P. Why modern terrorism? Differentiating types and distinguishing ideological motivations. In: Kegley Jr. CW, editor. The new global terrorism: characteristics, causes, controls. Upper Saddle River, NJ: Prentice Hall; 2003. p. 106–38.

[15] Country Reports on Terrorism 2006. U.S. Department of State. Online: <http://www.terrorisminfo.mipt.org/pdf/Country-Reports-Terrorism-2006.pdf)>; April 2006.

[16] Henderson H. In: Global terrorism: the complete reference guide. New York: Checkmark Books; 2001.

[17] Trotsky L. On the Ninetieth Anniversary of the *Communist Manifesto*. In: Marx K, Bender Frederic L, editors. The communist manifesto. New York: W.W. Norton & Company; 1988. p. 139–45.

[18] Courtois S, Werth N, Panne J, Paczkowski A, Bartosek K, Margolin J. In: The black book of communism: crimes, terror, repression, fourth printing. Cambridge: Harvard University Press; 2001.

[19] Nechaev S. Catechism of a Revolutionary. <www.geocities.com/countermedia/5.html>.

[20] Holms J, Burke T. In: Terrorism. New York: Pinnacle Books—Kensington Publishing Corporation; 2001.

[21] Friedman G. The geopolitics of the palestinians. Geopolitical intelligence report. Online: <www.stratfor.com>; (2007).

[22] Country Report on Terrorism 2007. U.S. Department of State. Online: <http://www.state.gov/s/ct/rls/crt/2006/82736.htm>; April 30, 2007.

[23] Hoffman B. Terrorism trends and prospects. In: Lesser I, Hoffman B, Arquilla J, Ronfeldt D, Zanini M, Jenkins B, editors. Countering the new terrorism. RAND Corporation; 1999.

[24] Urbancic FC. Acting Coordinator for Counterterrorism; Russ Travers, Deputy Director of the National Counterterrorism Center. Washington, DC; April 30, 2007. Online: <http://www.state.gov/s/ct/rls/rm/07/83999.htm>.

[25] National Counter Terrorism Center: Report: Online: <http://wits.nctc.gov/crn2/cgi-bin/cognos.cgi>; May 29, 2007.

[26] Christian Science Monitor, August 5, 2010.

[27] Pillar PR. Terrorism Goes Global: Extremist Groups Extend Their Reach Worldwide. Brookings Rev Fall 2001:34–7.

[28] Schmid A. Political Terrorism (1983), cited by Simonsen C. and Spindlove J. Terrorism today: the past, the players, the future. Upper Saddle River, NJ: Prentice Hall; 2000.

[29] Terrorism: Can You Trust Your Bathtub? Terrorism Research Center (September 12, 1996), downloaded 1/13/2003, <www.terrorism.com>.

[30] Whitaker B. The Definition of Terrorism. Guardian unlimited May 7, 2001, <www.guardian.co.uk/elsewhere/journalist/story/0%2C7792%2C487098%2C00.htm>.

[31] Jenkins B. International terrorism: the other World War. In: Kegley Jr CW, editor. The new global terrorism: characteristics, causes, controls. Upper Saddle River, NJ: Prentice Hall.

[32] Hubbard DG. In: Winning back the sky: a tactical analysis of terrorism. San Francisco: Stonybrook; 1986.

[33] Kruglanski AW, Fishman S. Terrorism between 'syndrome' and 'tool'. Curr Dir Psychol Sci February 2006;15(1):1–48.

[34] Crenshaw M. Questions to be answered, research to be done, knowledge to be applied. In: Reich W, editor. Origins of terrorism: psychologies, ideologies, theologies, states of mind. Cambridge, MA: Cambridge University Press; 1990. p. 247–60.

[35] Jenkins B, Kegley Jr. C. The characteristics, causes, and controls of the new global terrorism: an introduction. In: Kegley Jr. CW, editor. The new global terrorism: characteristics, causes, controls. Upper Saddle River, NJ: Prentice Hall; 2003. p. 1–14.

[36] Katz R. Director SITE Institute Testimony before the House Armed Services Committee Terrorism, Unconventional Threats and Capabilities Subcommittee United States House of Representatives. The Online Jihadist Threat. February 14, 2007. Online: <http://armedservices.house.gov/pdfs/TUTC021407/Katz_Testimony021407.pdf>; February 14, 2007.

[37] Web on Terror. Online: <http://expertlancer.com/web-of-terror-part-1-extremists-take-to-the-net/>; July 6, 2007.

[38] Mueller R. Statement Before the Senate Select Committee on Intelligence. Online: <http://www.fbi.gov/congress/congress07/mueller011107.htm>; January 11, 2007.

[39] Navanti Group, 2011. Number of Terrorist Attacks by Region 2005-2010. Online: <http://navantigroup.com/content/number-terrorist-attacks-region-2005-2010>.

[40] "Report on Terrorist Incidents – 2006": National Counterterrorism Center. Online: <http://wits.nctc.gov/reports/crot2006nctcannexfinal.pdf>; April 30, 2007.

[41] D'Arcy P. Terrorists, enemies of mankind. Law Enforcement Executive Forum, March 2002.

[42] Patterns of Global Terrorism, U.S. State Department Report, 2002.

[43] Countering the Changing Threat of International Terrorism, report from the National Commission on Terrorism, Washington, DC; June 2000, downloaded 1/24/2003, <www.fas.org/irp/threat/commission.html>.

[44] Urbancic FC. Acting Coordinator for Counterterrorism; Russ Travers, Deputy Director of the National Counterterrorism Center. Washington, DC; April 30, 2007. Online: <http://www.state.gov/s/ct/rls/rm/07/83999.htm>.

[45] Lee R, Perl R. Terrorism, the Future, and U.S. Foreign Policy, Congressional Research Service, downloaded 1/24/2003, <www.fas.org/irp/crs/IB95112.pdf>.

[46] Niksch L. North Korea's Nuclear Weapons Program, Congressional Research Service, downloaded 1/24/2003, <http://fas.org>.

[47] United States Institute for Peace Press, "Terror on the Internet: The New Arena, The New Challenges," 2006; Global Issues Report, "Electronic Jihad Group Coordinates Sophisticated Hacking Attacks," October 10, 2006.

[48] Countering the Changing Threat of International Terrorism, 2000.

[49] Brynjar Lia. (February 2007) "Al-Suri's Doctrines for Decentralized Jihadi Training." Online: <http://www.jamestown.org/news_details.php?news_id=217>.

[50] Fischer R, Berger F, Heininger B. Terrorism: nothing new—a predictive model for handling terrorist incidents. Law Enforcement Executive Forum March 2002 [23+].

[51] Hudson R, Majeska M. The sociology and psychology of terrorism: who becomes a terrorist and why? Federal Research Division, Library of Congress, Washington, DC; 1999, downloaded 1/27/2003, <www.neuromaster.com/LOsocpsyterrorism/spt_12.htm>.

[52] Countering the Changing Threat of International Terrorism.

[53] Spindlove JR. Terrorism and countering the threat. Law Enforcement Executive Forum March 2002.

[54] Newsru (April 26, 2005). Online: (May 29, 2007) <http://www.newsru.com/world/26apr2005/bomb.html>.

[55] Dirty Bomb Vulnerabilities Taff Report, Permanent Subcommittee on Investigations, United States Senate, Washington, D.C. 2007. Released in conjunction with the Permanent Subcommittee on Investigations July 12, 2007 hearing, p. 1.

[56] Komisar L. Tracking Terrorist Money—Too Hot for U.S. to Handle. Pacific News Service (October 2001), downloaded 1/24,/2003, <www.alternet.org/story.html?StoryID=11650>.

[57] Frank A. On the Terrorist Money Trail, downloaded 1/24/2003, <www.evesmag.com/terroristmoney.htm>.

[58] Resolution AGN/54/RES/1 (Washington D.C.: U.S. 54th General Assembly, 1985).

[59] Spindlove.

[60] Terrorism: assessing the threat. Secur Manage September 1996:40–2.

[61] Kushner H, Editor. Introduction. Am Behav Sci February 2003.

[62] Uner E. The threat of cyber terrorism is real; it's only a matter of when. Secur Prod February 3, 2003:26.

[63] The Shape of Things to Come. Secur Manage March 2003:20.

17 ⣿

Computer Technology and Information Security Issues

OBJECTIVES

The study of the chapter will enable you to:

1. Identify various computer products.
2. Discuss possible attacks on computer systems and software.
3. Discuss options for protecting computers and information from fraudulent use and theft.

Introduction

Computers and information systems have traditionally been treated as something that the security/loss-prevention director needs to consider as a vulnerability; however, the 21st century has brought about a revolution in security operations. The following discussion on computers and information systems security will focus primarily on the services provided by the traditional roles of security in protecting computers. However, the trend is for security technologies to rely on the very computers that they are designed to protect. For example, information technology has bought closed-circuit television (CCTV), primarily used for surveillance, of age. Technologies like biometrics[1] have made possible video monitoring in the areas of facial and physical characteristics recognition, fire and smoke detection, and advanced alarm monitoring. With this growing integration of technology and the security operation, the traditional dichotomy associated with security and information technology often creates problems.

In 1946 the U.S. Army developed ENIAC (Electronic Numerical Integrator and Calculator), the first viable full-scale computer. At that time, computers were mysterious boxes utilized by scientists and thought to be the top-secret weapons of generals. Today, scientific pocket calculators have greater computing power than ENIAC, and most kindergarten kids know how to use a computer[2] or some type of handheld personal digital assistant (PDA) computing device, particularly those designed for electronic games. Computers have become an important part of peoples' lives, becoming an integral part of the way we work, teach, learn, and even play.

In government and business, computers are used to process, store and transmit vast amounts of information. Information processing tasks that used to take days or weeks for workers to compile are handled by today's computers in mere minutes, translating into greater efficiencies and greater productivity. Moreover, information systems are becoming primary methods of communications. E-mail, instant messaging, voice-over Internet protocol (essentially using computers and the Internet for voice communications, until recently the exclusive capability of telephones and telephone companies) are common and in many cases essential means of effective and efficient communications. Cellphones, smart phones (e.g., iPhone, Blackberry, Android) and laptops, along with tablet computers and electronic book readers, are virtually ubiquitous in today's society.

The criminal justice sector also relies on computers. Since 1924 the Federal Bureau of Investigation (FBI) has been responsible for keeping the nation's fingerprint and criminal history records. In 1967 the National Crime Information Center (NCIC) was established. Today the FBI has a computer system they call the Investigative Data Warehouse (IDW), described as one-stop shopping, giving FBI agents, from anywhere in the world, almost instant access to a database containing more than 650 million records. The search capability of this system has been described as an "Uber-Google."[3]

In the private sector, banks, insurance agencies, and credit rating agencies also process enormous volumes of computer data. For example, in the early part of this decade it was estimated that TRW Data Systems of California collected, stored, and sold access to information containing the credit histories of more than 90 million Americans. Banks, department stores, jewelry stores, and credit card companies pay them a subscription fee to access such information on current and potential customers. Today, Choicepoint, acquired by Reed Elsevier in September 2008, is a leading information broker with personal files on more than 220 million people in the United States and Latin America. This data is for sale to government organizations and the private sector.[4] Likewise, every major insurance company in America collects and stores information on past, current, and future policyholders.

Telemarketing and mail order professionals similarly buy, sell, and repackage such information like so many tangible products. The countless pieces of junk mail stuffed in Americans' mailboxes each day attest to the proliferation of such information brokers. Information brokers sell personal data to companies who then target for mail campaigns people who might be interested in their products.

The Dow Jones News/Retrieval Service offers stock market quotations, reports on business and economic forecasts, plus profiles of companies and organizations. The Source not only provides news and stock market indexes but also provides games and other forms of entertainment to its subscribers. Each of these information services is available to anyone with a computer, laptop, iPad, smart phone or any other type of personal digital assistant (PDA) device.

However, as with all great advances, there is a downside. Computer technology is changing so fast that equipment and software are often outdated before or as soon as it is installed, having a negative impact on the profit margin of the company. This is especially true for microcomputers.[5]

Of greater importance for the security professional are the criminal activities associated with the misuse of computers and the technology supported by them. Early in the 21st century, one of the fastest growing problems in this arena is identify theft. Problems that did not exist 25 years ago are commonplace today. For example, 25 years ago, few people had any fear of computer viruses. Today several major firms are in the business of protecting not only company computers, but also the computers used at home, from destructive viruses.

CSO, CISO and CIO Interactions

Information and information systems have become so critical to the efficient operation of business and government that organizations have in place senior executives to direct strategic and tactical operations associated with the creation, processing, transmission, storage and protection of information. Virtually all major corporations and government organizations have in place chief information officers (CIO) and chief information security officers (CISO). These executives either hold a seat in the C-suite (a term used to refer to corporate and organizational positions of the chief executive level for a particular function, most commonly the chief executive officer (CEO), chief financial officer (CFO), chief technology officer (CTO) and in the security profession, the chief security officer (CSO)) or directly report to someone with "chief" responsibilities.

The CIO and CISO work closely with the CSO and in most organizations have distinctively separate responsibilities. Where the CIO is responsible for the delivery of information services capabilities to the company, its workforce and other stakeholders, the CISO is responsible for the security of those information systems and the information contained within. In more traditional companies, the CSO is responsible for determining the sensitivity of information and is responsible for the protection of information when it is not residing within information systems. More specifically, CSOs have been, and often still are, responsible for the protection of information when it is in forms other than electronic. For example, much information exists in the form of documents. These documents, when containing pages of sensitive information, require protection. This protection usually is accomplished with more traditional security methods such as locked containers, files and safes kept in secure or protected company areas where unauthorized persons are not allowed physical access. These traditional security methods help prevent compromise or theft of sensitive company or organization information. In some companies and organizations the CISO duties are assigned to the CSO; however, it is more common to see them separated or to see a CISO reporting to a CSO.

Furthermore, CSOs are often charged with the responsibility of working with the creators of information and intellectual property attorneys to determine and assign some level of sensitivity to information. Information has different degrees of value and sensitivity. Some information is routine business information with no particular sensitivity or value while other information may contain trade secrets or strategic data that possess high value to the organization and perhaps even provide the organization with a unique competitive advantage. To properly protect sensitive information it is essential to be able to identify that information that is truly sensitive and separate it from less valuable information, by virtue of a physical separation or a process of uniquely identifying (marking) that sensitive information so it is clear to the possessor just how sensitive that information is. Moreover, the CSO is generally charged with developing procedures for protecting information determined to be sensitive when not contained within information systems and with ensuring the workforce understands how to protect sensitive information.

Essentially, the CIO, CSIO and CSO are collectively responsible for protecting the confidentiality, integrity and availability of all company or organization information. Confidentiality of information is the process of ensuring only authorized persons have access to protected information and that same information is used only for authorized purposes. Integrity of information is the process of ensuring the information is not manipulated in an unauthorized way or corrupted, thus diminishing its value and utility to the organization. Availability of information is the process of information being made available for authorized business use to authorized persons when they need access to perform work on behalf of the company or organization. Properly maintaining information in these three conditions—confidentiality, integrity and availability—is particularly complex and difficult for information residing on electronic information systems.

IT and Security Cooperation

The importance of cooperation between the CIO, CISO and CSO is critical if the organization is going to successfully protect information and information systems. It is best expressed by the reported responses of CIOs to the 2003 *CIO Magazine* survey. According to this report, security that had once been on the bottom half of the CIO spending lists has now moved to the fourth highest priority. Only systems and process integration, and finding ways to lower cost, are at the top, ahead of security. And even these priorities are of concern to the CSO.[6] However, the global economic recession, which began in 2008 and as of early 2012 has shown some improvement (but with economic forecasts indicating slow growth over the next several years), has adversely impacted corporate IT spending. In late 2010, Gartner predicted information technology (IT) executives would shift their spending focus to IT infrastructure upgrades.[7] What impact that will have on security-related spending remains to be seen.

Types of Computer Systems

Regardless of the type of computer system a given agency or company is using, there are four common elements: input, processing, storage, and output. *Input* refers to entering data and programs into the computer. This can be accomplished by using a keyboard, mouse, scanner, voice recognition software, or telecommunications methods such as traditional phone lines or wireless transmissions. *Processing* transforms the input into machine instructions. These instructions then exist in executable form within the computer. Hardware components such as the central processing unit (CPU), memory, and basic input/output system (BIOS) affect the computer's ability to process the input. *Storage* is a generic term that refers to the areas of a computer and associated media that store information such as data and programs. Examples of storage include internal or main memory, tapes, zip drives, hard disks, CD-ROMs, and memory sticks. *Output* is any on-screen result or printed report generated by the computer. Output devices are printers, monitors, and communication data.[8]

Microcomputers, minicomputers, mainframe computers, and supercomputers are the four general categories of computer systems available today. What separates these categories from one another is how much information the computer can store, the processing speed of the system, and the size of the computer system.[9]

Microcomputers

These are the smallest and least expensive of the four computer categories. Microcomputers are designed primarily for individuals or small businesses. Such systems can fit either on or beside a person's desktop.[10] Within this category are two types of computers: personal computers (PCs) and workstations.[11]

Personal Computers

These machines can sit on a desk, stand on the floor, or are portable, and are either IBM- or Apple-compatible. Both systems can operate easy-to-use programs such as word processing, spreadsheets, and data management programs.[12]

Non-portable PCs require an AC outlet and weigh more than 20 pounds. These systems do not require special installation requirements (for example, extra air conditioning or heavy-duty wiring). With desktop and floor-standing computers, the user can add circuit boards to the system to add functionality, such as boards for modems, scanners, video capture systems, and fax machines. The following are non-portable PCs:

- *Desktops* are machines that can fit on a single table or desk. A potential difficulty with this type of system is how much space the cabinet "foot-print" occupies.[13]
- *Floor-standing computers* are those in which the system cabinet sits as a "tower" on the floor next to the desk.[14]
- *Luggable systems* weigh between 20 and 25 pounds. These systems contain all the components (monitor, computer, and keyboard) in one unit, sometimes including a printer as well. These machines are also called *transportable* because they are designed to be moved, but not to be used in transit.[15]

Portable computers do not require an AC outlet. Instead, these machines operate from a battery. Weight for portables ranges from ½ pound to 20 pounds. Portable systems are designed to be used in transit and have no special installation requirements. The following are portable PCs:

- *Laptop computers* weigh between 8 and 20 pounds. These systems have a flat display screen, which can display mono or color images.
- *Notebook computers* get their name from their size, which is roughly the size of a thick notebook, and weigh between 4 and 7.5 pounds. These machines can easily be tucked into a briefcase, backpack, or simply under a person's arm.[16] Essentially, notebook computers are a smaller version of laptop computers.
- *Sub-notebooks* weigh between 2.5 and 4 pounds.
- *Pocket PCs* weigh about 1 pound. These computers are also called *hand-helds* and are useful in specific situations. Pocket PCs may be classified as either electronic organizers, palmtop computers, personal digital assistants (PDAs), or personal communicators.[17] Personal communicators include smart phones that can function as a video camera, portable media player and an Internet client with email and browsing capability, in addition to providing traditional telephone capabilities.[18]

- *Pen computers* are often the size of a sub-notebook or pocket computer. These machines lack a keyboard or mouse but allow the user to enter data by writing directly on the screen with stylus or pen.[19]
- *Tablet computers* are mobile computers larger than a mobile phone or PDA and integrated into a flat touch screen operated by touching the screen rather than using a physical key board.[20] Apple's iPad is a prime example of a tablet computer.

Although these computers are used at home or during travel, most also have the ability to be used as remote terminals to access company information. Through the Internet, it is not unusual for company employees to access company records and email from home. Given the ability of hackers to access home computers that are "always on" the Internet, security executives need to consider how to protect proprietary systems from well-meaning employees who may need remote access to systems and data.

Workstations

Workstations look like desktop PCs but are more powerful. These systems cost between $10,000 and $150,000.[21] Essentially, a workstation is a high-end microcomputer.

Minicomputers

Minicomputers make up the middle class of computer size and power. They are popular with small- to medium-size businesses because they can be used as servers and do not require special installation. *Servers* are central computers that hold data and programs for many PCs or terminals, called *clients*, which are linked by a computer network. The entire network is called a client/server network.[22]

Mainframes

Mainframe systems occupy specially wired, air-conditioned rooms and are the oldest category of computers. Mainframe computers are capable of great processing speed and data storage, allowing multiple users to utilize the system simultaneously. Because of their costs (between $50,000 and $5 million), large organizations use these systems, operating them with a staff of professional programmers and technicians.[23]

Supercomputers

The largest and most powerful computers are called *supercomputers*. Such computers are high-capacity machines that also require special air-conditioned rooms and specially trained staff. They are the fastest calculating devices ever invented. To achieve this capability, cost (typically from $225,000 to more than $30 million) is set aside to achieve the maximum capabilities that technology has to offer. Because of the cost, these machines are used primarily by government, large companies, and universities.[24]

Networks

With increasing numbers of computers in the workplace, employees and employers want to be able to share computer resources. This sharing of resources typically includes proprietary or sensitive data, printers, and other types of applications. Because of this need, networks were developed. A network is just two or more computers connected together.[25]

Local Area Networks

Local area networks (LANs) consist of two or more computers physically connected with some type of wire or cable (normally coaxial or fiber optic) that forms a data path over which information is transferred. Communications to a computer on the LAN are instantly broadcast to all the computers connected to the LAN.[26]

The most popular LAN communication protocols are Ethernet, Token Ring, and ARCnet. The Xerox Corporation developed Ethernet. When using the Ethernet protocol, computers must ensure that there is no traffic on the network before they are allowed to transmit information. IBM developed both the Token Ring and ARCnet protocols. LANs using these protocols pass a special data frame (or token) around the network in a predetermined order to enable data transmission. Under ARCnet, the order of token movement is based on a network address; in Token Ring networks, it relies on the physical placement of devices.[27]

Because of the way LANs are wired and the protocols they use, communication is limited to a short distance. This is not a major limitation; organizations and businesses have discovered that as much as 80 percent of their communications occur within a limited geographic area. This geographic area is frequently within the same department, office, building, or group of buildings.[28]

Wireless LANs

New technology has led several major organizations to adopt wireless LANs (WLANs). The networks operate on the open air, eliminating hardwire applications and their limitations. While the systems offer added flexibility in connectivity because users are not tied to telephone or other hard lines, they present real problems for those assigned to the protection of assets. The federal government will not allow any government-funded agency to introduce wireless technology until security is improved.

Wide Area Networks

It was generally recognized in the 1970s and 1980s that computers in different locations need to talk with one other. This led to the development of wide area networks (WANs). WANs are more powerful networks that can function across wide geographic areas at greater speeds than LANs. Most WANs are connected via telephone lines, although a variety of other technologies, such as satellite links, are used as well. Because telephone lines were used in this system, WANs do not allow multiple computers to share the same communication line, as is possible with LANs.[29]

The Internet

For years WANs used the X.25 protocol developed by the Consultative Committee for International Telephone and Telegraph, whereas LANs utilized different protocols. Because LANs communicate with Ethernet, Token Ring, and ARCnet, and WANs use X.25, these networks cannot communicate directly with each other.

The Department of Defense started a network in the 1970s called ARPAnet. This system allowed LANs and WANs to communicate with one another by using a new communications rule called the Internet Protocol (IP) packet. Today ARPAnet has evolved into the Internet, which still uses the protocol developed for ARPAnet.[30]

IP sends information across networks in packets, with each packet containing between 1 and approximately 1,500 characters, creating two problems. First, most information transfers are longer than 1,500 characters. Second, when data exceeds 1,500 characters, IP breaks the information into packets. These individual packets are then transmitted, which can lead to further problems. Packets can get lost or damaged in transit, or may arrive out of sequence.[31]

The transmission control protocol (TCP) was developed to deal with the problems of IP. TCP divides the information into packets, sequentially numbers each packet, and inserts some error control information. Each sequentially numbered packet is then addressed to the recipient. IP then transports the information over the network. When the host computer receives the packets, TCP then checks for errors in transmitting. If errors occur, TCP asks for that particular packet to be resent. Once all the packets are received correctly, TCP will use the sequence numbers to reconstruct the original message.[32]

There are many services available on the Internet. Electronic mail (email) allows individuals to send and receive messages from anyone on the Internet. Telnet allows people to log on to a remote computer and use the resources of that system if they have a valid account. Finger services allow people to ask for information about a particular user. Usenet is a system of discussion groups in which individual articles are distributed throughout the world. File Transfer Protocol (FTP) allows people to copy or move files from one computer to another. Gophers provide a series of menus from which a person can access virtually any type of textual information. The World Wide Web (the Web or WWW) is a hypertext-based tool that allows people to retrieve and display data. Utilizing both graphics and hypertext (data linked to other data), the Web is one of the most popular tools on the Internet. This is only a sampling of the services provided by the Internet.[33]

As noted earlier, although the Web has made life easier, it has also brought with it many new problems. Anyone using the WWW is well aware of the spam problem, cookies, and viruses. These are minor problems compared to the possibility that someone could steal your identity by stealing information that you share while online. In the first half of 2006, Symantec reported 2,249 documented new vulnerabilities representing an increase of 18 percent over the previous period and the highest volume of vulnerabilities recorded for any reporting period.[34] Five years later (April, 2011), Symantec reported a massive increase in the threat volume to more than 286 million new threats identified in the previous year. This number represents a dramatic increase in the frequency and sophistication of attacks on enterprises.[35] It also demonstrates how expansive computer usage has become. From government and commerce to personal usage, global dependence on computers and information systems is massive.

The Database Problems

There is little doubt that the data (note: within in this chapter the authors will frequently use the terms information and date interchangeably depending upon the context of the situation or description) collected for business has become the backbone of most organizations. Data management resulted in the creation of data management personnel or IT departments. Because data is stored in computers the management and security of these systems has created more problems for security than any other threat in recent years. Some 30 years ago the security department simply controlled access to the computing center, restricting access to only those few who needed to work in the center. Today, control of access is much more complex. The task of safeguarding these assets has many parts. As previously mentioned, the three major aspects include: 1) integrity: making sure that data is changed only in intended ways, 2) confidentiality: making sure that only authorized individuals view the information, 3) availability: making sure the data is available when needed to authorized persons.

But even when proper measures are in place to assure the above, there are still two problems. First, even authorized users sometimes use data improperly (deliberately or accidentally). Second, unknown flaws in policy and its implementation can allow for unintended data access and data changes.

CIOs stress the importance of accountability in maintaining database integrity. This accountability should determine who did what to which data when, and by what means. The CSO generally agrees with this approach. The answer rests in a simple concept: because technical systems are involved in storing data, technical systems must be involved in *safeguarding* the data. Such a program should do the following:

- Send notification when someone changes data or permissions
- Keep a record of all changes to data or permissions
- Know what data was changed, when, and by whom
- Know who has viewed certain data and when
- Generate periodic reports on who accessed certain tables
- Investigate suspicious behavior on certain tables
- Know who modified a set of tables over a period of time
- Automate procedures across multiple servers[36]

The Need for Computer Security

What is computer security? People normally answer that it is protecting computers and information from some type of theft. While true, this is only part of the answer. Earlier in this chapter we mentioned the need to protect information residing on computers or within information systems in the context of the confidentiality, integrity and availability of such information. This too is a form of computer security as it requires protecting access to the computer allowing only authorized persons. Furthermore, computer security must also deal with other hazards such as natural disasters like fires, floods, accidents, and so forth, essentially physically protecting them from harm. In fact, the *American Heritage Talking Dictionary* defines *security* as freedom from risks and dangers.[37]

Unfortunately, the types of crimes committed on a grand scale are often also perpetrated on the small scale. Those computer crimes that startled people a few years ago for their uniqueness and scope are now being mirrored in many communities across the nation.[38] According to a *Newsweek* report, independent hackers account for 82 percent of all Web attacks. Seventy-five percent of the problems are from disgruntled employees; this includes independent hackers. Other attacks come from competitors accounting for between 25 and 35 percent of the problems. Among this group are domestic and foreign corporations and some foreign governments. The number of hacks in 2002 reached 87,000 worldwide with the United States being the biggest target, followed in number of attacks on Brazil, Britain, Germany, and Italy. By 2010, the number of cyber attacks on United States federal government networks alone exceeded 41,700 out of more than 107,400 reported to the United States Computer Emergency Readiness Team (US-CERT).[39] Clearly, cyber attacks continue to be a persistent problem.

According to an American Society for Industrial Security (ASIS) survey, sponsored by the ASIS Council on Safeguarding Proprietary Information, during the period of July 1, 2000 through June 30, 2001, U.S. companies lost up to $59 billion in proprietary information and intellectual property.[40] Furthermore, e-commerce online retailers also suffered losses of more than $2 billion from the cost of purchases made with stolen credit cards (identity theft). By 2010, on-line fraud cost retailers in the United States $2.7 billion.[41] Although lower than the $3.3 billion in losses in 2009, on-line fraud continues to be a serious problem. Common types of Web scams are listed below:

- Internet auctions
- Shop-at-home/catalog sales
- Internet access services
- Foreign money offers
- Internet info/adult services
- Business opportunities
- Computers
- Web site design[42]

Classic Methods for Committing Computer Crimes

An initial entry into a business's computers often requires virtually no expertise. For employees, it is a routine matter, especially if security measures are not used. For nonemployees, it may be as easy as dialing a published telephone number and then using an obvious password such as "system" or "test." Once connected to the computer, the criminal has a wide range of methods available to disrupt system activity or to observe, steal, or destroy information.

Data Manipulation or Theft
Changing data during or after input into a computer system is the simplest, safest, and most common method of committing computer crime. Any size business is vulnerable to it. It can be performed by anyone associated with or having access to the processes for creating,

recording, transporting, encoding, examining, checking, converting, or transforming the data that is eventually entered.[43] Data theft has become a major precursor for identity theft. Insiders often sell data files to an individual who then uses the information to "steal" identities. (See the following section on identity theft.)

Salami Technique

This descriptive term implies trimming off small amounts of money from many sources and diverting these slices into one's own or an accomplice's account. This form of crime is most common in banking environments with a large number of savings and/or checking accounts and automated financial processing. By creating a new program or altering an existing one, an employee can randomly deduct one to five cents from a few thousand different individual accounts. The accumulated sums can then be withdrawn by normal methods from his or her receiving account.[44]

Trojan Horse

Appropriately named after the hollow horse given to the city of Troy, Trojan horse programs initially appear legitimate and will behave as if they were doing what the computer operator expects. However, the Trojan horse contains either a block of undesired computer code or another computer program that allows it to do detrimental things to the system of which the operator is not aware, such as infecting a machine with a virus, worm, bomb, or trapdoor. Remember, a Trojan horse program appears innocent and attracts users by inviting them to load it as some type of software. In reality, Trojan horse programs are not software, but ruses designed to penetrate a computer system so that a program of the penetrator's choosing can become active.[45]

Viruses

According to the popular press and the world in general, a virus is any hidden computer code that copies itself to other programs. In the computer field, a *virus* is a set of unwanted instructions executed on a computer and resulting in a variety of effects. In the year 2000 there were over 50,000 known computer viruses.[46] By 2008, it was estimated the number of known computer viruses stood at in excess of one million.[47] The problem appears to be growing exponentially. The term *virus disruption* is used to categorize computer viruses.[48]

Viruses fall into one of four categories based on the type of damage that the virus inflicts. *Innocuous viruses*, the first category, cause no noticeable disruption or destruction in the computer system. When humorous text or a graphic message is displayed without causing any damage or loss of data, then a *humorous virus* (second category) has infected the system. Categories three and four cause damage to the data stored in the computer system. *Altering viruses* change system data subtly (for example, moving a decimal to a different place, or adding or deleting a digit). When sudden widespread destruction of data both on the computer system and on peripheral devices occurs, the machine is possibly infected with the fourth category, a *catastrophic virus*.[49]

Worms

Some people regard worms and viruses as the same type of program. Each has a replication mechanism, an activation mechanism, and an objective. Nevertheless, viruses and worms are very different kinds of programs. While viruses just infect programs, worms take over computer memory and deny its use to legitimate programs.[50]

Hostile Applets

A new danger exists when using the World Wide Web to obtain information. The danger is from so-called hostile applets that utilize a Java-enabled Web browser. Java is Sun Microsystems' scripting language. Just as viruses perform a variety of tasks without the user's knowledge, so do hostile applets. The effects can range from mild distraction to data loss.[51]

Bombs

Like the Trojan horse method, a bomb is a computer code inserted by a programmer into legitimate software. There are two types of bombs: time bombs and logic bombs. A date or time triggers a time bomb, whereas some event, perhaps the copying of a file, triggers a logic bomb. There are several advantages to using bombs. The built-in delay makes the program harder to trace. Perpetrators can plan the event for maximum effect, with the delay allowing the bomb to be copied into backup files. Also, some companies implant bombs in their software. If customers fall behind in payments, or if customers attempt to copy the program, the bomb is set off and the program stops or the system is halted.[52]

Trapdoors and Back Doors

Doors allow programmers extensive access to test systems while they are being developed, allowing programmers access that would normally be denied. There are two types: trapdoors and back doors. *Trapdoors* are intentionally created and are normally inserted during software development. These doors are supposed to be removed once the software is completed. Unintentional access to software code is referred to as a *back door*.[53]

Time Stealing

This is one of the most common forms of computer crime because people do not consider the cost of accessing a computer without authorization. Any access uses the computer's resources (hardware, memory, software, peripherals), which cost money. Time stealing is comparable to driving another person's car without his or her knowledge.[54]

Electronic Eavesdropping

Tapping, without authorization, into communication lines over which digitized computer data and messages are being sent is electronic eavesdropping. By using technologically advanced listening devices, eavesdropping can be done on traditional telephone lines and even satellite transmission networks. If data transmitted are not encoded, capturing and transforming the data is equivalent to using a clandestine tape recorder to record a standard telephone conversation.[55]

Software Piracy

Providing software for computers is big business. Software programs can cost from a few dollars to thousands of dollars. Because of this, some people are willing to copy software and resell it or give it away. This unauthorized copying of copyrighted computer programs is referred to as *software piracy*. It has been estimated that for each legitimate copy of a software package sold, between 4 and 30 additional copies are made illegally. Although most copied programs are not resold, they deny vendors and software developers' profits that they should have accrued legally.[56]

Scavenging Memory

Information contained in buffers or random access memory is kept until the space is written over or the machine is turned off. This fact allows a person gaining access to these areas to search for sensitive data that may be left from previous operations.[57]

War Driving

This self-attached term refers to hackers who drive around locating wireless network points of entry. With today's technology anyone with a laptop and powerful wireless card can enter a company's wireless network. As the range of the systems increases so do the threats from the war driver.[58]

Identity Theft

Using information stolen from computer databases, criminals are committing criminal acts that impact on the person whose identity is stolen. While using false identification is an old method of criminal activity, the ability to access all types of information on the computer has given this old problem an entirely new life.

Between November 1999 and September 2001, the Federal Trade Commission (FTC) received 94,100 complaints from victims of identity theft. By 2010, Reuters reported some 8.1 million people in the United States were victims of identity theft.[59] Complaints show the types of criminal activity as well as the consequences to the victims caused by the identity thief. The following suggests some of the more common losses:

- Cash theft using ATM machines
- Electronic check fraud
- Denial of credit
- Financial service charges for overdrafts
- Lost time in dealing with the aftermath
- Criminal investigation, arrest

A study by California Public Interest Research Group and the Privacy Rights Clearinghouse reported an average of 175 hours of lost time (over a month of activity) in attempting to correct errors caused by identity theft. Of the reports to the FTC on money losses, 200 individuals reported losses of between $5,000 and $10,000, with an additional 200 persons reporting losses of more than $10,000. The American Banking Association (ABA) reports that 29 percent

($197 million) of its check fraud losses were attributable to identity theft. Two major credit card associations reported to the U.S. General Accounting Office losses of $144.3 million in 2000, up 43 percent from 1996. A 2001 report by Celet Communications projected that losses to financial institutions from identity theft would exceed $8 billion by 2004.[60] In January of 2006, a Reuters news report concluded identity theft costs U.S. consumers 4 percent more in 2005 than the $54.4 billion it cost in 2004.[61] By 2010, the *New York Times* reported (February, 2011) the average consumer out-of-pocket cost due to identity fraud increased to $631 per incident, up 63% from $387 in 2009.[62]

Federal Computer Legislation

The Computer Fraud and Abuse Act of 1986 (CFAA) was the first truly comprehensive federal computer crime statute and was an extension of Federal Statute 18 U.S.C. 1030, enacted in 1984. The CFAA has been amended in 1986, 1988, 1989, and 1990. This law only covers federal interest computers. A *federal interest computer* is one that is owned, leased, or operated by or for the federal government, contains federally protected information, or is used in interstate commerce.[63]

This act contemplates six offenses: the unauthorized access of a computer to obtain information relating to national security with an intent to injure the United States or give advantage to a foreign nation, the unauthorized access of a computer to obtain protected financial or credit information, the unauthorized access into a computer used by the federal government, the unauthorized interstate or foreign access of a computer system with an intent to defraud, the unauthorized interstate or foreign access of computer systems that results in at least $1,000 aggregate damage or modifies or impairs medical records, and fraudulent trafficking in computer passwords affecting interstate commerce. Penalties range from $5,000 to $100,000 or twice the value obtained by the offense, whichever is higher, or imprisonment from 1 to 20 years or both. These violations are investigated by the FBI's National Computer Crime Squad (NCCS), which was authorized by the CFAA.[64]

The Computer Fraud and Abuse Act covers all phases of computer crime including hacking, misuse of passwords, and bulletin boards. Electronic trespassers now commit a felony when they enter a federally related computer with intent to defraud. The malicious damage felony violation applies to any hacker altering information in that computer. Preventing other legal users from accessing the computer is also defined as a felony.

The CFAA has a far-reaching new provision regarding electronic bulletin boards. It is now a misdemeanor for any bulletin board operator to provide "any password or similar information through which a computer may be accessed without authorization." This includes any sharing of information with other board users on how to break into computers.[65]

Patriot Act

Immediately following 9-11 the U.S. government passed the Patriot Act. Part of its mandate deals with computer records and the Internet. Combating identity fraud is one of the act's primary goals, covered in Title III. Because false identities were found on a number of terrorists involved in the World Trade Center attack, and the terrorists in financing their efforts used identity fraud, the government has required financial institutions to increase efforts to prevent theft of information that allows for identity theft.[66]

Computer Systems Protection

Security professionals must protect information contained within the computer system from damage or loss. The system might contain any of the following components: an electronic data processing (EDP) center, a LAN, a WLAN, a WAN, or a PC. Regardless of the type of system, the security professional's dilemma is how to balance convenience in using the system against protecting the system from disasters, systems failures, or unauthorized access. Disaster-recovery planning, identification and access control of software and data, encryption, and physical security are the four facets of computer protection.[67]

Disaster-Recovery Planning

Disasters such as fires, floods, and earthquakes are potential hazards to essential computer systems. Because these threats are unpredictable, businesses must develop contingency or disaster-recovery plans. Contingency planning requires more than an occasional emergency drill; such plans must cover all business functions, including but not limited to emergency response requirements, personnel resources, hardware backup, software and data file backup, and backup for related and special activities.[68]

Contingency Procedures

Every company should have procedures for dealing with emergencies, whether natural or man-made. Without such planning, the initial response might be a knee-jerk reaction that could lead to people being injured or killed and damage or destruction of data, software, and hardware. To guard against counterproductive knee-jerk reactions, companies must implement contingency planning. Prior comprehensive planning is the first line of defense against all types of disasters.[69]

Placing the Computer Center. As a rule computer centers should not be in a basement, below grade level, or on first-floor sites. This prevents the entry of surface water into the center. In addition to avoiding areas that are prone to flooding, computer centers should not be placed in sites along known geological fault lines. If this is not possible, make sure that the building is constructed using approved earthquake-proof practices.

Certain areas of any building present problems for security. First-floor sites are most vulnerable to forcible attack, surreptitious intrusion, civil commotion, or terrorist attack. The top floor also presents opportunities for illegal activities. People can enter the facility through skylights or by cutting through the roof.

Ideally, from a security standpoint, computer centers should be within a company-owned area at least 200 feet from the closest public access. If the building houses other types of businesses, then the computer center should be on a floor completely occupied by the company and the floor above and below the site should also be company occupied. If a new site is being selected, the preferred location is either rural or suburban.[70]

Fire Protection. Buildings housing computer centers should be of noncombustible construction to reduce the chance of fire. These facilities must be continuously monitored for temperature, humidity, water leakage, smoke, and fire. Most building codes today require that sprinkler systems be installed.

Remember that water and electrical equipment do not mix. It is preferable to install a dry pipe sprinkler system rather than a wet pipe system. Dry pipe systems only allow water into the pipes after heat is sensed. This avoids potential wet pipe problems, such as leakage. In addition, fast-acting sensors can be installed to shut down electricity before water sprinklers are activated. Sprinkler heads should be individually activated to avoid widespread water damage.

Another type of fire-suppression system uses chemicals instead of water. Once this system is utilized it must be recharged. FM-200 is similar to Halon, which is no longer available, but with no atmospheric ozone-depleting potential. Carbon dioxide flooding systems are also available but should never be used. Carbon dioxide suffocates fire by removing the oxygen from the room. While this effectively extinguishes most fires, it also suffocates people still in the affected area.

All chemical fire-suppression systems are relatively expensive and require long and complex governmental approval to install. Neither chemical fire-suppression system protects people from smoke inhalation, nor can they deal effectively with electrical fires. They are, however, the only fire-suppression systems that do not require computer equipment to be turned off, assuring the quickest possible return to normal operations.

There should be at least one 10-pound fire extinguisher within 50 feet of every equipment cabinet. At least one 5-pound fire extinguisher should also be installed for people unable to handle the larger units. These extinguishers should be filled with either FM-200 or carbon dioxide. None of these agents requires special cleanup.

Install at least one water-filled pump-type fire extinguisher to use for extinguishing minor paper fires. Employees should be trained and constantly reminded not to use water extinguishers on electrical equipment because of the possibility of electric shock to personnel and damage to the equipment. They should also be discouraged from using foam, dry chemical, acid-water, or soda-water extinguishers. The first two are hard to remove and the others are caustic and will damage computer components.[71]

Personnel Issues. Crisis management focuses on the swift and effective action of personnel. This means that anyone involved in the emergency response plan must be adequately trained and kept up to date on any changes in procedures. When ranking emergency response procedures, protection of life is the most important, followed by protection of property, and finally limitation of damage. One way to verify that employees are familiar with and have current knowledge of the contingency plan is to conduct periodic drills.[72]

Hardware Backup

Most people think contingency planning and hardware backup are the same thing. This is not the case. Hardware backup is only one element of contingency planning. In this phase, classifying possible disruptions is useful so that hardware backup strategies can be developed. There are three categories of disruptions: non-disasters, disasters, and catastrophes.

Non-disaster disruptions are normally system malfunctions or other failures. Disasters cause the entire facility to be inoperative for longer than one day. Catastrophes entail the destruction of the data-processing facility. In this last category, a new facility must be built or an existing alternate structure must be identified to be used as the computer center.[73]

Once the extent of the disruption is ascertained, the company must make arrangements for alternate locations in which to conduct their computer operations. Alternative locations are categorized into hot, warm, and cold sites. *Hot sites* are fully configured and ready to operate within several hours. *Warm sites* are partially configured but are missing the central computer. Because the central computer is missing, these sites are less expensive than hot sites. However, it may take several days or weeks to locate and install the main computer and any other missing equipment necessary for operation. Once the equipment is installed, these sites can be operational within several hours. The least expensive sites are referred to as *cold sites*. These locations are ready to receive equipment but do not have any components installed in advance. Cold sites take at least several weeks to become operational.

The major factors in choosing the right "temperature" of the three types of sites are the company's needs in terms of activation time and cost. All companies must also have a way of alerting personnel of a disruption and telling employees which site to report to for work. Computer personnel must also be trained to operate the hardware at the new site. Finally, the hardware must be compatible with the equipment damaged or destroyed.[74]

Software and Information Backup

Software includes operating systems (for example, DOS, Windows, and Unix), programming languages (C, Pascal, Ada, COBOL, and so forth), utilities (virus checkers, security programs, and batch files), and application programs (word processors, databases, accounting programs, and so forth). Keep in mind that if the hardware at the alternate site is not compatible with the computers at the company, then the software will not operate.

Information and software are both less tangible and more dynamic than hardware. To protect these elements, it is necessary to consider both the physical storage environment and the frequency of change in data. Backing up information and software can protect the company from loss. Regardless of the approach used, backing up data involves copying files onto machine-readable media. The backup media can be tapes, hard drives, CD-R-RW, DVDs, and off-site third-party providers.

Information and software should be stored at on- and off-site locations. Many large organizations employ a tiered strategy, using several levels of backups to achieve a balance of safety and convenience. Ideally, a business should have four sets of backup files, with one set of files staying on-site and three sets of files being stored off-site.

On-site files should be housed in a fire-resistant safe designed for computer media. These files are the most recently created backup files until replaced by newer generations. Next, there is an off-site local backup location. This location is normally within a half-mile radius of the computer site. Files at this site are stored in a fire-resistant vault and accessed daily for rotation. Backup files are retained there for one week. Once files leave the off-site local storage facility, then they are moved to an off-site remote location, which is a minimum of 5 miles from the computer center. This site also contains a fire-resistant vault designed for computer media and is accessed weekly. The remote location is used to retain remaining backup files in active use for more than 1 week. Finally, any permanent records that need to be retained for several years are removed to archival storage. Archival facilities should be more than 50 miles away from the original computer site. The vault should be fire-resistant and earthquake-resistant. From the security standpoint, as the storage facility becomes more remote, accessibility decreases and security increases.[75]

Backup for Related and Special Activities

Besides protecting computer hardware and software, source documents must be protected. Source documents contain information transformed into machine-readable data from which printouts are generated. The printouts are referred to as either *human-readable output* or *hard copies*. This output is used to help in furthering the business activities of the organization. Source documents should be copied or backed up in the event of loss or destruction so that the basic information can be reconstructed in an emergency. Backing up may take the form of duplicate copies, photocopies, microfilm, microfiche, or many other forms of media.[76]

Identification and Access Control of Software and Data

People used to believe that only highly skilled technicians could gain access to computers. This illusion has been shattered by many well-publicized news stories. Today many people believe that any individual possessing basic computer skills can break into a computer system. Because of this perception and the fact that it has occasionally been proven correct, organizations must now go to tremendous lengths to protect their software and data.[77]

Computer systems can use three methods to determine if a person has a legitimate right to access the system. The three categories are:

1. What a person has: cards, keys, and badges
2. What a person knows: personal identification numbers (PINs), passwords, and digital signatures
3. Who a person is: physical traits

Each of these authentication methods is designed to make impersonation difficult.[78]

WHAT A PERSON HAS
Some systems require that an employee insert cards, a key, or a badge into the machine before it will allow access to data. Credit cards, debit cards, cash-machine cards, and ID badges are examples of cards. Cards can contain either a magnetic strip or a computer chip. Cards containing a computer chip are referred to as *smart cards*. With this system, the operator must insert the card before the machine will allow that person to access any information. With a key-lock system, a person must unlock the computer to use the system. This is one of the most popular types of security features found on PCs. Most PCs have a key-lock installed that allows the authorized user to lock out the keyboard. When the system is locked, keyboard input is not recognized.[79]

Cards, keys, or badges can be lost, stolen, or counterfeited.[80] In addition, the key-locks on PCs can be disabled if a person can remove the case of the machine. This drastic method is seldom necessary, because most PC locks use the same type of key. If someone has a computer with a key-lock, then it is possible that his or her key can open or close the lock on an unauthorized computer.[81]

WHAT A PERSON KNOWS

PINs, passwords, and digital signatures fall under this category. These security features work with any computer system. PINs work in conjunction with various types of card systems (for example, ATM cards or phone cards). With this system one inserts a card and then enters the PIN, a security number known only to the user. Passwords are special words, codes, or symbols required to access a computer system. Passwords work in conjunction with access logs. Access logs keep track of who got in, how often they tried to enter, when they entered (date, time, and even location), and when they left, whereas passwords allow an operator into the system. To discourage the misuse of passwords, companies should require passwords to contain at least eight characters that could be any combination of symbols, capital and lowercase letters, and numbers. Easily guessed or obvious passwords should be discouraged. Finally, the company may assign passwords to employees that are meaningless numbers, letters, or both. If the system requires a high degree of security, then a password should only be used once. The last "what a person knows" category, digital signatures, is relatively new. This system uses a public/private key system. One person creates the signature with a public key, and the receiver reads it with a second, private key. The "signature" is a string of characters and numbers that a user signs to an electronic document.[82]

The two biggest pitfalls of the "knows" systems are associated with passwords and PINs. Passwords can be guessed. People have a tendency to use real words or dates (their name, birth date, friends' or children's names, user initials, social security numbers, and so forth). Some system operators even fail to replace the default password. PINs and passwords are frequently written down by employees in convenient places easily discovered by others.[83]

WHO A PERSON IS

Biometric methods are utilized in this category. *Biometrics* encompasses the science of measuring individual body characteristics. Fingerprints, hand geometry, retinal patterns, voice recognition, keystroke dynamics, signature dynamics, and lip prints are common methods used to identify authorized users. In each of these methods, the computer compares the item being scanned with a copy of the item stored in the computer's memory. If the compared items match, the computer allows access. If not, the person is denied entry. Biometric techniques are not usually found on PCs because they require expensive equipment to be connected to the computer. This equipment limits mobility, which restricts its use with portable computers.[84]

Encryption

The best way to protect any type of data is to encrypt it. This also happens to be one of the best ways to protect data on portable machines, like laptop computers. Encryption scrambles the information so that it is not usable unless the changes are reversed. Today there are at least five different methods for encrypting data. Data Encryption Standard (DES) is a 56-bit algorithm. This standard was first published in 1977 and is used to protect federal unclassified information (in this usage, *unclassified* means sensitive information not falling within the United States Government's national classification system including confidential, secret and top secret data that requires protection due to national security concerns). Commercial users have adopted it. DES is used in financial applications to protect electronic fund transfers and by the Internet to encrypt information.

In 1978 a 512-bit key was developed that uses the Rivest, Shamir, and Adleman (RSA) algorithm. Another encryption algorithm is Pretty Good Privacy (PGP). Using both the International Data Encryption algorithm (IDEA) and RSA algorithm, PGP is available over the Internet and has become somewhat of a de facto standard for encryption on the Internet.

A new algorithm called Skipjack has been developed by the National Security Agency (NSA) to replace DES. Placed in a computer chip, this algorithm is referred to as a *clipper chip*. An enhanced clipper chipset is called Capstone. These chips allow law enforcement and other agencies with access to the algorithm key to break encrypted information. The Communications Assistance for Law Enforcement Act (CALEA) preserves law enforcement's ability, pursuant to court order or other lawful authorization, to access communications and associated call-identifying information. CALEA mandates that law enforcement agencies have the legal right to break encryption algorithms.

One last system, developed but not fully deployed, utilizes both encryption and digital signatures to protect email. This system is called Privacy Enhanced Mail (PEM) and uses both DES and RSA algorithms to encrypt email messages.[85] In 1996 the U.S. government mandated that all exported data had to be set at 128-bit encryption. Internet Explorer, Firefox, Safari and Netscape are all capable of using 128-bit encryption.[86]

Physical Security

Physical security places barriers in the path of attackers to deter them from attacking, delay them if they decide to attack, and deny them access to high-value targets should they succeed in penetrating the security system. There are two methods of security planning: traditional planning and strategic planning. Traditional methods start from the outside perimeter and work inward, whereas strategic methods are applied in just the opposite way.[87]

Electronic Data Processing (EDP) Centers

EDP centers have the same physical security needs as any other business or industrial establishments. Most EDP centers use the traditional security approach, beginning with the protection of the grounds around the building, then proceeding to the building's perimeter, the building's interior, and the contents of the building.[88]

With an EDP center, the outer shell provides perimeter protection and includes walls, fences, or partitions. Entrance protection restricts entry points to the EDP center. Doors and other entry points should be restricted to locations essential for safe evacuation in an emergency. A receptionist or security officer should be stationed at each entry point during all hours that the department is working.

Compartmentalizing a computer center into clearly defined rooms according to function (control, central processor, test and maintenance, storage, media library, forms, printing, waste) provides additional security. It enables access to each area to be controlled and restricted to authorized personnel. Electronic access control mechanisms such as badge-reading locks should be installed. Badges should only be issued to personnel with a need to be in a given area; badges can also be time-stamped to restrict access to authorized times.

In all circumstances, the computer room should be limited to operations personnel. Protection of these critical areas should follow the principle of "Authorized Access Only." Only those persons specifically necessary to its operation are allowed into the computer room. This should be the only room where programs, data, and computer equipment are all brought together. Extremely tight control of this room is imperative if the integrity and confidentiality of the data and programs are to be preserved. Installation of intrusion detection devices to monitor these critical areas when not occupied is warranted. These devices are usually wired directly to the department or company security office station to alert, identify, and monitor the location of an intruder.[89]

Personal Computers

Security used to be much easier when we only had EDP centers. These centers were and still are centralized, containing mainframes or supercomputers. Today, there are a multitude of personal computer systems. These systems range from minicomputers to pocket PCs. In addition, many of these stand-alone PCs are connected to either LANs, WLANs, or WANs. Furthermore, the user community is mobile and needs access to ever-increasing online resources.[90] Because of this, traditional security methods are inappropriate and inadequate. To protect PCs, the strategic method, where protection starts from the computer and works toward the perimeter, is best.[91]

A company's personal computers, like all other corporate computers, should have access limited to authorized users only. If all the computers are in a central location, restrict entry to this area using methods similar to the security measures used in EDP centers.

With LAN and WLAN systems, begin security procedures by locking up everything that can be physically secured. With the strong trend toward concentrating control at hubs, the LAN and WLAN systems become increasingly vulnerable. With LANs make sure that the wiring closets are secured with an appropriate lock system. Another entry point for obtaining data from a LAN system is through the wiring itself. In most companies, the wiring is hidden in the ceiling, walls, or under the carpet, giving a wiretapper a choice of points of entry. All original, necessary wiring needs to be documented and diagrammed. By routinely checking the diagrams against existing wiring, new or suspicious additions will alert security to a potential problem.[92] With WLANs problems are even greater. More will be said about protecting these state-of-the art wireless systems later.

For any PCs placed on a person's desk, a lock-down system attaching the equipment to the desk must be installed. There are four types of lock-down systems: cages, plates, cables, and alarms. These various systems discourage theft of the equipment. Do not neglect to ensure that equipment covers are tamper resistant. Some criminals are now removing computer chips taken from inside computers' cases and reselling them.[93]

In a similar vein, portable computers have become a popular item to steal. During the Gulf War, a laptop computer was stolen from a military staff officer's automobile in England. This machine's hard drive contained detailed plans for Great Britain's participation in the war. A common scam in many hotels, motels, and airports is for a person in front of an individual with the targeted portable computer to slow the line. If the target puts the notebook computer down, an accomplice standing behind the intended victim picks up the notebook and walks away. The first line of defense against that theft is "street smarts." This basically means keeping the computer in one's constant physical possession.[94]

Besides protecting the computer itself, security must also be concerned with storage media, particularly removable media (discs, tapes, drives, and so forth). People do transport storage media between work and home even if company policy forbids the practice. Memory sticks, flash drives, CDs discs, etc., are small enough to fit into a shirt or coat pocket. Even if the work environment is secure, the home environment is not. Media can also be lost between work and home. If the information contained is sensitive or irreplaceable data or programs, such a loss could be catastrophic. Employees can also alter the data, taking it back to the office where it is used to update the central computer. This incorrect data would then affect the entire organization.[95] To help defend against unauthorized downloads of information, some organizations disable USB ports and disc drives on computer systems containing their most sensitive data.

Content Monitoring and Filtering

The following discussion is based on an article by D. E. Levine, "Content Monitoring and Filtering," *Security Technology & Design* (March 2003): 70–74.

Less than a decade ago, companies paid little attention to monitoring network use, whether LAN, WAN, WLAN, or the Internet. Today, with the widespread use of the Internet, companies cannot ignore looking at who is using the service and what they are doing while on the network. Traditional security wisdom devoted time to monitoring specific types of activity. Unfortunately, experience has shown that traditional solutions are not always effective.

Vulnerabilities due to remote access through the networks fall into several major categories:

- Hacking—These technologically experienced computer users keep finding ways to enter and misuse corporate data and systems.
- Voice systems—The interconnectivity of computers and telephone systems has opened opportunities for computer techies to abuse telephone and voice systems.
- Remote and traveling employees—While the "road warriors" need access to company data and computers, maintaining security while allowing such remote access is a challenge.
- Disgruntled employees—Although not a new threat, the computer provides such employees with new opportunities to strike back at employers.

In a recent survey by the American Management Association, 75 percent of its members reported regularly monitoring employee phone calls, email and Internet use.[96] The number one rule in such activities is to inform the employees of the company's policies. The courts have ruled that while employers have a right to monitor their own systems, employees have the right to be informed regarding such monitoring activities.

Possible Security Solutions

Traditionally, security has relied on a well-written company policy to enforce access controls and handle computer abuse problems. In recent years, monitoring of information has been added to the tools available. However, until recently most of the monitoring was restricted to looking for destructive or prohibited content that entered the company computer network. Today security and IT managers are just as concerned about what goes out. Software packages allow companies to block out entertainment, gaming, pornography and other non–work-related sites, while still allowing for Internet access for company-related work. Moreover, many companies have developed policies specifically addressing access to social media sites (e.g., Facebook, Twitter and Myspace). On one hand, companies are using social media sites for legitimate business purposes so allowing access within that context is generally permitted. However, access for personal (non-company business) use is generally prohibited.

Network Security Policy (NSP)

Although this may be old information for some, it is vital that companies have a clear statement of network use. The establishment of an NSP is critical in developing other solutions. Employees need to know their rights as well as the expectations of the company. Most NSPs define the problem, set the requirements, discuss solutions, and set out punishment for infractions.

While many NSPs are written from scratch, there are companies that sell model policies that can be modified. In some cases firms or consultants will gladly sell their services to assist an organization in the development of these policies.

Appropriate Use Policy (AUP)

Closely allied to the NSP is the AUP. This document aids the NSP by clearly delineating what the company believes is appropriate use of company computers, software, networks, and email.

Virus Scanning

Most computer users are at least familiar with the concept of scanning for viruses. Some form of antivirus software should protect every computer system or stand-alone computer. As noted earlier, there are thousands of virus threats every year, making it almost impossible for the end user to keep current. There are a number of companies providing virus-scanning software. Among these are: McAfee, Symantec, Computer Associates, Panda, and Trend Micro. It is important to remember that, while these programs are generally effective, no developer can claim to be 100 percent effective because new viruses appear regularly.

Email Filtering Software

Because of problems noted earlier, such as spam, email filtering software has become common in many company security programs. Both CSOs and CIOs are interested in this type of software, just as they are with the solutions discussed above. Most of the commercial software allows the end user to set rules or protocols that mail must meet. When the email fails to meet criteria it is blocked. Some of the more popular vendors include:

- IM Message Inspector (www.elronsoftware.com)
- Email Filter (www.surfcontrol.com)
- Imira Screening (www.ulead.com)
- Mailwasher (www.mailwasher.net)
- Eblaster 3.0 (www.spectorsoft.com)
- Eudora (www.eudora.com)

Web Monitoring Software

Just as the email software allows for monitoring and blocking of email that fails to meet certain criteria, the Web monitoring software allows the end user to monitor what Web sites employees are using and for how long. Some software allows the user to filter or restrict access to certain sites. Many schools use this type of software to block pornographic and other adult content Web sites from their systems.

Some of the more popular vendors in this area are:

- Websense (www.websense.com)
- IM Web Inspector (www.elronsoftware.com)
- Surfcontrol Web Filter (www.surfcontrol.com)

Spam Filtering

Some authorities estimate that between 50 percent and 70 percent of the email received each day is spam. This may be one of the biggest problems, or more accurately annoyances, associated with Web use. This specialized email filtering software blocks email based on key words, sender's address, mail content, or other specified criteria. Vendors include the following:

- Spam Killer (www.mcafee.com)
- Spam Assassin (spamassassin.org)

Computer Forensic Investigations

Another tool in combating a variety of computer crimes is investigations. There are individuals, primarily consultants, who specialize in forensic investigations associated with computer crimes. These individuals are often self-taught computer users from the public law enforcement sector, security, or IT.

These experts can trace email, viruses, and other computer transactions. When the "I Love You" virus was trailed, it led investigators to the Philippines and the originator. Investigators have traced mails from bomb threats, as well as viruses. The investigator follows the trail using information stored in the receiving computer to post offices that handled the transmission. Investigators commonly use tools such as *Whois* or *Better-Whois*. These database search engines look for databases of registrars that record online users and their Internet Protocol (IP). Ultimately, the investigator finds the initiating machine. The next step is to determine who used the machine at the time the message was drafted.

Unfortunately, the tracing is usually not this easy because most computer users know methods to send false trails. Spoofing, or making the email appear it is from someone else, is common. Re-mailing is also designed to cause investigators additional grief. Stealing email accounts is another means of protecting the criminal's identity. It would be nice if there was always an easily followed audit trail, but the reality is that smart programmers find ways to get around security protection. It takes time for the investigator to discover these new techniques.[97]

As noted earlier in this chapter, the use of WLANs is expanding. As these wireless systems, with their advantages, become more widespread, CIOs and CSOs are challenged to protect the information that is transmitted over the airwaves. Tools currently available to detect unauthorized access to the WLAN include vulnerability scanners (software) such as Ping and other well-known network discovery technologies. These software packages can detect points of access, but will not identify the perpetrator. However, point of access information is vital to security efforts, since hackers need open ports to operate.

Dealing with Identity Theft

From a security position, the recent increase in identity theft presents unique problems. Identify theft is defined as using the identity information of another person to commit fraud or engage in other unlawful activities. Criminals are stealing identities by raiding the databases of legitimate company customers. The schemes may be old-style or high-tech. For example, a simple theft of personal information by a help-desk worker resulted in thousands of individual identity thefts by accomplices. The worker used his position at a credit-checking firm to access credit reports. The worker sold the credit reports to accomplices, who then sold the social security numbers and names of the individuals to identity thieves. In another scam, an identity theft ring placed a cohort as a temporary employee at a company's world headquarters. The employee, using access codes needed for work, accessed executive records. With social security numbers, names, and birth dates, the ring obtained credit cards. When apprehended the ring had charged more than $100,000 to the cards. In yet another case, an employee of a major insurance firm stole 60,000 personnel records, selling them over the Internet. A simple ad announced the sale of thousands of names and social security numbers. The going price for an individual identity can be less than $100. The bottom line is that personal information is only as safe as the company securing it.

What do identity thieves do with the information? Among other things the following are their most frequent activities:

- Open new credit card accounts
- Take over existing credit card accounts
- Apply for loans
- Rent apartments
- Establish services with utility companies
- Write fraudulent checks
- Steal and transfer money from existing bank accounts
- File bankruptcy
- Obtain employment using the victim's name

Organizations can no longer ignore this problem. The number of victims has become too great and the federal government has taken an interest in protecting citizens' personal information.

What to Do?

The proper action depends on the level of potential victimization. Companies need to find ways to protect the information that they gather, whether company information or private client information. Users/customers need to know what to do if and when they become victims.

Safeguarding Corporate Information

John May, consultant and author specializing in identity theft, makes the following recommendations:

- Properly dispose of personal information. These documents should be shredded.
- Conduct proper background checks on all individuals with access to personal information.
- Limit the number of temporary agencies working within your organization.
- Develop guidelines on handling personal information.
- Train the staff on information security.
- Limit the use of social security numbers. Don't use social security numbers on identification cards, time cards, or paychecks.
- Control access to personal information to those who have a legitimate reason for access.
- Secure personal employee information in locked files or with proper password access or through file encryption.
- Implement and enforce password security measures.
- Change passwords on a regular basis.

Protecting Your Identity

John May also has suggestions for protecting your own identity. While there are no totally fail-safe programs, the following will reduce personal risks.

- Invest in a personal shredder. Shred all personal information, credit card statements, cancelled checks, preapproved credit card offers.

- Purchase a mailbox with a locking mechanism.
- Review your monthly bills promptly.
- Order a copy of your credit report at least once each year. (There are three major credit bureaus—TransUnion, Experian, and Equifax.)
- Keep a record of all your accounts—numbers, expiration dates, telephone numbers, and addresses.
- Opt out of preapproved credit card offers by calling 888-567-8688.
- Minimize the amount of information you carry in your wallet or purse. Don't carry a social security card.
- Cancel any seldom used cards. Limit the number of cards you use.
- Don't leave outgoing checks or paid bills in your residential mailbox. Take them to the post office.

Much of the preceding information is from Johnny R. May, "Feeling Vulnerable? Corporate and Personal Identity-Theft Protection Procedures," *Security Products* (March 2003): 30–31.

When You Are the Victim

The FTC recommends the following should you become a victim of identity theft:

- Contact the fraud departments of the three major credit bureaus to report the theft of your identity. Ask that a fraud alert be placed on your file and that no new credit be authorized without your personal consent.
- Contact the security department of those organizations where your accounts have been accessed. Close those accounts. Put passwords on any new accounts.
- File a report with the local police. Get a copy of the report for your own protection, showing the date and time the theft was reported.

The FTC has created a simple fraud affidavit that can be sent to all financial institutions to alert them of the potential of fraud from stolen identities.

Education/Training

Companies should educate employees about the problem of identity theft, how to prevent it and what to do if victimized. Orientation opportunities should be conducted until all employees understand the significance of the problem to individuals and the company. Awareness can be increased through traditional techniques such as posters, brochures, or booklets. The use of email alerts is also encouraged.

Other Data Resource Vulnerabilities

While the focus of this chapter has been on computers, there are other company assets that may also present vulnerabilities to data theft. Sharp Electronics reports that many IT and security managers did not recognize the potential risks associated with copiers, faxes, and scanners. Survey results indicate that 77 percent of the respondents did not know that copier/printers contained a hard drive. Sixty-five percent said that copiers/printers presented little or no risk to data security.

What most users, including security and IT personnel, do not realize is that document information in these devices remains in memory until memory needs eventually overwrite the data. Peter Cybuck, Senior Manager, Business Development, Sharp Electronics Corp., suggests the following to provide proper security for these vulnerabilities:

- Limit access to copy/print/fax/scanners to authorized users only.
- Install network-based software to monitor use and flag abuse.
- Protect your devices from hacking by using secured network interfaces.
- Automatically erase document data.
- Protect confidential information from accidental or intentional viewing and distribution.[98]

■ ■ CRITICAL THINKING ■

What problems are created when a company decides that laptop computers are a better option for employees than desktop versions? What are the advantages of such a program as well as potential pitfalls?

Summary

The world of computers and the information that is stored, processed, analyzed, and disseminated by them are constantly changing. With change come vulnerabilities. While the progress achieved in this dynamic field has improved the general state of the world, there are always those who use the technology for personal gain or criminal activity. The CIOs, CISOs, and CSOs must work together to protect the companies and the individuals that they serve.

Review Questions

1. Why is it safe to assume that computer crime will increase in the years ahead?
2. What are some of the vulnerabilities unique to computer systems?
3. What are LANs? What are WLANs?
4. What significance does the term *password* have in the area of computer security?
5. What is the World Wide Web (WWW)?
6. How are LANs, WLANs, and the WWW security issues?
7. What are some of the management principles basic to computer security?

References

[1] See <http://en.wikipedia.org/wiki/Biometric>. Biometrics (ancient Greek: *bios*= "life", *metron*= "measure") is the study of methods for uniquely recognizing humans based upon one or more intrinsic physical or behavioral traits.

[2]   Covington PA. In: Computers: the plain English guide, 3rd ed. Jackson, MI: QNS Publishing; 1991.

[3]   CBS Evening News, FBI's New Data Warehouse A Powerhouse; August 30, 2006.

[4]   From Wikipedia, the free encyclopedia; See: <http://en.wikipedia.org/wiki/Reed_Elsevier>.

[5]   Fischer RJ, Green G. In: Introduction to security, 6th ed. Boston: Butterworth-Heinemann; 1998.

[6]   Ware and Lorraine Cosgrow. The survey: what you have to say. CIO Magazine; April 1, 2003, downloaded 4/2/2003, <www.cio.com/archive/040103/results_content.html?printversion=yes>.

[7]   See: <http://www.cio.com.au/article/360432/cio_spending_priorities_shifting_gartner/>.

[8]   Carroll JM. In: Computer security, 4th ed. Boston: Butterworth-Heinemann; 1996. [Cobb, Covington]

[9]   Sawyer SC, Williams BK, Hutchinson SE. In: Using information technology: a practical introduction to computers and communications. Chicago: Irwin; 1995.

[10] Clontz Covington, Rothman S, Mosmann C. In: Computer uses and issues. Chicago: Science Research Associates; 1985.

[11] Sawyer, Williams, Hutchinson.

[12] Covington, Sawyer, Williams, Hutchinson.

[13] Sawyer, Williams, Hutchinson.

[14] Ibid.

[15] Covington, Sawyer, Williams, Hutchinson.

[16] Ibid.

[17] Sawyer, Williams, Hutchinson.

[18] See: <http://en.wikipedia.org/wiki/IPhone>.

[19] Sawyer, Williams, Hutchinson.

[20] See: <http://en.wikipedia.org/wiki/Tablet_computer>.

[21] Sawyer, Williams, Hutchinson.

[22] Covington, Rothmann, Mosmann, Sawyer, Williams, Hutchinson.

[23] Covington.

[24] Covington, Sawyer, Williams, Hutchinson.

[25] Amoroso E, Sharp R. In: PCWeek: Intranet and Internet firewall strategies. Emeryville, CA: Ziff-Davis Press; 1996. Cobb Covington, Hahn H, Stout R. In: The Internet complete reference. St. Louis: Osborne McGraw-Hill; 1994. Levine DE. Local area network security. In: Hutt AE, Bosworth S, Hoyt DB, editors. Computer security handbook 3rd ed. New York: John Wiley and Sons; 1995. p. 22.1–22.21.

[26] Amoroso, Sharp, Cobb, Covington, Levine. Local area network security.

[27] Amoroso, Sharp, Levine. Local area network security.

[28] Amoroso, Sharp, Cobb, Covington, Levine. Local area network security.

[29] Amoroso, Sharp, Covington, Hahn, Stout, Levine. Local area network security.

[30] Amoroso, Sharp, Hahn, Stout, Krol E. In: The whole Internet: users' guide and catalog. Sebastopol, CA: O'Reilly and Associates; 1992.

[31] Amoroso, Sharp, Krol.

[32] Hahn, Stout, Krol.

[33] Hahn, Stout.

[34] Symantec Report; Dtd. September 25, 2006, Vulnerabilities in Desktop Applications and Use of Stealth Techniques are on the Rise, Cupertino, Calif.

[35] Symantec Corp., Internet Security Threat Report, Vol. 16; See: <http://www.symantec.com/about/news/release/article.jsp?prid=20110404_03>.

[36] Mazer MS. Data access accountability: who did what to your data when? A Lumigent, Data Access Accountability Series, White Paper, Lumigent Technologies, Inc.; 2002, downloaded 4/4/2003, <www.lumigent.com>.

[37] American Heritage Talking Dictionary, SoftKey, 1996.

[38] Boni WC, Kovacich GL. In: I-way robbery. Boston: Butterworth-Heinemann; 1999. p. 28.

[39] InformationWeek, Federal Cyber Attack Rose 39% in 2010: See <http://www.informationweek.com/news/government/security/229400156>.

[40] See: <http://www.asisonline.org/newsroom/pressReleases/093002trends.xml>. New York, N.Y. (September 30, 2002).

[41] See: <http://ecommercejunkie.com/2011/01/18/online-fraud-losses-decline-in-2010/>.

[42] The Dark Side of the Internet. Newsweek, March 17, 2003: special insert.

[43] Clontz, Magel, Schweitzer.

[44] Clontz, Magel.

[45] Carroll, Clontz, Levine DE. Viruses and related threats to computer security. In: Hutt AE, Bosworth S, Hoyt DB, editors. Computer security handbook, 3rd ed. New York: John Wiley and Sons; 1995. p. 19.1–19.24. Simond F. In: Network security: *data and voice communications.* New York: McGraw-Hill; 1996.

[46] Computer virus information: See: <http://www.cknow.com/cms/vtutor/number-of-viruses.html>.

[47] See: <http://www.prlog.org/10814398-number-of-known-computer-viruses-exceeds-1-million.html>.

[48] Carroll, Clontz, Dunham K. Introduction to viruses, <www.iste.org/~iste/antivirus/intro.htm> Levine. Viruses and related threats; Sawyer, Williams, Hutchinson, Simond; 1996.

[49] Clontz, Levine. Viruses and related threats.

[50] Clontz, Levine. Viruses and related threats, Simond.

[51] Clontz, Hoffman H. Hostile applets: the dark side of Java. Comput Shopper, October 1996:80.

[52] Carroll, Clontz, Levine. Viruses and related threats; Sawyer, Williams, and Hutchinson, Simond.

[53] Clontz, Levine. Viruses and related threats.

[54] Clontz, Magel.

[55] Ibid.

[56] Clontz, Rothman, Mosmann.

[57] Kabay ME. Penetrating computer systems and networks. In: Hutt AE, Bosworth S, Hoyt DB, editors. Computer security handbook, 3rd ed. New York: John Wiley and Sons; 1995. p. 18.1–18.55. [Schweitzer]

[58] Mattox M. Worried about wireless? Secur Prod, February 2003. [30+]

[59] See: <http://wsau.com/news/articles/2011/feb/08/id-theft-down-28-percent-in-us-in-2010-survey/>.

[60] Green link: the threat of terrorism and the role of financial institutions. Secur Prod, February 2003:38–41.

[61] See: Reuters news report, January 31, 2006, Identity theft losses grow; Web a small factor, <http://news.com.com/Identity+theft+losses+grow,+Web+a+small+factor/2100-1029_3-6033610.html>.

[62] See: <http://bucks.blogs.nytimes.com/2011/02/09/the-rising-cost-of-identity-theft-for-consumers/>.

[63] Carroll, Clontz, Rasch MD. Legal lessons in the computer age; 1996, <www.securitymanagement.com/library/000122.html>.

[64] Carroll, Clontz. Federal bureau of investigation national computer crime squad; 1996, <www.fbi.gov/compcrim.htm>.

[65] Clontz, Magel.

[66] Mattox.

[67] Sawyer, Williams, Hutchinson.

[68] Clontz, Hutt AE. Contingency planning and disaster recovery. In: Hutt AE, Bosworth S, Hoyt DB, editors. Computer security handbook, 3rd ed. New York: John Wiley and Sons; 1995. p. 7.1–7.35. [Sawyer, Williams, Hutchinson]

[69] Clontz, Hutt.

[70] Carroll, Clontz.

[71] Carroll, Clontz, Platt FN. Computer facility protection. In: Hutt AE, Bosworth S, Hoyt DB, editors. Computer security handbook, 3rd ed. New York: John Wiley and Sons; 1995. p. 12.1–12.24.

[72] Carroll, Clontz, Hutt.

[73] Clontz, Hutt.

[74] Clontz, Hutt, Sawyer, Williams, Hutchinson.

[75] Clontz, Hutt.

[76] Ibid.

[77] Clontz, Sawyer, Williams, Hutchinson, Walsh ME. Software and information security. In: Hutt AE, Bosworth S, Hoyt DB, editors. Computer security handbook, 3rd ed. New York: John Wiley and Sons; 1995. p. 14.1–14.20.

[78] Clontz, Kabay, Sawyer, Williams, Hutchinson.

[79] Bologna GL. Computer crime and computer criminals. In: Hutt AE, Bosworth S, Hoyt DB, editors. Computer security handbook, 3rd ed. New York: John Wiley and Sons; 1995. p. 6.1–6.31. [Clontz, Kabay, Sawyer, Williams, Hutchinson]

[80] Kabay.

[81] Clontz.

[82] Carroll, Clontz, David JR. Security for personal computers. In: Hutt AE, Bosworth S, Hoyt DB, editors. Computer security handbook, 3rd ed. New York: John Wiley and Sons; 1995. p. 21.1–21.21. Hoyt DB. Security of computer data, records, and forms. In: Hutt AE, Bosworth S, Hoyt DB, editors. Computer security handbook, 3rd ed. New York: John Wiley and Sons; 1995. p. 15.1–15.24. [Kabay, Sawyer, Williams, Hutchinson]

[83] Clontz, David, Hoyt, Kabay.

[84] Clontz, Kabay, Sawyer, Williams, Hutchinson.

[85] Clontz, Freeh LJ. Impact of encryption on law enforcement and public safety (June 26, 1996), downloaded 7/25/1996, <www.fbi.gov/congress/encrypt/encrypt.htm> Levine. Viruses and related threats. Rothfeder J., Hacked! Are your company files safe? PC World; November 1996:170–82. Simond, Sussman V. Policing cyberspace. U.S. News & World Report. January 1995:55–60.

[86] Determining your browser and encryption level. http://cgi.scotiabank.com/simplify/browser.html

[87] Clontz, National Crime Prevention Institute. In: Understanding crime prevention. Boston: Butterworth-Heinemann; 1986.

[88] Clontz, Fischer, Green, National Crime Institute.

[89] Magel.

[90] Clontz, Simonds.

[91] Clontz, National Crime Institute.

[92] Clontz, Simonds.

[93] Carroll, Clontz, David.

[94] Carroll, Clontz, Stone.

[95] Clontz, Simonds.

[96] Moses J. Checking Employees' Phone, Email and Internet Usage, National Federation of Independent Business E-News; April 03, 2003, downloaded 4/4/2003, <www.NFIB.com>.

[97] Poole T, Hansen J. Tips for tracking the E-Mail trail. Secur Manage; January 2001:42–7.

[98] Cybuck P. Machine talk. Secur Prod; March 2003:34.

18 ::::

Selected Security Threats of the 21st Century

OBJECTIVES

The study of the chapter will enable you to:

1. Gain knowledge of specific security threats not assigned to chapter level coverage.
2. Know the extent of the specific security problems and identify some of the security strategies used to reduce company or personal exposure.

Introduction

The preceding chapters highlighted some of the most important aspects of the security and loss-prevention field. The discussions in the past few chapters centered on security in areas such as terrorism, retailing and transportation/cargo security, and on the devices and techniques used to reduce potential losses. This chapter examines several of the security threats common to many businesses. Although these threats are common to many businesses in the United States, they are not necessarily vulnerabilities, with one possible exception, that will be faced by every firm.

Economic/White-Collar Crime

Today, the concept of white-collar/economic crime is familiar to most of us. The deceptions of executives at Enron, WorldCom and Martha Stewart and Bernie Madoff made the news as we entered the 21st century. Along with the corporate misconduct of these enterprises, accounting giant Arthur Andersen was humiliated (and ultimately ruined) in its alleged failure to apply proper accounting practices in its dealing with Enron. The junk-bond market and insurance scams are regularly reported in the news media. Ivan Boesky and other names are familiar for their involvement in various financial scams. We know that white-collar crime is against the law, yet there is no legal definition for it, and it does not appear as part of the criminal code (although some of the crimes committed by white-collar criminals are codified—for example, embezzlement). The Federal Bureau of Investigation (FBI) has defined white-collar crime as "those illegal acts which are characterized by deceit, concealment, or violation of trust and which are not dependent upon the application or threat of physical force or violence. Individuals and organizations commit these acts to obtain money, property, or services; to avoid payment or loss of money or services; or to secure personal or business advantage."[1]

Government and the private sector keep declaring war on white-collar crime, yet they have failed to define what they are fighting. The problem is that white-collar crime is a social abstraction, not a legal concept. Morality and society's commitment to equal justice are more a part of the white-collar crime issue than is a true legal definition. While some white-collar crimes are violations of criminal laws, others are violations of regulatory statutes and general standards of moral conduct. The most recent addition is the Sarbanes-Oxley Act of 2002, which addresses securities fraud. Corporate executives and auditors now face prison time if they knowingly submit false financial reports, alter or destroy records or fail to maintain a proper audit trail. Furthermore, most major corporations (including for-profit and non-profit) have instituted corporate ethics programs designed to ensure all employees, including those with fiduciary responsibilities, understand their obligation to comply with company policy, defined standards of conduct, government regulations and the law.

Still, the way corporate "wrongdoers" are handled appears to be lax, as borne out by the frequent lack of harsh penalties faced by executives at Enron, Arthur Andersen and Martha Stewart. A good example is the Credit Suisse First Boston settlement. Accused of a pervasive scheme to siphon tens of millions of dollars from customers' trading profits in 2000, the Securities and Exchange Commission noted that the company had failed to observe "high standards of commercial honor" and fined them $100 million dollars. Still, no one was charged with a crime.[2] Conversely, after pleading guilty to 11 federal crimes including securities fraud, money laundering, and theft from an employee benefit plan, Bernie Madoff was sentenced to 150 years in prison and $170 billion in restitution. Clearly, his penalty was deservingly harsh.[3] Executives generally do not go to jail. According to a University of California Irvine and New York's St. John's University study, executives who stole more than $100,000 got an average of 36.4 months in prison, while burglars received an average of 55.6 months, car thieves, 38 months, and first-time drug offenders, 64.9 months. Ironically the total cost from all bank robberies in 1992 totaled $35 million, about 1 percent of the estimated cost of Charles Keating's fraud at Lincoln Savings and Loan.[4]

While we are familiar with traditional white-collar crime activities, the newest entries, compliments of a shrinking global environment, are transnational white-collar scams. Criminals, terrorists, warlords, and drug-barons can reach freely around the globe. According to Michael Robert, editor of *Threat Level,* a newsletter devoted to corporate security issues, "Mafia-like organizations are springing up like daisies all through what used to be the Warsaw pact."[5] Perhaps one of the best-known activities in this area is the Nigerian money scam. The Nigerian money scam is a fraudulent scheme designed to extract money from residents of affluent nations, primarily Europe and North America.

Definitions

Edwin H. Sutherland made the first mention of white-collar crime as a concept in 1939 at an American Sociological Society conference. He defined white-collar crime as "an offense committed by a person of respectability and high social status in the course of his [or her] occupation." In 1970 Herbert Edelhertz presented a newer and perhaps more descriptive definition: an "illegal act or series of acts committed by nonphysical means and by traditional notions of deceit, deception, manipulation, concealment or guile to obtain money or properties, to avoid the payment or loss of money or property, or to obtain a business or personal advantage."[6] Edelhertz's definition has been widely accepted because it relies more on the actions of the perpetrator than it does on the economic social status. By including regulatory issues as well, there is also no restriction to criminal offenses in his definition. Experience has shown that white-collar offenses are not limited to the rich and powerful and that the offenses need not arise out of the context of one's occupation.

Regardless of which of the definitions one chooses for white-collar crime, the concept today has the following common elements:

1. It is an illegal act committed in the context of a lawful occupation.
2. It generally involves deceit, deception, manipulation, and breach of trust.
3. It does not rely on physical force.
4. It has the acquisition of money, property, or power as the primary goal.

Impact

Although the crime itself may be nonviolent, the end result could be violent (an industrial plant knowingly allows carcinogenic waste to pollute water; a pharmaceutical corporation knowingly sells a weight-loss drug that has potentially harmful effects on the heart). Thus, the nonviolent statement is only appropriate to the means of delivery or action necessary, not to the end results. As Gilbert Geis says, "corporate criminals deal death not deliberately, but through inadvertence, omission and indifference."[7]

The exact extent of white-collar and economic crime is difficult to establish. A lack of standard classifications and definitions coupled with limited reporting of workplace crime contributes to the problem of accurate measurement. Over the past 25 years, five studies have identified the need for indexes to measure economic crime and its true impact on society accurately. But there is still no progress! "Security executives told the Hallcrest research staff that their companies' incident or crime loss reporting system was incomplete or nonexistent." One key security executive with one of the nation's largest corporations said, "We probably know of only 1 fraud out of every 10 that is occurring or has occurred."[8]

White-collar crime will continue to be a growing problem into the 21st century. A relatively recent $1.1 million loss was the result of collusion between the contracting company vice president and the security firm vice president. Even security is not immune from white-collar crime problems.

Burglary and Robbery

Burglary and robbery have remained, in terms of the number of incidents, two of the most persistent crime problems in the United States. All types of establishments are subject to their attack, but certain businesses are more prone to these crimes than are others. Obviously banks and retail stores have more to fear from robbery than do manufacturing firms. Before discussing these two problems, we should define them; far too often people use the terms burglary and robbery interchangeably. *Burglary* is committing a crime through stealth by entering a building or other structure. In most cases, the crime committed is larceny (theft), but burglars also commit rape and arson among other felonies.

Robbery, on the other hand, is a crime of force or threat of force. While burglars do not like to confront people, robbers work by creating fear or by forcing people to give up their property. Although burglary may involve different kinds of crimes, robbery is strictly a crime comprising only two offenses—theft and assault and battery.

Statistics presented in the Uniform Crime Report (UCR) indicate that the growth of these two crimes has decelerated; however, both are difficult crimes to deal with. Robbery has the potential of physical harm that all firms must consider. Burglary has the problem of insolvability because less than 2 percent of all burglary cases are solved, and the recovery rate for stolen property is less than 4 percent.

Burglary

According to the FBI's UCR, burglary accounted for over 23.8% of all property crimes in 2010 with an average loss of $2,119 per incident.[9] Even though the number of incidents has decreased over the past two decades, burglary costs Americans billions of dollars annually. Because burglary is the most frequently occurring crime, it is essential that every business take particular care to protect itself against this form of crime.

The Attack

A burglary attack on a retail establishment is similar to such attacks on other types of facilities except that the job of the burglar is simplified by the ease with which the premises can be inspected before the attack. The physical layout, store routines, police patrols, and internal inspections (if any) can be easily assessed by a burglar posing as a customer. For a more exhaustive survey, the burglar might even pose as a building or fire inspector and make a minute examination of alarm installations, safe location, interior lock construction, and every other detail of the defenses.

Assuming a weak point has been found, the burglar enters through a door or window, through the roof, or possibly from a neighboring occupancy. Most successful burglaries (and more than 90 percent of detected attempts are successful) are made without forced entry. Curiously enough, the greatest numbers are made through the front door or main entrance.

A considerably smaller number of burglaries involve the stay-in who gathers the loot then breaks out and is gone before the guards or police can respond to an alarm.

Merchandise as Target

Most burglaries involve the theft of high-value merchandise, although any goods will do in the absence of the big-ticket items. Police reports repeatedly show the most astonishing variety of goods that have been targeted by enterprising thieves. Anything from unassembled cardboard cartons to bags of flour is fair game. Obviously such merchandise would hardly be considered high-risk assets, but it cannot be overlooked because possible burglars of loot come in all sexes, races, and sizes and what may seem to one as cumbersome and unprofitable may be remarkably appealing to another.

Cash as Target

Cash is naturally the most sensitive asset and the most eagerly sought, but because it is usually secured in some manner, it represents the greatest challenge to the burglar.

Stores keeping supplies of cash on hand are particularly susceptible, especially before payday. If such stores customarily cash payroll checks as a service to their employees, a burglar can assume that adequate cash must be on hand in anticipation of the next day's demands. Particular care must be taken in such cases if there is no way to handle cash needs other than to store it on the premises overnight. Every other avenue should be explored before the decision is made to keep substantial supplies of cash on hand overnight.

Because any cash will normally be held in a safe, it requires some degree of expertise to get at it. Boring, jimmying, blasting, and even carrying away the entire safe are the methods most commonly used. Few burglars are sufficiently skilled to enter a safe by manipulation of the combination, so applied violence is the normal approach. Unfortunately, even in cases where an attack is unsuccessful, the damage to the container is likely to be severe enough to require its replacement. This is equally true in all areas where entry is made or attempted. The high cost of repairs of damage to buildings and equipment can be almost as harmful as the loss in cash or merchandise.

Physical Defense against Burglary

Burglary defense has been widely studied for years, and several strategies have been developed to combat this crime. The most common approach is referred to as target hardening. Simply speaking, this means that the basic security precautions outlined in the preceding chapters are applied to the facility, including attention to door construction, locks, alarms, and surveillance devices.

It is always wise to let potential burglars know that the facility is well protected. In most cases, a burglar will avoid facilities where the chance of getting caught is great, and instead seek out other locations that are not as well protected.

Another defense against burglary that has met with some success is "reducing the value of merchandise." This approach usually includes marking company property with company identification tags or recording serial numbers that are easily traced. Because many fencing operations will turn away merchandise that is well marked and thus identifiable, burglars prefer merchandise that has no identifiable numbers.

In addition, some communities have developed "sting" operations—that is, fencing operations run by law enforcement officials. These operations require large outlays of cash but have proven effective in closing down major theft rings. Firms suffering substantial losses from theft should consider working with local officials in developing sting operations.

Alarms

Alarms of some kind can make the difference between an especially effective program and one that is providing only minimal protection. The type of alarm providing the best results is a matter of some disagreement, but there is little disagreement over the effectiveness of having some kind of system, however simple. Several alarm systems are covered in Chapter 10.

Many stores report satisfaction with local alarm systems. They feel that the sound of the signaling device scares off the burglar in time to prevent the looting of the premises. Many police and security experts feel such alarms are ineffective because the response to the signal is only by chance and because such a device serves to warn the intruder rather than aid in capture. Because most managers are less interested in apprehending thieves than they are in preventing theft and because local systems are inexpensive and can be installed anywhere, they continue to be widely used. Certainly they are preferable to no alarm system at all. With the relatively inexpensive dial alarm systems and national central station paid services, many retailers and homeowners have replaced older, less reliable systems with new state-of-the art technology.

Whether the outer perimeter should be fitted with alarms and whether space coverage alarms should be used is a matter peculiar to each facility. This must be carefully studied and determined by the manager. But it is always important to have the safe fitted with an alarm in some way.

Safes

The location of the safe within the premises will depend on the layout and the location of the store. Many experts agree that the safe should be located in a prominent, well-lighted position readily visible from the street where it can be seen easily by patrols or by passing city police. In some premises—especially where no surveillance by passing patrols can be expected—it is generally recommended that the safe be located in a well-secured and alarm-rigged inner room that shares no walls with the exterior of the building. Floors and ceilings should also be reinforced. The safe should be further protected by a capacitance alarm. The classification of the safe and the complexity of its alarm protection will be dictated by the amount of cash it will be expected to hold. This should be computed on the maximum amount to be deposited in the safe and the frequency with which such maximums are stored therein.

Basic Burglary Protection

Obviously the amount of protection and the investment involved in it will depend on the results of a careful analysis of the risks involved in each facility. There is no single way—no magic solution to the problem of burglary. Every system and each element of that system must be tailored to the individual premises.

Managers must evaluate the incidence of burglary in their neighborhoods as well as the efficiency and response time of the police. They must consider the nature of the construction of the building and the type of traffic in the area. Consideration must also be given to the ease with which merchandise can be carried off and by what probable routes. Managers must consider the reductions in insurance premiums provided by various security measures. The advice of city police in antiburglary measures and their experience in analogous situations should be sought out. In short, retailers, home owners and other businesses owe it to themselves to learn as much as they can about coping with the problem of burglary and dealing with it as forcefully, economically, and energetically as possible.

Robbery

Although the number of robbery incidents has declined, robbery remains a serious crime because of the potential harm to its victims. Firearms were used in 41.4 percent of the robberies for which the FBI's Uniform Crime Reporting Program received additional information in 2010. Furthermore, in another 16.7 percent of robberies, cutting instruments and other dangerous weapons were used.[10] Deaths in convenience store holdups make this point all too clear! However, convenience stores have taken a proactive approach to robbery reduction. According to the Bureau of Justice Statistics, the overall rate of robbery has remained relatively low since 1973.[11] Attention to the following points is credited with most of the reduction:

- Employee training programs.
- Proper cash control, extensive use of "drop" safes with signs noting that there is no more than $50 in the register during the day and $30 at night.
- Visibility measures, including proactive work with police agencies offering police substations in conjunction with store operations. Other programs allow police to write up reports at convenience stores.
- Use of technology, for example, closed-circuit TV (CCTV) and silent panic alarms.

Aside from private individuals, retailers and banks take the brunt of these attacks. Because less than a third of all robbers are arrested, retailers are obliged to take measures to protect themselves from this most dangerous crime.

The Nature of Robbery

The incidence of robbery varies only slightly according to geographical location, and the occurrence of such incidents is fairly constant throughout the week. The daily peak of robberies is at 10:30 P.M. Those stores remaining open all night, however, suffer the majority of robberies between midnight and 3:00 A.M. More than 90 percent of the money stolen is taken from cash registers, and 8 out of 10 robbers are armed with handguns. Eight out of 10 assailants are generally under 30, and of these, the vast majority is male. Although weapons are used (or the threat of harm), violence is generally carried out in less than 5 out of 100 cases. Eighty percent of the fatal cases occur in situations in which store personnel did nothing to motivate the attack. And while a lone robber carries out 60 percent of all robberies, 75 percent of the death or injury occurrences take place in situations involving two or more robbers.

We thus find that the robber is most apt to be a young man working alone, carrying a pistol, and threatening employees with bodily harm unless they hand over the money from cash registers. He rarely carries out his threats—probably because the employees wisely comply with his demands—unless accompanied by a partner, in which case he is very likely to cause death or injury without provocation. The deaths of convenience store employees during holdups with multiple assailants clearly illustrate this trend.

Robbery Targets

Robbers are typically criminals making a direct assault on people responsible for cash or jewelry. Their targets may be messengers or store employees taking cash for deposit in the bank or bringing in cash for the day's business. Robbers frequently enter stores a few minutes before closing and hold up the managers for the day's receipts, or they may break into stores before opening and wait for the first employees, whom they force to open safes or cash rooms. In some cases, managers or members of their families have been kidnapped or threatened in order to coerce access to company cash.

Delivery trucks and warehouses can also be targets for robbers. These latter cases require several holdup people working together, and they are a very real threat and do occur. The majority of cases, however, are still carried out by the lone bandit attacking the cash register.

Businesses most frequently attacked are supermarkets, drugstores, jewelry stores, liquor stores, gas stations, and all-night restaurants or delicatessens.

Cash Handling

Probably the major cause of robbery is the accumulation of excessive amounts of cash. Such accumulations not only attract robbers who have noticed the amount of cash handled but they are also more damaging to business if a robbery does occur.

It is essential that only the cash needed to conduct the day's business be kept on hand, and most of that should be stored in safes. Cash should never be allowed to build up in registers, and regular hourly checks should be conducted to audit the amounts each register has on hand.

Every store should make a careful, realistic study of actual cash needs under all predictable conditions, and the manager should then see to it that only that amount plus a small reserve is on hand for the business of the day. Limits should also be set on the maximum amount permitted in each register, and cashiers should be instructed to keep cash only to that maximum. Overages should be turned in as often as necessary and as unobtrusively as possible.

In large-volume stores, a three-way safe is an invaluable safety precaution. Such safes provide a locked section for storage of 2 or 3 hours' worth of money that may be called on for check cashing or other cash outlays. The middle section has a time lock that is not under the control of any of the store personnel and has a slot for the deposit of armored-car deliveries or cash buildups. Cash register trays are stored in the bottom compartment.

Movement of cash buildups from registers to the safes should be accomplished one at a time, and every effort should be made to conceal the transfer. This is particularly important at closing time when registers are being emptied. The sight of large amounts of cash can prove irresistibly tempting to a potential robber.

Cash-Room Protection

If the store has a cash room, it must be protected from assault. Basic considerations as to location and alarming were outlined in the discussion on antiburglary measures earlier in this chapter, but an additional precaution to protect against robbery should also be noted. The room itself must be secure against unauthorized entry during business hours. If possible, a list naming those persons authorized to enter the cash room and under what conditions should be drawn up. It should be made clear that there are to be no deviations under any circumstances from these authorizations.

The door to the room must be secure itself and securely locked. If fire regulations indicate the need for additional doors, they should be equipped with panic locks and have alarms fitted to them. The entrance should be under the control of a designated person who, through a peephole or other viewing device, can check persons wishing to enter. In larger facilities, an employee at a desk outside the room can control entrance, although this arrangement should be studied carefully before implementation because a robber could force such an employee to permit entrance unless the position is protected from the threat of attack.

Opening Routine
Because the potential for robbery is always present when opening or closing the store, it is important that a carefully prescribed routine be established to protect against that possibility.

At opening, the employee responsible for this routine should arrive accompanied by at least one other employee. The second employee should stand well away from the entrance while the manager prepares to enter. The manager should check the burglar alarm, turn it off, and enter the store. The interior, including washrooms, offices, back rooms, and other spaces that might offer concealment to robbers, should also be checked. After a specific, pre-established time, the manager should reappear in the main entrance and signal the second person with a predetermined code.

This code, which should be changed periodically, should be given both in voice and in some hand signal or gesture such as adjusting clothing or smoothing hair. One coded reply and gesture should indicate that all is clear; another should indicate trouble. It is important that this code be innocent and reasonable sounding to allay any suspicions on the part of robbers (if any are present).

If assistants get the danger sign, they should reply with something to the effect that they will return after getting a paper, checking the car, or buying a cup of coffee. The assistants should then proceed slowly and deliberately to a predetermined phone and call the police. In the past, experienced security experts all stress the value of a card with the number of the police station typed on one side and coins to make the call taped to the back. For lack of change, stores have been robbed. However, with today's communications capabilities, virtually everyone carries a cell phone. Store personnel should consider maintaining local police and other emergency numbers programed in their cell phone contact file for quick access.

Transporting Cash
Because a basic principle in robbery protection is minimizing cash on hand, money will necessarily have to be moved off the premises to a bank from time to time. Ideally this should be done by an armored-car service. If this cannot be done because of cost or unavailability of such service, efforts should be made to get a police escort for the store employee acting as messenger.

In any event, messengers so assigned must be instructed to change their routes; they should also be sent out at different times of day and if possible on different days of the week. They should regularly change the carriers in which the money is stored. It might be anything from a brown paper bag to a toolbox. They should stay on well-populated streets and move at a predetermined speed. The bank should be notified of their departure and given a close estimate of the time of arrival.

Closing Routine

Shortly before closing time, the manager should make a check of all spaces, similar to that conducted during opening. An assistant should watch unobtrusively for any unusual activity on the part of departing customers. When all registers have been emptied and the money locked up, the assistant should wait in the parking lot and watch as the manager completes the closing routine. When the manager has checked the alarm and locked up, the assistant will be free to go. If the manager signals trouble during any of this routine, the assistant should follow the same routine as at opening.

Other Routines

Because any entry into a store leads to a potential for robbery, it is important that the manager never try to handle it alone. There have been a number of instances where robbers have called managers and reported damage or a faulty alarm ringing in order to lure the managers into opening the store. They are then in a position to be coerced into opening the safe or at least into providing a bypass of the alarm system. If so notified, the manager should phone the police and/or the appropriate repair people and wait for them to arrive before getting out of the car. A manager should never try to handle the matter without some backup present.

Employee Training

A fully cooperative effort by all employees is essential to any robbery prevention program. Properly indoctrinated employees will know how to use a silent, "hands off" alarm if such an installation is deemed advisable. They will learn to question persons loitering in unauthorized areas. They will be alert to suspicious movements by customers and will know what to do and how to do it if they are aware that a holdup is in progress in some other part of the store. They will remain cool and make mental notes of the robbers' appearance for later identification. They will cooperate with the robbers' demands to an acceptable minimum. They will not under any circumstances try to fight back or resist. If possible, they will hand over the lesser of two packs of money if that choice is open, but they will never do so if there is the slightest chance that the robbers will notice. They will remain alert to the possibility of robbery, and if they are involved in one, they will make careful observations for assistance in apprehending the criminals.

Robbery Prevention

By incorporating some of the ideas already discussed in this chapter, several companies have developed robbery prevention programs. These are comprehensive plans designed to make stores less attractive targets for robbers, to protect employees, and to assist officials in apprehension.

The primary goal of these programs is to make the store less attractive to potential robbers. One successful strategy has been to publicize the fact that the store has a robbery prevention program and does not keep large amounts of money in cash registers and that employees do not have the combination to the drop safe. Today it is not unusual to see large signs stating these facts placed in plain view in many stores. This program became essential to several major convenience store and gas station chains during the early 1970s when a rash of robberies left several employees dead, and the importance of such programs again dramatically came to public attention with the rash of slayings associated with convenience stores in the 1990s.

Cash register location has also been carefully studied. In general, it is advisable to locate cashiers in areas where they are visible from the street but where the register drawer and cash transactions are not easily observed by passersby.

Robbery prevention programs also stress the well-being of the employee. No cash loss is worth the death of an employee. Employees should be instructed to cooperate and not do anything that would make the robber resort to violence. Although they should cooperate, employees should also be trained to be observant and to note any features and actions of the robber that might later help officials apprehend the robber. After a robbery has occurred, employees are often supplied with forms that ask for specific details about the robber's appearance and voice and show pictures of various weapon types as an aid to identification. While it may not be possible to describe a robber in the police jargon (for example, 6'2", 200 lb, light complexion, dark brown hair), it is possible to use descriptive aids that will allow the police to develop such a description. Because many people are not practiced enough to guess someone's specific height, such information is often not reliable. Employees can be trained, however, to match a person's height to reference points in the store or to other people. For example, employees might notice how the robber's height compared to their own or might observe that the robber came up to a certain point on a doorway. Weight and build might be better described by comparing the robber with another person—perhaps the police officer or store manager. In addition, employees must realize that the best information concerning the robbery is the escape mode. Did the robber leave on foot, on a motorcycle, or in a car? Only one employee should try to follow the robber after allowing the robber the time to get far enough away so that the employee is not in jeopardy.

As in the case of burglary, there is no absolute solution to the problem of robbery, but this does not mean that firms must accept robbery as a given of business. It has been shown that robbery attempts can be reduced by careful management procedures and by certain physical security measures.

Labor Disputes

If a facility is faced with a deteriorating labor-management relationship during negotiation of a new union contract, it may be necessary to review those security measures designed to protect personnel and property during unusual periods.

With the passage of the National Labor Relations Act and the Norris LaGuardia Act in the early 1930s, followed by similar legislation in many industrial states, a change in strike profiles began that has continued to the present. Whereas violence occurs with significant frequency, it is not generally planned or used in a tactical way by modern management or organized labor. Still, in specific disputes and local situations, violence may be intentionally generated.

Spontaneous violence occurs in almost every strike situation except the most mild. Dealing promptly and effectively with these sudden crises is critical in preventing the entire mood and complexion of the strike from becoming violent.

Security's Role

Security directors should review every element of the company's strike plans with their departments to ensure that the staff is totally familiar with the job to be performed and that every member of the staff is adequately trained to perform as necessary. They should be certain that all security personnel are aware of their role to protect personnel and property only and that they not participate in any way in the elements of the dispute, which is strictly between management and the participating union.

In clarifying policy and getting approval of specific plans of action from responsible management, foresight is indicated. The earlier the protection system is established, the better. Many elements will require time and considerable concentration from those executives who will later be occupied with handling the labor problem and therefore may not be available for consultation.

A checklist of some possible measures is listed below as a guide to the development of a more thorough plan that will apply more specifically to a given facility.

1. Secure all doors and gates not being used during the strike and see that they remain secured.
2. Remove all combustibles from the area near the perimeter—both inside and outside.
3. Remove any trash and stones from the perimeter that could be used for missiles.
4. Change all locks and padlocks on peripheral doors of all buildings to which keys have been issued to striking employees.
5. Recover keys from employees who will go out on strike.
6. Nullify all existing identification cards for the duration and issue special cards to workers who are not striking.
7. Check all standpipe hoses, fire extinguishers, and other fire-fighting equipment after striking workers have walked out.
8. Test sprinkler systems and all alarms—both fire and intrusion—after striking workers have walked out.
9. Consider construction of barriers for physical protection of windows, landscaping, and lighting fixtures.
10. Move property most likely to be damaged well back from the perimeter.
11. Be certain all security personnel are familiar with the property line and stay within it at all times when they are on duty.
12. Guards are not to be armed, nor are guards to be used to photograph, tape, or report on the conduct of the strikers. The only reports will relate to injury to personnel or property.
13. Notify employees who will continue to work to keep the windows of their automobiles closed and their car doors locked when they are moving through the picket line.

14. Consider the establishment of a shuttle bus for nonstriking employees.
15. Establish, in advance, which vendors or service persons will continue to service the facility, and make arrangements to provide substitute services for those unwilling to cross the picket line.
16. Keep lines of communication open.
17. Because functional organizational lines may be radically changed during the walkout, find out who is where and who is responsible for what.

Right to Picket

Federal and state courts, along with the Labor Management Relations Act, protect picketing on public property as an exercise of the right of free speech and free assembly. Although picketing is protected, there are established rules restricting both the purposes and methods used by picketing unions. The employer, nonstriking employees, customers, suppliers, service personnel, management staff, and visitors have a right to enter and leave the business without obstruction. There is no right to picket on private property.

It is important to remember that union officials and pickets have the right to talk to anyone entering or leaving company facilities. On the other hand, the nonunion individuals, truckers, managers, etc. also have the right to not engage in conversation. There is also no law against passing out pamphlets or leaflets on public property.

The responsibilities of security generally do not involve pickets unless they extend onto company property. Pickets are usually the responsibility of public law enforcement.

Pre-strike Planning

The best place to begin handling a strike is long before one is anticipated. The company should have comprehensive written strike and work stoppage procedures. These policies should be reviewed annually. Education of all management personnel, especially security, is a must.

Law Enforcement Liaison

Because a portion of the responsibility in dealing with strikes falls on the shoulders of local law enforcement agencies, it is essential that security and police personnel understand each other's mutually supportive roles. At least one company person, usually security management or someone from human resources, is assigned as a liaison to ensure a common understanding of company roles and law enforcement expectations. Such relationships should be cultivated over time and not wait until a strike situation. Similar relationships should be developed with local fire, telephone, city services, etc.

Espionage

Espionage usually brings about thoughts of spies sneaking into a company's private vaults and copying or stealing formulas or products. Government spies, typified by James Bond, working in glamorous settings, retrieve government secrets. The reality is somewhat different, but in a world of increasing corporate competition and computer-based data storage, the problem of espionage is increasing. The spy's tools may have transitioned to the computer and other sophisticated technology, but many of the people are Cold War participants, now working for private firms.

Still, the issue of what constitutes espionage can be gray. Competitive intelligence (which if not properly conducted can also be a form of industrial espionage) involves legal means of data collection. This ranges from analyzing publicity documents, schmoozing representatives to research on securities filings and news reports. It does not involve illegal means. The following quotation from a 2001 *Business Week* publication illustrates the point well: "Companies that engage in corporate spying see a payoff in increased revenue, costs avoided, and better decision-making."[12]

Research and development investment in the United States was about $174 billion for 1996. By 2010 research and development investment in the U.S. had more than doubled to nearly $380 billion, or 2.62% of nominal GDP.[13] In 1994 the FBI reported that its economic counter-intelligence unit had information that nearly 50 percent of research and development firms had a trade-secret theft, and 57 percent reported repeat or multiple thefts. An employee most often accomplished the thefts. Director of the FBI Louis Freeh noted that the cost to U.S. business for all types of espionage was more than $100 billion annually.[14] Corporations have gathered information on products, pricing, research, and corporate strategies for years from media reports, public financial statements, and other open-source research materials. In today's market more and more companies seem to be turning to the dark area of industrial espionage. According to James Chandler, executive director of the National Intellectual Property Law Institute, "We have seen industry after industry collapse after losses of intellectual property." An American Society for Industrial Security (ASIS) report found that 700 cases of theft of intellectual property at 310 companies were discovered during 1993.

Although espionage is not a new threat, the fall of the Soviet Union in the 1990s has prompted increased activity. The 2005 case of Hewlett-Packard executives spying on employees and each other and members of the press may be the current pinnacle of unethical practices. In this case the espionage (spying) was used against their own people in an attempt to find out who was leaking company information to reporters. In a recent Amercian Management Association and ePolicy Institute survey of 526 companies 76% reported that they monitor website connections used by workers, 26% have fired workers for "misusing" the Internet and 6% have fired workers for misusing the office phones.[15] Espionage (spying) can involve the disclosure or theft of many types of information. We tend to think of espionage involving information classifed under national security legislation and relating to political or military secrets. However, economic espionage where industrial, trade, and economic secrets are compromised or stolen is a real and growing threat and is practiced by private individuals and organizations as well as nation states.

Probably the most embarrassing case in recent years occurred in 1994, when the FBI arrested Aldrich Ames on espionage charges. Ames had been working for the CIA for more than 31 years and spying for the Soviet government since 1986. Ames passed secret CIA information on Russian agents working for the U.S. government to the KGB, resulting in the deaths of some compromised agents. Ames was paid millions of dollars by his KGB contacts during the years that he provided them with intelligence information. The case shows that any organization can be victimized.

One of the biggest cases of industrial espionage was filed in December 1996 against a purchasing chief for General Motors after he defected to Volkswagen. The former GM employee is accused of taking thousands of pages of confidential documents, slides, computer disks, and price lists. Specifically, the former GM employee is accused along with two other GM associates of taking plans for a superefficient car factory. VW recently opened a plant similar to the GM plan.[16] VW has also been the target of espionage. In September 1996 the company discovered a night-vision video camera and satellite uplink transmitter hidden in its Wolfsburg, Germany, testing facility.[17]

The full extent of the problem is not known, because many corporations do not disclose problems. Companies fear that disclosure of problems will hurt their image with stockholders and customers. A recent survey of Fortune 1000 companies, conducted by ASIS, estimates that the theft of intellectual property may amount to more than $300 billion each year.[18]

Those cases that are known tell a story of great financial gain for the agency or company acquiring information, while the victim suffers bankruptcy in some cases. Ellery Systems Inc. of Boulder, Colorado, a software developer, went out of business when an employee allegedly transferred 122 computer files with source codes from Ellery Systems computers to his own start-up company. In 1994 Genetics, Inc. was the victim of a theft of a patented drug formula. FBI agents posing as Russian agents discovered the theft. The two thieves contacted the agents in an attempt to sell the formula.[19]

To counter these espionage trends, the U.S. House Judiciary Committee approved the Economic Espionage Act of 1996; the legislation was sponsored by the U.S. Justice Department.[20] The Department of Justice has prosecuted several major cases since the introduction of the act. Insiders played a major role in the espionage cases providing information for pay to interested third parties. In one case an insider sold information to rival company Four Pillars Enterprises, Taiwan. The insider received a 6-month sentence while Four Pillars Enterprises was fined $5 million. This was the first case tried under the new law but not the last.

Following the implementation of the 1996 Act, the U.S. government has successfully prosecuted offenders with the most recent case occurring in 2007, where a naturalized American citizen of Chinese heritage was convicted of conspiring to export restricted defense technology to China. Chi Mak was employed by Power Paragon, a California defense contractor, part of L-3 Communications.[21] The actual data he shared with foreign nationals was export controlled. The United States Arms Export Control Act places restrictions on the sale of certain technologies with military applications. Such technology sales or transfers must have prior approval by the United States Department of State.[22]

In a 2004 presentation before the International Security Managers Association in Scottsdale, Arizona, Deputy Director Bruce Gebhardt highlighted the importance of dealing with economic espionage. Over $250 billion per year is lost in the theft of trade secrets and technologies. Counterfeiting of U.S. products by overseas competition costs at least the same. The number of countries now spying against U.S. companies has increased, with many countries using students and business executives to gather information. However, approximately 75% of the cases involve company insiders.[23]

Piracy

While certainly not a new problem, the copying of copyrighted materials by unauthorized means is a problem that has grown with the increase in the amount of material produced for public consumption. A police raid of a private home in Queens, N.Y., in 1996 discovered more than 85,000 pirated cassettes. The operation was copying a Mariah Carey tape with the confiscated value of $82,000 in cassettes and $30 million for 3 million labels.

According to the Recording Industry Association of America (RIAA), this one operation alone cost the recording industry more than $300 million in 1995. Overall the U.S. industry loses $200 billion from counterfeiting, according to the Better Business Bureau.[24] This is up from $60 billion in 1988. In 1982 the loss was only $5.5 billion. However, the problem has become so prevalent and complex that in 2010 the Government Accounting Offfice (GAO) concluded it is difficult, if not impossible, to quantify the economic-wide impact.[25]

Advances in technology have made possible high-speed, high-quality duplication by pirates. Compact disks (CDs), either audio or video, are the medium most often copied. Even home computer owners have the ability to download and copy their own audio (CD) and video (DVD) with burners. In a statement made in November 2002, the RIAA chairman and CEO pointed out the growing problem of piracy of CDs in Mexico. The Mexican Supreme Court recently relocated to a quieter area in an effort to get away from the loud music being played by street vendors who were selling pirated CDs.[26] Russia and China top the United States Trade Representatives list of piracy hot spots in 2007. Russia has pledged to take action against CD and physical piracy as it is seeking admission to the World Trade Organization.[27]

In related events Kazaa, a file-trading firm, was charged with American copyright violations through the use of the Internet. An Australian firm, the company has to date avoided the fate of parallel company Napster. The use of international offices to continue offering its product has upset the International Federation of the Phonographic Industry, which represents global music industry interests. The World Intellectual Property Organization (WIPO) tried to eliminate nation hopping when it enacted the WIPO Copyright Treaty, which outlines copyright legislation that all countries should adopt. The problem is that many European Union nations and Italy did not sign on.[28]

Piracy refers to the illegal duplication and distribution of recordings. It takes three specific and often confused forms:

- *Counterfeiting.* The unauthorized recording of the prerecorded sounds, as well as the unauthorized duplication of original artwork, label, trademark, and packaging.
- *Pirating.* The unauthorized duplication of sounds or images only from a legitimate recording.
- *Bootlegging.* The unauthorized recording of a musical broadcast on radio, television, or a live concert. Bootlegs are also known as underground recordings.

With the growth in this type of criminal activity, security measures are a must for any company involved in the production of audio and visual products for sale. The United States Customs Office seized $100 million in counterfeit products in 2003, almost double the figure seized in 2001.[29]

Besides the high-publicity piracy, other violations of copyrights and patent laws are less known to the public. In fact, many individuals who use the Internet and download documents may be committing violations of the copyright and patent laws. While there have been guidelines in the copying and distribution of copyrighted materials for years, recent use of materials from the Internet has prompted the federal government to take a closer look at electronic media and the potential abuse of copyright laws.

Of interest and some recognition, however, is the production of illegal copies of computer software. Much like the battle to control file swapping of music, production and sale of illegal copies of operating systems and computer games is big business. While Microsoft sells its Microsoft Office CD-ROM for approximately $800, pirated copies sell for much less, for as little as $10. The problem costs Microsoft an estimated $300 million each year.[30]

In the past several years counterfeiting of products has also come to represent a number of health and safety risks. Counterfeit pharmaceuticals have caused physical harm. An alert on Cialis sold on several websites was issued in March 2006 by the Danish Medicines Agency when counterfeit products were advertised on the Internet.[31] The pet industry is also plagued by counterfeit pet health care products. And most recently counterfeit auto parts have been making their way into the system. Brake pads that look like the real thing are made from compressed grass. It has been reported that there are enough counterfeit parts to assemble a complete car. In still another area, a shipment of counterfeit extension cords was seized by U.S. Customs. Not so lucky were Verizon customers who purchased new batteries that later overheated. In China, 60 infants were reportedly killed by fake baby formula.[32]

The worst violators of copyright law are China and Russia, according to a recent report by the International Intellectual Property Alliance (IIPA). The IIPA reported losses from industry members that topped $15.25 billion in 2006 from only the top 60 worst pirating nations. If the United States and the rest of the world were included, the figure could be as much as $35 billion annually. The only recourse available to the IIPA is through the U.S. Trade Representative, which could take action against countries by taking away certain duty-free trade privileges.[33]

Summary

Although the security concerns we have covered are not present in every business, they are of significant interest to private individuals where the impact of the crime is often felt in increased costs for merchandise and services. With the growing use of the Internet and associated technology, many of these crimes are changing direction to make use of the reach of the Internet or to exploit computer technology. In order to combat these crimes, security and law enforcement will need to enhance their tools by becoming proficient at monitoring and investigating crimes using computers and related technology.

■ ■ CRITICAL THINKING ■

Why should we as individuals have any concern over the illegal copying of products, particularly DVDs and CDs? After all, the only one hurt is the large corporation and they lose only small amounts compared to their total profits.

Review Questions

1. What distinguishes robbery from burglary?
2. What is target hardening? List several target-hardening techniques.
3. Describe the concept of robbery prevention.
4. List some of the measures that should be taken prior to a labor strike.
5. What is the latest problem associated with espionage?
6. What is the difference between piracy and bootlegging in the recording industry?

References

[1] U.S. Department of Justice, Federal Bureau of Investigation. In: White collar crime: a report to the public. Washington, DC: U.S. Department of Justice; 1989.

[2] Leaf C. White-collar criminals: enough is enough, downloaded 1/17/2003, <www.doublestandards.org/leaf1.html>.

[3] 10 White Collar Crime Cases That Made Headlines – Criminal Justice USA: <http://www.criminaljustice-usa.com/blog/2011/10-white-collar-crime-cases-that-made-headlines/>.

[4] Leaf C.

[5] The new white collar crime: enemy within? SC Magazine, downloaded 1/17/2003, <www.scmagazine.com/scmagazine/enemy>.

[6] Edelhertz as quoted in Green GS. Occupational crime. Chicago: Nelson Hall; 1990. p. 11.

[7] Gilbert Geis as quoted in Green, p. 16.8. Cunningham W, Strauchs JJ, Van Meter CW. The hallcrest report II: private security trends 1970–2000. Boston: Butterworth-Heinemann; 1990. p. 28.

[8] Cunningham W, Strauchs JJ, Van Meter CW. In: The hallcrest report II: private security trends 1970–2000. Boston: Butterworth-Heinemann; 1990. [p. 28]

[9] http://www.fbi.gov/about-us/cjis/ucr/crime-in-the-u.s/2010/crime-in-the-u.s.-2010/property-crime/burglarymain

[10] http://www.fbi.gov/about-us/cjis/ucr/crime-in-the-u.s/2010/crime-in-the-u.s.-2010/violent-crime/robberymain

[11] Bureau of Justice Statistics. National crime victimization survey: violent crime trends, 1973–2001, downloaded 1/17/2003, <www.ojp.usdoj.gov/bjs/glance/tables/viortrdtab.htm>.

[12] The corporate spy. The Integrity News, November 26, 2001; x(28).

[13] http://en.wikipedia.org/wiki/Research_%26_development

[14] Economic espionage plots loom. Security August 1996:14–5.

[15] As quoted in "The Changing Rules of Corporate Spy Games," <www.csmonitor.com/2006/0925/p02s01-usec.htm>.

[16] Maynard M. Execs indicted in theft of GM secrets. USA Today, December 11, 1996:1.

[17] Industrial security. Security October 1996:31.

[18] Wallace WA. Industrial espionage experts, downloaded 4/5/2003, <www.newhaven.edu/california/CJ625/p6.html>.

[19] Bernstein D. Pilfering trade secrets. Infosecurity News May/June 1996:23–5.

[20] Industrial security. Security October 1996:31.

[21] http://en.widipedia.org/wiki/Chi_Mak

[22] http://en.widipedia.org/wiki/Arms_Export_Control_Act

[23] <www.fbi.gov/pressrel/speeches/gebhardt011204.htm>, downloaded 7/15/2007.

[24] Better Business Bureau.

[25] Today @ PCWorld, April 14, 2010. <http://www.pcworld.com/article/194203/government_says_data_estimating_piracy_losses_is_unsubstantiated.html>.

[26] Glasner J. No More Music Piracy, *Por Favor*. Wired News November 22, 2002 [downloaded 4/5/2003, <www.wired.com/news/digiwood/0,1412,56522.00.html>].

[27] Russia, China Once Again Top USTR List of Piracy Hot Spots, <www.riaa.com/newsitem.php?news_year>, downloaded 7/16/2007.

[28] King B. Kazaa taunts record biz: catch Us. Wired News September 25, 2002 [downloaded 4/5/2003, <www.wired.com/news/mp3/0,1285,55356.00.html>].

[29] <www.bosbbb.org/counterfeit/index/html>, downloaded 7/16/2007.

[30] Kamber M. Street scene: mexico's CD pirates. Wired News July 22, 2000 [downloaded 4/5/2003, <www.wired.com/news/culture/0,1284,37482.00.html>].

[31] Warning: Illegal copy of the medicinal product Cialis sold on the Internet, <www.dkma.dk/visUKLSArtikel.asp?artikelID=8728>, downloaded 7/16/2007.

[32] Hazardous Counterfeit Products, <www.cbsnews.com/stories/2004/06/25/eveningnews/consumer/main626211.shtml>.

[33] Nystedt D. China and Russia Top List of Worst Copyright Violators. InfoWorld, <www.infoworld.com/archives/article/07/02/12/HNwortcopyrightv... 2/13/2007>.

19 Security: The Future

OBJECTIVES

The study of this chapter will enable you to:

1. Consider the future and what challenges and opportunities will face the security profession.
2. Know what resources might be available to the security profession.
3. Discuss the critically important relationship between the public sector and private industry.
4. Understand the importance of pandemic planning and brand protection management.

Introduction

In the 21st century, most of the world is concerned about terrorism. In particular the United States and other developed nations have focused resources to wage war against terrorist networks. While the federal government focuses on homeland security, with the cooperation of state and local governments, local communities must still contend with crime prevention.

Despite the efforts of law enforcement agencies, community programs and private security, the crime problem continues to be a major concern for most U.S. citizens. Today, few would argue that the criminal justice system alone has been totally effective in reducing crime. In addition, declining funding for public law enforcement—as funds are diverted to homeland security and the war on terrorism—has forced most law enforcement administrators to reduce services even in a time when the federal government is asking for more. Many such policies forbid law enforcement personnel from responding to or investigating certain types of crime. In fact, most local law enforcement agencies are triaging their responses, focusing and responding primarily to the most violent crimes in their communities. Hope is not lost, however, and to fill this void, criminal justice administrators and scholars have called for greater involvement from the private sector. This is where the security profession comes into play on the crime prevention field. A collaborative partnership among the security profession, public law enforcement, and citizens at large is necessary to protect the homeland and reduce the fear of terrorism and crime. The private sector has embraced this challenge and has responded willingly, including participating in volunteer federal programs such as the Customs Trade Partnership against Terrorism (CTPAT).

The overall perceptions of the terrorist threat and crime are important, but the economic impact on businesses is equally significant to security management and the resiliency of the business in the long run. The problem stems from the nation's inability to find a set of accurate metrics regarding criminal behavior. Not only has the problem of economic crime continued to grow in size, it has also become increasingly sophisticated, aided by the use of computers (what is commonly referred to as *cybercrime*). For example, and since 2008, the recent rash of insider trading and Ponzi schemes are a result of major corporate and government governance failures. The growth of identity theft is an indicator of problems yet to come. As such, it is imperative that the security industry come up with proactive solutions to help defeat these kinds of white collar and cybercrimes, which naturally includes a reporting mechanism back to federal law enforcement agencies.

The Aftermath of September 11, 2001

In January 2003 *Security Management* asked security managers, senior business executives, Wall Street analysts, and others for their perspective on how the events of September 11, 2001 had impacted security. In general, their report shows that security has become more visible and important to the business enterprise. This is true not only in the corporate view of security but also in the security technology and services sector. There is no doubt that security now plays an important role as part of company's business resiliency strategy.

A study conducted for the American Society for Industrial Security (ASIS) International by Westat, Inc., ASIS found that almost 75 percent of the security executives surveyed reported no changes in the structure of security within their companies. However, almost half reported an increase in budgets. The increased budgets did not result in additional staff in 66 percent of the organizations. Some of the increased budgets went to enhance and in some cases supplant security through new technologies, but most respondents did not place equipment at the top of their lists.[1]

Where was the money spent? For most companies, funds were used to improve business resiliency and to update security policies that addressed issues raised in the events of 9/11. Companies with established policies spent time *dusting them off* and revising them for the current state of threats to the business enterprise. For others, policies were nonexistent and had to be created. Policy concerns focused on suspicious packages, bioterrorism concerns, access control issues, followed by technology.[2] Business resiliency and incident management also became key focus areas for companies, to include practical exercises with the support of executive management to ensure the resiliency of the enterprise in the aftermath of a crisis.

Private Security Resources

Growth in private security is shown by increases in expenditures, employment numbers, and value of equipment. In addition, many new firms have entered the business, established firms are showing strong growth, and some Fortune 500 companies as well as foreign firms are buying security firms.

In a time when public law enforcement is experiencing little or no growth, private security directors are experiencing greater demands on their resources and professional expertise. More than half of local and central station alarm firms reported annual budget increases. The private security industry will continue to grow commensurate with the need to secure the business enterprise as a result of homeland security concerns. Businesses have also gone global with supply chain security now being a major remit for security managers.

To fill the gap left by public law enforcement and to help alleviate some of the public's fears, private sector security has offered its services in a variety of areas, including private patrols of residential areas, security for courtrooms and correctional facilities, traffic control, and so on.

The private sector has a tremendous impact on crime and the criminal justice system. Therefore, it is absolutely necessary to completely understand the private sector system. In the future, the private sector will bear more of the burden for crime prevention and asset protection, while public law enforcement will narrow the focus of police services to crime control. Most of the security managers surveyed by the researcher were willing to accept this growing role and continue to embrace this mission even to this day.

Interactions and Cooperation

Although *Hallcrest II,* in 1990, rated interaction and cooperation between the public and private sector as good, it also noted (as did the RAND report and the Private Security Advisory Council [PSAC]) that certain impediments to interaction and cooperation still existed. These impediments included role conflict, negative stereotypes, lack of mutual respect, and minimal knowledge on the part of law enforcement about private security. As noted when these topics were discussed in earlier chapters, both private and public sector administrators are cooperating to try to overcome these problems.

We presented some discussion of these areas earlier in this text. The trend by the public sector to contract services or develop hybrid security operations, with greater cooperation between the public and private sectors, continues in the 21st century. According to the late Robert Trojanowicz:

> *One question that need not be asked is whether the trend will persist. We are already too far down the road to turn back. Therefore, the ultimate question is not whether this change is good or bad, but whether these changes will occur piecemeal and poorly or thoughtfully and well.*[3]

Limitations of Security

A great burden has been placed on the private sector. It goes without saying that some of the expectations will very likely not be met. It is impossible to be *everything to everyone*, especially in an area that is changing so rapidly and dynamically.

Consider, for example, the changes in technology. Security's main purpose is to prevent losses, whether traditional, such as burglary; new age, such as cybercrime of hacking and identity theft; or military, as in reduction of vulnerabilities to terrorist attacks. Historically, losses were reduced through the use of various physical security devices that served as delaying devices to would-be intruders. As the targets for theft changed, security modified its protective strategies and adopted new security technologies.

Today, security in many companies is truly a full loss-prevention operation, and within such operations, state-of-the-art and enterprise-wide security systems are coupled with traditional security methods. Integration of security systems with data processing and other company operations are becoming commonplace as logical and physical security converge. As discussed earlier, the relationship between the chief security officer (CSO) and the chief information officer (CIO) will need to be reinforced and clearly defined. In short, it is important that the CSO and CIO stay very much aligned regarding how to mitigate threats against the business enterprise.

How effective are security devices today? Security managers would be wise to remember that no system is completely effective. A determined criminal can find ingenious methods of defeating any system. For example, consider the progressive development of lock technology. It began with a simple doorknob lock with a spring latch. As we know today, this combination offers virtually no protection from a person who wants to gain entry. The simplest and least destructive method of gaining access would be to card the lock. If this option is removed by installing a dead latch, the intruder will have to choose another means of access. Since the latch does not extend more than half an inch into the doorframe, a burglar could spring the door or twist the doorknob off to gain access to the locking mechanism. Installing a good dead bolt lock may solve this problem.

Each step that security takes requires a burglar to use more time, but the burglar is still not stopped. The burglar chooses to either cut the dead bolt or twist the collar mechanism off the dead bolt lock. Security then responds by installing a dead bolt with a slip collar (or recessed system) and a dead bolt with a case-hardened core.

In most cases, if the door and frame are of solid construction (that is, metal), burglars will not attack them. But some might want company assets enough to find another method of entry. The keying device then becomes a target; key cores can be drilled or popped. Security managers who have been thinking ahead will purchase locks with case-hardened drill plates, which will foil both drilling and popping. However, there is still the chance of picking the lock. But security can anticipate this possibility and install a pick-resistant lock.

Has security now stopped the potential intruder? In reality most thieves would not bother to assault this level of protection; but if the target is sufficiently valuable, other alternatives exist. The thief's imagination is the only limiting factor.

With today's high-tech security devices and reliance on computers, a security manager can virtually eliminate any target's probability of being attacked. The modern age has brought "Star Wars" devices such as computer access-control systems based on fingerprints or voice identification, laser beam alarm systems and communication devices, and sophisticated listening devices and their counterparts, sweeping devices. Lighting technology has made it possible to illuminate areas to daylight levels with reduced costs. The list goes on, but the question remains: Why does crime continue to plague us?

The answer could lie in the fact that when professional thieves are sufficiently motivated, they will simply find a way—the professional thief being one whose job it is to steal. Or it may be because a security manager has failed to convince executives to pay for a security operation that appears to them to be an added expenditure that will not add to the profit margin. Return on investment (ROI) and value-add analysis must become part of the working vocabulary of the security manager and CSO.

High technology has made the operation of most businesses much easier. The photocopier, the fax, the computer, and 21st-century communication devices make information transfer simpler and more efficient. This convenience has at the same time created new problems for security in the areas of increased information and physical security risks, as discussed in this book.

Evolving Trends

Although the security field is now composed of more than 1.1 million personnel, with the majority employed in contract security, the trend will continue toward fewer and smaller proprietary security operations and even fewer proprietary unionized firms will hire contract security or security consultants for special-purpose projects. Contract firms will continue to merge. As noted in earlier discussions, Securitas is now the largest contractor in the United States and continues to grow by buying other firms. Hybrid security operations are now commonplace.

The testing industry, geared to pre-employment, drug, and psychological testing, will continue to offer services sought after by the private sector. Advocates of privacy standards will continue to restrict the use of these types of tests.

The use of outside investigators and security specialists will increase. The trend will be toward a manager of security services using security consultants for special projects. However, given recent corporate scandals, such as the pretexting investigation into Hewlett-Packard, which brought the CEO to testify before Congress, more contract management and rules of engagement will be required in order to mitigate liability and ethical issues.

Liability costs, though remaining relatively high, will continue in the downward trend that began in 1986. The downward trend may be attributable to better prepared security managers who are learning the value of risk management, premise liability mitigation and workplace privacy.

Drug testing will continue to be a major issue for corporations. Until the nation is able to control the influx of drugs over the border, this issue will remain a legal and safety problem for many companies. The cost associated with screening can be prohibitive, and though the public relations aspect of the Employee Assistance Program (EAP) cannot be disputed, there are those who are now asking whether the company can afford to spend thousands of dollars on each employee who needs rehabilitation. The question remains, what is enough?

In addition to its role in fighting the increase of economic crime in general, security will continue to face an increase in computer- and information-related crime. As noted earlier in this text, estimates of the cost of computer crime range from $1 billion to $200 billion annually. On the international scale, there is more concern about security issues. The Common Market has mandated that security laws be reviewed and upgraded to bring commonality to the European Union. Many companies have established "Situation Watch" programs designed to track escalating economic, social, political situations and weather patterns that could negatively impact a company's employee interests and supply chain in a given region.

Terrorism without a doubt is currently the most talked-about issue facing the nation, especially for companies operating on an international scale, referred to as *multinationals*. The terror of 9/11 made it abundantly clear that terrorism is truly global and can negatively affect the ecosystem of multinational companies. The continuing attacks, such as the Madrid train bombing and the Mumbai and London bombings, keeps the terrorism menace in the forefront of our minds. The 1991 Gulf War caused a growth in concern among most multinational companies. Since 2003 the war with Iraq has also had its impact on money markets around the world. Companies can protect their assets from terrorists but have little control over the impact of nations' decisions in their efforts to control terrorist states. The key in planning antiterrorist security operations for the 21st century will be providing adequate protection without spending money for unnecessary security.

Convergence: Integration of Logical and Physical Security Technologies

Technology will continue to drive the integration of computers and biometrics with CCTV and access control protocols. Telephone, wireless systems and Internet protocols will be used increasingly to connect security systems safety devices and building management. Micro-processing chips dedicated to security purposes will be embedded in equipment, resulting in smart cameras and employee identification cards. The field of robotics will continue to grow, particularly in the area of corrections and guard operations within confined areas. Miniaturization is a growing trend. CCTV cameras the size of small packages are already on the market. Electronic article surveillance (EAS) will continue to grow as prepared packaging becomes more prominent.

Still, integration of security technology is a continuing process rather than a single event. Though the industry has come a long way from the independent systems of the early 1990s, consultants and security practitioners continue to argue over the true extent that integration has arrived. Elliot Boxerbaum, CPP and president of Security/Risk Management Consultants, recently stated that "security [will become] just another piece of the puzzle, but at the same time its importance to the whole is increased."[4]

True integration creates a myriad of potential issues. First among these is the process of giving computer "techies" control over systems that were once the domain of security professionals. The real end result should be cooperation between the CIO and CSO as well as firm employee policies in place that could result in termination of employment for violating information technology practices.

The Future

A number of years ago, Ira Somerson, president of Loss Management Consultants, provided an excellent overview of the future in his article "The Next Generation." Security professionals must be futurists and ready to initiate and adapt to change based on the larger forces shaping the future given the turbulent world we all live in today. The following are the trends that Mr. Somerson noted. These are very much true as the author revises this section in 2012:

- Expansion of global markets and production will continue 24 hours per day across the globe.
- Multinational and multicultural workforces will be the norm.

- Downsizing will continue.
- Technology will move us into a "telehumanic age."
- Qualified workers will be difficult to find.
- Religious extremists will become more prevalent.
- Biotechnology will become a powerful force.
- Employee loyalty will continue to erode.[5]

The future for the security field is very positive. Professionalization is being pursued; large sums of money are being spent to improve security and loss-prevention operations. For the first time since World War II, security has found its place in business. In addition, limitations of the criminal justice system have been identified, and the resultant void is being filled with some success by the private sector. The future in private security is exceptionally bright and lucrative for the talented person who wants to do something special, especially if that person has had sound educational preparation. It cannot be underestimated that security is *going global*. It's imperative for security professionals to think and act globally, with those having proven international experience competing for the best jobs. As such, international experience and exposure through personal and business travel will be important hiring criteria for the security director of the future. In short, if you don't have it, it will be hard for you to compete for the top job!

Technology

The future for technological advances in security is also bright. In just 45 years, the security business has seen the arrival of CCTV, microwave detection systems, small portable radios and cameras, magnetic sensors, laser technology, and truly integrated computer and security systems. Security has taken advantage of these technological improvements through necessity—the realization that for every new piece of security equipment there is a person who, given the challenge, will find a way to subvert, bypass, or defeat it. Today's battles with criminals are far more often battles of intellect than they are of muscle. We may eventually see security measures such as force fields, holographic security officers, and laser pistols. The future is limited only by the security technician's imagination.

Combinations of biometric, electronic, and computer technologies will allow security and other management personnel to follow employees from the time they enter the facility until the time they leave it. Global positioning systems (GPS) are now available that will allow monitors to track vehicle position within yards on a geographic earth grid. The satellite technology that has made this possible also allows people with access to monitor activities across the globe.

By combining GPS and radio frequency identification (RFID) technologies, it is now possible for retailers to track merchandise in real time. Where is that shipment of Viagra? We can find out relatively easily. A "geofence" may also be created that would signal an alarm should a load deviate from a chosen route.

In the area of intellectual technology, the field continued its rocket-like growth from a $17 billion industry in 2001 to more than $45 billion in 2006. The strongest market will likely continue to be in security hardware and then software, with a 16 percent growth rate.[6] The demand for technology is the result of increased interest in integration of all security operating systems to include access control, surveillance, databases, GPS, and RFID technologies.

Guards equipped with wireless handheld tablets will be able to monitor surveillance cameras from remote locations, submit reports over WLANs, and access email and other data in real time. It will also allow them to "call up" CCTV cameras and floor plans while on scene handling a crisis situation. It is also likely that GPS technology will allow control centers to monitor guard locations for enhanced safety and accountability. Workplace privacy issues will always be a consideration when it comes to surveillance technology in the workplace.

Managers equipped with similar technology will be able to provide real-time updates or alerts to officers in the field without radio broadcasts.[7] "Smart cameras" will replace traditional fixed CCTV and even today's state-of-the art integrated systems. These almost stand-alone cameras will not rely on hardwire systems or computer support to make adjustments. Rather, wireless technology coupled with memory chips will allow these cameras to make decisions based on pre-programmed scenarios and send information back to the central station or to remote computer receivers carried by security officers.

The issues associated with identity theft may remain, but with the increased use of "smart cards" in place of traditional credit/debit cards, credit card operations may no longer need to retain database information on customers. Removing critical information from customer databases will mean less critical information will be available to identity thieves.

WLANs may be improved through "mesh-networking." Both Microsoft and Intel are looking at ways to make the WLAN individual systems a coherent network. Issues include ways to provide proper security to wireless systems.[8] The introduction of the iPhone is just a first step in this process.

The future will also bring increased use of biometrics and spending as part of the layered access security protocol.

Leadership and Management

As Tom Peters notes in his writings, the manager of the future must be able to predict and manage change.[9] In fact, Peters would probably prefer the word "leader" to "manager." The security leader of the 21st century will need to be flexible. As Thomas Sege, chairman and CEO of Varian Research, notes, "You have to approach things much more flexibly. Change can come from all directions. You can't chart one course and hold to it. Course corrections are going to have to be made."[10]

The author submits that if you are flexible you are still too rigid, as today's environment requires the security manager to be fluid in dealing with events and incidents, in addition to being a superior communicator and problem solver. All businesses will continue to stress the benefits of cost containment management. Profit margins must be improved, and security will be expected to offer its fair share (ROI). The security manager will thus become a security program educator and salesperson. Security managers must learn to sell security as though it were any other product. Because CEOs think in terms of profit, a good security manager must find ways of selling security as a cost-effective return on investment.

For example, perhaps better lighting is needed in a particular workplace. All the CEO sees is the cost for new lighting, but the security manager suggests it might reduce theft, vandalism, and accidents. Will possible savings in these areas offset the cost? Probably not in the first year! But why not project the cost of the project over five or more years? Also consider the savings in electricity costs from installing more efficient fixtures and the possible reduction in insurance rates. Now the CEO is interested. All of a sudden the investment is at worst breaking even—and it might even save the company money. The bottom line will be more important than ever.

Security managers will find that safety, architecture, consulting and engineering, human resources, compliance, transportation, telecommunications, information processing, and marketing will demand more from the security operation and that to develop future-oriented security programs, they must take an interest in these areas.

The security manager of the future must carefully consider terms such as *integration, rightsizing, interfacing, hybrid security systems, ROI,* and *reengineering.* The current trend is toward downsizing and the increasing use of technology to replace or at least augment existing operations, but the future may go beyond this trend. Even as this book is being completed, more and more professional employees are completing their work assignments using technology from their home offices. From a security point of view these trends offer new opportunities as well as challenges.

The Challenge of Theft

Arguably, the greatest threat for security managers will continue to be employee theft. Worker loyalty is at an all-time low. Ethical issues and a gross sense of entitlement will become important for security managers to understand as they struggle to protect company information as well as physical assets. Employee theft will continue to cost businesses about $40 billion annually. This challenge will become more complex as cybercrime continues to grow.

Education and Professionalization

Experts predict that by the year 2020 the majority of new jobs will require postsecondary education. Yet today, 20 percent of the U.S. workforce reads at no better than an eighth-grade level. This means that unless something changes, the U.S. workers of the future will not be able to read basic materials necessary to perform at the required level. The staff at *Security* predicts that in the future, professional organizations representing security will approach colleges, universities, and community colleges and ask for help with programs that train security officers and security professionals.[11] ASIS has already moved in this direction with the development of master's level education through Webster University. As mentioned earlier, the wise security professional will complement his or her accredited degrees and professional certifications with international experience and exposure.

Pandemic Planning and Brand Protection Management

Counterfeiting has become a problem for multinational companies, especially those with a recognizable brand. Make no mistake, the act of counterfeiting is universally held to be a crime around the world. Counterfeiters know that brand is important in a consumer's choice and therefore target companies with a recognizable brand in the marketplace. In short, they look to duplicate or fake products of companies who operate with high margins and product run rates and generally target products that are relatively easy to duplicate. The World Customs Organization (WCO) estimates that 5 to 10 percent of all global merchandise is counterfeit. Even more alarming, the World Health Organization (WHO) estimates that 10 percent of all medicines are counterfeit.[12,13]

In the federal sector, for example, a 2007 study estimates that 30 out of every million $100 bills are fakes. The National Research Council believes that within the next 10 years, even low-skilled amateurs will be able to duplicate our current currency. They are recommending that the government incorporate complex starburst patterns that copies cannot duplicate. Other suggestions include inks that change color as they're exposed to various temperatures.[14] Regardless of the sector, counterfeiting of all types is big business. It is predicted that this illegal enterprise will continue to grow because it is generally much more profitable than dealing in drugs and much less risky.

As the world's population tops 7 billion, the need for pandemic planning becomes even more important. In short, pandemic planning is mitigating the threat of disease such as a viral outbreak found in large population densities throughout the world. A good example of this is the avian flu outbreak in Southeast Asia. Often security management includes the environmental health and safety remit. Therefore, it is important for the security manager to embrace this role and develop plans along with medical experts to help mitigate the threat to the business enterprise.

Certification and Standards

The discussions presented earlier in this text point toward the need for some type of mandated minimum standards for the security industry. Recent developments led by ASIS point the way for some type of industry-imposed regulations similar to the British system. If security is to continue to take the place of law enforcement, the field must present a professional image. The image can exist only when outsiders can view the field with the respect that comes from established standards. The proliferation of certifications, if properly controlled, is a move in the right direction. Federal legislation mandating standards will hopefully lead to base-level standards for the security industry.

Summary

As stated, the future for security is bright and can be lucrative for the educated professional. The more education, certification and training become a regular part of the security occupation, the greater the professional development and acceptance of that occupation. It is obvious that the public sector cannot afford to pay for all the protection needed in our modern and turbulent world. The only means of providing security as defined in Chapter 1 is through the use of private resources that keep the costs of products down and provide an environment wherein our free-enterprise system can prosper.

Current events have definitely changed the security landscape, but as we focus on the war on terrorism, the question that needs to be answered is, "How far will the American public let the federal government go in mandating security measures for the public's own good?" Years of private initiative in protecting one's own property are at stake. Only time will tell where the line will be drawn. As noted in Chapter 1, there are efforts to create federal identification databases on each person. Whether this goal will be accomplished is questionable. If it is achieved, how will it be controlled?

■ ■ CRITICAL THINKING ■

Who should have the responsibility of maintaining an individual's security: the individual or the government? What are the consequences of deciding in favor of either option?

Review Questions

1. List several major trends for security in the 21st century. Explain their significance from a management perspective.
2. Explain the need for security managers to develop into educators and salespeople for security programs. Give an example of how you might present a security program to your CEO.
3. What role should security play in the prevention of terrorism in the 21st century?
4. What is the future of public/private cooperation in the 21st century?

References

[1] Harowitz S. The new centurions. Secur Manage January 2003:52.

[2] Anderson T. A year of reassessment. Secur Manage January 2003:61–5.

[3] Trojanowicz R. Public and private justice: preparing for the 21st century. Crim Justice Alumni Newsl Fall/Winter 1989;V(1):2. [Michigan State University]

[4] Lasky S. The age of convergence. Secur Technol Des March 2003:6.

[5] Somerson I. The next generation. Secur Manage January 1995:27–30.

[6] IT Security Market to Hit $45 Billion by 2006. Secur Beat February 25, 2003.

[7] Technology Training Boosts Level of Service for Barton. Access Control Secur Syst January 2003:18.

[8] Jones M. Wireless meshes with its future. Infoworld March 2003:10.

[9] Peters T. In: Thriving on Chaos: a handbook for a management revolution. New York: Harper & Row Publishers; 1987.

[10] Exploring Security Trends, special report. Security 1989:2.

[11] Ibid., pp. 5–6.

[12] World Customs Organization (WCO) <www.wcomd.org>; 2009.

[13] World Health Organization (WHO) <www.who.int>; 2009.

[14] <www.whbf.com/Global/story.asp?S56143306>.

Appendix A
Security Journals, Magazines, and Newsletters

United States

Access Control and Security Systems	www.securitysolutions.com
Beyond Computing	http://beyondcomputingmag.com
Campus Law Enforcement Journal	www.iaclea.org
CIO	www.cio.com
CSO	www.csoonline.com
Economist	www.economist.com
Eweek	www.eweek.com
Federal Computer Week	www.fcw.com
Homeland Security News	www.homelandsecurity.org/bulletin
Information Technology Security Magazine	www.information-security-magazine.com
InfoSecurity	www.infosecurity-magazine.com
InfoWorld	www.infoworld.com
Journal of Economic Management (Economic Crime Institute)	www.jecm.org
Journal of Healthcare Protection Management	www.iahss.org
Library and Archival Security	www.tandf.co.uk/journals/WLAS
Lipman Report	www.guardsmark.com/library/library_sec.asp
Locksmith Ledger	www.locksmithledger.com
Polygraph Journal	www.polygraph.org/section/polygraph-journals
Protection of Assets Manual	www.protectionofassets.com/
SCMagazine	www.scmagazine.com
Security Journal	www.palgrave-journals.com/sj/
Security Management	www.securitymanagement.com
Security Management Weekly	www.securitymanagement.com
Security	www.securitymagazine.com
Security Products	www.secprodonline.com
Security Technology Executive	www.securityinfowatch.com

Appendix B
World Wide Web Sites

The following World Wide Web sites are useful resources for security professionals. The list is certainly not comprehensive, but contains resources that the author has used on a regular basis. The list does not include on-line journals (see Appendix A). If readers believe there are other valuable sites, please contact the author via email at rjfish@macomb.com or via fax to Assets Protection Associates, Inc., (309) 837-4305.

www.aais.org	American Association of Insurance Services
www.salary-surveys.erieri.com	ERI- Salary Surveys.
www.truckline.com	American Trucking Association
www.asisonline.org	American Society for Industrial Security International
www.bls.gov	U.S. Department of Labor, Bureau of Labor Statistics
www.brookings.edu	Brookings Institute
www.cfenet.com	Association of Certified Fraud Examiners
www.fbi.gov	Federal Bureau of Investigation
www.iasir.org	International Association of Security & Investigative Regulators
www.theiacp.org	International Association of Chiefs of Police
www.ilj.org	Institute of Law and Justice
www.retailmerchants.com	National Retail Merchants Association
www.josephsoninstitute.org	Josephson Institute
www.lexis-nexis.com	Lexis-Nexis
www.ojp.usdoj.gov/nij	National Institute of Justice
www.naaa.org	National Alarm Association of America
www.ncjrs.org	National Criminal Justice Reference Service
www.newslibrary.com	News Library
www.policefoundation.org	Police Foundation
www.state.gov	U.S. Department of State
www.usdoj.gov	U.S. Department of Justice
www.westpub.com	ThomsonWest Group
www.dhs.gov	Department of Homeland Security
www.wcomd.org	World Customs Organization
www.whoin.org	World Health Organization

Appendix C
Security Surveys

Table C.1 Security Vulnerability Survey

Facility _____ Survey date_____

Address _____ Facility manager _____

  Telephone no. _____

1. GENERAL FUNCTION   Leased
  Owned

_____

No. employees assnd. _____

Operating   Weekdays   Saturday   Sunday
Hours:

  Opens _____ Opens _____ Opens _____
  Closes _____ Closes _____ Closes _____

Address & phone of police jurisdiction: _____

Area evaluation:

_____

II. BUILDING & PERIMETER

_____ 1. Type of construction?
_____ 2. Door construction (hinges, hinge pins, solid core, etc.)?
_____ 3. Total number of perimeter entrances?
_____ 4. Are all exits & entrances supervised?
  If not, how controlled?
_____ 5. Are there perimeter fences?
  Type?
  Height?
  Distance from bldg.?
  Cleared areas?
  Barbed wire top?
  Roof or wall areas close to fence?
_____ 6. Are there any overpasses or
  subterranean passageways?

(Continued)

503

Table C.1   (Continued)

_____ 7. Height of windows from ground?
            Adequately protected?
_____ 8. Any roof openings or entries?
_____ 9. Any floor grates, ventilation openings?
_____10. Any materials stored outside
            bldg.? How controlled?
_____11. Adjacent occupancy?
            Comments:

III. VEHICULAR MOVEMENT

_____ 1. Is employee parking within perimeter fence?
_____ 2. Are cars parked abutting interior fences?
_____ 3. Are cars parked adjacent to loading docks, bldg. entrances, etc.?
_____ 4. Do employees have access to cars during work hours?
_____ 5. Vehicle passes or decals?
_____ 6. Are guards involved in traffic control?
            Comments:

IV. LIGHTING

_____ 1. Is perimeter lighting provided?
            Adequate?
_____ 2. Is there an emergency lighting system?
_____ 3. Are all doorways sufficiently lighted?
_____ 4. Is lighting in use during all night hours?
_____ 5. Is lighting directed toward perimeter?
_____ 6. Is lighting adequate for parking area?
_____ 7. How is lighting checked?
_____ 8. Is interior night lighting adequate for surveillance by night guards (or by municipal law enforcement
            agents)?
_____ 9. Are guard posts properly illuminated?
            Comments:

V. LOCKING CONTROLS

_____ 1. Does the facility have adequate control and records for all keys?
_____ 2. Is a master key system in use?
_____ 3. How many master keys are issued?
_____ 4. Are all extra keys secured in a locked container?
_____ 5. Total number of safes?
_____ 6. Last time combination(s) changed?
_____ 7. If combination is recorded, where is it stored?
_____ 8. Total number of employees processing combination?
_____ 9. Review procedures for securing sensitive items (i.e., monies, precious metals, high dollar value items,
            narcotics, etc.)?
_____10. Who performs locksmithing function for the facility?
_____11. Is a key inventory periodically taken?
_____12. Are locks changed when keys are lost?
            Comments:

(Continued)

Table C.1 (Continued)

VI. ALARMS

_____ 1. Does the facility utilize any alarm devices?
Total number of alarms?
Type Location Manufacturer Remarks
_____ 2. Are alarms of central station type connected to police department or outside guard service?
_____ 3. Is authorization list or personnel authorized to "open & close" alarmed premise up to date?
_____ 4. Are local alarms used on exit doors?
_____ 5. Review procedure established on receipt of alarm?
_____ 6. Is closed-circuit TV utilized?
Comments:

VII. GUARDS/SECURITY CONTROLS

_____ 1. Is a guard service employed to protect this facility?
If yes. Name:____ No. of guards____
No. of posts_____
_____ 2. Are after hours security checks conducted to assure proper storage of classified reports, key controls, monies, checks, etc.?
_____ 3. Is a property pass system utilized?
_____ 4. Are items of company property clearly identified with a distinguishing mark that cannot be removed?
_____ 5. Review guard patrols & frequency?
_____ 6. Are yard areas and perimeter areas included in guard coverage?
_____ 7. Are all guard tours recorded?
_____ 8. Are package controls exercised re packages brought on or off premises?
_____ 9. Does facility have written instructions for guards?
_____10. What type of training do guards receive?
_____11. Are personnel last leaving building charged with checking doors, windows, cabinets, etc.? Record or identity:
_____12. Are adequate security procedures followed during lunch hours?
Comments:

VIII. EMPLOYEE AND VISITOR CONTROLS

_____ 1. Is a daily visitors register maintained?
_____ 2. Is there a control to prevent visitors from wandering in the plant?
_____ 3. Do employees use identification badge?
_____ 4. Are visitors issued identification passes?
_____ 5. What types of visitors are on premises during down hours and weekends?
_____ 6. Does any company's employees other than _____ have access to facility?
List Company Names   Type Service Performed
_____ 7. Are controls over temporary help adequate?
Comments:

IX. PRODUCT CONTROLS (Shipping and Receiving)

_____ 1. Are all thefts or shortages or other possible problems (i.e., anonymous letters, crank calls, etc.) reported immediately?
_____ 2. Inspect and review controls for shipping area.

(Continued)

Table C.1   (Continued)

_____ 3. Inspect and review controls for receiving area.

_____ 4. Supervision in attendance at all times?

_____ 5. Are truck drivers allowed to wander about the area?
Is there a waiting area segregated from product area?
Are there toilet facilities nearby? Water cooler?
Pay telephone?

_____ 6. Are shipping or receiving doors used by employees to enter or leave facility?

_____ 7. What protection is afforded loaded trucks awaiting shipment?

_____ 8. Are all trailers secured by seals?

_____ 9. Are seal numbers checked for correctness against shipping papers? "In" and "Out"

_____10. Are kingpin locks utilized on trailers?

_____11. Is a separate storage location utilized for overages, shortages, damages?

_____12. Is parking (employees and visitor vehicles) prohibited from areas adjacent to loading docks or emergency exit doors?

_____13. Is any material stored in exterior of building? If so how protected?

_____14. Are trailers or shipments received after closing hours? If so how protected?

_____15. Are all loaded trucks or trailers parked within fenced area?

_____16. Review facility's product inventory control.

Loss   Breakage   Returns

Average
Monthly

_____17. Review controls over breakage.
Comments:

X. MONEY CONTROLS

_____ 1. How much cash is maintained on the premises?

_____ 2. What is the location and type of repository?

_____ 3. Review cashier function.

_____ 4. What protective measures are taken for money deliveries to facility?
To bank?

_____ 5. If armored car service utilized, list name and address.
_____

_____ 6. Does facility have procedure to control cashing of personal checks?

_____ 7. Are checks immediately stamped with restricted endorsement?

_____ 8. Are employee payroll checks properly accounted for and stored in a locked container (including lunch hours) until distributed to the employee or his supervisor?
Comments:

XI. PROPRIETARY INFORMATION

_____ 1. What type of proprietary information is possessed at this facility?

_____ 2. How is it protected?

_____ 3. Is "_____ Restricted" marking used?

_____ 4. Are safeguards followed for paper waste, its collection and destruction?

_____ 5. Are desk and cabinet tops cleared at end of day?

_____ 6. Is management aware of need for protecting proprietary information?
Comments:

(Continued)

Table C.1   (Continued)

XII. OTHER VULNERABILITIES

_____ 1. Trash pickups (hours of pickups, control of contractor, physical controls)?
_____ 2. Scrap operations (physical controls of material and area, control over scrap pickups, etc.)?
_____ 3. Other?
Comments:

XIII. PERSONNEL SECURITY

_____ 1. Are background investigations conducted on employees handling products?
Handling cash?
Engaged in other sensitive duties?
Supervisory position?
All employees?
_____ 2. If so, who conducts background investigation?
_____ 3. Are new employees given any security or other type of orientation?
_____ 4. Do newly hired employees execute a corporate briefing form for inclusion in their personnel file?
_____ 5. Are exit interviews conducted of terminating employees?
_____ 6. Is a program followed to ensure return of keys, credit cards, ID cards, manuals and other company property?

General Comments

Source: Charles A. Sennewald, *Effective Security Management* (Los Angeles: Security World Publishing, 1978).

Table C.2   Small Store Security Survey

To illustrate how even small facilities can be big when it comes to loss prevention, the following store survey is a comprehensive example. It is formatted in a question and answer style.

SMALL BUSINESS SECURITY SURVEY

Name of Business:
Street Address of Business:
Manager's Name:
Business Phone:
Type of Goods and Services:

MANAGEMENT SECURITY

Employee Screening	Yes No N/A	Comments
1. Previous employers		
a. Write them?	Yes No N/A	
b. Call on phone?	Yes No N/A	
c. Personal inquiry?	Yes No N/A	
2. Education		
a. Write them?	Yes No N/A	
b. Call on phone?	Yes No N/A	
c. Personal inquiry?	Yes No N/A	

(Continued)

Table C.2 (Continued)

3. Criminal history check (state and region)	
a. Is there a felony conviction record?	Yes No N/A
4. Are personal references checked?	Yes No N/A
5. Are Department of Motor Vehicles records checked?	Yes No N/A

EMPLOYEE AWARENESS

6. Shoplifting	
a. Is there a shoplifting prevention program in place?	Yes No N/A
b. Do employees know what to observe for?	Yes No N/A
c. Do employees know what to do and what not to do around suspected shoplifters?	Yes No N/A
d. Do you know if there are any community-sponsored programs relating to shoplifting?	Yes No N/A
7. Robbery	
a. Is a robbery prevention plan in place?	Yes No N/A
b. Are employees trained if a robbery occurs?	Yes No N/A
8. Credit cards	
a. Are employees familiar with the different types of credit cards?	Yes No N/A
b. Do employees know the proper identification needed for the use of credit cards?	Yes No N/A
c. Do employees know what to do if confronted with fraudulent, altered, or stolen cards?	Yes No N/A
9. Are personal check procedures in place for cashing checks and verifying identification?	Yes No N/A

EMPLOYEE ACCESS CONTROL

10. Is a key log maintained and are keys audited periodically?	Yes No N/A
11. Are keys turned in during extended absences?	Yes No N/A
12. Are keys retrieved when employees separate or are transferred?	Yes No N/A
13. When security is breached, are locks rekeyed?	Yes No N/A
14. Are locks rekeyed every 3 to 5 years regardless of known security violations?	Yes No N/A
15. Are keys marked "Do not duplicate?"	Yes No N/A
16. Is someone responsible for locking up premises and ensuring no security risks exist (open doors, windows, stay behinds)?	Yes No N/A

CASH AND DEPOSIT CONTROLS

17. Is cash in registers kept to a minimum at all times?	Yes No N/A
18. Are robbery alarm actuators hidden from view?	Yes No N/A
19. When handling large sums of money, is more than one employee present at all times?	Yes No N/A
20. Are all checks, money orders, and traveler's checks stamped "for deposit only" immediately upon receipt?	Yes No N/A
21. Are all cash receipts deposited daily?	Yes No N/A
22. Is there an established routine for transportation of cash receipts to the bank to help deter a robbery?	Yes No N/A

(Continued)

Table C.2   (Continued)

INTERNAL CONTROLS

23. Are surprise audits or physical inventories conducted periodically?	Yes No N/A
24. Is valuable equipment marked with the store identification?	Yes No N/A
25. Is a record kept of serial numbers on valuable equipment?	Yes No N/A
26. Are cash registers emptied and left open after closing?	Yes No N/A
27. Are relations with the local police department maintained?	Yes No N/A

PERIMETER DOORS

28. Are all unused doors secured?	Yes No N/A
29. Are doors of high quality and securely fastened in place?	Yes No N/A
30. If doors contain glass, do wire or bars protect them?	Yes No N/A
31. Are hinge pins protected so they cannot be pulled?	Yes No N/A
32. Is the latch bolt protected with a latch guard so that it cannot be pushed back, opened with a thin instrument, and/or pried?	Yes No N/A
33. Is the lock a double cylinder type?	Yes No N/A
34. Are the locks and door hardware in good working condition?	Yes No N/A
35. Are padlock hasps installed so that the screws cannot be removed?	Yes No N/A
36. Are the padlock hasps heavy duty enough?	Yes No N/A

WINDOWS

37. Do heavy screens or bars protect easily accessible windows?	Yes No N/A
38. Are unused windows permanently closed?	Yes No N/A
39. Are bars or screens mounted securely?	Yes No N/A
40. Are window locks designed or located so they cannot be opened by just breaking the glass?	Yes No N/A
41. Are window displays avoided so as to not obstruct view into the store?	Yes No N/A
42. Are metal window grates padlocked across window displays used as protection?	Yes No N/A
43. Is all valuable merchandise removed from unprotected window displays at night?	Yes No N/A

OTHER OPENINGS

44. Is there a lock on manholes, skylights, and roof hatches that give direct access to the building?	Yes No N/A
45. Are the sidewalk doors or grates securely in place so that the entire frame cannot be pried off?	Yes No N/A
46. Are accessible skylights protected by bars, screens, or intrusion alarm?	Yes No N/A
47. Are the roof doors in good condition and securely locked?	Yes No N/A
48. Are all air conditioning ducts, ventilation shafts, and fan openings protected against unauthorized entrance?	Yes No N/A
49. Are transoms properly locked or protected by bars or screens?	Yes No N/A
50. Do fire exits meet fire code for egress?	Yes No N/A

SAFES

51. Is the safe designed for burglary protection as well as fire protection?	Yes No N/A
52. Is the safe located and well-lighted so that the police can view it from outside at night?	Yes No N/A
53. If there is a built-in vault, are the walls as well as the door secure?	Yes No N/A

(Continued)

Table C.2   (Continued)

54. Has the combination been changed if persons have separated and no longer need it?	Yes No N/A

LIGHTING

55. Are all areas where an intrusion might occur lighted by street lights or the store's lights?	Yes No N/A
56. Are blind alleys, where a burglar might work unobserved, protected with adequate illumination?	Yes No N/A
57. Is the interior of the store lighted from rear to silhouette an intruder?	Yes No N/A
58. Are low-mounted lights free of being compromised by easily unscrewing a bulb, vandalism, etc.?	Yes No N/A
59. Is lighting adequate to allow a policeman, security officer, or passerby to readily detect an intruder attempting unauthorized entry?	Yes No N/A
60. Are lights controlled by automatic timer or manually operated?	Yes No N/A
61. If one exterior light goes out, will other existing lights compensate?	Yes No N/A

ALARMS

62. Is there a protective alarm system for the premises?	Yes No N/A
63. To detect intrusion, is there an off-premises central station that sends an alarm to the police department?	Yes No N/A
64. To prevent burglary, is there a local alarm that rings into the premises to ward off intruders?	Yes No N/A
65. Is there a hold-up alarm for robberies?	Yes No N/A
66. Are access points into the building (doors, windows, vents) protected by an alarm?	Yes No N/A
67. Is the alarm system maintained regularly to verify operating condition?	Yes No N/A

FIRE SAFETY

68. Do local firefighters visit annually for inspections?	Yes No N/A
69. Are the type and number of fire extinguishers adequate?	Yes No N/A
70. Are extinguishers inspected monthly to verify they are in good working order?	Yes No N/A
71. Are all fire exits correctly marked?	Yes No N/A
72. Are all fire exits and extinguishers unobstructed?	Yes No N/A
73. If a sprinkler system is used, is it tested annually for effectiveness?	Yes No N/A

Source: Robert J. Fischer and Richard Janoski, ***Loss Prevention and Security Procedures*** (Boston: Butterworth-Heinemann, 2000).

Index

Printed by Printforce, the Netherlands